Lecture Notes in Computer Science　　　6410

Commenced Publication in 1973
Founding and Former Series Editors:
Gerhard Goos, Juris Hartmanis, and Jan van Leeuwen

Advanced Research in Computing and Software Science

Subline of Lectures Notes in Computer Science

Dimitrios M. Thilikos (Ed.)

Graph-Theoretic Concepts in Computer Science

36th International Workshop, WG 2010
Zarós, Crete, Greece, June 28-30, 2010
Revised Papers

 Springer

Volume Editor

Dimitrios M. Thilikos
National and Kapodistrian University of Athens
Department of Mathematics
Panepistimioupolis, 15784 Athens, Greece
E-mail: sedthilk@math.uoa.gr

Library of Congress Control Number: 2010938221

CR Subject Classification (1998): G.2.2, I.2.8, E.1, F.2, I.3.5, C.2

LNCS Sublibrary: SL 1 – Theoretical Computer Science and General Issues

ISSN   0302-9743
ISBN-10   3-642-16925-2 Springer Berlin Heidelberg New York
ISBN-13   978-3-642-16925-0 Springer Berlin Heidelberg New York

springer.com

© Springer-Verlag Berlin Heidelberg 2010
Printed in Germany

Typesetting: Camera-ready by author, data conversion by Scientific Publishing Services, Chennai, India
Printed on acid-free paper   SPIN: 06/3180   5 4 3 2 1 0

Preface

The *36th International Workshop on Graph-Theoretic Concepts in Computer Science* (WG 2010) took place in Zarós, Crete, Greece, June 28–30, 2010. About 60 mathematicians and computer scientists from all over the world (Australia, Canada, Czech Republic, France, Germany, Greece, Hungary, Israel, Japan, The Netherlands, Norway, Poland, Switzerland, the UK, and the USA) attended the conference.

WG has a long tradition. Since 1975, WG has taken place 21 times in Germany, four times in The Netherlands, twice in Austria, twice in France and once in the Czech Republic, Greece, Italy, Norway, Slovakia, Switzerland, and the UK.

WG aims at merging theory and practice by demonstrating how concepts from graph theory can be applied to various areas in computer science, or by extracting new graph theoretic problems from applications. The goal is to present emerging research results and to identify and explore directions of future research. The conference is well-balanced with respect to established researchers and young scientists.

There were 94 submissions, two of which where withdrawn before entering the review process. Each submission was carefully reviewed by at least 3, and on average 4.5, members of the Program Committee. The Committee accepted 28 papers, which makes an acceptance ratio of around 30%. I should stress that, due to the high competition and the limited schedule, there were papers that were not accepted while they deserved to be.

The program also included two excellent invited talks: the first one was given by Dimitris Achlioptas (Department of Computer Science, UC Santa Cruz) on "Algorithmic Barriers from Phase Transitions in Graphs" and the second one was given by Erik D. Demaine (MIT Computer Science and Artificial Intelligence Laboratory) on "Algorithmic Graph Minors and Bidimensionality." This volume contains the abstracts of both talks.

I wish to thank *all* those who contributed to the success of WG 2010. While it is impossible to enumerate them all, this includes the authors for submitting high-quality papers, the reviewers and the members of the Program Committee for their detailed work, the speakers for their well-prepared talks, all the participants for their enthusiasm, and the personnel of the hotel "Idi" for the pleasant conference environment and facilities.

I am grateful to all the students of the Department of Mathematics of the National and Kapodistrian University of Athens that helped with the organization. Among people from the Department of Mathematics, University of Crete, I should thank Christos Kourouniotis for his link to the Anogia Academic Village and Mihalis Kolountzakis for his valuable advice. But most of all, I am indebted to the conference secretary Marina Vassilaki for her professionality and diligence during the preparation and the management of the conference.

All material of the conference, including photos and videos of the invited talks, can be found at the homepage of WG 2010 (http://www.math.uoa.gr/wg2010/). Special thanks to Charalampos Tampakopoulos for his programming, development, and hosting services.

Finally, I should also thank the EasyChair team for the wonderful conference management system.

September 2010   Dimitrios M. Thilikos
                                                         Δημήτριος Μ. Θηλυκός

WG 2010

Ζαρός, Κρήτη

The Long Tradition of WG

WG 1975 U. Pape – Berlin, Germany
WG 1976 H. Noltemeier – Göttingen, Germany
WG 1977 J. Mühlbacher – Linz, Austria
WG 1978 M. Nagl, H.J. Schneider – Castler Feuerstein, Germany
WG 1979 U. Pape – Berlin, Germany
WG 1980 H. Noltemeier – Bad Honnef, Germany
WG 1981 J. Mühlbacher – Linz, Austria
WG 1982 H.J. Schneider, H. Göttler – Neuenkirchen, Germany
WG 1983 M. Nagl, J. Perl – Haus Ohrbeck near Onasbrück, Germany
WG 1984 U. Pape – Berlin, Germany
WG 1985 H. Noltemeier – Castle Schwanberg near Würzburg, Germany
WG 1986 G. Tinhofer, G. Schmidt – Bernried near Munich, Germany
WG 1987 H. Göttler, H.J. Schneider – Kloster Banz near Bamberg, Germany
WG 1988 J. van Leeuwen – Amsterdam, The Netherlands
WG 1989 M. Nagl – Castle Rolduc, The Netherlands
WG 1990 R.H. Möhring – Berlin, Germany
WG 1991 G. Schmidt, R. Berghammer – Fischbachau near Munich, Germany
WG 1992 E.W Mayr – Wiesbaden-Naurod, Germany
WG 1993 J. van Leeuwen – Utrecht, The Netherlands
WG 1994 G. Tinhofer, E.W. Mayr, G. Schmidt – Herrsching near Munich, Germany
WG 1995 M. Nagl – Aachen, Germany
WG 1996 G. Ausiello, A. Marchetti-Spaccamela – Como, Italy
WG 1997 R.H. Möhring – Berlin, Germany
WG 1998 J. Hromkovič, O. Sýkora – Smolenice Castle, Slovak Republic
WG 1999 P. Widmayer – Ascona, Switzerland
WG 2000 D. Wagner – Konstanz, Germany
WG 2001 A. Brandstädt, Boltenhagen near Rostock, Germany
WG 2002 L. Kučera – Český Krumlov, Czech Republic
WG 2003 H.L. Bodlaender – Elspeet, The Netherlands
WG 2004 J. Hromkovič, M. Nagl – Bad Honnef, Germany
WG 2005 D. Kratsch – Metz, France
WG 2006 F.V. Fomin – Bergen, Norway
WG 2007 A. Brandstädt, D. Kratsch, H. Müller – Dornburg near Jena, Germany
WG 2008 H. Broersma, T. Erlebach – Durham, UK
WG 2009 C. Paul, M. Habib – Montpellier, France
WG 2010 D.M. Thilikos – Zarós, Crete, Greece

WG 2010 Organization

Program Chair

Dimitrios M. Thilikos
Department of Mathematics,
National and Kapodistrian
University of Athens, Greece

Program Committee

Fedor V. Fomin	University of Bergen, Norway
Pierre Fraigniaud	CNRS and University Paris Diderot, France
Gregory Z. Gutin	Royal Holloway, University of London, UK
Frédéric Havet	INRIA Sophia-Antipolis, France
Giuseppe F. Italiano	University of Rome Tor Vergata, Italy
Kazuo Iwama	Kyoto University, Japan
Jan Kratochvíl	Charles University, Czech Republic
Bojan Mohar	Simon Fraser University, Canada
David Peleg	Weizmann Institute of Science, Israel
Prabhakar Ragde	University of Waterloo, Canada
Dieter Rautenbach	Ilmenau University of Technology, Germany
Saket Saurabh	Institute of Mathematical Sciences, India
Ingo Schiermeyer	Freiberg University of Mining and Technology, Germany
Maria Serna	Technical University of Catalonia, Spain
Martin Skutella	Technical University of Berlin, Germany
Dimitrios M. Thilikos	National & Kapodistrian University of Athens, Greece
Jan van Leeuwen	Utrecht University, The Netherlands
Peter Widmayer	Federal Institute of Technology Zurich, Switzerland
Gerhard J. Woeginger	Eindhoven University of Technology, The Netherlands

Organizing Committee

Dimitrios M. Thilikos (Chair)
Marina Vassilaki (Secretary)

Archontia Giannopoulou, Athanassios Koutsonas,
Ignasi Sau, Konstantinos Stavropoulos,
Charalampos Tampakopoulos, Dimitris Zoros

Organization Entities

Anogia Academic Vilage
Department of Mathematics, National and Kapodistrian University of Athens

External Reviewers of WG 2010

Ittai Abraham, Faisal Abu-Khzam, Noa Agmon, Carme Álvarez, David Avis, Jørgen Bang-Jensen, Robert Benkoczi, Philip Bille, Jean Blair, Andreas Bley, Hans L. Bodlaender, Andreas Brandstädt, Hajo Broersma, Sergio Cabello, Sourav Chakraborty, Markus Chimani, Nathann Cohen, Derek Corneil, Robert Crowston, Marek Cygan, Ajoy K. Datta, Guoli Ding, Yann Disser, Benjamin Doerr, Frederic Dorn, Michael Dracopoulos, Feodor Dragan, Amalia Duch, Vida Dujmović, Zdeněk Dvořák, Louis Esperet, Uriel Feige, Andreas Emil Feldmann, Michael Fellows, Stefan Felsner, Holger Flier, Radoslav Fulek, Anna Galluccio, Leszek Gąsieniec, Serge Gaspers, Archontia Giannopoulou, Boris Goldengorin, Petr Golovach, Martin Golumbic, Fabrizio Grandoni, Jiong Guo, Michel Habib, Mohammad, Taghi Hajiaghayi, Johannes Hatzl, Yinnon Haviv, Pinar Heggernes, Pavol Hell, Christoph Helmberg, Petr Petr Hliněný, Andreas Holmsen, Takashi Horiyama, Tomas Hruz, Thore Husfeldt, Hiro Ito, Bart Jansen, Jeannette Janssen, Mark Jones, Tibor Jordan, Hirotsugu Kakugawa, Marcin Kamiński, Mamadou Moustapha Kanté, Alexis Kaporis, Jan-Philipp Kappmeier, Menelaos I. Karavelas, Ján Katrenic, Jun Kawahara, Ken-ichi Kawarabayashi, Judith Keijsper, Jonathan Kelner, Eun Jung Kim, Ralf Klasing, Bettina Klinz, Christian Knauer, Anja Kohl, Mikko Koivisto, Stavros Kolliopoulos, Athanassios Koutsonas, Daniel Král, Ilia Krasikov, Dieter Kratsch, Matthias Kriesell, Sven Krumke, Michael Lampis, Monique Laurent, Mathieu Liedloff, Giuseppe Liotta, Daniel Lokshtanov, Vadim Lozin, John Maharry, Monaldo Mastrolilli, Yasuko Matsui, Jannik Matuschke, Jens Maue, Klaus Meer, George Mertzios, Matúš Mihalák, Zoltán Miklós, Neeldhara Misra, Valia Mitsou, Matthias Mnich, Janina Müttel, Shin-Ichi Nakano, N.S. Narayanaswamy, Hung Ngo, Rolf Niedermeier, Prajakta Nimbhorkar, Nicolas Nisse, Jan Obdrzálek, Yoshio Okamoto, Michael Okun, Sang-il Oum, Attila Pór, Christophe Paul, Daniël Paulusma, Rudi Pendavingh, Geevarghese Philip, Preyas Popat, Ian Post, Andrzej Proskurowski, Arash Rafiey, Venkatesh Raman, Bert Randerath, Friedrich Regen, Peter Rossmanith, Zdeněk Ryjáček, Ignasi Sau, Mathias Schacht, Diego Scheide, Ildikó Schlotter, Marcel Schöngens, Philipp Matthias Schäfer, Jean-Sébastien Sereni, Kazuhisa Seto, Akiyoshi Shioura, Michiel Smid, Bettina Speckmann, Daniel Spielman, Rastislav Sramek, Juraj Stacho, Konstantinos Stavropoulos, Michal Stern, Hisao Tamaki, Suguru Tamaki, Wolfgang Thomas, Ioan Todinca, Csaba Toth, Ryuhei Uehara, Takeaki Uno, Pim van 't Hof, Leo van Iersel, Erik Jan van Leeuwen, Johan M.M. van Rooij, Martin Vatshelle, Yngve Villanger, Kristina Vušković, Koichi Wada, Oren Weimann, Andreas Wiese, Thomas Wolle, Koichi Yamazaki, Hiroki Yanagisawa, Yuichi Yoshida, Dimitris Zoros, Vadim Zverovich, Anna Zych.

Table of Contents

Algorithmic Barriers from Phase Transitions in Graphs

Dimitris Achlioptas

Department of Computer Science,
UC Santa Cruz
optas@cs.ucsc.edu

Abstract. For a number of optimization problems on random graphs and hypergraphs, e.g., k-colorings, there is a very big gap between the largest average degree for which known polynomial-time algorithms can find solutions, and the largest average degree for which solutions provably exist. We study this phenomenon by examining how sets of solutions evolve as edges are added. We prove in a precise mathematical sense that, for each problem studied, the barrier faced by algorithms corresponds to a phase transition in the problems solution-space geometry. Roughly speaking, at some problem-specific critical density, the set of solutions shatters and goes from being a single giant ball to exponentially many, well-separated, tiny pieces. All known polynomial-time algorithms work in the ball regime, but stop as soon as the shattering occurs. Besides giving a geometric view of the solution space of random instances our results provide novel constructions of one-way functions.

D.M. Thilikos (Ed.): WG 2010, LNCS 6410, p. 1, 2010.

Algorithmic Graph Minors and Bidimensionality

Erik D. Demaine

MIT Computer Science and Artificial Intelligence Laboratory
edemaine@mit.edu

Abstract. Graph Minor Theory, developed by Robertson and Seymour over two decades, provides powerful structural results about a wide family of graph classes (anything closed under deletion and contraction). In recent years, this theory has been extended and generalized to apply to many algorithmic problems. Bidimensionality theory is one approach to algorithmic graph minor theory. This theory provides general tools for designing fast (constructive, often subexponential) fixed-parameter algorithms, kernelizations, and approximation algorithms (often PTASs), for a wide variety of NP-hard graph problems for graphs excluding a fixed minor. For example, some of the most general algorithms for feedback vertex set and connected dominating set are based on bidimensionality. Another approach is "deletion and contraction decompositions", which split any graph excluding a fixed minor into a bounded number of small-treewidth graphs. For example, this approach has led to some of the most general algorithms for graph coloring and the Traveling Salesman Problem on graphs. I will describe these and other approaches to efficient algorithms through graph minors.

D.M. Thilikos (Ed.): WG 2010, LNCS 6410, p. 2, 2010.

Complexity Results for the Spanning Tree Congestion Problem

Yota Otachi[1], Hans L. Bodlaender[2], and Erik Jan van Leeuwen[3]

[1] Graduate School of Information Sciences, Tohoku University, Sendai 980-8579, Japan, JSPS
Research Fellow
otachi@dais.is.tohoku.ac.jp
[2] Institute of Information and Computing Sciences, Utrecht University, P.O. Box 80.089, 3508
TB Utrecht, The Netherlands
hansb@cs.uu.nl
[3] Department of Informatics, University of Bergen, P.O. Box 7803, 5020 Bergen, Norway
E.J.van.Leeuwen@ii.uib.no

Abstract. We study the problem of determining the *spanning tree congestion* of
a graph. We present some sharp contrasts in the complexity of this problem. First,
we show that for every fixed k and d the problem to determine whether a given
graph has spanning tree congestion at most k can be solved in linear time for
graphs of degree at most d. In contrast, if we allow only one vertex of unbounded
degree, the problem immediately becomes NP-complete for any fixed $k \geq 10$. For
very small values of k however, the problem becomes polynomially solvable. We
also show that it is NP-hard to approximate the spanning tree congestion within
a factor better than 11/10. On planar graphs, we prove the problem is NP-hard in
general, but solvable in linear time for fixed k.

1   Introduction

Spanning tree congestion is a relatively new graph parameter, which was formally defined by Ostrovskii [21] in 2004. Prior to Ostrovskii [21], Simonson [25] studied the
same parameter under a different name to approximate the cutwidth of outerplanar graphs.
Although several graph theoretical results have been presented [7, 16–18, 20, 22] after
Ostrovskii [21], so far, no results on the complexity of the problem were known. In this
paper, we present the first such results. The parameter is defined as follows. Let G be a
graph and T a spanning tree of G. The *detour* for an edge $\{u, v\} \in E(G)$ is the unique u–v
path in T. We define the *congestion* of $e \in E(T)$, denoted by $cng_{G,T}(e)$, as the number of
detours that contain e. The *congestion of G in T*, denoted by $cng_G(T)$, is the maximum
congestion over all edges in T. The *spanning tree congestion* of G, denoted by $stc(G)$, is
the minimum congestion over all spanning trees of G. We denote by STC the problem
of determining whether a given graph has spanning tree congestion at most given k. If
k is fixed, we denote the problem by k-STC.

The name of the parameter comes from the following analogy [7]: Edges of G are
roads, and edges of T are those roads which are cleaned from snow after snowstorms.
For an edge $h \in E(T)$, it is natural to define the congestion of h as the number of detours
passing through h. Clearly, the congestion of the busiest roads should be minimized. The

D.M. Thilikos (Ed.): WG 2010, LNCS 6410, pp. 3–14, 2010.

tree spanner problem [6] is a variant of the problem, which minimize the dilation, that is, the length of the longest detours. Several pairs of congestion and dilation problems are known [23]. The most famous pair is the cutwidth problem and the bandwidth problem.

The rest of the paper is organized as follows. Section 2 provides some definitions and basic facts. In Section 3, we study the problem for planar graphs, and show that STC for planar graphs is NP-complete, and k-STC for planar graphs is solvable in linear time. In Section 4, we show that k-STC can be solved in linear time for $1 \leq k \leq 3$. In Section 5, we show that k-STC can be solved in linear time also for graphs of bounded degree. In Section 6, we show that k-STC is NP-complete for edge weighted graphs if $k \geq 10$. Using the result of Section 6, we show in Section 7 that for $k \geq 10$, k-STC is NP-complete for simple unweighted graphs with only one vertex of unbounded degree. In the last section, we conclude the paper and show the approximation hardness of the spanning tree congestion. Due to space limitation, some proofs are omitted.

2   Preliminaries

We extend the notion of spanning tree congestion to edge weighted graphs, by defining the congestion of an edge as the sum of the weights of edges whose detours pass through the edge. We denote by $w(F)$ the sum of weights of edges in F for an edge set $F \subseteq E(G)$.

Let G be a connected graph. For $S \subseteq V(G)$, we denote by $G[S]$ the subgraph induced by S. For an edge $e \in E(G)$, we denote by $G - e$ the graph obtained by the deletion of e from G. For $A, B \subseteq V(G)$, we define $E_G(A, B) = \{u, v \in E(G) \mid u \in A, v \in B\}$. For $S \subseteq V(G)$, we define the *boundary edges* of S, denoted by $\theta_G(S)$, as $\theta_G(S) = E_G(S, V(G) \setminus S)$. Using this notation, we can redefine $cng_{G,T}(e)$ as $cng_{G,T}(e) = |\theta_G(A_e)|$, where A_e is the vertex set of one of the two components of $T - e$. From this redefinition through boundary edges, we can see that *c-cut trees* defined by Fekete and Kremer [12] and spanning trees of congestion at most c are equivalent.

For an edge e in a tree T, we say that e separates A and B if $A \subseteq A_e$ and $B \subseteq B_e$, where A_e and B_e are the vertex sets of the two components of $T - e$. Clearly, if T is a spanning tree of G and $e \in E(T)$ separates A and B, then $cng_{G,T}(e) \geq |E(A, B)|$ (if G is weighted, $cng_{G,T}(e) \geq w(E(A, B))$). If e separates A and B, we also say that e *divides* $A \cup B$ into A and B.

From the definition of the spanning tree congestion, the following proposition holds.

Proposition 2.1. *The spanning tree congestion of G equals the maximum spanning tree congestion of its biconnected components.*

Ostrovskii [21] showed the following lower bound on the spanning tree congestion of graphs.

Lemma 2.2 ([21]). *Let G be a graph, $u, v \in V(G)$. If G has k edge disjoint u–v paths, then $stc(G) \geq k$.*

Let G be a graph. We say that a graph H is obtained from G by an *edge subdivision* if $V(H) = V(G) \cup \{w\}$ and $E(H) = E(G) \setminus \{\{u, v\}\} \cup \{\{u, w\}, \{w, v\}\}$ for some edge $\{u, v\} \in E(G)$ and a new vertex w. We say that H is a *subdivision* of G if H can be obtained from G by a finite sequence of edge subdivisions. If H is a subdivision of a subgraph of G, then H is a *topological minor* of G.

The concept of treewidth was introduced by Robertson and Seymour in their project of Graph Minor Theory (see [24] for example). A *tree decomposition* of a graph G is a pair $(\mathcal{X}, T)$, where T is a tree and $\mathcal{X} = \{X_i \mid i \in V(T)\}$ is a collection of subsets of $V(G)$ such that

- $\bigcup_{i \in V(T)} X_i = V(G)$,
- for each edge $\{u, v\} \in E(G)$, there is a *node* $i \in V(T)$ such that $u, v \in X_i$, and
- for each $v \in V(G)$, the set of nodes $\{i \mid v \in X_i\}$ forms a subtree of T.

The elements in $\mathcal{X}$ are called *bags*. The *width* of a tree decomposition $(\mathcal{X}, T)$ equals $\max_{i \in V(T)} |X_i| - 1$. The *treewidth* of G, denoted by $tw(G)$, is the minimum width over all tree decompositions of G.

3 Spanning Tree Congestion of Planar Graphs

Ostrovskii [22] has asked whether STC can be solved in polynomial time for planar graphs. By combining a number of known results, we answer this question negatively (assuming $P \neq NP$), and show that k-STC can be solved in linear time for planar graphs. Our results follow easily from some known results for the tree spanner problem. Let G be a graph and T a spanning tree of G. If $dist_T(u, v) \leq k$ for any $\{u, v\} \in E(G)$, then T is a *tree k-spanner* [6]. We denote by $tsp(G)$ the minimum number k such that G has a tree k-spanner. For planar graphs, the following results are known.

Lemma 3.1 ([12]). *It is NP-complete to decide $tsp(G) \leq k$ for planar graphs G and integers k.*

Lemma 3.2 ([11]). *For every fixed k, $tsp(G) \leq k$ can be decided in linear time for planar graphs G.*

A *dual graph* G^* of a planar graph G is a graph that has the vertex set $\mathcal{F}(G)$, the faces of a certain embedding of G, and in which two vertices $f, f' \in \mathcal{F}(G)$ are adjacent in G^* if and only if the two faces f and f' have a common edge in G. It is known that a graph G is planar if and only if G is a dual graph of a planar graph (see e.g. [10]). Since a cut in G corresponds to a cycle in G^*, the following relation holds.

Lemma 3.3 ([12]). *For any planar graph G, $stc(G) = tsp(G^*) + 1$.*

A planar embedding of a planar graph can be constructed in linear time by an algorithm proposed by Hopcroft and Tarjan [15]. From a planar embedding of a planar graph G, we can easily construct geometrically a dual graph G^* (see e.g. [19]). Note that $G = (G^*)^*$. Thus, from Lemma 3.3, we can have the conclusions of this section.

Theorem 3.4. *It is NP-complete to decide $stc(G) \leq k$ for planar graphs G and integers k.*

Theorem 3.5. *For every fixed k, $stc(G) \leq k$ can be decided in linear time for planar graphs G.*

4   Linear Time Solvability of k-STC for $1 \leq k \leq 3$

In this section, we show that k-STC can be solved in linear time for $1 \leq k \leq 3$. First, we give characterizations for graphs of spanning tree congestion one and two.

Theorem 4.1. *For a connected graph G, $stc(G) = 1$ if and only if G is a tree.*

Proof. If G is a tree, then clearly $stc(G) = 1$. Assume G has a cycle C. Then, for any two vertices in C, G has two edge disjoint paths between them. Thus, by Lemma 2.2, G cannot have any cycle.   □

A graph G is a *cactus graph* if no two cycles in G have a common edge.

Theorem 4.2. *For a connected graph G, $stc(G) = 2$ if and only if G is not a tree but a cactus graph.*

Proof. Clearly, every biconnected component of a cactus graph G is either a cycle or a single edge, and thus, G has spanning tree congestion at most two. It is easy to verify that a biconnected graph G has no vertex pair u, v such that G contains three edge disjoint u–v paths if and only if G is either a cycle or a single edge. Thus, from Proposition 2.1 and Lemma 2.2, the theorem holds.   □

Obviously, the recognition of trees and cactus graphs can be done in linear time, by using standard depth first search techniques (see e.g. [8]). For $k = 3$, we need the following lemma.

Lemma 4.3. *For a graph G, if $stc(G) \leq 3$, then G is planar.*

Proof. Suppose $stc(G) \leq 3$ and G is not planar. From Kuratowski's Theorem (see e.g. [10]), G has either K_5 or $K_{3,3}$ as a topological minor. If G has K_5 as a topological minor, then clearly G contains two vertices such that G has at least four edge disjoint paths between them. From Lemma 2.2, we have $stc(G) \geq 4$, which is a contradiction. Thus, G contains $K_{3,3}$ as a topological minor. Let G' be this topological minor, and $X = \{x_1, x_2, x_3\}, Y = \{y_1, y_2, y_3\} \subset V(G')$ be the two sets corresponding to the two color classes of $K_{3,3}$. By Lemma 7.2 edge subdivisions do not change the spanning tree congestion. Thus, $stc(G') = stc(K_{3,3})$. Moreover, by Hruska's result that shows $stc(K_{m,n}) = m + n - 2$ [16], we can conclude $stc(G') = 4$. Now we need the following two propositions.

Proposition 4.4. *Let H be a connected graph and H' a connected subgraph of H. If a spanning tree S of H has a spanning tree S' of H' as a subgraph, then $cng_H(S) \geq cng_{H'}(S')$.*

Proof. Let $e \in E(S') \subseteq E(S)$. Assume e divides $V(H)$ into A and B, and $V(H')$ into A' and B'. Clearly, $A' \subseteq A$ and $B' \subseteq B$. Thus, $cng_{H,S}(e) = |E(A, B)| \geq |E(A', B')| = cng_{H',S'}(e)$.   □

Proposition 4.5. *Let H be a connected graph, S a spanning tree of H, and $A, B \subset V(H)$. If H has p edge disjoint paths $P_1, \ldots, P_p$ between A and B, and $e \in E(S)$ separates A and B, then $cng_{H,S}(e) \geq p$. Moreover, if e does not belong to any P_i, then $cng_{H,S}(e) \geq p + 1$.*

Proof. For each P_i, there exists at least one edge e_i such that the detour of e_i in S passes through the edge e. Since the paths $P_1, \ldots, P_p$ are edge disjoint, $cng_{H,S}(e) \geq p$. Since e itself is the detour for e, $cng_{H,S}(e) \geq p + 1$ if $e \notin \{e_i \mid 1 \leq i \leq p\}$. □

We will show that $cng_{G,T}(e) > 3$ for any spanning tree T of G. If T has a spanning tree T' of G' as a subgraph, then by Proposition 4.4, $cng_G(T) \geq cng_{G'}(T') \geq 4$. Thus T contains no such a subgraph T'. This implies that T contains an edge $e \notin E(G')$ that divides $X \cup Y$ into two nonempty sets, say A and B. Since there are nine edge disjoint paths between X and Y in G', there exist at least three edge disjoint paths between $A \subset X \cup Y$ and $B = (X \cup Y) \setminus A$ in G'. Proposition 4.5 implies $cng_{G,T}(e) \geq 4$ since $e \notin E(G')$. □

From Theorem 3.5 and Lemma 4.3, 3-STC can be solved in linear time, with the linear time algorithm for recognizing planar graphs [15]. This proves the following theorem.

Theorem 4.6. *For $1 \leq k \leq 3$, k-STC can be solved in linear time.*

5 Linear Time Solvability of *k*-STC for Graphs of Bounded Degree

In this section, we show that k-STC can be solved in linear time for graphs of bounded degree. To this end, we use Courcelle's theorem and a connection between the spanning tree congestion and the treewidth. Courcelle [9] showed that every problem expressible in MS_2 can be solved in linear time for graphs of bounded treewidth, where MS_2 is a graph logic in the monadic second-order logic (see also [14]). In MS_2, we are allowed to use the incident relation inc, the membership relation $\in$, and variables over vertices, edges, vertex sets, and edge sets.

Theorem 5.1. *For graphs of bounded treewidth, k-STC can be solved in linear time.*

Proof. We show that k-STC is expressible in MS_2. The proof is omitted. □

We can show that the treewidth of a graph of bounded degree is linear in its spanning tree congestion. (The proof is omitted.)

Lemma 5.2. *For any connected graph G, $tw(G) \leq \max\{stc(G), \Delta(G)(stc(G) - 1)/2\}$. Moreover, this bound is tight.*

Lemma 5.2 can be proved similarly to results on the edge remember number, reported in [3]. The upper bound improves on an earlier bound by Kozawa, Otachi, and Yamazaki [17]. A tight example for the bound of Lemma 5.2 is a cycle. By using expanders, we can even show that any upper bound must depend linearly on both the maximum degree and the spanning tree congestion of the graph (we omit the proof). This gives strong evidence that our bound cannot be improved upon. Combining the above facts, we can obtain the main result of this section.

Theorem 5.3. *For graphs of bounded degree, k-STC can be solved in linear time.*

Proof. Let G be a graph of bounded degree and $\Delta(G) = d$. Since k and d are constants, we can check whether $tw(G) \leq \max\{k, d(k-1)/2\}$ in linear time by Bodlaender's algorithm [2]. If the output of the algorithm is "no," then $stc(G) > k$ from Lemma 5.2. Otherwise, G has bounded treewidth. Hence, from Theorem 5.1, we can determine whether $stc(G) \leq k$ in linear time. □

6   Weighted k-STC is NP-Complete for $k \geq 10$

In this section, we prove the following hardness result.

Theorem 6.1. *For any fixed $k \geq 10$, k-STC is NP-complete for edge weighted graphs.*

Clearly, the problem belongs to NP. To show NP-completeness, we present a reduction from (3, B2)-SAT. The problem (3, B2)-SAT is a restricted version of the 3-SAT problem, which is a well-known NP-complete problem [13]. An instance (U, C) of (3, B2)-SAT consists of a set U of n distinct Boolean variables and a collection C of m clauses such that each clause has exactly three literals, and each literal occurs exactly twice. Berman, Karpinski, and Scott [1] showed the NP-completeness of (3, B2)-SAT. In their construction of a hard instance of (3, B2)-SAT, every clause has exactly three variables, that is, there is no clause like $(u, u, *)$, $(\bar{u}, \bar{u}, *)$, or $(u, \bar{u}, *)$. Thus, in what follows, we assume that instances of (3, B2)-SAT satisfy this condition as well.

The constructions in our proof are inspired by the proof of Cai and Corneil [6] for the NP-completeness of the Weighted Tree Spanners problem. Let $k \geq 10$ be a fixed integer. For an arbitrary instance (U, C) of (3, B2)-SAT, we construct an edge weighted graph G_C such that C is satisfiable if and only if $stc(G_C) \leq k$. Let $a = \lceil k/2 \rceil + 1$ and $b = \lfloor k/2 \rfloor - 3$. Each edge in G_C has a weight which will be either a, b, or 1. For example, if $k = 10$, then the weight of an edge is six, two, or one. Clearly, the following proposition holds.

Proposition 6.2. *For $k \geq 10$, $a + b + 2 = k$, $2b + 6 \leq k$, $2a > k$, $6b > k$, and $4b + 4 > k$.*

From an instance (U, C) of (3, B2)-SAT, the graph G_C is constructed as follows (see Fig. 1):

1. Take a vertex x, *literal vertices* u_i and $\bar{u}_i$ for each variable $u_i \in U$, and *clause vertices* c_i for each clause $c_i \in C$.
2. Connect x to all literal vertices by *literal edges* of weight b.
3. For each variable $u_i \in U$, create a path of length two between u_i and $\bar{u}_i$ such that edges in the path, which are called *bridge edges*, have weight a and the center vertex of the path is a new vertex y_i.
4. For each clause $c_i = \{l_p, l_q, l_r\} \in C$, connect the clause vertex c_i to the literal vertices l_p, l_q, and l_r by *clause edges* of unit weight.

Clearly, the above construction can be done in polynomial time.

Now, we show the following useful properties of a spanning tree of G_C with small congestion.

Lemma 6.3. *Let T be a spanning tree of G_C. If $cng_{G_C}(T) \leq k$, then*

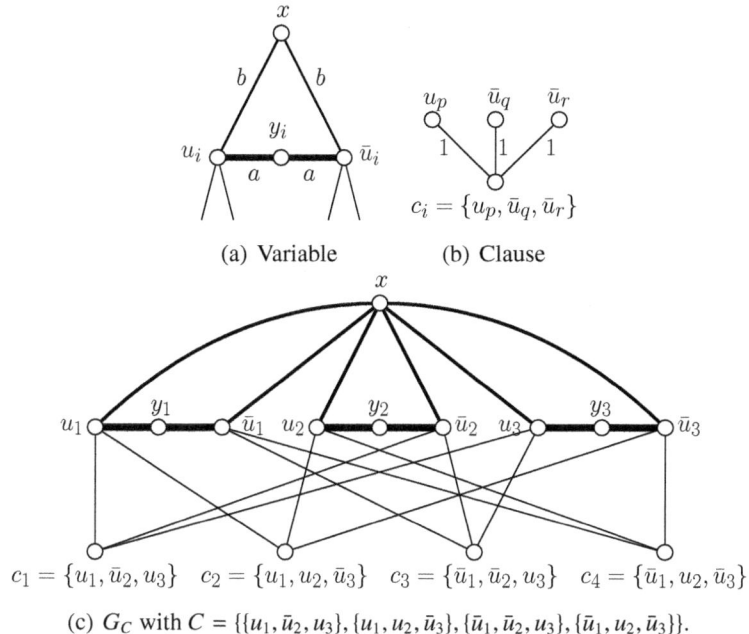

(a) Variable   (b) Clause

$c_i = \{u_p, \bar{u}_q, \bar{u}_r\}$

$c_1 = \{u_1, \bar{u}_2, u_3\}$ $c_2 = \{u_1, u_2, \bar{u}_3\}$ $c_3 = \{\bar{u}_1, \bar{u}_2, u_3\}$ $c_4 = \{\bar{u}_1, u_2, \bar{u}_3\}$

(c) G_C with $C = \{\{u_1, \bar{u}_2, u_3\}, \{u_1, u_2, \bar{u}_3\}, \{\bar{u}_1, \bar{u}_2, u_3\}, \{\bar{u}_1, u_2, \bar{u}_3\}\}$.

Fig. 1. Gadgets, and a constructed graph

1. *All bridge edges are contained in T;*
2. *Each clause vertex is a leaf of T;*
3. *For each variable, exactly one of its two literal edges is contained in T.*

Proof (of the first property). Since y_i has degree two, at least one of $\{u_i, y_i\}$ and $\{\bar{u}_i, y_i\}$ must be in T. If $\{\bar{u}_i, y_i\}$ is not in T, then $cng_{G_C,T}(\{u_i, y_i\}) = w(\theta(\{y_i\})) = 2a > k$. The other case is almost the same.   □

Proof (of the second property). Assume T has the first property. By way of contradiction, suppose some clause vertex $c_i = \{l_p, l_q, l_r\}$ has degree larger than one in T. Let u_p, u_q, u_r be the variables corresponding to the literals l_p, l_q, l_r, respectively. We divide the proof into two cases depending on the degree of c_i in T. Recall that all bridge edges are in T from the first property.

Case 1: $deg_T(c_i) = 3$. The three neighbors of c_i in T are l_p, l_q, and l_r. Let e be the unique literal edge in the unique c_i–x path in T. Then, e separates $\{x\}$ and $\{u_p, \bar{u}_p, u_q, \bar{u}_q, u_r, \bar{u}_r\}$. Thus, $cng_{G_C,T}(e) \geq w(E(\{x\}, \{u_p, \bar{u}_p, u_q, \bar{u}_q, u_r, \bar{u}_r\})) = 6b > k$.

Case 2: $deg_T(c_i) = 2$. Without loss of generality, we assume that the two neighbors of c_i in T are l_p and l_q. Then, at most one of the literal edges of u_p and u_q can be in T. From the above case, we can assume that no clause vertex has degree three in T.

First, assume that none of the literal edges of u_p and u_q are in T. Let $e = \{x, l_s\}$ be the unique literal edge in the unique c_i–x path in T. Then, $l_s \notin \{u_p, \bar{u}_p, u_q, \bar{u}_q\}$, and e separates $\{x\}$ and $\{u_p, \bar{u}_p, u_q, \bar{u}_q, u_s, \bar{u}_s\}$. Thus, $cng_{G_C,T}(e) \geq 6b > k$.

Next, assume that one of the literal edges of u_p and u_q, say e, is in T (see Fig. 2). Let us consider the clause vertices adjacent to at least one of the literal vertices $u_p, \bar{u}_p, u_q$, and $\bar{u}_q$ in G_C. If a clause vertex c_z ($\neq c_i$) is adjacent to two vertices in $\{u_p, \bar{u}_p, u_q, \bar{u}_q\}$ in T, then T has a cycle. Hence, if $c_z \neq c_i$ has degree two in T, and one of the two neighbors of c_z is in $\{u_p, \bar{u}_p, u_q, \bar{u}_q\}$, then another neighbor, say l_s, is not in $\{u_p, \bar{u}_p, u_q, \bar{u}_q\}$. In such a case, e separates $\{x\}$ and $\{u_p, \bar{u}_p, u_q, \bar{u}_q, u_s, \bar{u}_s\}$, and thus, $cng_{G_C,T}(e) \geq 6b > k$ (see Fig. 2(a)). Therefore, every clause vertex (except for c_i) that has at least one of $\{u_p, \bar{u}_p, u_q, \bar{u}_q\}$ as a neighbor in T is a leaf of T. Let C_1 be the set of such leaf clauses. Since every clause has exactly three variables, each $c \in C_1$ has at most two neighbors in $\{u_p, \bar{u}_p, u_q, \bar{u}_q\}$ in G_C. Hence, $cng_{G_C,T}(e) = w(\theta(\{u_p, \bar{u}_p, u_q, \bar{u}_q\} \cup \{c_i\} \cup C_1)) \geq 4b + |C_1| + 1$ (see Fig. 2(b)). Since $cng_{G_C}(T) \leq k < 4b + 4$, we can conclude that $|C_1| \leq 2$. It is easy to see that $cng_{G_C,T}(e) \geq 4b + 5 > k$ if $|C_1| \leq 2$ (see Fig. 3).   □

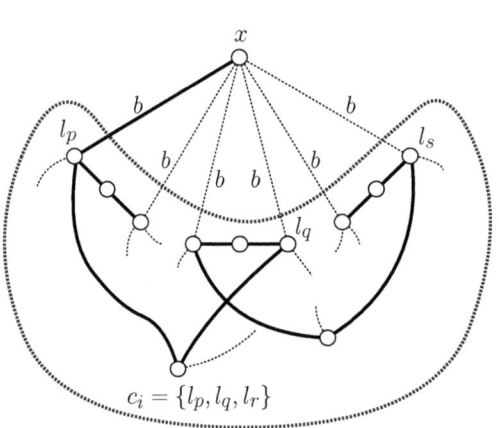

(a) Another clause vertex of degree two.

(b) No other clause vertex of degree two.

Fig. 2. A clause vertex c_i of degree two

Proof (of the third property). Assume T has the first and the second properties. Since T is a tree and contains all bridge edges, at most one of $\{x, u_i\}$ and $\{x, \bar{u}_i\}$ can be in T for each $u_i \in U$. Suppose T contains none of them. Since any clause vertex is a leaf of T, there is no path between u_i and x.   □

The next two lemmas show that C is satisfiable if and only if $stc(G_C) \leq k$, thus proving Theorem 6.1.

Lemma 6.4. *If $stc(G_C) \leq k$ then C is satisfiable.*

Proof. Let T be a spanning tree of G_C such that $cng_{G_C}(T) \leq k$. From Lemma 6.3, (1) T contains all bridge edges, (2) every clause vertex is a leaf of T, and (3) T contains exactly one literal edge for each variable. From the third property, we can define a truth assignment ξ_T by setting $\xi_T(u_i) = $ **true** if $\{x, u_i\} \in E(T)$ and $\xi_T(u_i) = $ **false** if

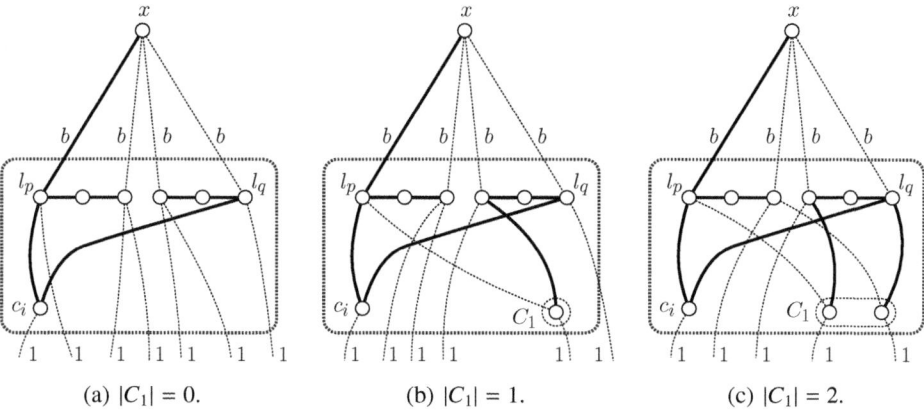

(a) $|C_1| = 0$.   (b) $|C_1| = 1$.   (c) $|C_1| = 2$.

Fig. 3. The cases of $|C_1| \leq 2$

$\{x, \bar{u}_i\} \in E(T)$. We show that ξ_T satisfies C. It suffices to show that for every $c_j \in C$, the unique neighbor l_i of c_j is adjacent to x. If l_i is not adjacent to x, then $cng_{G_C,T}(\{l_i, y_i\}) \geq a + b + 3 > k$ (see Fig. 4). This contradicts $cng_{G_C}(T) \leq k$. $\qquad\square$

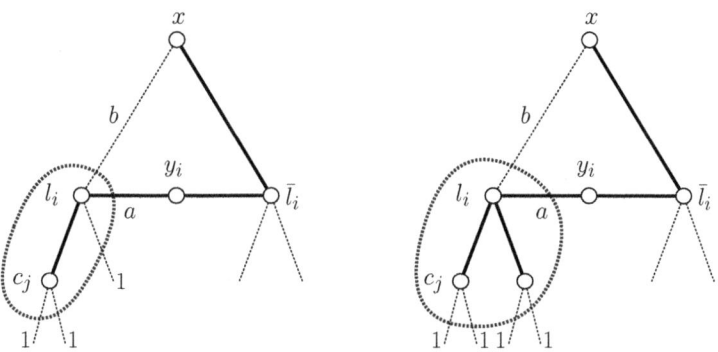

Fig. 4. Unsatisfied clauses

Lemma 6.5. *If C is satisfiable then $stc(G_C) \leq k$.*

Proof. Let ξ be a satisfying truth assignment for C. We say that a literal vertex l_i is a *true vertex* if l_i becomes **true** by the assignment ξ. We construct a spanning tree T of G_C as follows:

1. Take all bridge edges.
2. Take all literal edges incident to true vertices.
3. For each clause, take an arbitrary clause edge incident with a true vertex.

Clearly, T is a spanning tree of G_C. We show that $cng_{G_C}(T) \leq k$.

Let $u_i \in U$. Without loss of generality, we assume that $\{x, u_i\} \in E(T)$. Then T contains edges $\{x, u_i\}$ and $\{u_i, y_i\}$, $\{\bar{u}_i, y_i\}$. From the construction of T, T may contain any clause edge incident with u_i, but cannot contain any clause edge incident with $\bar{u}_i$. See Fig. 5. Clearly, the edge $\{u_i, y_i\}$ and $\{\bar{u}_i, y_i\}$ have the same congestion, and $cng_{G_C,T}(\{\bar{u}_i, y_i\}) = w(\theta(\{\bar{u}_i\})) = a + b + 2 = k$. If a clause edge incident with u_i is contained in T, then the edge has congestion $3 \le k$. Obviously, $cng_{G_C,T}(\{x, u_i\}) = w(\theta(\{u_i, \bar{u}_i\} \cup N_T(u_i) \setminus \{x\})) \le 2b + 6 \le k$ (see Fig. 5). $\qquad\square$

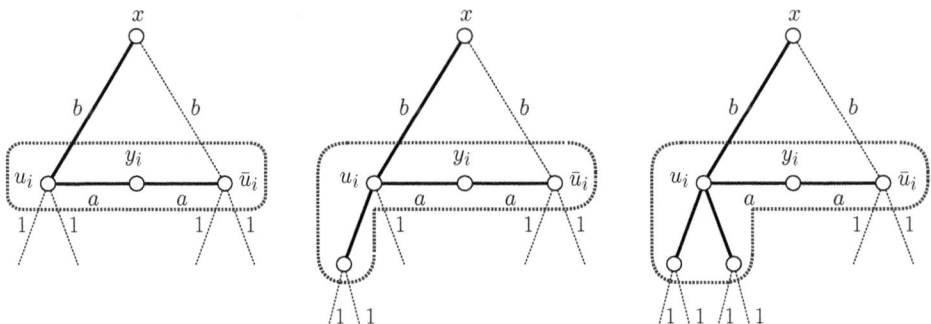

Fig. 5. A spanning tree of congestion at most k

7   Unweighted k-STC is NP-Complete for $k \ge 10$

Extending the result in the previous section, we prove the main theorem of the paper, that is, NP-completeness of k-STC for unweighted graphs. We need the following two lemmas. The proofs are omitted.

Lemma 7.1. *An edge e of weight $w \in \mathbb{Z}^+$ can be replaced by w parallel edges of unit weight without changing the spanning tree congestion.*

Lemma 7.2. *Edge subdivisions do not change the spanning tree congestion of unweighted graphs.*

Combining the above two lemmas, we can conclude that an edge $\{u, v\}$ of weight w can be replaced by w internally disjoint u–v paths of length two that consist of unweighted edges, without changing the spanning tree congestion. It is easy to see that this replacement can be done in $O(w)$ time. Thus, we have the following corollary.

Corollary 7.3. *Let G be an edge weighted graph such that the weight of every edge of G is a positive integer, and the maximum weight of the edges is w. Then G can be transformed into unweighted simple graph G' in $O(w \cdot |E(G)|)$ time, such that $stc(G) = stc(G')$.*

Now, we prove the main theorem of the paper.

Theorem 7.4. *For any fixed $k \ge 10$, k-STC is NP-complete for simple unweighted graphs that have only one vertex of unbounded degree.*

Proof. Let (U, C) be an instance of (3, B2)-SAT, and G_C the corresponding graph constructed in the previous section. From Corollary 7.3, we can construct a simple unweighted graph G_C' in polynomial time such that $stc(G_C') = stc(G_C)$. Clearly, $stc(G_C') \leq k$ if and only if C is satisfiable.

We show that the vertices other than x have bounded degree. The new vertices added by subdivisions have degree two. Clause vertices have degree three in G_C. Since clause vertices are only incident to unit weight edges, they have degree three in G_C'. Since every y_i is incident to two bridge edges of weight $a = \lceil k/2 \rceil + 1$, y_i has degree $2a \leq k+3$ in G_C'. Literal vertex l_i is incident to two clause edges, one bridge edge, and one literal edge that have weight one, a, and $b = \lfloor k/2 \rfloor - 3$, respectively. Thus, $deg_{G_C'}(l_i) = a + b + 2 = k$. Hence, the maximum degree of G_C' is bounded by $k + 3$, which is a constant. □

8   Concluding Remarks

We have proved that for fixed k, the problem of determining whether the spanning tree congestion of a given graph is at most k is solvable in linear time for planar graphs, graphs of bounded treewidth, and graphs of bounded degree. We also show that the problem can be solved in linear time for any graph if $1 \leq k \leq 3$. On the other hand, we show that if the input graph has one vertex of unbounded degree, then the problem becomes NP-complete for $k \geq 10$. The complexity of k-STC remains open for $4 \leq k \leq 9$.

Since the problem is hard in general, an approximation algorithm with good approximation ratio is required. We say that a polynomial time algorithm for spanning tree congestion is a c_1-approximation algorithm for positive number c_1 if there is a positive integer c_2 such that for any input graph G, the output k of the algorithm satisfies $k \leq c_1 \cdot stc(G) + c_2$. Using NP-hardness of 10-STC, the following constant lower bound on the approximation ratio can be shown (the proof is omitted).

Theorem 8.1. *There is no polynomial time c_1-approximation algorithm for the spanning tree congestion of simple unweighted graphs such that $c_1 < 11/10$, unless $P = NP$.*

We also considered the complexity of STC or k-STC on some restricted graph classes. It is known that the tree spanner problem is NP-hard for chordal graphs [4] and chordal bipartite graphs [5]. It would be interesting to determine the complexity of STC or k-STC for these graph classes.

References

1. Berman, P., Karpinski, M., Scott, A.D.: Approximation hardness of short symmetric instances of MAX-3SAT, ECCC TR03-049 (2003)
2. Bodlaender, H.L.: A linear-time algorithm for finding tree-decompositions of small treewidth. SIAM J. Comput. 25, 1305–1317 (1996)
3. Bodlaender, H.L.: A partial k-arboretum of graphs with bounded treewidth. Theoret. Comput. Sci. 209, 1–45 (1998)
4. Brandstädt, A., Dragan, F.F., Le, H.-O., Le, V.B.: Tree spanners on chordal graphs: complexity and algorithms. Theoret. Comput. Sci. 310, 329–354 (2004)
5. Brandstädt, A., Dragan, F.F., Le, H.-O., Le, V.B., Uehara, R.: Tree spanners for bipartite graphs and probe interval graphs. Algorithmica 47, 27–51 (2007)

6. Cai, L., Corneil, D.G.: Tree spanners. SIAM J. Discrete Math. 8, 359–387 (1995)
7. Castejón, A., Ostrovskii, M.I.: Minimum congestion spanning trees of grids and discrete toruses. Discuss. Math. Graph Theory 29, 511–519 (2009)
8. Cormen, T.H., Leiserson, C.E., Rivest, R.L., Stein, C.: Introduction to Algorithms, 3rd edn. MIT Press, Cambridge (2009)
9. Courcelle, B.: The monadic second-order logic of graphs III: Tree-decompositions, minor and complexity issues. Theor. Inform. Appl. 26, 257–286 (1992)
10. Diestel, R.: Graph Theory, 3rd edn. Springer, Heidelberg (2005)
11. Dragan, F.F., Fomin, F.V., Golovach, P.A.: Spanners in sparse graphs. In: Aceto, L., Damgård, I., Goldberg, L.A., Halldórsson, M.M., Ingólfsdóttir, A., Walukiewicz, I. (eds.) ICALP 2008, Part I. LNCS, vol. 5125, pp. 597–608. Springer, Heidelberg (2008)
12. Fekete, S.P., Kremer, J.: Tree spanners in planar graphs. Discrete Appl. Math. 108, 85–103 (2001)
13. Garey, M.R., Johnson, D.S.: Computers and Intractability: A Guide to the Theory of NP-Completeness. Freeman, New York (1979)
14. Hliněný, P., Oum, S., Seese, D., Gottlob, G.: Width parameters beyond tree-width and their applications. J. Comput. 51, 326–362 (2008)
15. Hopcroft, J., Tarjan, R.: Efficient planarity testing. J. ACM 21, 549–568 (1974)
16. Hruska, S.W.: On tree congestion of graphs. Discrete Math. 308, 1801–1809 (2008)
17. Kozawa, K., Otachi, Y., Yamazaki, K.: On spanning tree congestion of graphs. Discrete Math. 309, 4215–4224 (2009)
18. Law, H.-F.: Spanning tree congestion of the hypercube. Discrete Math. 309, 6644–6648 (2009)
19. Lawler, E.: Combinatorial Optimization: Networks and Matroids. Holt, Rinehart and Winston (1976)
20. Löwenstein, C., Rautenbach, D., Regen, F.: On spanning tree congestion. Discrete Math. 309, 4653–4655 (2009)
21. Ostrovskii, M.I.: Minimal congestion trees. Discrete Math. 285, 219–226 (2004)
22. Ostrovskii, M.I.: Minimum congestion spanning trees in planar graphs. Discrete Math. 310, 1204–1209 (2010)
23. Raspaud, A., Sýkora, O., Vrťo, I.: Congestion and dilation, similarities and differences: A survey. In: SIROCCO 2000, pp. 269–280. Carleton Scientific (2000)
24. Robertson, N., Seymour, P.D.: Graph minors. X. Obstructions to tree-decomposition. J. Combin. Theory Ser. B 52, 153–190 (1991)
25. Simonson, S.: A variation on the min cut linear arrangement problem. Math. Syst. Theory 20, 235–252 (1987)

MAX-CUT and Containment Relations in Graphs

Marcin Kamiński*

Département d'Informatique
Université Libre de Bruxelles
Marcin.Kaminski@ulb.ac.be

Abstract. We study MAX-CUT in classes of graphs defined by forbidding a single graph as a subgraph, induced subgraph, or minor. For the first two containment relations, we prove dichotomy theorems. For the minor order, we show how to solve MAX-CUT in polynomial time for the class obtained by forbidding a graph with crossing number at most one (this generalizes a known result for K_5-minor-free graphs) and identify an open problem which is the missing case for a dichotomy theorem.

Keywords: MAX-CUT, subgraph, induced subgraph, minor.

1 Introduction

MAX-CUT is a classical problem in combinatorial optimization and have been studied in different contexts – heuristics, approximation algorithms, exact algorithms, polyhedra. Here we suggest to look at the computational complexity of the problem in different classes of graphs. We focus on graphs obtained by forbidding a single graph as a subgraph, induced subgraph, or minor.

A *cut* in a graph is a partition of the vertex set into two disjoint sets. The *value* of a cut is the total weight of edges whose endpoint belong to two different parts of the cut. A cut of maximum value in G is called a *maximum cut* (or *max-cut*) and the value is denoted by $\mathbf{mc}(G)$. Notice that there is a one to one correspondence between a cut and the set of edges whose endpoint belong to two different parts of the cut. For convenience, we will sometimes refer to this set of edges as a cut as well.

The algorithmic MAX-CUT problem is to determine $\mathbf{mc}(G)$ given an input graph G. A cardinality variant of MAX-CUT is called SIMPLE MAX-CUT. (It is MAX-CUT in which all the weights on edges are equal.) Clearly, if MAX-CUT is solvable in polynomial time for some class of graphs, so it SIMPLE MAX-CUT. Also, if SIMPLE MAX-CUT is NP-complete for some class of graphs, so is MAX-CUT.

In this paper we consider simple, undirected, and real-weighted graphs. The terminology used is standard; for notions not defined here, we refer the reader to [15]. K_k is the complete graph on k vertices and P_k is the induced path on k vertices. $G \cup H$ denotes the disjoint union of G and H.

Our contribution. We look at the classes of graphs defined by forbidding a single graph H as a subgraph, induced subgraph, or minor.

---

* Chargé de Recherches du FRS-FNRS.

D.M. Thilikos (Ed.): WG 2010, LNCS 6410, pp. 15–26, 2010.

1. For H-subgraph-free graphs, we show that both SIMPLE MAX-CUT and MAX-CUT are solvable in polynomial time, if H is a forest every connected component of which is a tree with at most one vertex of degree 3; and are NP-complete otherwise.

2. For H-induced-subgraph-free graphs, we show that SIMPLE MAX-CUT is solvable in polynomial time, if H is an induced subgraph of P_4; and is NP-complete otherwise. (For MAX-CUT this containment relation is rather uninteresting.)

Our contribution here is to notice that such dichotomy theorems hold and to put known algorithmic and hardness results into this framework. Dichotomy theorems of this type are rather rare. We are aware of just one result of this type. For CHROMATIC NUMBER, Král et al. proved that the class of H-induced-subgraph-free graphs admits a polynomial-time algorithm, if H is an induced subgraph of P_4 or of $P_3 \cup K_1$; and the problem is NP-complete otherwise [25]. No such theorem is known for instance for STABLE SET. In fact, P_5-induced-subgraph-free graphs is the unique minimal class defined by a single forbidden induced subgraph for which the computational complexity of STABLE SET is unknown [26].

The case of minors is a bit different. Revisiting known results, we can show that MAX-CUT is solvable in polynomial time for H-minor-free graphs, for planar H; and SIMPLE MAX-CUT is NP-hard, when H is at "vertex-distance" at least 2 to planarity (becomes planar only when two of its vertices are removed). A remaining open problem is to determine the computational complexity of (SIMPLE) MAX-CUT, when H is a strict apex, which means a graph at "vertex-distance" 1 to planarity. Perhaps those classes admit a polynomial-time algorithm for MAX-CUT. We show that this indeed is a case when H is at "edge-distance" 1 to planarity (becomes planar only when one of its edges is removed). Clearly, graphs at "edge-distance" 1 to planarity are also at "vertex-distance" 1.

3. For H-minor-free graphs, we show that MAX-CUT is solvable in polynomial time, if H is a graph that can be drawn on the plane with at most one crossing.

This generalizes previous work on MAX-CUT for planar graphs [27], [23] and K_5-minor-free graphs [6].

2   Previous Work

Maximum Cut

MAX-CUT was among the twenty one problems whose NP-hardness was established in the foundational paper *"Reducibility among combinatorial problems"* by Richard Karp [24]. Since then it has been extensively studied and became one of the classical problems in the field of combinatorial optimization.

An early result that is of interest to us is a polynomial-time algorithm for MAX-CUT in planar graphs. It was first discovered by Orlova and Dorfman [27] and then independently by Hadlock [23]. The main idea of the solution is to fix an embedding of the planar input graph, take the dual, and find a pairing of vertices of odd degree in the dual graph using matching.

Grötschel and Pulleyblank introduced – by means of a polyhedral definition – the class of *weakly bipartite* graphs and showed how to solve MAX-CUT in this class [18]. Both planar and bipartite graphs are weakly bipartite, thus their result generalizes [27] and [23], as well as a trivial polynomial-time algorithm for bipartite graphs. (Notice that a maximum cut in a bipartite graph contains all its edges.) Later, Guenin proved that weakly bipartite graphs are exactly these that do not contain an odd-K_5-minor[1] [19], [20], [21] (see also [31] for a short proof).

The result of Guenin implies that K_5-minor-free graphs are weakly bipartite (since every odd-K_5-minor is a K_5-minor) and therefore there is a polynomial-time algorithm for MAX-CUT in this class. Before the characterization of weakly bipartite graphs was known, Barahona showed how to solve MAX-CUT in polynomial time in the class of graphs without a K_5-minor [6]. The paper uses a decomposition theorem for K_5-minor-free graphs due to Wagner [32]. The same paper also proves that MAX-CUT is NP-hard for K_6-minor-free graphs.

MAX-CUT is also solvable in polynomial-time in the class of graphs of bounded orientable genus [1]. This results was already attributed to Barahona by [18] but the preprint to which [18] refers apparently has never been published.

MAX-CUT is also known to be NP-complete on unit disk graphs [14] and solvable in polynomial time on graphs without long odd cycles (= a class of graphs with no odd cycles longer than k, for some $k \geq 3$) [17], on line graphs [5] (see also [22] for a simple proof), on graphs of bounded tree-width [8], [9]. Also, there exists a PTAS for MAX-CUT in classes of graphs with a forbidden minor [10].

Graphs with No Single-Crossing Minor

A graph is a *single-crossing* graph when it can be drawn on the plane with at most one crossing. $K_{3,3}$ and K_5 are examples of single-crossing graphs.

Wagner proved that a graph is K_5-minor-free if and only if it can be constructed from planar graphs and copies of the four-rung Möbius ladder glued together along cliques of size ≤ 3 [32]. He also showed that a graph is $K_{3,3}$-minor-free if and only if it can be constructed from planar graphs and copies of K_5 glued together along cliques of size ≤ 2 (possibly removing an edge after pasting along it).

Robertson and Seymour proved a more general theorem, describing the structure of graphs with a forbidden single-crossing minor [30]. They can be obtained from planar graphs and graphs of bounded tree-width (the bound depends on the forbidden single-crossing graph) by pasting them along cliques of size at most 3 and (possibly) removing some of the edges of those cliques afterwards.

---

[1] An odd minor is a restriction of the standard minor relation.

This structural result was made algorithmic by Demaine et al.; they gave an $\mathcal{O}(n^4)$ algorithm for finding this decomposition [11]. This was subsequently used to give parameterized algorithms with better dependence on the parameter [13] and approximation algorithms with better approximation ratio in classes defined by forbidding a single-crossing graph as a minor [11].

3   Forbidden Subgraph

In this section, we study classes of graphs defined by forbidding a single graph as a subgraph. Let us start with a simple lemma.

Lemma 1. *Let e be an edge of a graph G and G' the graph obtained by subdividing edge e twice. Then, $mc(G') = mc(G) + 2$.*

Proof. A double subdivision of e replaces e with 3 edges; let them be called e_L, e', and e_R. Let us say that a cut in G' is *good* if it contains both e_L and e_R. Notice that there is a maximum cut in G' which is good. Also, there exists a one-to-one correspondence between cuts in G and good cuts in G': cuts have the same edges f, for $f \neq e$, and e belongs to the cut in G iff e' belongs to the cut in G'. This correspondence makes the value of the cut in G' bigger by 2 than the cut in G.

The following lemma is a consequence of this subdivision property.

Lemma 2. SIMPLE MAX-CUT *is* NP-*complete in the following two classes of graphs:*

- *graphs not containing cycles of length at most k, for every $k \geq 3$;*
- *graphs not containing a pair of vertices of degree at least 3 at distance at most k, for every $k \geq 1$.*

Proof. Let us take a graph G, double subdivide all its edges, and then repeat the operation $\lceil \log_3 k \rceil$ times more, each time applying the operation to the outcome of the previous operation. It is easy to see that the graph G' obtained after the series of subdivisions has no cycles of length at most k, and has no pair of vertices of degree at least 3 at distance at most k. Notice that $\mathbf{mc}(G') = \mathbf{mc}(G) + 2(\lceil \log_3 k \rceil + 1)$ from Lemma 1. Since SIMPLE MAX-CUT is NP-hard in the class of all graphs, so it is in the two classes mentioned in the theorem.

We will need one more hardness result from [33].

Lemma 3. *[33]* SIMPLE MAX-CUT *is* NP-*complete in the class of graphs with maximum degree 3.*

We will need the following theorem from [7] (see also [28]) due to Bienstock, Robertson and Seymour.

Theorem 1. *[7] For every forest F, every graph with path-width $\geq |V(F)| - 1$ has a minor isomorphic to F.*

Lemma 4. *Let H be a forest whose every connected component is a tree with at most one vertex of degree three. A graph contains H as a minor if and only if it contains H as a subgraph.*

Proof. The backward implication is clear. We will suppose that H is connected; the forward implication will follow by induction on the number of connected components of H. Let G contain H as a minor. Let us fix a model of H in G. We will build a subgraph T. For each pair of adjacent vertices in H, we select an edge of G whose endpoints are in the two different bags corresponding to the vertices of H. Now, for each bag of the model of H that contains at least two (at most three) endpoints of the edges already in T, let us add to T a tree spanning these endpoints inside the bag. Clearly, T is a tree and T contains H as a subgraph. Hence, G contains H as a subgraph.

Lemma 5. *[9] For every constant t, there exists a polynomial-time algorithm solving* MAX-CUT *in the class of graphs of tree-width at most t.*

Theorem 2. *Both the* SIMPLE MAX-CUT *and* MAX-CUT *problems in the class of H-subgraph-free graphs are:*

- *solvable in polynomial time, if H is a forest whose every connected component is a tree with at most one vertex of degree three;*
- NP-*hard, otherwise.*

Proof. For the first part, let H be a forest whose every connected component is a tree with at most one vertex of degree three. The class of H-subgraph-free graphs is also H-minor-free by Lemma 4. Since H is a forest, from Theorem 1, the path-width of H-subgraph-free graphs is at most $|V(H)| - 2$. Therefore also their tree-width is at most $|V(H)| - 2$. From Lemma 5, MAX-CUT is solvable in polynomial on H-subgraph-free graphs.

For the second part, assume that H is not a forest whose every connected component is a tree with at most one vertex of degree three. Then, H contains a vertex of degree at least 4, or a cycle, or a pair of vertices of degree 3 in the same connected component. In the first case, the SIMPLE MAX-CUT is NP-complete in this class of H-subgraph-free graphs by applying Lemma 3, and in the two last cases by applying Lemma 2 with $k = |H|$.

4 Forbidden Induced Subgraph

In this section, we study classes of graphs defined by forbidding a single graph as an induced subgraph. We start with some useful definitions.

A *co-bipartite* graph is the complement of a bipartite graph. A *split graph* is one whose vertex set can be partitioned into a clique and an independent set. The class of split graphs was first studied in [16] where a characterization of these graphs was proved.

Lemma 6. *[16] The class of split graphs is the class of $(2K_2, C_4, C_5)$-induced-subgraph-free graphs.*

We will need two results from [9] (see also [8] for the conference version) that we state as the following lemma.

Lemma 7. *[9]* SIMPLE MAX-CUT *is solvable in polynomial time in the class of* P_4-*induced-subgraph-free graphs and is* NP-*hard in the class of split graphs and in the class of co-bipartite graphs.*

Lemma 8. *Let H be a tree containing a vertex of degree at least 3. Then,* SIMPLE MAX-CUT *is* NP-*hard in the class of H-induced-subgraph-free graphs.*

Proof. A tree with a vertex of degree at least 3 has stability number at least 3. Co-bipartite graphs have stability number at most 2 and hence are H-induced-subgraph-free. The NP-hardness follows from Lemma 7.

Lemma 9. SIMPLE MAX-CUT *is* NP-*hard in the class of* P_k-*induced-subgraph-free graphs, for all* $k \geq 5$.

Proof. Split graphs are P_k-induced-subgraph-free graphs, for all $k \geq 5$. The NP-hardness follows from Lemma 7.

Theorem 3. *The* SIMPLE MAX-CUT *problem in the class of H-induced-subgraph-free graphs is:*

○ *solvable in polynomial time, if H is an induced subgraph of* P_4;
○ NP-*hard, otherwise.*

Proof. The first part of the theorem follows from Lemma 7.

For the second part, suppose H is not an induced subgraph of P_4. If it contains a cycle, then MAX-CUT is NP-complete in the class of H-induced-subgraph-free graphs by applying Lemma 2 with $k = |H|$. (Shortest cycle in a graph is necessarily induced.) We can assume that H is a forest. If it has a vertex of of degree 3,T then from Lemma 8, SIMPLE MAX-CUT is NP-hard. Thus, we can assume that H is a forest of paths. If one of the paths (connected components of H) contains more than 5 vertices, then the class of P_5-induced-subgraph-free graphs is contained in the class of H-induced-subgraph-free graphs and therefore SIMPLE MAX-CUT is NP-hard in H-induced-subgraph-free graphs from Lemma 9.

We can assume that H is a forest of induced subgraphs of P_4. If two of the connected components of H are not singletons, then H contains $2K_2$ and SIMPLE MAX-CUT is NP-hard in H-induced-subgraph-free graphs from Lemmas 6 and 7. Also, if H has three connected components, then the stability number of H is at least 3, and since co-bipartite graphs have stability number at most 2, the NP-hardness follows from Lemma 7. If H has two components, and one has at least 3 vertices, SIMPLE MAX-CUT is NP-hard for the same reason. Otherwise, H is an induced subgraph of P_4.

A similar theorem for MAX-CUT is perhaps less interesting but we include it here for completeness.

Theorem 4. *The* MAX-CUT *problem in the class of H-induced-subgraph-free graphs is:*

○ *solvable in polynomial time, if H is clique on at most two vertices;*
○ NP-*hard, otherwise.*

Proof. The first part of the theorem is trivial since K_2-induced-subgraph-free graphs are edgeless. For the second part, notice that MAX-CUT is NP-hard whenever SIMPLE MAX-CUT is. It remains to show that MAX-CUT is NP-hard for three classes of graphs: $K_1 \cup K_1$-induced-subgraph-free, $K_1 \cup K_2$-induced-subgraph-free, and P_3-induced-subgraph-free. However, each of these three classes contains the class of cliques and MAX-CUT is NP-hard on cliques. Indeed, every graph can be embedded in a clique using weights 0 and 1.

The techniques we use in this and the previous section have been dveloped by Alekseev and Lozin and applied to different problems in the context of boundary graph classes; see for example [2], [4], [3].

5   Forbidden Minor

In this section, we study classes of graphs defined by forbidding a single graph as a minor.

Definitions

Let G and H be two graphs with disjoint vertex sets, K_G and K_H cliques of size $k \geq 0$ in G and H respectively. A k-*sum* $G \oplus H$ is the graph obtained from G and H by identifying vertices of K_G and K_H (according to some bijection between the cliques) and then possibly removing some edges between the vertices of the identified clique.

A *single-crossing* graph is one that can be drawn on the plane with at most one crossing of edges. $K_{3,3}$ and K_5 are examples of single-crossing graphs. A graph H is an *apex* graph if it has a vertex v such that $H - v$ is planar. Clearly, single-crossing graphs are apex graphs. A graph H is a k-*apex* (for $k \geq 1$) if it has a set of vertices S of cardinality k such that $H \setminus S$ is planar. An apex is a 1-apex. We say that a k-apex is *strict*, when it is not $(k-1)$-apex, for $k > 1$, or planar for 1-apex.

Single-Crossing-Minor-Free Classes

Now we focus on graph classes defined by forbidding a single-crossing graph as a minor. We need to introduce a variant of MAX-CUT, called RESTRICTED MAX-CUT, that allows to specify for some vertices of the input graph, to which part of the cut they must belong.

First we need a decomposition theorem for graphs excluding a single-crossing as a minor. This algorithmic version is due to Demaine et al. [11] (see also [12] for the conference version).

Theorem 5. *[11] For a single-crossing H, there exists a constant c_H such that every H-minor-free graph G can be decomposed in time $O(n^4)$ into a series of clique-sum operations $G = G_1 \oplus \ldots \oplus G_m$, where each G_i ($1 \leq i \leq m$) is a minor of G and is either planar or its tree-width is at most c_H, and each $\oplus$ is an k-sum, for $0 \leq k \leq 3$.*

When the graph is weighted, the edges of graphs G_i that also exist in G have the same weights as in G; the edges of graphs G_i that do not exist in G are assigned weight 0.

To be able to use the decomposition theorem we need to analyze how a solution to MAX-CUT propagates through clique sums.

Lemma 10. *Let G and H be two graphs and $G \oplus H$ be their k-sum, for $0 \leq k \leq 3$. Given solutions to the instances of RESTRICTED MAX-CUT on G defined by considering all possible assignments of vertices from the k-clique to different parts of the cut, one can find in time $O(1)$ weights w^* on H such that $\mathbf{mc}_w(G \oplus H) = \mathbf{mc}_{w^*}(H) + T_\emptyset$, where $T_\emptyset$ is the value of RESTRICTED MAX-CUT on G when all vertices of the k-clique are required to belong to the same part of the cut.*

Proof. We will consider different cases depending on k.

CASE $k = 0, 1$. If $k = 0$, then $G \oplus H$ is a disjoint union of G and H. If $k = 1$, $G \oplus H$ is obtained from G and H by identifying one vertex. In both cases $\mathbf{mc}_w(G \oplus H) = \mathbf{mc}_w(H) + \mathbf{mc}_w(G)$. Setting $w^* = w$, we get $\mathbf{mc}_{w^*}(H) = \mathbf{mc}_w(H)$. Clearly $T_\emptyset = \mathbf{mc}_w(G)$. Hence, $\mathbf{mc}_w(G \oplus H) = \mathbf{mc}_w(H) + \mathbf{mc}_w(G) = \mathbf{mc}_{w^*}(H) + T_\emptyset$.

CASE $k = 2$. Let e_0 be the edge of the 2-clique. Let T_{e_0} be the value of RESTRICTED MAX-CUT on G when we require edge e_0 to be in the cut (= the endpoints of e_0 are forced to be in two different parts of the cut). Let $w^*(e_0) = T_{e_0} - T_\emptyset$; and for all edges $e \in E(H)$ different than e_0, $w^*(e) = w(e)$. Now, it is easy to verify that $\mathbf{mc}_w(G \oplus H) = \mathbf{mc}_{w^*}(H) + T_\emptyset$.

CASE $k = 3$. Let e_0, e_1, e_2 be the three edges of the 3-clique. Notice that a maximum-cut in G (in H, and in $G \oplus H$ as well), will always contain an even number of the edges of a 3-clique – either none, or exactly two. For two distinct edges $f, g \in \{e_0, e_1, e_2\}$, let $T_{f,g}$ be the value of RESTRICTED MAX-CUT on G when we require f and g to belong to the cut (= the common endpoint of edges f, g is forced to be in the other part of the cut than their "private" endpoints).

Let $w^*(e_j) = T_{e_{j-1}, e_{j+1}} - T_\emptyset$, for $0 \leq j \leq 2$, where all indices are taken modulo 3; and for all edges $e \in E(H)$ different than e_0, e_1, e_2, $w^*(e) = w(e)$. Now, it is easy to verify that $\mathbf{mc}_w(G \oplus H) = \mathbf{mc}_{w^*}(H) + T_\emptyset$.

We need to tailor the previous lemma to our needs.

Lemma 11. *Let c be a constant and G be a planar graph or a graph of tree-width at most c. Let H be a graph and $G \oplus H$ be a k-sum, for $0 \leq k \leq 3$ of G and H. One can find in polynomial time weights w^* on H and a constant d such that $\mathbf{mc}_w(G \oplus H) = \mathbf{mc}_{w^*}(H) + d$.*

Proof. To use Lemma 10 we need to show how to compute solutions to the instances of RESTRICTED MAX-CUT on G defined by considering all possible assignments of vertices from the k-clique to different parts of the cut.

Notice that increasing the weight of an edge by a large number M ($\geq$ than the sum of the weights of all the edges in the graph) will force the endpoints of this edge to belong to two different parts of every maximum cut in the new graph.

MAX-CUT in a graph with an edge contracted corresponds to RESTRICTED MAX-CUT in the original graph with the two endpoints of the contracted edge required to belong to the same part of the cut. (After contraction we remove loops; for multiple edges, we also remove them but the weight of the edge that remains equals to the sum of the weights of all the parallel ones.)

Hence, we can simulate RESTRICTED MAX-CUT by instances of MAX-CUT. Notice that edge contraction preserves planarity and does not increase tree-width. Since G is planar or of bounded tree-width, MAX-CUT can be solved in polynomial time [27], [23], [9]. Also, since $k \leq 3$, we only need to consider a constant number of different MAX-CUT instances on G.

Theorem 6. *Let H be a single-crossing graph. MAX-CUT can be solved in polynomial time in the class of H-minor-free graphs.*

Proof. First we apply Theorem 5 and find a decomposition of the input graph $G = G_1 \oplus \ldots \oplus G_m$. We will be processing graphs G_i from left to right. Let $G_1^* = G_1$.

We apply Lemma 11 to $G_i^* \oplus G_{i+1}$, $i = 1, \ldots, m-1$. Hence, there is a constant d and weights w^* such that $\mathbf{mc}_w(G_i^* \oplus G_{i+1}) = \mathbf{mc}_{w^*}(G_i) + d$. Let us denote G_i with the new weights by G_i^*. Notice that every G_i^* is either planar or a graph of tree-width at most c_H (c_H is the constant from Theorem 5), so Lemma 11 can be applied.

Finally, we conclude that there is a constant d' such that $\mathbf{mc}(G_m^*) + d' = \mathbf{mc}_w(G_1 \oplus \ldots \oplus G_m) = \mathbf{mc}_w(G)$.

H-Minor-Free Classes

Now we will look at classes of graphs defined by forbidding a single graph as a minor. We will need the following theorem due to Robertson and Seymour.

Theorem 7. *[29] For every planar graph H, there is a number w such that every planar graph with no minor isomorphic to H has tree-width $\leq w$.*

The following lemma is a consequence of Lemma 5 and Theorem 7.

Lemma 12. *Let H be a planar graph. MAX-CUT can be solved in the class of H-minor-free graphs in polynomial time.*

We mentioned in Section 2 that Barahona proved in [6] that SIMPLE MAX-CUT is NP-hard on K_6-minor-free graphs. Here is the precise statement of his result.

Theorem 8 (Theorem 5.1 in [6]). *Let G be a graph with a vertex v such that $G - v$ is a cubic planar graph.* SIMPLE MAX-CUT *is NP-complete in the class of such graphs.*

This class of graphs is in fact K_6-minor-free. However, as easily seen, it does not contain any strict k-apex graph (for $k \geq 2$) as a minor. Let us state it as a lemma.

Lemma 13. *Let H be a strict k-apex graph, for $k \geq 2$.* SIMPLE MAX-CUT *is NP-hard in the class of H-minor-free graphs.*

Considering the computational complexity of (SIMPLE) MAX-CUT in the class of H-minor-free graphs, we find an interesting situation. If H is planar, then MAX-CUT is solvable in polynomial time (Lemma 12); if H is a strict k-apex, for $k \geq 2$, then SIMPLE MAX-CUT is NP-complete (Lemma 13). What happens when H is an apex graph? – this is the missing case in a dichotomy theorem.

Graphs in classes of bounded orientable genus and graph classes obtained by excluding some single-crossing are H-minor-free for some apex graph H. The fact that MAX-CUT is solvable in polynomial time in those classes of graphs provide grounds for the following conjecture.

Conjecture. Let H be an apex graph. (SIMPLE) MAX-CUT is solvable in polynomial time in the class of H-minor-free graphs.

References

1. Vondrak, J., Galluccio, A., Loebl, M.: Optimization via enumeration: a new algorithm for the max cut problem. Mathematical Programming 90(2), 273–290 (2001)
2. Alekseev, V.E.: On easy and hard hereditary classes of graphs with respect to the independent set problem. Discrete Applied Mathematics 132(1-3), 17–26 (2003)
3. Alekseev, V.E., Boliac, R., Korobitsyn, D.V., Lozin, V.V.: Np-hard graph problems and boundary classes of graphs. Theor. Comput. Sci. 389(1-2), 219–236 (2007)
4. Alekseev, V.E., Korobitsyn, D.V., Lozin, V.V.: Boundary classes of graphs for the dominating set problem. Discrete Mathematics 285(1-3), 1–6 (2004)
5. Arbib, C.: A polynomial characterization of some graph partitioning problems. Inf. Process. Lett. 26(5), 223–230 (1988)
6. Barahona, F.: The Max-Cut problem on graphs not contractible to K_5. Oper. Res. Lett. 2(3), 107–111 (1983)
7. Bienstock, D., Robertson, N., Seymour, P.D., Thomas, R.: Quickly excluding a forest. J. Comb. Theory, Ser. B 52(2), 274–283 (1991)
8. Bodlaender, H.L., Jansen, K.: On the complexity of the maximum cut problem. In: Enjalbert, P., Mayr, E.W., Wagner, K.W. (eds.) STACS 1994. LNCS, vol. 775, pp. 769–780. Springer, Heidelberg (1994)
9. Bodlaender, H.L., Jansen, K.: On the complexity of the maximum cut problem. Nordic J. of Computing 7(1), 14–31 (2000)
10. Demaine, E.D., Hajiaghayi, M.T., Kawarabayashi, K.: Algorithmic graph minor theory: Decomposition, approximation, and coloring. In: FOCS, pp. 637–646. IEEE Computer Society, Los Alamitos (2005)

11. Demaine, E.D., Hajiaghayi, M.T., Nishimura, N., Ragde, P., Thilikos, D.M.: Approximation algorithms for classes of graphs excluding single-crossing graphs as minors. J. Comput. Syst. Sci. 69(2), 166–195 (2004)
12. Demaine, E.D., Hajiaghayi, M.T., Thilikos, D.M.: 1.5-approximation for treewidth of graphs excluding a graph with one crossing as a minor. In: Jansen, K., Leonardi, S., Vazirani, V.V. (eds.) APPROX 2002. LNCS, vol. 2462, pp. 67–80. Springer, Heidelberg (2002)
13. Demaine, E.D., Hajiaghayi, M.T., Thilikos, D.M.: Exponential speedup of fixed-parameter algorithms for classes of graphs excluding single-crossing graphs as minors. Algorithmica 41(4), 245–267 (2005)
14. Díaz, J., Kamiński, M.: Max-Cut and Max-Bisection are NP-hard on unit disk graphs. Theoretical Computer Science 377, 271–276 (2007)
15. Diestel, R.: Graph Theory, Electronic edn. Springer, Heidelberg (2005)
16. Földes, S., Hammer, P.L.: Split graphs. Congress. Numer., 311–315 (1978)
17. Grötschel, M., Nemhauser, G.L.: A polynomial algorithm for the Max-Cut problem on graphs without long odd cycles. Math. Prog. 29, 28–40 (1984)
18. Grötschel, M., Pulleyblank, W.R.: Weakly bipartite graphs and the Max-Cut problem. Oper. Res. Lett. 1, 23–27 (1981)
19. Guenin, B.: A characterization of weakly bipartite graphs. In: Bixby, R.E., Boyd, E.A., Ríos-Mercado, R.Z. (eds.) IPCO 1998. LNCS, vol. 1412, pp. 9–22. Springer, Heidelberg (1998)
20. Guenin, B.: A characterization of weakly bipartite graphs. Electronic Notes in Discrete Mathematics 5, 149–151 (2000)
21. Guenin, B.: A characterization of weakly bipartite graphs. J. Comb. Theory, Ser. B 83(1), 112–168 (2001)
22. Guruswami, V.: Maximum cut on line and total graphs. Discrete Applied Mathematics 92(2-3), 217–221 (1999)
23. Hadlock, F.: Finding a maximum cut of a planar graph in polynomial time. SIAM J. Comput. 4(3), 221–225 (1975)
24. Karp, R.M.: Reducibility among combinatorial problems. In: Miller, R.E., Thatcher, J.W. (eds.) Complexity of Computer Computations, pp. 85–104. Plenum Press, New York (1972)
25. Král, D., Kratochvíl, J., Tuza, Z., Woeginger, G.J.: Complexity of coloring graphs without forbidden induced subgraphs. In: Brandstädt, A., Van Bang, L. (eds.) WG 2001. LNCS, vol. 2204, pp. 254–262. Springer, Heidelberg (2001)
26. Lozin, V.V., Mosca, R.: Maximum independent sets in subclasses of p_5-free graphs. Inf. Process. Lett. 109(6), 319–324 (2009)
27. Orlova, G., Dorfman, Y.: Finding the maximum cut in a graph. Tekhnicheskaya Kibernetika (Engineering Cybernetics) 10, 502–506 (1972)
28. Robertson, N., Seymour, P.D.: Graph minors. I. Excluding a forest. J. Comb. Theory, Ser. B 35(1), 39–61 (1983)
29. Robertson, N., Seymour, P.D.: Graph minors. V. Excluding a planar graph. J. Comb. Theory, Ser. B 41, 92–114 (1986)
30. Robertson, N., Seymour, P.D.: Excluding a graph with one crossing. In: Robertson, N., Seymour, P.D. (eds.) Graph Structure Theory. Contemporary Mathematics, vol. 147, pp. 669–676. American Mathematical Society, Providence (1991)

31. Schrijver, A.: A short proof of Guenin's characterization of weakly bipartite graphs. J. Comb. Theory, Ser. B 85(2), 255–260 (2002)
32. Wagner, K.: Über eine Eigenshaft der ebenen Komplexe. Math. Ann. 114, 570–590 (1937)
33. Yannakakis, M.: Node- and edge-deletion NP-complete problems. In: STOC, pp. 253–264. ACM, New York (1978)

The Longest Path Problem is Polynomial on Cocomparability Graphs

Kyriaki Ioannidou and Stavros D. Nikolopoulos

Department of Computer Science, University of Ioannina
GR-45110 Ioannina, Greece
{kioannid,stavros}@cs.uoi.gr

Abstract. The longest path problem is the problem of finding a path of maximum length in a graph. As a generalization of the Hamiltonian path problem, it is NP-complete on general graphs and, in fact, on every class of graphs that the Hamiltonian path problem is NP-complete. Polynomial solutions for the longest path problem have recently been proposed for weighted trees, ptolemaic graphs, bipartite permutation graphs, interval graphs, and some small classes of graphs. Although the Hamiltonian path problem on cocomparability graphs was proved to be polynomial almost two decades ago [9], the complexity status of the longest path problem on cocomparability graphs has remained open until now; actually, the complexity status of the problem has remained open even on the smaller class of permutation graphs. In this paper, we present a polynomial-time algorithm for solving the longest path problem on the class of cocomparability graphs. Our result resolves the open question for the complexity of the problem on such graphs, and since cocomparability graphs form a superclass of both interval and permutation graphs, extends the polynomial solution of the longest path problem on interval graphs [18] and provides polynomial solution to the class of permutation graphs.

Keywords: Longest path problem, cocomparability graphs, permutation graphs, polynomial algorithm, complexity.

1 Introduction

The problem of finding a path of maximum length in a graph (Longest Path Problem) generalizes the Hamiltonian path problem and thus it is NP-complete on general graphs; in fact, it is NP-complete on every class of graphs that the Hamiltonian path problem is NP-complete. It is thus interesting to study the longest path problem on classes of graphs $\mathcal{C}$ where the Hamiltonian path problem is polynomial, since if a graph $G \in \mathcal{C}$ is not Hamiltonian, it makes sense in several applications to search for a longest path of G. Although the Hamiltonian path problem has been extensively studied in the past two decades, only recently did the longest path problem start receiving attention [11,12,13,23,25,26,27].

The Hamiltonian path problem is known to be NP-complete in general graphs [14,15], and remains NP-complete even when restricted to some small

D.M. Thilikos (Ed.): WG 2010, LNCS 6410, pp. 27–38, 2010.

classes of graphs such as split graphs [16], chordal bipartite graphs, split strongly chordal graphs [21], directed path graphs [22], circle graphs [7], planar graphs [15], and grid graphs [19,24]. On the other hand, it admits polynomial time solutions on some known classes of graphs; such classes include interval graphs [1,8], circular-arc graphs [8], biconvex graphs [2], and cocomparability graphs [9]. Note that the problem of finding a longest path on proper interval graphs is easy, since all connected proper interval graphs have a Hamiltonian path which can be computed in linear time [3].

Polynomial time solutions for the longest path problem are known only for small classes of graphs. Specifically, a linear-time algorithm for finding a longest path in a tree was proposed by Dijkstra early in 1960, a formal proof of which can be found in [5]. Recently, through a generalization of Dijkstra's algorithm for trees, Uehara and Uno [25] solved the longest path problem for weighted trees and block graphs in linear time and space, and for cacti in $O(n^2)$ time and space, where n is the number of vertices of the input graph. Polynomial algorithms for the longest path problem have been also proposed on bipartite permutation and ptolemaic graphs having $O(n)$ and $O(n^5)$ time complexity, respectively [23,26]. Furthermore, Uehara and Uno in [25] solved the longest path problem on a subclass of interval graphs in $O(n^3(m + n\log n))$ time, and as a corollary they showed that a longest path on threshold graphs can be found in $O(n + m)$ time and space. Recently, Ioannidou et al. [18] showed that the longest path problem has a polynomial solution on interval graphs by proposing an algorithm that runs in $O(n^4)$ time, answering thus the question left open in [25] concerning the complexity of the problem on interval graphs.

In this paper we present a polynomial-time algorithm for solving the longest path problem on the class of cocomparability graphs, an important and well-known class of perfect graphs [16]. The Hamiltonian path problem on cocomparability graphs has been proved to be polynomial [9], while the status of the longest path problem on such graphs was unknown; actually, the status of the longest path problem was unknown even on the smaller class of permutation graphs. Thus, our result resolves the open question for the complexity of the problem on cocomparability graphs, and since cocomparability graphs form a superclass of both interval and permutation graphs, extends the polynomial solution of the longest path problem on interval graphs [18], and also provides polynomial solution to the class of permutation graphs.

2   Theoretical Framework

For basic definitions in graph theory refer to [4,16,20]. A *simple path* (resp. *antipath*) of a graph G is a sequence of distinct vertices $v_1, v_2, \ldots, v_k$ such that $v_i v_{i+1} \in E(G)$ (resp. $v_i v_{i+1} \notin E(G)$), for each i, $1 \le i \le k - 1$, and is denoted by $(v_1, v_2, \ldots, v_k)$; throughout the paper all paths and antipaths are considered to be simple. We denote by $V(P)$ the set of vertices in the path (antipath) P, and define the *length* of the path (antipath) P to be the number of vertices in P, i.e., $|P| = |V(P)|$. We call *right endpoint* of a path (antipath) $P = (v_1, v_2, \ldots, v_k)$ the last vertex v_k of P. Moreover, if $P = (v_1, v_2, \ldots, v_{i-1}, v_i, v_{i+1}, \ldots, v_j, v_{j+1}, v_{j+2}, \ldots, v_k)$ is a path (antipath) of a graph and $P_0 = (v_i, v_{i+1}, \ldots, v_j)$ is a subpath (subantipath) of P, we shall denote the path (antipath) P by $P = (v_1, v_2, \ldots, v_{i-1}, P_0, v_{j+1}, v_{j+2}, \ldots, v_k)$.

2.1 Partial Orders and Cocomparability Graphs

A *partial order* will be denoted by $\mathcal{P} = (V, <_{\mathcal{P}})$, where V is the finite ground set of elements or vertices and $<_{\mathcal{P}}$ is an irreflexive, antisymmetric, and transitive binary relation on V. Two elements $a, b \in V$ are comparable in $\mathcal{P}$ (denoted by $a \sim_{\mathcal{P}} b$) if $a <_{\mathcal{P}} b$ or $b <_{\mathcal{P}} a$; otherwise, they are said to be incomparable (denoted by $a \parallel b$). An *extension* of a partial order $\mathcal{P} = (V, <_{\mathcal{P}})$ is a partial order $L = (V, <_L)$ on the same ground set that *extends* $\mathcal{P}$, i.e., $a <_{\mathcal{P}} b \Rightarrow a <_L b$, for all $a, b \in V$. The *dual partial order* $\mathcal{P}^d$ of $\mathcal{P} = (V, <_{\mathcal{P}})$ is a partial order $\mathcal{P}^d = (V, <_{\mathcal{P}^d})$ such that for any two elements $a, b \in V$, $a <_{\mathcal{P}^d} b$ if and only if $b <_{\mathcal{P}} a$.

The graph G, edges of which are exactly the comparable pairs of a partial order $\mathcal{P}$ on $V(G)$, is called the *comparability graph* of $\mathcal{P}$, and is denoted by $G(\mathcal{P})$. The complement graph $\overline{G}$, whose edges are the incomparable pairs of $\mathcal{P}$, is called the *cocomparability graph* of $\mathcal{P}$, and is denoted by $\overline{G}(\mathcal{P})$. Alternatively, a graph G is a cocomparability graph if its complement graph $\overline{G}$ has a transitive orientation, corresponding to the comparability relations of a partial order $\mathcal{P}_{\overline{G}}$. Note that a partial order $\mathcal{P}$ uniquely determines its comparability graph $G(\mathcal{P})$ and its cocomparability graph $\overline{G}(\mathcal{P})$, but the reverse is not true, i.e., a cocomparability graph G has as many partial orders $\mathcal{P}_{\overline{G}}$ as the number of the transitive orientations of $\overline{G}$. Also, the class of cocomparability graphs is hereditary.

Let G be a comparability graph, and let $\mathcal{P}_G$ be a partial order which corresponds to G. The graph G can be represented by a directed covering graph with layers $H_1, H_2, \ldots, H_h$, in which each vertex is on the highest possible layer. That is, the maximal vertices of the partial order $\mathcal{P}_G$ are on the highest layer H_h, and for every vertex v on layer H_{i-1} there exists a vertex u on layer H_i such that $v <_{\mathcal{P}_G} u$; such a layered representation of G (respectively $\mathcal{P}_G$) is a called the *Hasse diagram* of G (respectively $\mathcal{P}_G$) [9].

Let $\sigma = (V(G), <_\sigma)$ be a partial order on the vertices of a comparability graph G, such that for any two vertices $v, u \in V(G)$, $v <_\sigma u$ if and only if $v \in H_i$, $u \in H_j$, and $i < j$; hereafter, we equivalently denote $v <_\sigma u$ by $u >_\sigma v$. For simplicity sometimes we shall write $v =_\sigma u$, for vertices $v, u \in V(G)$ which belong to the same layer H_i; we write $v \neq_\sigma u$ to denote that vertices $v, u \in V(G)$ belong to different layers. Also, $v \leq_\sigma u$ implies that either $v <_\sigma u$ or $v =_\sigma u$; again, we equivalently denote $v \leq_\sigma u$ by $u \geq_\sigma v$. Throughout the paper, such an ordering σ is called a *layered ordering* of G. Note that, the partial order σ is an extension of the partial order $\mathcal{P}_G$; in particular, it holds $v <_{\mathcal{P}_G} u$ if and only if $v <_\sigma u$ and $vu \in E(G)$, for any two vertices $u, v \in V(G)$.

Since a comparability graph G does not uniquely determine a partial order, hereafter we will represent a comparability graph G by its Hasse diagram and we will denote the partial order $(V(G), <_{\mathcal{P}_G})$ to which the Hasse diagram of G

corresponds by $\mathcal{P}_G$; that is, the vertices which are on the highest layer H_h of the Hasse diagram are the maximal vertices of the partial order $\mathcal{P}_G$, and for two vertices $u, v \in V(G)$, $v <_{\mathcal{P}_G} u$ if $v \in H_{i-1}$, $u \in H_i$ and $uv \in E(G)$. Thus, we will say that $\mathcal{P}_G$ is the partial order which *corresponds* to the comparability graph G. Note that vertices in the Hasse diagram satisfy the following property: for any three vertices $v, u, w \in V(G)$ such that $v \in H_i$, $u \in H_j$, $w \in H_k$, and $i < j < k$ (or, equivalently, $v <_\sigma u <_\sigma w$), if $vu \in E(G)$ and $uw \in E(G)$, then $vw \in E(G)$.

The following definition and results where given by Damaschke *et al.* in [9], based on which they prove the correctness of their algorithm for finding a Hamiltonian path of a cocomparability graph; note that their algorithm uses the bump number algorithm which is presented in [17].

Definition 1. *(Damaschke et al. [9]): Let G be a comparability graph, and let $\mathcal{P}_G$ be the partial order which corresponds to G. A path $P = (v_1, v_2, \ldots, v_k)$ of the cocomparability graph $\overline{G}$ is monotone if $v_i <_{\mathcal{P}_G} v_j$ implies $i < j$.*

Lemma 1. *(Damaschke et al. [9]): Let G be a comparability graph, and let $\mathcal{P}_G$ be the partial order which corresponds to G. Let $P = (v_1, v_2, \ldots, v_k)$ be a Hamiltonian path of the cocomparability graph $\overline{G}$ such that v_1 is a minimal element of $\mathcal{P}_G$. Then there exists a monotone Hamiltonian path P' of $\overline{G}$ starting with v_1.*

Theorem 1. *(Damaschke et al. [9]): Let G be a cocomparability graph. Then, G has a Hamiltonian path if and only if G has a monotone Hamiltonian path.*

It appears that the above two results hold not only for Hamiltonian paths of a cocomparability graph $\overline{G}$, but also for any path of $\overline{G}$. Indeed, let P be a path of $\overline{G}$ and let $\overline{G'} = \overline{G}[V(P)]$ be the subgraph of $\overline{G}$ induced by the vertices of P. Also, let $\mathcal{P}_{G'}$ be the partial order which corresponds to G' such that $\mathcal{P}_G$ is an extension of $\mathcal{P}_{G'}$, i.e., for any two vertices $u, v \in V(\overline{G})$, if $u <_{\mathcal{P}_G} v$ and $u, v \in V(\overline{G'})$, then $u <_{\mathcal{P}_{G'}} v$. Then, since P is a Hamiltonian path of $\overline{G'}$, from Theorem 1 there exists a monotone path P' of $\overline{G'}$ (with respect to $\mathcal{P}_{G'}$) such that $V(P') = V(P)$. From Definition 1 it is easy to see that P' is also a monotone path of $\overline{G}$ (with respect to $\mathcal{P}_G$), since $\mathcal{P}_G$ is an extension of $\mathcal{P}_{G'}$.

Additionally, since a path P of a cocomparability graph $\overline{G}$ is an antipath of the comparability graph G, and since our algorithm for computing a longest path of a cocomparability graph $\overline{G}$ computes in fact a longest antipath of the comparability graph G, we restate the above definition and results and whenever P denotes a path of a cocomparability graph $\overline{G}$, we refer to P as an antipath of the comparability graph G.

We first restate Definition 1 as follows: an antipath $P = (v_1, v_2, \ldots, v_k)$ of a comparability graph G is monotone if $v_i <_{\mathcal{P}_G} v_j$ implies $i < j$, where $\mathcal{P}_G$ is the partial order which corresponds to G. We next restate Lemma 1 and Theorem 1 in a form stronger than the one stated in [9].

Lemma 2. *Let G be a comparability graph, and let $\mathcal{P}_G$ be the partial order which corresponds to G. Let $P = (v_1, v_2, \ldots, v_k)$ be an antipath of G such that v_1 is a minimal element of $V(P)$ in $\mathcal{P}_G$. Then there exists a monotone antipath P' of G starting with vertex v_1 such that $V(P') = V(P)$.*

Theorem 2. *Let G be a comparability graph. If P is an antipath of G, then there exists a monotone antipath P' of G such that $V(P') = V(P)$.*

The following lemma holds.

Lemma 3. *Let G be a comparability graph, and let σ be the layered ordering of G. Let $P = (v_1, v_2, \ldots, v_k)$ be an antipath of G, and let $v_\ell \notin V(P)$ be a vertex of G such that $v_1 \leq_\sigma v_\ell <_\sigma v_k$ and $v_\ell v_k \in E(G)$. Then there exist two consecutive vertices v_{i-1} and v_i in P, $2 \leq i \leq k$, such that $v_{i-1}v_\ell \notin E(G)$ and $v_\ell <_\sigma v_i$.*

2.2   Normal Antipaths on Comparability Graphs

Our algorithm for computing a longest antipath P of a comparability graph G uses a specific type of antipaths, which we call *normal* antipaths.

Definition 2. *Let G be a comparability graph, and let σ be a layered ordering of G. The antipath $P = (v_1, v_2, \ldots, v_k)$ of G is called normal, if v_1 is a leftmost (i.e., minimal) vertex of $V(P)$ in σ, and for every i, $2 \leq i \leq k$, the vertex v_i is a leftmost vertex of $N_{\overline{G}}(v_{i-1}) \cap \{v_i, v_{i+1}, \ldots, v_k\}$ in σ.*

Based on Lemma 3 and Definition 2, we prove the following result.

Lemma 4. *Let G be a comparability graph, and let σ be the layered ordering of G. Let $P = (v_1, v_2, \ldots, v_k)$ be a normal antipath of G, and let v_ℓ, and v_j be two vertices of P such that $v_\ell <_\sigma v_j$ and $v_\ell v_j \in E(G)$. Then $\ell < j$, i.e., v_ℓ appears before v_j in P.*

Recall that, if $\mathcal{P}_G$ is the partial order corresponding to a comparability graph G, and σ is the layered ordering of G, then $v_\ell <_{\mathcal{P}_G} v_j$ if and only if $v_\ell <_\sigma v_j$ and $v_\ell v_j \in E(G)$, for any two vertices $v_\ell, v_j \in V(G)$. Therefore, the definition of a monotone antipath can be paraphrased as follows: an antipath $P = (v_1, v_2, \ldots, v_k)$ of a comparability graph G is monotone if $v_\ell <_\sigma v_j$ and $v_\ell v_j \in E(G)$ implies that v_ℓ appears before v_j in P. Then, from Lemma 4 we obtain the following result.

Corollary 1. *Let G be a comparability graph. If P is a normal antipath of G, then P is a monotone antipath of G.*

Note that the inverse of Corollary 1 is not always true; for example, see the antipath P in Figure 1. In [9], for proving that for any Hamiltonian path P of a cocomparability graph $\overline{G}$ there exists a monotone Hamiltonian path of $\overline{G}$, Damaschke *et al.* first show that there exists a path $P' = (v_1, v_2, \ldots, v_{|V(\overline{G})|})$ of $\overline{G}$ such that v_1 is a minimal vertex of either $\mathcal{P}_G$ or $\mathcal{P}_G^d$. Using the same arguments, we obtain the following lemma.

Lemma 5. *Let G be a comparability graph, and let $\mathcal{P}_G$ be the partial order which corresponds to G. If P is an antipath of G, then there exists an antipath P' of G such that $V(P') = V(P)$ which starts with a minimal vertex of $V(P)$ in $\mathcal{P}_G$.*

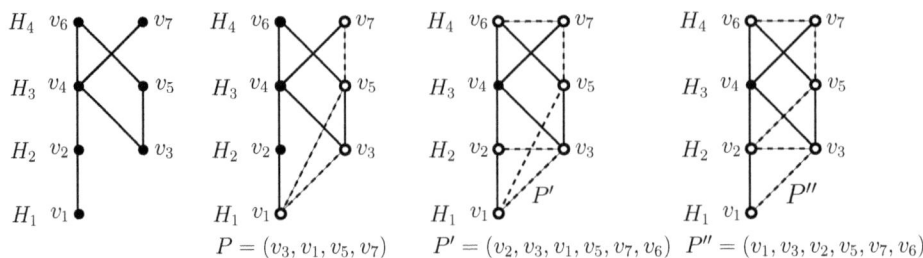

$$P = (v_3, v_1, v_5, v_7) \qquad P' = (v_2, v_3, v_1, v_5, v_7, v_6) \quad P'' = (v_1, v_3, v_2, v_5, v_7, v_6)$$

Fig. 1. Illustrating a Hasse diagram of a comparability graph G, an antipath P of G which is neither normal nor longest, an antipath P' of G such that $|P'| > |P|$ which is not normal, and a normal antipath P'' of G such that $V(P'') = V(P')$

The following result is central for the correctness of our algorithm.

Lemma 6. *Let P be a longest antipath of a comparability graph G. Then, there exists a normal antipath P' of G such that $V(P') = V(P)$.*

Figure 1 illustrates a Hasse diagram of a comparability graph G. The antipath $P = (v_3, v_1, v_5, v_7)$ of G is not normal, and there exists no normal antipath $\widehat{P}$ of G such that $V(\widehat{P}) = V(P)$; however, note that P is monotone. Also, P is not a longest antipath of G, since there exists an antipath $P' = (v_2, v_3, v_1, v_5, v_7, v_6)$ of G such that $|P'| > |P|$. Also, P' is not a normal antipath of G and there exists a normal antipath $P'' = (v_1, v_3, v_2, v_5, v_7, v_6)$ of G such that $V(P'') = V(P')$; note that it is easy to see that P'' is a longest antipath of G.

3   The Algorithm

Our algorithm, which we call Algorithm LP_Cocomparability, computes a longest path P of a cocomparability graph G by computing a longest antipath P of the comparability graph $\overline{G}$.

Let G be a comparability graph and let $H_1, H_2, \ldots, H_k$ be the layers of its Hasse diagram. For simplifying our description, we add a dummy vertex u_0 to G such that u_0 belongs to a layer H_0 and $u_0 u_i \in E(G)$ for every i, $1 \leq i \leq n$; let G' be the resulting graph. Note that, G' is a comparability graph having a Hasse diagram with layers $H_0, H_1, H_2, \ldots, H_k$, and let σ be a layered ordering of G', where $V(G') = \{u_0, u_1, u_2, \ldots, u_n\}$. It is easy to see that u_0 does not participate in any longest antipath P of G' such that $|P| \geq 2$. In general, a longest antipath P of G' which does not contain the vertex u_0 is also a longest antipath of G. Algorithm LP_Cocomparability computes a longest antipath of G' which is a longest antipath of the original graph G as well. Hereafter, we consider comparability graphs G having assumed that we have already added the dummy vertex u_0. Thus, the antipaths we compute in G are also antipaths of the graph $G \setminus \{u_0\}$.

We next give some definitions and notations necessary for the description of the algorithm. Let $L_j = (v_1, v_2, \ldots, v_k)$ be an arbitrary ordering of the vertices $v_1, v_2, \ldots, v_k$. We denote by $V(L_j)$ the set $\{v_1, v_2, \ldots, v_k\}$ and by $|L_j|$ the cardinality of the set $V(L_j)$. For every vertex $v_z \in L_j$, we denote by $L_j(v_z)$ the ordering $(v_1, v_2, \ldots, v_{z-1}, v_{z+1}, v_{z+2}, \ldots, v_{|L_j|}, v_z)$, and for every index r, $0 \leq r \leq |L_j|$, we denote by $L_j^r(v_z)$ the ordering containing the first r vertices of $L_j(v_z)$; thus:

- $L_j = (v_1, v_2, \ldots, v_k)$,
- $L_j(v_z) = (v_1, v_2, \ldots, v_{z-1}, v_{z+1}, v_{z+2}, \ldots, v_{|L_j|}, v_z)$,
- $L_j^r(v_z) = (v_1, v_2, \ldots, v_r)$ if $1 \leq r \leq z-1$,
- $L_j^r(v_z) = (v_1, v_2, \ldots, v_{z-1}, v_{z+1}, v_{z+2}, \ldots, v_{r+1})$ if $z \leq r \leq |L_j| - 1$,
- $L_j^0(v_z) = \emptyset$, and $L_j^{|L_j|}(v_z) = L_j(v_z)$.

Definition 3. *Let G be a comparability graph, let $H_0, H_1, H_2, \ldots, H_k$ be the layers of its Hasse diagram, let $V(G) = \{u_0, u_1, u_2, \ldots, u_n\}$, and let σ be the layered ordering of G. For every triple p, i, and j, where $1 \leq i \leq j \leq k$ and $u_p \in H_{i-1}$, we define the graph $G(u_p, i, j)$ to be the subgraph $G[S]$, where $S = \{u_x : u_x \in H_\ell, i \leq \ell \leq j\} \setminus \{u_x : u_p u_x \notin E(G)\}$.*

Definition 4. *Let L_j be an ordering of the set $H_j \cap V(G(u_p, i, j))$. We define the graph $G_{u_z}^r(u_p, i, j)$, where $u_z \in L_j$ and $0 \leq r \leq |L_j|$, to be the subgraph $G[S]$, where $S = V(G(u_p, i, j-1)) \cup L_j^r(u_z)$ if $i < j$, and $S = L_j^r(u_z)$ if $i = j$.*

Note that, since the dummy vertex u_0 is adjacent to every other vertex of G, the graph $G(u_p, 1, j)$, $1 \leq j \leq k$, is the subgraph $G[S]$ of G induced by the set $S = \{u_x : u_x \in H_\ell, 1 \leq \ell \leq j\}$. Additionally, $G_{u_z}^{|L_j|}(u_p, i, j) = G(u_p, i, j)$, and if $i < j$, then $G_{u_z}^0(u_p, i, j) = G(u_p, i, j-1)$.

Figure 2 illustrates examples of the graphs defined in Definitions 3 and 4. In particular, the figure to the left illustrates a Hasse diagram of a comparability graph G with layers $H_0, H_1, \ldots, H_5$. The figure in the middle illustrates the subgraph $G(v_1, 2, 4)$ of G induced by the vertices $\{v_3, v_6, v_7, v_8, v_9, v_{10}\}$. The figure to the right illustrates the subgraph $G_{v_9}^2(v_1, 2, 4)$ of G, if we consider the ordering $L_4 = (v_8, v_9, v_{10})$ for the vertices of $H_4 \cap V(G(v_1, 2, 4))$. The subgraph $G_{v_9}^2(v_1, 2, 4)$ of G is induced by the vertices $\{v_3, v_6, v_7, v_8, v_{10}\}$, and it is actually an induced subgraph of $G(v_1, 2, 4)$.

Notation 1. *For every vertex $u_t \in V(G_{u_z}^r(u_p, i, j))$, if $u_t \in H_j$, then we denote by $f(u_t)$ the smallest index such that $f(u_t) < j$, for which there exists a vertex u_x of $G_{u_z}^r(u_p, i, j)$ such that $u_x \in H_{f(u_t)}$ and $u_x u_t \notin E(G)$; in the case where no such index $f(u_t)$ exists, we set $f(u_t) = j$.*

Notation 2. *For every vertex $u_y \in V(G_{u_z}^r(u_p, i, j))$ we denote by $P(u_y; G_{u_z}^r(u_p, i, j))$ a longest normal antipath of $G_{u_z}^r(u_p, i, j)$ with right endpoint the vertex u_y, and by $\ell(u_y; G_{u_z}^r(u_p, i, j))$ the length of $P(u_y; G_{u_z}^r(u_p, i, j))$.*

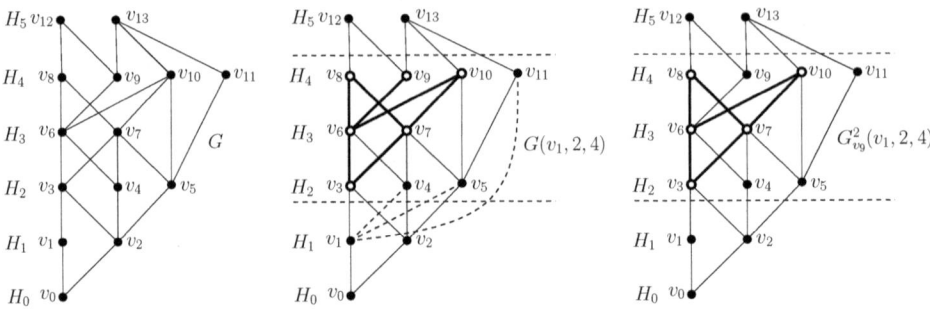

Fig. 2. Illustrating a Hasse diagram of a comparability graph G and the induced subgraphs $G(v_1, 2, 4)$ and $G_{v_9}^2(v_1, 2, 4)$ of G

Note that, if P is a longest normal antipath of $G(u_p, i, j)$ with right endpoint the vertex u_y, i.e., $P = P(u_y; G(u_p, i, j))$, then P is not necessarily a longest antipath of $G(u_p, i, j)$. However, if P is a longest antipath of $G(u_p, i, j)$, then from Lemma 6 there exists in $G(u_p, i, j)$ a normal antipath P' such that $V(P') = V(P)$; let u_y be the right endpoint of the normal antipath P'. Thus, there exists a longest normal antipath $P' = P(u_y; G(u_p, i, j))$ which is also a longest antipath in $G(u_p, i, j)$ for some vertex $u_y \in V(G(u_p, i, j))$.

Given a comparability graph G, Algorithm LP_Cocomparability (presented in Figures 3 and 4) computes for every induced subgraph $G(u_p, i, j)$ and for every vertex u_y of $G(u_p, i, j)$, the length $\ell(u_y; G(u_p, i, j))$ and the corresponding antipath $P(u_y; G(u_p, i, j))$, and outputs the maximum among the values $\{\ell(u_y; G(u_0, 1, k)) : u_y \in V(G(u_0, 1, k))\}$, and the corresponding normal antipath $P(u_y; G(u_0, 1, k))$. We prove that $P(u_y; G(u_0, 1, k))$ is a longest antipath of G.

4   Correctness and Time Complexity

Let G be a comparability graph, let $H_0, H_1, H_2, \ldots, H_k$ be the layers of its Hasse diagram, and let σ be the layered ordering of G. We prove the following results.

Lemma 7. *Let L_j be an ordering of the set $H_j \cap V(G(u_p, i, j))$, let $P = (P_1, v_\ell, P_2)$ be a normal antipath of $G_{u_z}^r(u_p, i, j)$, and let v_ℓ be the last vertex of $L_j^r(u_z)$. Then, P_1 and P_2 are normal antipaths of $G_{u_z}^r(u_p, i, j)$.*

Lemma 8. *Let L_j be an ordering of the set $H_j \cap V(G(u_p, i, j))$, and let u_t be the last vertex of $L_j^r(u_z)$. Let P_1 be a normal antipath of $G_{u_z}^{r-1}(u_p, i, j)$ with right endpoint a vertex u_x such that $u_x \in H_\ell$, $f(u_t) \leq \ell \leq j - 1$, and $u_t u_x \notin E(G)$. Let P_2 be a normal antipath of $G_{u_z}^{r-1}(u_x, \ell+1, j)$ with right endpoint a vertex u_y such that $u_y \in H_h$, $\ell+1 \leq h \leq j$, and $V(P_1) \cap V(P_2) = \emptyset$. Then, $P = (P_1, u_t, P_2)$ is a normal antipath of $G_{u_z}^r(u_p, i, j)$ with right endpoint the vertex u_y.*

ALGORITHM LP_COCOMPARABILITY

Input: a comparability graph G where $V(G) = \{u_0, u_1, u_2, \ldots, u_n\}$, the layers $H_0, H_1, H_2, \ldots, H_k$ of its Hasse diagram, and a layered ordering σ of G.

Output: a longest normal antipath of G.

1. **for** $j = 1$ to k
2.   **for** $i = j$ **downto** 1
3.   **for** every vertex $u_p \in H_{i-1}$
4.   let L_j be an ordering of $H_j \cap V(G(u_p, i, j))$
5.   **for** every vertex $u_z \in L_j$
6.   **for** $r = 1$ to $|L_j|$
7.   let u_t be the last vertex of $L_j^r(u_z)$
8.   **for** every vertex $u_y \in V(G_{u_z}^r(u_p, i, j))$ and $y \neq t$ {*initialization*}
9.   **if** $r = 1$ **then**
10. $\ell(u_y; G_{u_z}^0(u_p, i, j)) \leftarrow \ell(u_y; G(u_p, i, j-1));$
11. $P(u_y; G_{u_z}^0(u_p, i, j)) \leftarrow P(u_y; G(u_p, i, j-1));$
12. $\ell(u_y; G_{u_z}^r(u_p, i, j)) \leftarrow \ell(u_y; G_{u_z}^{r-1}(u_p, i, j));$
13. $P(u_y; G_{u_z}^r(u_p, i, j)) \leftarrow P(u_y; G_{u_z}^{r-1}(u_p, i, j));$
14.   **end_for**
15.   **if** $i = j$ **then**   {*case $i = j$*}
16. $\ell(u_t; G_{u_z}^r(u_p, j, j)) \leftarrow |L_j^r(u_z)|;$
17. $P(u_t; G_{u_z}^r(u_p, j, j)) \leftarrow L_j^r(u_z);$
18.   **if** $i \neq j$ **then**
19. $\ell(u_t; G_{u_z}^r(u_p, i, j)) \leftarrow 1;$ {*initialization for $u_y = u_t$*}
20. $P(u_t; G_{u_z}^r(u_p, i, j)) \leftarrow (u_t);$
21.   **execute** process$(G_{u_z}^r(u_p, i, j));$
22.   **end_for**
23. $\ell(u_z; G(u_p, i, j)) \leftarrow \ell(u_z; G_{u_z}^{|L_j|}(u_p, i, j));$ {*for the vertex $u_z \in L_j$*}
24. $P(u_z; G(u_p, i, j)) \leftarrow P(u_z; G_{u_z}^{|L_j|}(u_p, i, j));$
25.   **end_for**
26.   **for** every vertex $u_y \in V(G(u_p, i, j))$ and $u_y \notin L_j$ {*for $u_y \notin L_j$*}
27. $\ell(u_y; G(u_p, i, j)) \leftarrow \ell(u_y; G_{u_z}^{|L_j|}(u_p, i, j));$
28. $P(u_y; G(u_p, i, j)) \leftarrow P(u_y; G_{u_z}^{|L_j|}(u_p, i, j));$
29.   **end_for**
30.   **end_for**
31.   **end_for**
32. **end_for**

33. **compute** the $max\{\ell(u_y; G(u_0, 1, k)) : u_y \in G(u_0, 1, k)\}$ and the corresponding antipath $P(u_y; G(u_0, 1, k));$

Fig. 3. The algorithm for finding a longest antipath of G

PROCESS($G_{u_z}^r(u_p, i, j)$)

```
procedure bridge(G^r_{u_z}(u_p, i, j))
if f(u_t) < j then          {u_t is the last vertex of L^r_j(u_z)}
   for h = f(u_t) + 1 to j
      for ℓ = f(u_t) to h - 1
         for every vertex u_x ∈ H_ℓ ∩ V(G^{r-1}_{u_z}(u_p, i, j)) and u_x u_t ∉ E(G)
            for every vertex u_y ∈ H_h ∩ V(G^{r-1}_{u_z}(u_x, ℓ + 1, j))
               w_1 ← ℓ(u_x; G^{r-1}_{u_z}(u_p, i, j));   P'_1 ← P(u_x; G^{r-1}_{u_z}(u_p, i, j));
               w_2 ← ℓ(u_y; G^{r-1}_{u_z}(u_x, ℓ + 1, j));   P'_2 ← P(u_y; G^{r-1}_{u_z}(u_x, ℓ + 1, j));
               if w_1 + w_2 + 1 > ℓ(u_y; G^r_{u_z}(u_p, i, j)) then
                  ℓ(u_y; G^r_{u_z}(u_p, i, j)) ← w_1 + w_2 + 1;
                  P(u_y; G^r_{u_z}(u_p, i, j)) ← (P'_1, u_t, P'_2);

procedure append(G^r_{u_z}(u_p, i, j))
for ℓ = f(u_t) to j          {u_t is the last vertex of L^r_j(u_z)}
   for every vertex u_x ∈ H_ℓ ∩ (V(G^{r-1}_{u_z}(u_p, i, j)) and u_x u_t ∉ E(G)
      w_1 ← ℓ(u_x; G^{r-1}_{u_z}(u_p, i, j));   P'_1 ← P(u_x; G^{r-1}_{u_z}(u_p, i, j));
      if w_1 + 1 > ℓ(u_t; G^r_{u_z}(u_p, i, j)) then
         ℓ(u_t; G^r_{u_z}(u_p, i, j)) ← w_1 + 1;
         P(u_t; G^r_{u_z}(u_p, i, j)) ← (P'_1, u_t);
```

return (the value $\ell(u_y; G^r_{u_z}(u_p, i, j))$ and the antipath $P(u_y; G^r_{u_z}(u_p, i, j))$, for every vertex $u_y \in V(G^r_{u_z}(u_p, f(u_t) + 1, j))$ if $f(u_t) < j$, and for $u_y = u_t$ if $f(u_t) = j$);

Fig. 4. The procedure process()

Lemma 9. *For every induced subgraph $G(u_p, i, j)$ of G, and for every vertex $u_y \in V(G(u_p, i, j))$, the value $\ell(u_y; G(u_p, i, j))$ computed by Algorithm LP_Cocomparability is equal to the length of a longest normal antipath of $G(u_p, i, j)$ with right endpoint the vertex u_y and, also, the corresponding computed antipath $P(u_y; G(u_p, i, j))$ is a longest normal antipath of $G(u_p, i, j)$ with right endpoint the vertex u_y.*

Let P be a longest antipath of G such that $|P| \geq 2$. From Lemma 6 we may assume that P is a longest normal antipath of G and let u_y be its right endpoint. Also, P belongs to the graph $G \setminus \{u_0\}$. Since $G(u_0, 1, k) = G \setminus \{u_0\}$ and since Algorithm LP_Cocomparability computes the maximum among the lengths $\{\ell(u_y; G(u_0, 1, k)) : u_y \in V(G(u_0, 1, k))\}$ and the corresponding antipath P', from Lemma 9 we obtain that $|P'| = |P|$. Therefore, we obtain the following.

Theorem 3. *Algorithm LP_Cocomparability computes a longest path of a cocomparability graph in polynomial time.*

References

1. Arikati, S.R., Pandu Rangan, C.: Linear algorithm for optimal path cover problem on interval graphs. Inform. Proc. Lett. 35, 149–153 (1990)
2. Asdre, K., Nikolopoulos, S.D.: The 1-fixed-endpoint path cover problem is polynomial on interval graphs. Algorithmica, doi:10.1007/s00453-009-9292-5
3. Bertossi, A.A.: Finding Hamiltonian circuits in proper interval graphs. Inform. Proc. Lett. 17, 97–101 (1983)
4. Brandstädt, A., Le, V.B., Spinrad, J.P.: Graph Classes: A Survey. SIAM, Philadelphia (1999)
5. Bulterman, R., van der Sommen, F., Zwaan, G., Verhoeff, T., van Gasteren, A., Feijen, W.: On computing a longest path in a tree. Inform. Proc. Lett. 81, 93–96 (2002)
6. Chang, M.S., Peng, S.L., Liaw, J.L.: Deferred-query: An efficient approach for some problems on interval graphs. Networks 34, 1–10 (1999)
7. Damaschke, P.: The Hamiltonian circuit problem for circle graphs is NP-complete. Inform. Proc. Lett. 32, 1–2 (1989)
8. Damaschke, P.: Paths in interval graphs and circular arc graphs. Discrete Math. 112, 49–64 (1993)
9. Damaschke, P., Deogun, J.S., Kratsch, D., Steiner, G.: Finding Hamiltonian paths in cocomparability graphs using the bump number algorithm. Order 8, 383–391 (1992)
10. Deogun, J.S., Steiner, G.: Polynomial algorithms for hamiltonian cycle in cocomparability graphs. SIAM J. Computing 23, 520–552 (1994)
11. Feder, T., Motwani, R.: Finding large cycles in Hamiltonian graphs. In: Proc. of the 16th Annual ACM-SIAM Symp. on Discrete Algorithms (SODA), pp. 166–175. ACM, New York (2005)
12. Gabow, H.N.: Finding paths and cycles of superpolylogarithmic length. In: Proc. of the 36th Annual ACM Symp. on Theory of Computing (STOC), pp. 407–416. ACM, New York (2004)
13. Gabow, H.N., Nie, S.: Finding long paths, cycles and circuits. In: Hong, S.-H., Nagamochi, H., Fukunaga, T. (eds.) ISAAC 2008. LNCS, vol. 5369, pp. 752–763. Springer, Heidelberg (2008)
14. Garey, M.R., Johnson, D.S.: Computers and Intractability: A Guide to the Theory of NP-completeness. W.H. Freeman, New York (1979)
15. Garey, M.R., Johnson, D.S., Tarjan, R.E.: The planar Hamiltonian circuit problem is NP-complete. SIAM J. Computing 5, 704–714 (1976)
16. Golumbic, M.C.: Algorithmic Graph Theory and Perfect Graphs. Annals of Discrete Mathematics, vol. 57. North-Holland Publishing Co., Amsterdam (2004)
17. Habib, M., Möhring, R.H., Steiner, G.: Computing the bump number is easy. Order 5, 107–129 (1988)
18. Ioannidou, K., Mertzios, G.B., Nikolopoulos, S.D.: The longest path problem has a polynomial solution on interval graphs. In: Královič, R., Niwiński, D. (eds.) MFCS 2009. LNCS, vol. 5734, pp. 403–414. Springer, Heidelberg (2009)
19. Itai, A., Papadimitriou, C.H., Szwarcfiter, J.L.: Hamiltonian paths in grid graphs. SIAM J. Computing 11, 676–686 (1982)
20. McKee, T.A., McMorris, F.R.: Topics in Intersection Graph Theory. Society for Industrial and Applied Mathematics, Philadelphia (1999)
21. Müller, H.: Hamiltonian circuits in chordal bipartite graphs. Discrete Math. 156, 291–298 (1996)

22. Narasimhan, G.: A note on the Hamiltonian circuit problem on directed path graphs. Inform. Proc. Lett. 32, 167–170 (1989)
23. Takahara, Y., Teramoto, S., Uehara, R.: Longest path problems on ptolemaic graphs. IEICE Trans. Inf. and Syst. 91-D, 170–177 (2008)
24. Uehara, R.: Simple geometrical intersection graphs. In: Nakano, S.-i., Rahman, M. S. (eds.) WALCOM 2008. LNCS, vol. 4921, pp. 25–33. Springer, Heidelberg (2008)
25. Uehara, R., Uno, Y.: Efficient algorithms for the longest path problem. In: Fleischer, R., Trippen, G. (eds.) ISAAC 2004. LNCS, vol. 3341, pp. 871–883. Springer, Heidelberg (2004)
26. Uehara, R., Valiente, G.: Linear structure of bipartite permutation graphs and the longest path problem. Inform. Proc. Lett. 103, 71–77 (2007)
27. Zhang, Z., Li, H.: Algorithms for long paths in graphs. Theoret. Comput. Sci. 377, 25–34 (2007)

Colorings with Few Colors: Counting, Enumeration and Combinatorial Bounds[*]

Petr A. Golovach[1], Dieter Kratsch[2], and Jean-Francois Couturier[2]

[1] School of Engineering and Computing Sciences, Durham University,
South Road, Durham, DH1 3LE, United Kingdom
petr.golovach@durham.ac.uk
[2] Laboratoire d'Informatique Théorique et Appliquée,
Université Paul Verlaine - Metz, 57045 Metz Cedex 01, France
{kratsch,couturier}@univ-metz.fr

Abstract. We provide exact algorithms for enumeration and counting problems on edge colorings and total colorings of graphs, if the number of (available) colors is fixed and small. For edge 3-colorings the following is achieved: there is a branching algorithm to enumerate all edge 3-colorings of a connected cubic graph in time $O^*(2^{5n/8})$. This implies that the maximum number of edge 3-colorings in an n-vertex connected cubic graph is $O^*(2^{5n/8})$. Finally, the maximum number of edge 3-colorings in an n-vertex connected cubic graph is lower bounded by $12^{n/10}$. Similar results are achieved for total 4-colorings of connected cubic graphs. We also present dynamic programming algorithms to count the number of edge k-colorings and total k-colorings for graphs of bounded pathwidth. These algorithms can be used to obtain fast exact exponential time algorithms for counting edge k-colorings and total k-colorings on graphs, if k is small.

1 Introduction

Graph coloring is one of the classical subjects in graph theory. The four color conjecture asking whether every planar graph can be vertex-colored using at most 4 colors has been triggering the research in graph theory for more than a century. From an algorithmic point of view, for many coloring type problems, like vertex coloring, edge coloring and total coloring, the existence problem asking whether the graph has a coloring with a given number of colors is NP-complete. Even more, these coloring problems remain NP-complete when the question is whether there is a coloring of the input graph with a fixed (and small) number of colors [12,13,20]. (For Definitions see Section 2.)

[*] The first author has been supported by EPSRC under project EP/G043434/1. The second and third author have been supported by ANR Blanc AGAPE (ANR-09-BLAN-0159-03).

D.M. Thilikos (Ed.): WG 2010, LNCS 6410, pp. 39–50, 2010.

Exact algorithms to solve NP-hard problems are a challenging research subject in graph algorithms. Many papers on exact exponential time algorithms have been published in the last decade. One of the major results is the $O^*(2^n)$ inclusion-exclusion algorithm to compute the chromatic number of a graph first presented at FOCS 2006 by Björklund, Husfeldt [2] and Koivisto [14].[1] This approach can also be used to establish a $O^*(2^n)$ algorithm to count the k-colorings and to compute the chromatic polynomial of a graph. It also implies a $O^*(2^m)$ algorithm to count the edge k-colorings and a $O^*(2^{n+m})$ algorithm to count the total k-colorings of the input graph.

The existence problem asking whether a graph has a k-coloring for a fixed and small value of k also attracted a lot of attention. For vertex-colorability the fastest algorithm for $k = 3$ has running time $O^*(1.3289^n)$ and was proposed by Beigel and Eppstein [1], and the fastest algorithm for $k = 4$ has running time $O^*(1.7272^n)$ and was given by Fomin et al. [7]. They also established algorithms for counting vertex k-colorings for $k = 3$ and 4 [7]. The existence problem for an edge 3-coloring is considered in [1,15] and the currently fastest algorithm with the running time $O(1.344^n)$ is due to Kowalik [15]. Very recently Björklund et al. showed how to detect whether a d-regular graph admits an edge d-coloring in time $O^*(2^{(d-1)n/2})$ [3].

Combinatorial bounds on the maximum number of combinatorial objects in any n-vertex graph, as e.g. maximal independent sets or k-colorings, are of interest in combinatorics. Such upper bounds can sometimes be achieved via algorithms to enumerate all these objects. A well-known example is a branching algorithm to enumerate all maximal independent sets of a graph that can be used to establish an $O^*(3^{n/3})$ upper bound for the number of maximal independent sets in an n-vertex graph. Originally in 1965 Moon and Moser showed by an inductive proof that the maximum number of maximal independent sets in an n-vertex graph is $3^{n/3}$ [18]. Another interesting example is the upper bound of $O(1.7159^n)$ on the number of minimal dominating sets of an n-vertex graph established via a branching enumeration algorithm and its Measure & Conquer analysis by Fomin et al. [10].

Our Results. For edge 3-colorings we achieve the following enumeration algorithm and related combinatorial bounds.

- There is a branching algorithm to enumerate all edge 3-colorings of a connected cubic graph of running time $O^*(2^{5n/8}) = O(1.5423^n)$ using polynomial space (Subsection 3.1).
- The maximum number of edge 3-colorings in an n-vertex connected cubic graph is at most $O^*(2^{5n/8}) = O(1.5423^n)$ (Subsection 3.1).
- The maximum number of edge 3-colorings in an n-vertex connected cubic graph is lower bounded by $12^{n/10} = \Omega(1.2820^n)$ (Subsection 3.3).

For the counting problem of edge k-colorings of graphs we achieve the following algorithms.

---

[1] As has recently become standard, we write $f(n) = O^*(g(n))$ if $f(n) \le p(n) \cdot g(n)$ for some polynomial $p(n)$.

- The edge k-colorings of a graph given with a path decomposition of width at most p can be counted in time roughly $O^*((\binom{k}{\lfloor k/2 \rfloor})^{p+1})$ by a dynamic programming algorithm using exponential space (Section 4).[2]
- The number of edge 3-colorings of a graph can be counted in time $O^*(3^{n/6})$ $= O(1.201^n)$ and exponential space (Section 4).
- The number of edge 4-colorings of a graph can be counted in time $O^*(6^{n/3})$ $= O(1.8172^n)$ and exponential space (Section 4).

Note that our algorithm to count edge 3-colorings in time $O(1.201^n)$ improves upon the $O(1.344^n)$ time of Kowalik's polynomial space branching algorithm solving the decision problem [15].

For total k-colorings of graphs we achieve the following results.

- There is a branching algorithm to enumerate the total 4-colorings of a connected cubic graph in time $O^*(2^{13n/8}) = O(3.0845^n)$, implying that the maximum number of total 4-colorings in an n-vertex connected cubic graph is at most $O^*(2^{13n/8}) = O(3.0845^n)$ (Subsection 3.2).
- The number of total k-colorings of a graph given with a path decomposition of width at most p can be counted in time roughly $O^*((k \cdot \binom{k-1}{\lfloor (k-1)/2 \rfloor})^{p+1})$ (Section 4).[2]
- The number of total 4-colorings of a graph G can be counted in time $O(12^{n/6})$ $= O(1.5131^n)$ (Section 4).

Let us emphasize that edge 3-colorings and total 4-colorings exist only for graphs of maximum degree at most 3. Furthermore the largest number of such colorings for connected graphs is achieved by the n-vertex path. To avoid such trivial cases, it is natural to study these problems on connected cubic graphs; a well-known class of graphs for which upper bounding the number of colorings (of both types) is a non trivial and challenging task.

Furthermore we achieved similar results for the $L(2,1)$-labeling problem of graphs. Due to space restrictions only a short summary is given in Section 5. For the same reason some proofs will be omitted.

2   Preliminaries

We consider finite undirected graphs without loops or multiple edges. The vertex set of a graph G is denoted by $V(G)$ and its edge set by $E(G)$, or simply by V and E if this does not create confusion. For a set of edges $S \subseteq E(G)$, $G - S$ is the graph obtained from G by removing the edges of S. We denote by $\deg_G(v)$ the *degree* of a vertex v. For a vertex v, $E_G(v)$ is the set of edges incident with v. We may omit indices if the graph under consideration is clear from the context. The maximum degree of a graph G is denoted by $\Delta(G)$. Let r be a positive integer. A graph G is called r-*regular* if all vertices of G have degree r. A *cubic* graph is a 3-regular graph.

---

[2] See Theorem 3 for a precise estimation of the running time.

Let k be a positive integer. A *vertex (edge) k-coloring* of a graph G is an assignment $c\colon V(G) \to \{1,\dots,k\}$ ($c\colon E(G) \to \{1,\dots,k\}$ respectively) of a positive integer (*color*) to each vertex (edge) of G such that adjacent vertices (edges) receive distinct colors. A *total k-coloring* of a graph G is a mapping $c\colon V(G) \cup E(G) \to \{1,\dots,k\}$ such that adjacent vertices and adjacent edges have different colors, and for any edge e incident to vertex v, $c(v) \neq c(e)$. The *chromatic number* of G (denoted by $\chi(G)$) is the minimum k such that there is a vertex k-coloring of G. The *chromatic index* (or *edge chromatic number*) is the smallest k for which an edge k-coloring of G exists. We denote the chromatic index by $\chi'(G)$. The *total coloring number* $\tau(G)$ is the minimum k for which there is a total k-coloring of G.

It is easy to see that $\chi'(G) \geq \Delta(G)$ and $\tau(G) \geq \Delta(G) + 1$ for every graph G. Let us recall that by the well-known theorem of Vizing $\Delta(G) \leq \chi'(G) \leq \Delta(G)+1$ [21]. Also it is known that for any graph G with maximum degree at most three, there is a total 5-coloring of G [19].

3   Enumeration Algorithms for Cubic Graphs

3.1   Edge 3-Colorings

We present a branching algorithm to enumerate all edge 3-colorings of a given connected cubic graph. Our branching algorithm consists of a recursive procedure `EnumCol` and two auxiliary subroutines `Color` and `Extend`.

Procedure EnumCol(S,c);

1 Extend(S,c);
2 **if** *the graph $H = G - S$ contains a component F s.t. F is not a cycle and not an isolated vertex* **then**
  | *choose a vertex $v \in V(F)$ s.t. $\deg_F(v) < 3$ and edge $e \in E(F)$ incident with v; let $\{\alpha,\beta\} = \{1,2,3\} \setminus \{c(S \cap E_G(v))\}$;set $S' = S, c' = c$;*
  | **if** $Color(S,c,e,\alpha) = true$ **then** EnumCol(S,c);
  | **if** $Color(S',c',e,\beta) = true$ **then** EnumCol(S',c');
3 **if** *the graph $H = G - S$ contains a component F s.t. F is an odd cycle and there is $\alpha \in \{1,2,3\}$ s.t. for all $v \in V(F)$, $c(S \cap E_G(v)) = \alpha$* **then** Halt;
4 **if** *there is an edge $\{u,v\}$ in $H = G - S$ s.t. $c(S \cap E_G(u)) \neq c(S \cap E_G(v))$* **then**
  | *let $\alpha \in \{1,2,3\} \setminus c(S \cap (E_G(u) \cup E_G(v)))$;*Color$(S,c,\{u,v\},\alpha)$;
  | EnumCol(S,c);
5 **if** *the graph $H = G - S$ contains a component F s.t. F is a cycle and there is $\alpha \in \{1,2,3\}$ s.t. for all $v \in V(F)$, $c(S \cap E_G(v)) = \alpha$* **then**
  | *let $e \in E(F)$ and let $\{\beta,\gamma\} = \{1,2,3\} \setminus \{\alpha\}$;set $S' = S, c' = c$;*
  | Color(S,c,e,β);EnumCol(S,c);
  | Color(S',c',e,γ);EnumCol(S',c');
6 **if** $S = E(G)$ **then** Output(c)

The procedure `EnumCol` takes as input a connected cubic graph G and a set of colored edges $S \subseteq E(G)$, i.e. for each $e \in S$, the assigned color $c(e) \in \{1,2,3\}$ is given. The procedure enumerates all edge 3-colorings of G which are extensions of the given edge coloring of S. Note that if v is an isolated vertex in $G - S$ then all edges incident to v are colored, and thus v is of no importance for extending the partial edge coloring c.

The subroutine `Color` takes as input a set of colored edges S with a partial edge coloring c, an edge $\{u, v\} \in E(G) \setminus S$ and a color $\alpha \in \{1, 2, 3\}$. It tries to extend c by assigning to edge $\{u, v\}$ the color α. If possible the subroutine returns $true$, otherwise it returns $false$.

Subroutine Color$(S, c, \{u, v\}, \alpha)$;
if $\alpha \notin c(S \cap (E_G(u) \cup E_G(v)))$ **then**
    ⌊ set $c(\{u, v\}) = \alpha$, $S = S \cup \{e\}$;**Return(***true***)**;
else Return(*false***)**

The subroutine `Extend` tries to extend a given edge coloring c of S by reduction (i.e. without branching) if there is a vertex incident with two already colored edges.

Subroutine Extend(S, c);
while *there is a vertex* $v \in V(G)$ *s.t.* $|S \cap E(v)| = 2$ **do**
    ⌊ let $\{\alpha, \beta\} = S \cap E(v)$, $\gamma \in \{1, 2, 3\} \setminus \{\alpha, \beta\}$ *and* $e \in E(v) \setminus S$;
      if *Color*$(S, c, e, \gamma) = false$ **then Halt**

To enumerate all edge 3-colorings of a connected cubic G we choose any edge $e \in E(G)$, set $S = \{e\}$ and call the procedure `EnumCol` consecutively for $c(e) = 1$, $c(e) = 2$ and $c(e) = 3$. If the aim is to enumerate all edge 3-colorings up to permutations of colors then we choose two adjacent edges $e_1, e_2 \in E(G)$, set $S = \{e_1, e_2\}$, $c(e_1) = 1$, $c(e_2) = 2$ and call `EnumCol`.

Theorem 1. *Our algorithm enumerates all edge 3-colorings of a connected cubic graph G in time $O^*(2^{5n/8})$.*

Proof. To prove the correctness of the algorithm we consider Procedure `EnumCol`. By the step 1 we try to extend the coloring c of S without branching looking for vertices incident with two already colored edges. By the step 2 we choose a vertex v with exactly one incident colored edge, and then branch using an uncolored edge e incident with v. Clearly, we have two possibilities to color this edge. Notice that e is an edge of the component of $H = G - S$ which is not a cycle. If for the current input G, S, c the steps 1 and 2 are not applicable anymore, then since G is a connected graph, every vertex is incident to at least one edge of S. Hence the maximum degree of $G - S$ is at most two, which implies that $G - S$ is a union of paths, cycles and isolated vertices. Furthermore no component can be a path since such a component would be colored by applying step 1 successively to an end vertex of the path. Thus in steps 3–5 all non empty components of $H = G - S$ are cycles. Suppose F is a component of H which is a cycle. If F is an odd cycle and all vertices of F are incident with colored edges from S which are colored by the same color, then all edges of F have to be colored by the two remaining color, but this is impossible. This case is checked in the step 3. If F has at least two vertices which are incident with edges of S colored by different colors, then F contains two adjacent vertices u and v with the same property. Since there is only one possibility to color $\{u, v\}$, all edges of F can be colored without branching, and it is done in step 4 and by a recursive call of EnumCol. In step 5 we consider the final case when all non empty components of H are even cycles, and for each cycle F, all vertices of F are incident with the edges of S colored by the same color. Then two edge colorings of F are possible and each is generated via a recursive call of EnumCol.

Now we estimate the running time. Consider the graph G_S which is the subgraph of G with edge set S and the set of all end vertices of S as vertex set. Notice that by each successive execution of step 2 and step 1 for the next recursive call of EnumCol, we add at least two vertices to the graph G_S by our choice of the edges for the branching in step 2. It follows from the fact that after a call of the subroutine Extend all non isolated vertices of $G - S$ have degree at least 2 in this graph, and then the Extend subroutine colors edges along two paths in $G - S$ starting at the vertex v chosen for the branching and ending at a degree 3 vertex in $G - S$ for each of them. Observe also that if these two paths reach the same vertex then the subroutine extends the coloring along a new path starting at this vertex until we reach a new vertex of degree 3. This means that the depth of the search tree generated by branchings in step 2 only, and not considering steps 3–6, is at most $\frac{n-2}{2}$ (recall that G_S has two vertices when EnumCol is called first). The branching on the steps 3–5 is done only when all non empty components of H are even cycles, and for each cycle, only one binary branching is done. Let us estimate the number of such cycles. Suppose that the set S constructed by EnumCol is such that all non empty components of $H = G - S$ are cycles. Since G is a connected graph and by the choice of the edges for the branching, G_S is a connected graph with the vertex set $V(G)$. Notice that G_S has only vertices of degree one or three. Let n_1 be the number of vertices of degree one, and let n_3 be the number of vertices of degree three. Clearly, G_S has $\frac{1}{2}(n_1 + 3n_3)$ edges, and since it is connected, $\frac{1}{2}(n_1 + 3n_3) \geq n_1 + n_3 - 1$. Hence, $\frac{n+2}{2} \geq n_1$ and therefore the number of vertices of degree 2 in $H = G - S$ is at most $\frac{n+2}{2}$ implying that H has at most $\frac{n+2}{8}$ even length cycles. Therefore, the depth of the overall search tree is at most $\frac{n-2}{2} + \frac{n+2}{8}$. Since for each branching, we consider two cases, the number of leaves in the search tree is at most $2^{\frac{n-2}{2} + \frac{n+2}{8}}$, and the running time of the algorithm is $O^*(2^{5n/8})$.   □

Using the fact that edge 3-colorings correspond to leaves of the search tree, we have the following corollary.

Corollary 1. *Let G be a connected cubic graph with n vertices. Then the number of different edge 3-colorings of G is at most $3 \cdot 2^{(5n-6)/8}$.*

3.2   Total 4-Colorings

Using an algorithm similar to the one of the previous subsection it is possible to enumerate the total 4-colorings of connected cubic graphs. The major ingredient to be added to the above algorithm is that a vertex is colored as soon as one of its incident edges is colored.

Theorem 2. *All total 4-colorings of a connected cubic graph G can be enumerated in time $O^*(2^{13n/8})$.*

Proof. We discuss those properties that are different from the algorithm in the previous subsection.

Notice that if $\{u, v\}$ is a colored edge and the vertex u is colored too, then there are at most two possibilities to color v. Thus whenever a vertex is colored (except the first two) at most two colors are possible. If a vertex v is colored and it is incident with one colored edge, then there are at most two possibilities to color the remaining edges incident to v. Also, if a vertex v is colored and it is incident with two colored edge, then there is at most one possibility to color the remaining edge incident to v.

Suppose that S is the set of colored edges. Let F be a component of $G - S$ and assume that F is a cycle. Since we color vertices as soon as at least one incident edge is colored, all vertices of F are colored. If two adjacent vertices are colored by the same color we stop since there is no total 4-coloring extending the current partial total coloring. Assume that F is an odd cycle. Thus there is a vertex $u \in V(G)$ which is adjacent to vertices $v, w \in V(G)$ which are colored by different colors. In this case there is at most one possibility to extend the total coloring by coloring either $\{u, v\}$ or $\{u, w\}$, and therefore there is at most one possibility to color the edges of F. Assume that F is an even cycle. Then there are at most two possibilities to color the edges of F, and two possibilities may exist only if the vertices of the cycle are colored alternately by two colors.

The time analysis is similar to the one in the proof of Theorem 1. The number of possibilities to color the edges (using 4 colors) is $O^*(2^{5n/8})$ since the same recurrences apply. Furthermore there are at most two possibilities to color a vertex which contributes a factor of 2^n and implies the stated running time. □

3.3   Lower Bounds for Edge 3-Colorings

Let us note that the enumeration algorithm of Subsection 3.1 actively uses the fact that the considered graphs are connected and have no vertices of degree one or two. Particularly, our upper bound for the number of edge 3-colorings does not apply to all graphs of maximum degree 3. For example, for an n-vertex path P_n, the number of edge 3-colorings is $3 \cdot 2^{n-2}$, and for the disjoint union of $\frac{n}{6}$ copies of $K_{3,3}$, the number of edge 3-colorings is $12^{n/6}$. Now we give lower bound for the number of edge 3-colorings of connected cubic graphs.

We consider a complete bipartite graph $K_{3,3}$. Let e_1 and e_2 be edges of this graph. We replace these edges by paths of length 3 with middle vertices a_1, a_2 and b_1, b_2 respectively. Denote the obtained graph by H. We call vertices a_1, a_2, b_1, b_2 *roots* of H. Let $n = 10r$ be a positive integer. We construct r copies of H denoted by $H_1 \ldots, H_r$, and denote by $a_1^{(i)}, a_2^{(i)}, b_1^{(i)}, b_2^{(i)}$ the roots of H_i for $1 \leq i \leq r$. Assume that $b_1^{(0)} = b_1^{(r)}$ and $b_1^{(0)} = b_1^{(r)}$. For $1 \leq i \leq r$, we add edges $\{b_1^{(i-1)}, a_1^{(i)}\}$ and $\{b_2^{(i-1)}, a_2^{(i)}\}$. Let us call the resulting connected cubic graph G. It is possible to prove the following proposition.

Proposition 1. *The connected cubic graph G has n vertices and at least $12^{n/10}$ different edge 3-colorings.*[3]

4   Dynamic Programming Counting Algorithms

We establish dynamic programming algorithms (needing exponential space) to count the number of edge k-colorings and total k-colorings on graphs of bounded pathwidth. This allows the design of exact algorithms to count the edge and total k-colorings of graphs of bounded degree. It also implies a faster edge 3-coloring algorithm. Notice that since $\chi'(G) \geq \Delta(G)$ and $\tau(G) \geq \Delta(G) + 1$, it is sufficient to consider our problems for graphs of bounded maximum degree: $\Delta(G) \leq k$ for edge k-colorings and $\Delta(G) \leq k - 1$ for total k-colorings.

First we summarize a few fundamentals on treewidth and pathwidth.

A *tree decomposition* of a graph G is a pair (X, T) where T is a tree whose vertices we will call *nodes* and $X = (\{X_i \mid i \in V(T)\})$ is a collection of subsets of $V(G)$ (called *bags*) such that

1. $\bigcup_{i \in V(T)} X_i = V(G)$,
2. for each edge $\{v, w\} \in E(G)$, there is an $i \in V(T)$ such that $v, w \in X_i$, and
3. for each $v \in V(G)$ the set of nodes $\{i \mid v \in X_i\}$ forms a subtree of T.

The *width* of a tree decomposition $(\{X_i \mid i \in V(T)\}, T)$ equals $\max_{i \in V(T)} \{|X_i| - 1\}$. The *treewidth* of a graph G, denoted tw(G), is the minimum width over all tree decompositions of G.

A tree decomposition (X, T) of a graph G with T being a path is called a *path decomposition* of G. The *pathwidth* of G is the minimum width over all path decompositions of G. The pathwidth is denoted by pw(G). For a path decomposition (X, P), we assume that the path P has nodes $1, \ldots, r$ in the given order. For $1 \leq i \leq r$, by G_i we denote the graph induced by the set $X_1 \cup \cdots \cup X_i$. It is well known that every path decomposition (X, P) can be easily converted (in linear time) to a *nice* path decomposition of same width (and with a linear size of P), such that nodes of P are of two types:

1. *Introduce* nodes i with $X_i = X_{i-1} \cup \{v\}$ for some vertex $v \in V(G)$.
2. *Forget* nodes i with $X_i = X_{i-1} \setminus \{v\}$ for some vertex $v \in V(G)$.

We assume here that $X_0 = \emptyset$. Nice path decompositions are used in the design of our dynamic programming algorithm.

Nowadays dynamic programming algorithms on path or tree decompositions are often used to establish exact exponential time algorithms (see e.g. [8,9,11]). We discuss this approach for for graphs of maximum degree $d \geq 3$. This approach relies on upper bounds for the pathwidth. Fomin and Høie [11] proved that

---

[3] It was pointed to us by Artem Pyatkin that it is possible to improve this lower bound, if we consider the graph with $n = 2r$ vertices obtained by joining two cycles C_r by a perfect matching (vertices joined in the cyclical order). This graph has at least $\frac{3}{4} \cdot 2^{n/2}$ different edge 3-colorings.

for any $\varepsilon > 0$, there exists an integer n_ε such that for every graph G with maximum vertex degree at most three and with $|V(G)| > n_\varepsilon$, $\mathrm{pw}(G) \leq (\frac{1}{6} + \varepsilon)|V(G)|$. It should be noted that the proof of this fact is constructive and a path decomposition of width at most $(\frac{1}{6} + \varepsilon)|V(G)|$ can be constructed in polynomial time. Fomin and Høie pointed out in [11] that such path decompositions can be used for constructing fast exact algorithm for the graphs of maximum degree three. They demonstrated it for the problems MAXIMUM INDEPENDENT SET, MAX-CUT and MINIMUM DOMINATING SET. This technique was also used for the vertex 3- and 4-coloring problems in [7]. The best known upper bound is the following.

Proposition 2 ([8]). *For any $\varepsilon > 0$, there exists an integer n_ε such that for every graph G with $|V(G)| = n > n_\varepsilon$,*

$$\mathrm{pw}(G) \leq \frac{1}{6}n_3 + \frac{1}{3}n_4 + \frac{13}{30}n_5 + \frac{23}{45}n_6 + n_{\geq 7} + \epsilon n,$$

where n_i is the number of vertices of degree i in G for any $i \in \{3, 4, 5, 6\}$ and $n_{\geq 7}$ is the number of vertices of degree at least 7. Moreover, a path decomposition of the corresponding width can be computed in polynomial time.

This bound can be combined with dynamic programming algorithms for edge k-coloring and total k-coloring on graphs of bounded pathwidth.

Theorem 3. *For an n-vertex graph with pathwidth at most p,*

1. *all edge k-colorings can be counted in time $O((\binom{k}{\lfloor k/2 \rfloor})^{p+1} \cdot k \cdot k! \cdot \log k \cdot n^2)$,*
2. *all total k-colorings can be counted in time $O((k \cdot \binom{k-1}{\lfloor (k-1)/2 \rfloor}))^{p+1} \cdot k \cdot k! \cdot \log k \cdot n^2)$,*

if the path decomposition is given.

Proof. To describe the dynamic programming algorithms, we describe what we store in the tables corresponding to the bags of the path decomposition. We consider a nice path decomposition of a graph G with maximum degree at most k and pathwidth at most p with bags $X_1, \ldots, X_r$. For $1 \leq i \leq r$, we denote by G_i the subgraph of G induced by $X_1 \cup X_2 \cup \cdots X_i$. For $1 \leq i \leq r$ and $0 \leq j \leq k$, let $Z_i^{(j)} \subseteq X_i$ be the set of vertices in the bag X_i having degree j in G_i. Assume that $Z_i^{(j)} = \{z_1^{(j)}, \ldots, z_{p_j}^{(j)}\}$.

At first we consider edge colorings. The table of data for a bag X_i stores entries often called characteristics which contain collections of sets $\{S_1^{(j)}, \ldots, S_{p_j}^{(j)}\}$ for $1 \leq j \leq k$ and an integer σ such that

- $S_t^{(j)} \subseteq \{1, \ldots, k\}$ and $|S_t^{(j)}| = j$ for $1 \leq t \leq p_j$ and $1 \leq j \leq k$, and
- there are σ edge k-colorings of G_i such that edges of G_i incident with $z_t^{(j)}$ are colored by the colors from $S_t^{(j)}$ for $1 \leq t \leq p_j$ and $1 \leq j \leq k$.

The first claim of the proposition follows from the observation that the table contains at most

$$\prod_{j=1}^{k} \binom{k}{j}^{p_j} \leq \binom{k}{\lfloor k/2 \rfloor}^{p+1}$$

entries. Constructions of the tables for introduce and forget nodes are straightforward, and we omit these descriptions here. It remains to notice that the number of all edge k-colorings of G equals to the sum of all integers σ over all entries of the table for the node r.

The proof of the second claim is similar. For a node i, we keep collections of pairs $\{(\alpha_1^{(j)}, S_1^{(j)}), \ldots, (\alpha_{p_i}^{(j)}, S_{p_j}^{(j)})\}$ for $1 \leq j \leq k-1$ and an integer σ such that

- $\alpha_t^{(j)} \in \{1, \ldots, k\}$,
- $S_t^{(j)} \subseteq \{1, \ldots, k\} \setminus \{\alpha_t^{(j)}\}$ and $|S_t^{(j)}| = j$ for $1 \leq t \leq p_j$ and $0 \leq j \leq k-1$, and
- there are σ total k-coloring of G_i such that $z_t^{(j)}$ is colored by $\alpha_t^{(j)}$ and edges of G_i incident with $z_t^{(j)}$ are colored by the colors from $S_t^{(j)}$ for $1 \leq t \leq p_j$ and $0 \leq j \leq k-1$.

It remains to note that the table contains at most

$$\prod_{j=0}^{k-1} \left(k \cdot \binom{k-1}{j} \right)^{p_j} \leq \left(k \cdot \binom{k-1}{\lfloor (k-1)/2 \rfloor} \right)^{p+1}$$

entries.

This implies that the overall number of entries stored in tables of bags is $O(\binom{k}{\lfloor k/2 \rfloor}^{p+1} \cdot n)$ in the first and $O((k \cdot \binom{k-1}{\lfloor (k-1)/2 \rfloor})^{p+1} \cdot n)$ in the second algorithm.

The additional factors in the stated running times (though of little importance for our further applications in which $k > 0$ is a small integer) are discussed here for edge coloring only. When computing the entries for an insert node X_i obtained from a fixed entry of X_{i-1} with $v \in X_i \setminus X_{i-1}$, there are at most $k!$ possibilities to color the edges with endpoint v and another endpoint in X_{i-1}, and it takes time $O(k)$ to compute and verify the validity of an entry. Furthermore the algorithm stores in each entry the number of valid partial edge colorings in G_i. In a unit-cost RAM model the necessary arithmetic operations can be done in time $O(1)$. In a more realistic log-cost RAM model there is another factor $n \log k$ since the number of edge k-colorings in an n-vertex graph is at most k^n. □

The above dynamic programming algorithm is simpler and has a better running time for a small number of colors than the known dynamic programming algorithms for edge colorings on graphs of bounded treewidth [4,22,23].

Using our algorithms of Theorem 3 and upper bounds on the maximum pathwidth of graphs of a given maximum degree, we can obtain exact algorithms for counting edge k-colorings and total $(k+1)$-colorings on graphs of maximum degree k. Combined with the fact that graphs of maximum degree larger than k have neither edge k-colorings nor total $(k+1)$-colorings this implies algorithms for all graphs. To mention a few:

Theorem 4. *For any $\varepsilon > 0$,*

1. *all edge 3-colorings of a graph can be counted in time $O^*(3^{(1/6+\varepsilon)n})$,*
2. *all edge 4-colorings of a graph can be counted in time $O^*(6^{(1/3+\varepsilon)n})$,*
3. *all total 4-colorings of a graph can be counted in time $O^*(12^{(1/6+\varepsilon)n})$.*

Theorem 4 implies a $O^*(3^{n/6}) = O(1.2010^n)$ time exponential space algorithm to count the edge 3-colorings of graphs improving upon the $O(1.344^n)$ running time of Kowalik's polynomial space algorithm for the edge 3-coloring decision problem [15].

5 Conclusion

Results similar to those for edge and total k-colorings presented in previous sections can be obtained for $L(2, 1)$-labelings of graphs. Due to space restrictions, we give only a short summary of our results.

An $L(2, 1)$-*labeling* of a graph G of span k is a function $f \colon V(G) \to \{0, \ldots, k\}$ such that for any adjacent vertices u and v, $|f(u) - f(v)| \geq 2$, and for any two vertices u and v at distance two, $f(u) \neq f(v)$. Fiala et al. proved that it is NP-complete to decide whether a given graph has an $L(2, 1)$-labeling of span k for any $k \geq 4$ [6]. Exact algorithms for the $L(2, 1)$-labeling problem with span k are given by Král [16] (for the more general Channel Assignment problem) and Kratochvíl et al. [17]. For $k = 4$, an algorithm with running time $O(1.3161^n)$ was given in [17].

Since cubic graphs have no $L(2, 1)$-labeling of span k for $k \leq 4$, we are interested in $L(2, 1)$-labelings of span 5. One of the basic observations is the asymmetry within the colors. Coloring a vertex x by 0 or 5 makes only two colors unavailable for a neighbor of x. Coloring a vertex x by 1, 2, 3 or 4 makes three colors unavailable for each neighbor of x.

By a classical branching algorithm we can prove the following theorem.

Theorem 5. *All $L(2, 1)$-labelings of span 5 for a given connected cubic graph with n vertices can be enumerated in time $O(1.8613^n)$, and the number $L(2, 1)$-labelings of span 5 of any connected cubic graph is $O(1.8613^n)$.*

We can also show that the maximum number of $L(2, 1)$-labelings of span 5 in a connected cubic graph is lower bounded by $2^{n/6}$.

Finally for any $\varepsilon > 0$, the number of $L(2, 1)$-labelings of span 4 of n-vertex graphs can be counted in time $O^*(6^{(1/15+\varepsilon)n})$.

References

1. Beigel, R., Eppstein, D.: 3-coloring in time $O(1.3289^n)$. Journal of Algorithms 54, 168–204 (2005)
2. Björklund, A., Husfeldt, T.: Inclusion-exclusion algorithms for counting set partitions. In: Proceedings of the 47th Annual IEEE Symposium on Foundations of Computer Science (FOCS 2006), pp. 575–582. IEEE, Los Alamitos (2006)

3. Björklund, A., Husfeldt, T., Kaski, P., Koivisto, M.: Narrow sieves for parameterized paths and packings, arXiv:1007.1161v1 (2010)
4. Bodlaender, H.L.: Polynomial algorithms for graph isomorphism and chromatic index on partial k-trees. J. Algorithms 11, 631–643 (1990)
5. Eppstein, D.: Improved algorithms for 3-coloring, 3-edge-coloring, and constraint satisfaction. In: Proceedings of the 12th Annual ACM-SIAM Symposium on Discrete Algorithms (SODA), pp. 329–337. SIAM, Philadelphia (2001)
6. Fiala, J., Kloks, T., Kratochvíl, J.: Fixed-parameter complexity of lambda-labelings. Discrete Applied Mathematics 113, 59–72 (2001)
7. Fomin, F.V., Gaspers, S., Saurabh, S.: Improved exact algorithms for counting 3- and 4-colorings. In: Lin, G. (ed.) COCOON 2007. LNCS, vol. 4598, pp. 65–74. Springer, Heidelberg (2007)
8. Fomin, F.V., Gaspers, S., Saurabh, S.: On two techniques of combining branching and treewidth. Algorithmica 54, 181–207 (2009)
9. Fomin, F., Grandoni, F., Kratsch, D.: Some new techniques in design and analysis of exact (exponential) algorithms. Bulletin of the EATCS 87, 47–77 (2005)
10. Fomin, F.V., Grandoni, F., Pyatkin, A., Stepanov, A.: Combinatorial bounds via Measure and Conquer: Bounding minimal dominating sets and applications. ACM Transactions on Algorithms 5(1), Article 9 (2008)
11. Fomin, F.V., Høie, K.: Pathwidth of cubic graphs and exact algorithms. Inf. Process. Lett. 97, 191–196 (2006)
12. Garey, M.R., Johnson, D.S.: Computers and Intractability: A guide to the Theory of NP-completeness. Freeman, New York (1979)
13. Holyer, I.: The NP-completeness of edge-coloring. SIAM J. Comput. 10, 718–720 (1981)
14. Koivisto, M.: An $O(2^n)$ Algorithm for graph coloring and other partitioning problems via inclusion-exclusion. In: Proceedings of the 47th Annual IEEE Symposium on Foundations of Computer Science (FOCS 2006), 2nd edn., pp. 583–590. IEEE, Los Alamitos (2006)
15. Kowalik, L.: Improved edge-coloring with three colors. Theoret. Comp. Sci. 410, 3733–3742 (2009)
16. Král, D.: An exact algorithm for the channel assignment problem. Discrete Applied Mathematics 145, 326–331 (2005)
17. Kratochvíl, J., Kratsch, D., Liedloff, M.: Exact algorithms for L(2,1)-labeling of graphs. In: Kucera, L., Kucera, A. (eds.) MFCS 2007. LNCS, vol. 4708, pp. 513–524. Springer, Heidelberg (2007)
18. Moon, J.W., Moser, L.: On cliques in graphs. Israel J. Math. 3, 23–28 (1965)
19. Rosenfeld, M.: On the total coloring of certain graphs. Israel Journal of Mathematics 9, 396–402 (1971)
20. Sánchez-Arroyo, A.: Determining the total colouring number is NP-hard. Discrete Mathematics 78, 315–319 (1989)
21. Vizing, V.G.: On an estimate of the chromatic class of a p-graph. Diskret. Anal., 25–30 (1964) (in Russian)
22. Zhou, X., Nishizeki, T.: Optimal parallel algorithm for edge-coloring partial k-trees with bounded degrees. In: Proceedings of the International Symposium on Parallel Architectures, Algorithms and Networks, pp. 167–174. IEEE, Los Alamitos (1994)
23. Zhou, X., Nakano, S., Nishizeki, T.: Edge-coloring partial k-trees. J. Algorithms 21, 598–617 (1996)

On Stable Matchings and Flows

Tamás Fleiner*

Budapest University of Technology and Economics,
Department of Computer Science and Information Theory,
Magyar tudósok körútja 2. H-1117, Budapest, Hungary
fleiner@cs.bme.hu

Abstract. We describe a flow model that generalizes ordinary network flows the same way as stable matchings generalize the bipartite matching problem. We prove that there always exists a stable flow and generalize the lattice structure of stable marriages to stable flows. Our main tool is a straightforward reduction of the stable flow problem to stable allocations.

Keywords: Stable marriages; stable allocations; network flows.

1 Introduction

In the stable marriage problem of Gale and Shapley [6], there are n men and n women and each person ranks the members of the opposite gender by an arbitrary strict, individual preference order. A *marriage scheme* in this model is a set of marriages between different men and women. Such a scheme is *unstable* if there exists a *blocking pair*, that is, a man m and a woman w in such a way that m is either unmarried or m prefers w to his wife, and at the same time, w is either unmarried or prefers m to her partner. A marriage scheme is *stable* if it is not unstable, that is, not blocked by any pair. It is a natural problem to find a stable marriage scheme if it exists at all. Nowadays, it is already folklore that for any preference rankings of the n men and n women, a stable marriage scheme does exist. This theorem was proved first by Gale and Shapley in [6]. They constructed a special stable marriage scheme with the help of a finite procedure, the so-called deferred acceptance algorithm. It also turned out that for the existence of a stable scheme, it is not necessary that the number of men is the same as the number of women or that for each person, all members of the opposite gender are acceptable: the deferred acceptance algorithm is so robust that it works properly in these more general settings.

Several interesting properties about the structure of stable marriage schemes are known. Donald Knuth [7] attributes to John Conway the observation that

---

* Research is supported by OTKA grant K69027 and the MTA-ELTE Egerváry Research Group (EGRES).

stable marriages have a lattice structure: if each man picks the better assignment out of two stable marriage schemes then another stable marriage scheme is created in which each women receives the worse out of the two husbands.

There are further known extensions of the stable marriage problem. Baïou and Balinski proved in [1] that if each edge of the underlying bipartite graph has a nonnegative capacity and each vertex has a nonnegative quota then the accordingly modified deferred acceptance algorithm shows that there always exists a so called stable allocation. An allocation is an assignment of nonnegative values to the edges that do not exceed the corresponding capacities such that the total allocation of no vertex exceeds its quota. (That is, a "marriage" can be formed with an "intensity" different from 0 and 1 and each participant has an individual upper bound on his/her total "marriage intensity".) An allocation is stable if any unsaturated edge e has a saturated end vertex v such that no edge e' incident to v and preferred by v less than e has a positive value. Beyond proving the existence of stable assignments, Baïou and Balinski used flow-type arguments to speed up the deferred acceptance algorithm in [1]. Later, Dean and Munshi came up with an even faster algorithm for the same problem [3] that also has to do with network flows.

It is fairly well-known that the bipartite matching problem can be formulated in the more general network flow model, and the alternating path algorithm for maximum bipartite matchings is a special case of the augmenting path algorithm of Ford and Fulkerson for maximum flows. However, it seems that the question whether there exists a flow generalization of the stable marriage theorem has not been addressed so far. This very problem is in the focus of our present work. In section 2, we formulate the stable flow problem and state a result from [1] by Baïou and Balinski on stable allocations. Section 3 contains the stable flow theorem, a generalization of the Gale-Shapley theorem to flows. Our reduction of the stable flow problem to the stable allocation problem resembles to the reduction of the maximum flow problem to the maximum b-matching problem. Actually, our construction has to do also with the one that Cechlárová and Fleiner used in [2] to extend the stable roommates model to a multiple partner model. Section 4 is devoted to certain structural results on stable flows, in particular we generalize the lattice structure of stable marriages. To achieve this, we lean on the construction we used for the reduction. The interested reader can find the extended version of our work with the proofs and with an application showing a certain "linking property of flows" in [5].

It turned out that our model is closely related to so-called "supply chains" well-known in the Economics literature. Prior to our work, Ostrovsky had a related result in [8]. There, he considers only acyclic networks, but instead of the Kirchhoff law, he requires a less restrictive property that he calls "same side substitutability" and "cross side complementarity". In [8] the author proves the existence of a "chain stable network" and justifies that these "chain stable networks" form a lattice under a natural partial order. Ostrovsky's results are very close to ours and these cry for a common generalization. This will be subject of a future work.

2   Preliminaries

Recall that by a *network* we mean a quadruple (D, s, t, c), where $D = (V, A)$ is a digraph, s and t are different nodes of D and $c : A \to \mathbb{R}_+$ is a function that determines the capacity $c(a)$ of each arc a of A. (Sometimes it is assumed that no arc enters vertex s and no arc leaves vertex t. Though this assumption would allow a simpler proof, we do not require it for the reason that the result is significantly more general this way. Still, if the reader finds it difficult to follow the argument, it might be convenient to consider the source-sink case and skip the irrelevant parts.) A *flow* of network (D, s, t, c) is a function $f : A \to \mathbb{R}$ such that capacity condition $0 \le f(a) \le c(a)$ holds for each arc a of A and each vertex v of D different from s and t satisfies the Kirchhoff law: $\sum_{uv \in A} f(uv) = \sum_{vu \in A} f(vu)$, that is, the amount of the incoming flow equals the amount of the outgoing flow for v. Note that there is no conceptual difference between s and t: both are ordinary vertices that are exempt from the Kirchhoff law. (It seems that many people do not realize this. The reason perhaps is that when we teach network flows, we used to emphasize that the role of s and t are so different: one is "the source" and the other is a "the sink". To convince the sceptic, it is illuminative to find a formula for the minimum value of an st flow in a network. It is not 0 in general.)

A *network with preferences* is a network (D, s, t, c) along with a preference order $\le_v$ for each vertex v, such that $\le_v$ is a linear order on the arcs that are incident to v. (Note that preference orders $\le_s$ and $\le_t$ do not play a role in the notion of stability. Moreover, we shall never have to compare an incoming and an outgoing arc of the same vertex, so we may think that for each vertex v there is a preference order on the incoming arcs and another one on the outgoing ones.) For a given network with preferences, it is convenient to think that vertices of D are "players" that trade with a certain product. An arc uv of D from player u to player v with capacity $c(uv)$ represents the possibility that player u can supply at most $c(uv)$ units of product to player v. A "trading scheme" is described by a flow f of the network, as for any two players u and v, flow $f(uv)$ determines the amount of product that u sells to v. Everybody in the market would like to trade as much as possible, that is, each player v strives to maximize the amount of flow through v. In particular, if flow f allows player v to receive some more flow (that is, there are products on the market that v can buy) and v can also send some more flow (i.e. some player would be happy to buy more products from v) then flow f does not correspond to a stable market situation.

Another instability occurs when $vw \le_v vu$ (player v prefers to sell to w rather than to u) and flow f is such that w would be happy to buy more product from v (that is $f(vw) < c(vw)$ and w has some extra selling capacity), moreover $f(vu) > 0$ (v sells a positive amount of products to u). In this situation, v would send flow rather to w than to u, hence a stable market situation does not allow the above situation. A similar instability can be described if we talk about entering arcs instead of outgoing ones, that is, if we exchange the roles of buying and selling.

To formalize our concept of stability we need a few definitions. For a network (D, s, t, c) and flow f we say that arc a is f-*unsaturated* if $f(a) < c(a)$, that is, if it is possible to send some extra flow thorough P. A *blocking walk of flow* f is an alternating sequence of incident vertices and arcs $P = (v_1, a_1, v_2, a_2, \ldots, a_{k-1}, v_k)$ such that all the following properties hold.

$$\text{arc } a_i \text{ points from } v_i \text{ to } v_{i+1} \text{ for } i = 1, 2, \ldots, k-1 \text{ and} \tag{1}$$

$$\text{vertices } v_2, v_3, \ldots, v_{k-1} \text{ are different from } s \text{ and } t \tag{2}$$

$$\text{each arc } a_i \text{ is } f\text{-unsaturated and} \tag{3}$$

$$v_1 \in \{s, t\} \text{ or there is an arc } a' = v_1 u \text{ such that } f(a') > 0 \text{ and } a_1 <_{v_1} a' \tag{4}$$

$$v_k \in \{s, t\} \text{ or there is an arc } a^* \text{ to } v_k \text{ such that } f(a^*) > 0 \text{ and } a_{k-1} <_{v_k} a^*. \tag{5}$$

So directed walk P is blocking if each player that corresponds to an inner vertex of P is happy and capable to increase the flow along P, moreover v_1 can send extra flow either because $v_1 = s$ or $v_1 = t$ is a terminal node or because v_1 may decrease the flow toward some vertex u that v_1 prefers less than v_2, and at last, v_k can receive some extra flow either because either $v_k \in \{s, t\}$ or v_k can refuse some flow arriving from w whom v_k ranks below v_{k-1}. (As we mentioned before, there is no difference between the roles of s and t in the network: none of them have to obey the Kirchhoff law and both of them can send or receive flow. If the reader is uncomfortable with the idea that the target node sends flow to the source then consider the case where no arc enters s and no arc leaves t. This assumption simplifies some of the proofs.) We say that an f-unsaturated path $P = (v_1, v_2, \ldots, v_k)$ is f-*dominated at* v_1 if (4) does not hold, and P is f-dominated at v_k if (5) does not hold.

A flow f of a network with preferences is *stable* if no blocking walk exists for f. In the *stable flow problem* we have given a network with preferences and our task is to find a stable flow if such exists.

A special case of the stable flow problem is the stable allocation problem of Baïou and Balinski [1]. The *stable allocation problem* is defined by finite disjoint sets W and F of workers and firms, a map $q : W \cup F \to \mathbb{R}$, a set E of edges between W and F along with a map $p : E \to \mathbb{R}$ and for each worker or firm $v \in W \cup F$ a linear order $<_v$ on those pairs of E that contain v. We shall refer to pairs of E as "edges" and hopefully it will not cause ambiguity. Quota $q(v)$ denotes the maximum of total assignment that worker or firm v can accept and capacity $p(wf)$ of edge $e = wf$ means the maximum allocation that worker w can be assigned to firm f along e. An *allocation* is a nonnegative map $g : E \to \mathbb{R}$ such that $g(e) \le p(e)$ holds for each $e \in E$ and for any $v \in W \cup F$ we have

$$g(v) := \sum_{vx \in E} g(vx) \le q(v) , \tag{6}$$

that is the total assignment $g(v)$ of player v cannot exceed quota $q(v)$ of v. If (6) holds with equality then we say that player v is g-*saturated*. An allocation is *stable* if for any edge wf of E at least one of the following properties hold:

$$g(wf) = p(wf) \text{(the particular employment is realized with full capacity)}, \quad (7)$$

worker w is g-saturated and w does not prefer f to any of his employers

(we say that wf is g-*dominated at* w), $\quad (8)$

firm f is g-saturated and f does not prefer w to any of its employees

(that is, edge wf is g-*dominated at firm* f). $\quad (9)$

If g_1 and g_2 are allocations and $w \in W$ is a worker then we say that *allocation* g_1 *dominates allocation* g_2 *for worker* w (in notation $g_1 \leq_w g_2$) if one of the following properties is true:

$$\text{either } g_1(wf) = g_2(wf) \text{ for each } f \in F \quad (10)$$

$$\text{or } \sum_{f' \in F} g_1(wf') = \sum_{f' \in F} g_2(wf') = q(w), \text{ and}$$
$$g_1(wf) < g_2(wf) \text{ and } g_1(wf') > 0 \text{ implies that } wf' <_w wf. \quad (11)$$

That is, if w can freely choose his allocation from $\max(g_1, g_2)$ then w would choose g_1 either because g_1 and g_2 are identical for w or because w is saturated in both allocations and g_1 represents w's choice out of $\max(g_1, g_2)$. By exchanging the roles of workers and firms, one can define domination relation $\leq_f$ for any firm f, as well.

For any stable allocation problem, one can design a network (D, s, t, c) such that $V(D) = \{s, t\} \cup W \cup F$, $A(D) = \{sw : w \in W\} \cup \{ft : f \in F\} \cup \{wf : wf \in E\}$ and $c(sw) = q(w)$, $c(ft) = q(f)$ and $c(wf) = p(wf)$ for any worker w and firm f. That is, we consider the underlying bipartite graph, orient its edges from W to F, add new vertices s and t, with an arc from s to each worker-node and an arc from each firm-node to t, and capacities are given by the original edge-capacities and the corresponding quotas. Preference orders $<_v$ on the arcs incident to v are induced by the preference order on the corresponding edges incident to v, or, if there is no such edge, then it is a trivial linear order. It is straightforward to see from the definitions that g is a stable allocation if and only if there exists a stable flow f such that $g(e) = f(\boldsymbol{e})$ holds for each edge $e \in E$, where $\boldsymbol{e}$ is the arc that corresponds to edge e. The stable allocation problem was introduced by Baïou and Balinski as a certain "continuous" version of the stable marriage problem in [1]. It turned out that a natural extension of the deferred acceptance algorithm of Gale and Shapley [6] works for the stable allocation problem and the structure of stable allocations is similar to that of stable marriages. Beyond stating the existence of stable allocations, the theorem below describes some structural properties of them. The interested reader finds a proof based on Tarski's fixed point theorem in [5].

Theorem 1 (See Baïou and Balinski [1])

1. If stable allocation problem is described by W, F, E, p and q then there always exists a stable allocation g. Moreover, if p and q are integral, then there exists an integral stable allocation g.

2. If g_1 and g_2 are stable allocations and $v \in W \cup F$ then $g_1 \leq_v g_2$ or $g_2 \leq_v g_1$ holds.

3. Stable allocations have a natural lattice structure. I.e., if g_1 and g_2 are stable allocations then $g_1 \vee g_2$ and $g_1 \wedge g_2$ are stable allocations, where

$$(g_1 \vee g_2)(wf) = \begin{cases} g_1(wf) \ if \ g_1 \leq_w g_2 \\ g_2(wf) \ if \ g_2 \leq_w g_1 \end{cases} \ and \tag{12}$$

$$(g_1 \wedge g_2)(wf) = \begin{cases} g_1(wf) \ if \ g_1 \leq_f g_2 \\ g_2(wf) \ if \ g_2 \leq_f g_1 \end{cases} \tag{13}$$

In other words, if workers choose from two stable allocations then we get another stable allocation, and this is also true for the firms' choices. Moreover, it is true that

$$(g_1 \vee g_2)(wf) = \begin{cases} g_1(wf) \ if \ g_1 \geq_f g_2 \\ g_2(wf) \ if \ g_2 \geq_f g_1 \end{cases} \ and \tag{14}$$

$$(g_1 \wedge g_2)(wf) = \begin{cases} g_1(wf) \ if \ g_1 \geq_w g_2 \\ g_2(wf) \ if \ g_2 \geq_w g_1 \end{cases} \tag{15}$$

That is, in stable allocation $g_1 \vee g_2$ where each worker picks his better assignment, each firm receives the worse out of the two. Similarly, in $g_1 \wedge g_2$ the choice of the firms means the less preferred situation to the workers.

3 Stable Flows

Our goal in this section is to prove a generalization of Theorem 1. The "natural" approach to achieve this would be an appropriate generalization of the deferred acceptance algorithm of Gale and Shapley. The difficulty is that though the Gale-Shapley algorithm can handle quota function q, somehow it has problems with ensuring the Kirchhoff law.

Theorem 2. *If network (D, s, t, c) and preference orders $<_v$ describe a stable flow problem then there always exists a stable flow f. If capacity function c is integral then there exists an integral stable flow.*

Note that it is possible to prove Theorem 2 by a mixture of the deferred acceptance algorithm and the augmenting path algorithm. That is, starting from s or from t, we follow "first choice walks" until they arrive to s or t and we augment along them with observing the capacity constraints. If a new path collides with an earlier one then some amount of flow is refused by the receiving vertex and we try to reroute the flow excess from the starting point of the refused arc. We have a stable flow as soon as we cannot find an augmenting path between the terminals.

Our proof of Theorem 2 follows a different approach for two reasons. On one hand, it seems that in the area of stable matchings neither the reduction of one problem to another one nor the use of graph terminology is routine. We demonstrate here that these methods may be fruitful. On the other hand, the "deferred augmentation" algorithm we sketched above does not give much information about the rich structure of stable flows that we shall deduce from the lattice property of stable allocations.

With the help of the given stable flow problem, we shall define a stable allocation problem. For each vertex v of D calculate

$$M(v) := \min \left(\sum_{xv \in A(D)} c(xv), \sum_{vx \in A(D)} c(vx) \right),$$

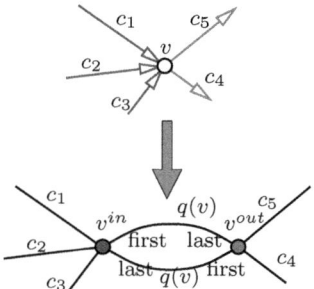

that is, $M(v)$ is the minimum of total capacity of those arcs of D that enter and leave v. So $M(v)$ is an upper bound on the amount of flow that can flow through vertex v. Choose $q(v) := M(v) + 1$. Construct graph G_D as follows. Split each vertex v of D into two distinct vertices v^{in} and v^{out}, and for each arc uv of D add edge $u^{out}v^{in}$ to G_D.

For each vertex v of D different from s and t add two parallel edges between v^{in} and v^{out}: to distinguish between them we will refer them as $v^{in}v^{out}$ and $v^{out}v^{in}$. Let $p(v^{in}v^{out}) = p(v^{out}v^{in}) := q(v)$, $p(u^{out}v^{in}) := c(uv)$ and $q(v^{in}) = q(v^{out}) := q(v)$. To finish the construction of the stable allocation problem, we need to fix a linear preference order for each vertex of G_D. For vertex v^{in} let $v^{in}v^{out}$ be the most preferred and $v^{out}v^{in}$ be the least preferred edge (if these edges are present), and the order of the other edges incident to v^{in} are coming from the preference order of v on the corresponding arcs. For vertex v^{out} the most preferred edge is $v^{out}v^{in}$ and the least preferred one is $v^{in}v^{out}$ (if it makes sense), and the other preferences are coming from $<_v$.

The proof of Theorem 2 is a consequence of the following Lemma that describes a close relationship between stable flows and stable allocations.

Lemma 1. *If network (D, s, t, c) and preference orders $<_v$ describe a stable flow problem then $f : A(D) \rightarrow \mathbb{R}$ is a stable flow if and only if there is a stable allocation g of G_D such that $f(uv) = g(u^{out}v^{in})$ holds for each arc uv of D.*

Proof. Assume first that g is a stable allocation in G_D. This means that none of the $v^{in}v^{out}$ edges is blocking, so either $g(v^{in}v^{out}) = p(v^{in}v^{out}) = q(v)$ or $v^{in}v^{out}$ must be g-dominated at v^{out}, hence v^{out} is assigned to $q(v^{out}) = q(v)$ amount of allocation. As $q(v)$ is more than the total capacity of arcs leaving v, $g(v^{in}v^{out}) > 0$ or $g(v^{out}v^{in}) > 0$ must hold. So v^{out} must have exactly $q(v)$ amount of allocation whenever $v^{in}v^{out}$ is present. An exchange of in and out shows that the presence of $v^{out}v^{in}$ implies that v^{in} has exactly $q(v^{in}) = q(v)$ allocation. These observations directly imply that the Kirchhoff law holds for f at each node different from s and t. The capacity condition is also trivial for f, hence f is a flow of D. Observe that by the choice of q, neither s nor t is g-saturated hence no edge is g-dominated at s or at t.

Assume that walk $P = (v_1, v_2, \ldots, v_k)$ blocks flow f. As P is f-unsaturated, each edge $v_i^{out} v_{i+1}^{in}$ of G_D must be g-dominated at v_i^{out} or at v_{i+1}^{in}. Walk P is blocking, hence either $v_1 \in \{s, t\}$, and hence $v_1^{out} v_2^{in}$ cannot be dominated at v_1 or there is a $v_1 u$ arc with positive flow value such that $v_1 u > v_1 v_2$. In both cases, edge $v_1^{out} v_2^{in}$ has to be g-dominated at v_2^{in}. It means that $g(v_2^{in} v_2^{out}) > 0$. As arc $v_2 v_3$ is f-unsaturated, it follows that edge $v_2^{out} v_3^{in}$ must be g-dominated at v_3^{in}. This yields that $g(v_3^{in} v_3^{out}) > 0$. Again, arc $v_3 v_4$ is f-unsaturated, hence edge $v_3^{out} v_4^{in}$ has to be g-dominated at v_4^{in}, and so on. At the end we get that $v_{k-1}^{out} v_k^{in}$ is g-dominated at v_k^{in}. If $v_k \in \{s, t\}$ then it is impossible as both these vertices are g-unsaturated. Otherwise by the blocking property of P there is an arc $w v_k$ with positive flow and $v_{k-1} v_k <_{v_k} w v_k$, hence again, $v_{k-1}^{out} v_k^{in}$ cannot be g-dominated at v_k^{in}. The contradiction shows that no path can block f.

Assume now that f is a stable flow of D. We have to exhibit a stable allocation g of G_D such that f is the "restriction" of g. To determine g, our real task is to find the $g(v^{in} v^{out})$ and $g(v^{out} v^{in})$ values, as all other values of g are determined directly by f: $g(u^{out} v^{in}) = f(uv)$. The stable allocation we look for might not be unique. In what follows, we shall construct the *canonical representation* g_f of f.

Let S be the set of those vertices u of D such that there exists an f-unsaturated directed path $P = (v_1, v_2, \ldots, v_k = u)$ that is not f-dominated at v_1. As no path can block f, neither s, nor t belongs to S. To determine g_f, for each vertex $v \neq s, t$ allocate the remaining quota of v to $v^{in} v^{out}$ or to $v^{out} v^{in}$ depending on whether $v \in S$ or $v \notin S$ holds. More precisely, define

$$g_f(v^{in} v^{out}) = \begin{cases} q(v) - \sum_{x \in V(D)} f(vx) & \text{if } v \in S \\ 0 & \text{if } v \notin S \end{cases} \quad \text{and} \quad (16)$$

$$g_f(v^{out} v^{in}) = \begin{cases} q(v) - \sum_{x \in V(D)} f(xv) & \text{if } v \notin S \\ 0 & \text{if } v \in S. \end{cases} \quad (17)$$

By the definition of q, both $g_f(v^{in} v^{out})$ and $g_f(v^{out} v^{in})$ are nonnegative. If $v \in S$ then the amount of total allocation of v^{out} is $q(v) = q(v^{out})$ by (16), and for $v \notin S$ the amount of total allocation of v^{in} is $q(v) = q(v^{in})$ by (17). So if $v \neq s, t$ then the total allocation of v^{in} and v^{out} is $q(v)$ by the Kirchhoff law. The total allocations of s^{in}, s^{out} and t^{in}, t^{out} is less than $q(s)$ and $q(t)$ respectively, by the choice of q. That is, g_f is an allocation on G_D.

To justify the stability of g_f, we have to show that no blocking edge exists. We have seen earlier, that the presence of $v^{in} v^{out}$ in G_D means that v^{out} g-dominates $v^{in} v^{out}$. Similarly, each edge $v^{out} v^{in}$ is g_f-dominated at v^{in}. Assume now that $g_f(v^{out} u^{in}) < p(v^{out} u^{in}) = c(vu)$ holds.

If there is an f-unsaturated path P that is not f-dominated at its starting node and ends with arc vu then $u \in S$ by the definition of S, hence $g_f(u^{out} u^{in}) = 0$. Moreover, if some edge $w^{out} u^{in}$ with $v^{out} u^{in} <_{u^{in}} w^{out} u^{in}$ would have positive allocation then path P would block f, a contradiction. As u^{in} has $q(u^{in})$ amount of total allocation, edge $v^{out} u^{in}$ is g_f-dominated at u^{in}.

The last case is when any f-unsaturated path that ends with arc vu is f-dominated at its starting vertex. In particular, $v \notin S$, so $g_f(v^{in} v^{out}) = 0$.

Moreover, f-unsaturated path (v, u) must be f-dominated at v, hence $v \notin \{s, t\}$ and $v^{out}u^{in}$ is g_f-dominated at v^{out} as v^{out} has $q(v) = q(v^{out})$ amount of allocation. The conclusion is that $g := g_f$ is a stable allocation, just as we claimed.

At this point, we are ready to prove our main result.

Proof (Proof of Theorem 2). There is a stable allocation for G_D by Theorem 1, hence there is a stable flow for D due to the first part of Theorem 1. If c is integral then $q(v)$ is an integer for each vertex v of D hence p is integral for G_D. The integrality property of stable allocations in the first part of Theorem 1 shows that there is an integral stable allocation g of G_D that describes an integral stable flow f of D.

At the end of this section let us point out a weakness of our stability concept. The motivation behind the notion is that we look for a flow that corresponds to an equilibrium situation where the players represented by the vertices of the network act in a selfish way. This equilibrium situation occurs if no coalition of the players can block the underlying flow f, and this blocking is defined by a certain f-unsaturated path (or cycle through s or t) along which the players are capable and prefer to increase the flow. However, in some sense an f-unsaturated cycle C per se causes instability because the players of C mutually agree to send some extra flow along C. So it is natural to define flow f of network (D, s, t, c) with preferences to be *completely stable* if f is stable and there exists no f-unsaturated cycle in D whatsoever. If f is a stable flow then we can "augment" along f-unsaturated cycles, and hence we can construct a flow $f' \geq f$ such that there no longer exists an f'-unsaturated cycle. But unfortunately flow f' might not be stable any more because we might have created a blocking walk by the cycle-augmentations.

In fact, there exist networks with preferences that do not have a completely stable flow. One example is on the figure: each arc has unit capacity, preferences are indicated around the vertices: lower rank is preferred to the higher.

As no arc leaves subset $U := \{a, b, c\}$ of the vertices, no flow can leave U, hence no flow enters U. In particular, arc sa has zero flow. If we assume indirectly that f is a completely stable flow then cycle abc cannot block, hence there must be a unit flow along it. But now path sa is blocking, a contradiction.

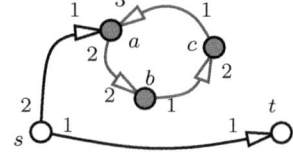

Stable flows have a blocking cycle

4   The Structure of Stable Flows

It is well-known about the stable marriage problem that in each stable marriage scheme, the same set of participants get married. That is, if someone does not get a marriage partner in some stable scheme then this very person remains single in each stable marriage schemes. A generalization of this is the rural hospital theorem of Roth [9] (see also Theorem 5.13 in [10]). It is about the college model, where instead of men we work with colleges, women correspond to students and each college has a quota on the maximum number of students. In the college admission problem, it is true that if a certain college c cannot fill up its quota in a stable admission scheme then c receives the same set of students in any stable admission scheme. (The phenomenon is named after the assignment problem of medical interns to hospitals.)

It seems that the rural hospital theorem cannot be generalized to the stable flow problem. It may happen in a network that a certain vertex transmits different amounts of flow in two stable flows.

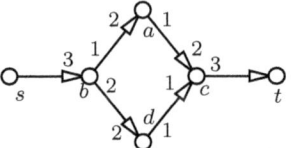

An example is shown in the figure where each arc has unit capacity. There are two stable flows: one is along path $sbact$ and the other follows path $sbdct$. So in one stable flow, vertex a transmits unit flow and no flow passes through a in the other one.

Network for a stable flow

There is however a consequence of the rural hospital theorem that can be generalized, namely, that the size of a stable matching is always the same. We have seen that the stable allocation problem is a special case of the stable flow problem, and from the construction it is apparent that the size of a stable matching (more precisely the total amount of assignments in a stable allocation) equals the value of the corresponding flow.

Theorem 3. *If network* (D, s, t, c) *and preference orders* $<_v$ *describe a stable flow problem and* f_1 *and* f_2 *are stable flows then the value of* f_1 *and* f_2 *are the same. More generally,* $f_1(a) = f_2(a)$ *for any arc of* D *that is incident to* s *(or to* t*).*

Proof. Lemma 1 implies that there exist stable allocations g_1 and g_2 of G_D that correspond to stable flows f_1 and f_2, respectively. The value of a flow is the net amount that leaves s in D, or, in G_D one can calculate it as the difference of total allocation of s^{out} and s^{in}. This means that the second part of the theorem implies the first one.

As there is no edge between s^{out} and s^{in}, the choice of $q(s)$ implies that both s^{out} and s^{in} are g_1-unsaturated. Hence property (11) can hold neither for s^{in} nor for s^{out}. But Theorem 1 implies that g_1 and g_2 are $\leq_{s^{out}}$ and $\leq_{s^{in}}$-comparable. So property (10) must be true for both flows g_1 and g_2 for vertices $v = s^{out}$ and $v = s^{in}$. This shows the second part of the Theorem for s. The argument for t is analogous to the above one.

As we have seen in Theorem 1, stable allocations have a lattice structure. Based on the connection of stable allocations and stable flows described in Lemma 1, we can prove that stable flows of a network with preferences also form a natural lattice. So assume that f is a stable flow in network $(D, s, t, c,)$ with preferences and let stable allocation g_f of G_D be the canonical representation of f as in the proof of Lemma 1.

Observe that any vertex $v \neq s, t$ of D, exactly one of $g_f(v^{in}v^{out})$ and $g_f(v^{out}v^{in})$ is positive by the choice of q and g_f. For stable flow f, we can classify the vertices of D different from s and t: v is an f-*vendor* if $g_f(v^{in}v^{out}) > 0$ and v is an f-*customer* if $g_f(v^{out}v^{in}) > 0$. If v is an f-vendor then no edge $v^{out}u^{in}$ can be g_f-dominated at v^{out} (as $g_f(v^{in}v^{out}) > 0$), hence player v sends as much flow to other vertices as much they accept. Similarly, if v is an f-customer then no edge $u^{out}v^{in}$ can be g_f-dominated at v^{out}, that is, player v receives as much flow as the others can supply her.

To explore the promised lattice structure of stable flows, let f_1 and f_2 two stable flows with canonical representations g_{f_1} and g_{f_2}, respectively. From Theorem 1 we know that stable allocations form a lattice, so $g_{f_1} \vee g_{f_2}$ and $g_{f_1} \wedge g_{f_2}$ are also stable allocations of G_D, and by Theorem 2, these stable allocations define stable flows $f_1 \vee f_2$ and $f_1 \wedge f_2$, respectively. How can we determine these latter flows directly, without the canonical representations? To answer this, we translate the lattice property of stable allocations on G_D to stable flows of D.

Theorem 3 shows that stable flows cannot differ on arcs incident to s or t, so on these arcs $f_1 \vee f_2$ and $f_1 \wedge f_2$ are determined. However, vertices different from s and t may have completely different situations in stable flows f_1 and f_2. The two colour classes of graph G_D are formed by the v^{in} and v^{out} type vertices, respectively. So, by Theorem 1, $g_{f_1} \vee g_{f_2}$ can be determined such that (say) each vertex v^{out} selects the better allocation and each vertex v^{in} receives the worse allocation out of the ones that g_{f_1} and g_{f_2} provides them. Similarly, for stable allocation $g_{f_1} \wedge g_{f_2}$ the "in"-type vertices choose according to their preferences and the "out"-type ones are left with the less preferred allocations. This means the following in the language of flows. If we want to construct $f_1 \vee f_2$ and v is a vertex different from s and t then either all arcs entering v will have the same flow in $f_1 \vee f_2$ as in f_1, or for all arcs a entering v we have $(f_1 \vee f_2)(a) = f_2(a)$ holds. A similar statement is true for the arcs leaving v. To determine which of the two alternatives is the right one, the following rules apply:

- If v is an f_1-vendor and an f_2-customer then v chooses f_2. If v is an f_2-vendor and an f_1-customer then v chooses f_1. That is, each vertex strives to be a customer.
- If v is an f_1-vendor and an f_2-vendor and v transmits more flow in f_1 than in f_2 (i.e. $0 < g_{f_1}(v^{in}v^{out}) < g_{f_2}(v^{in}v^{out})$) then v chooses f_1. That is, vendors prefer to sell more.
- If v is an f_1-customer and an f_2-customer and v transmits more flow in f_1 than in f_2 (i.e. $0 < g_{f_1}(v^{out}v^{in}) < g_{f_2}(v^{out}v^{in})$) then v chooses f_2. That is, customers prefer to buy less.
- Otherwise v is a customer in both f_1 and f_2 or v is a vendor in both flows and v transmits the same amount in both flows (i.e. $g_{f_1}(v^{out}v^{in}) = g_{f_2}(v^{out}v^{in})$ and $g_{f_1}(v^{in}v^{out}) = g_{f_2}(v^{in}v^{out})$). In this situation, v chooses the better "selling position" and gets the worse "buying position" out of stable flows f_1 and f_2.

Clearly, for the construction of $f_1 \wedge f_2$, one always has to choose the "other" options than the one that the above rules describe.

The lattice structure of stable flows defines a partial order on stable flows: $f_1 \preceq f_2$ if and only if $f_1 \vee f_2 = f_2$ holds, or equivalently, if $f_1 \wedge f_2 = f_1$ is true. By to the above rules, this means that each f_1-customer v is an f_2-customer, such that v buys at least as much in f_1 as in f_2. Each f_2-vendor u is an f_1-vendor and u sells at most as much in f_1 as in f_2. If w plays the same role (vendor or customer) in both flows and transmits the same amount then v prefers the selling position of f_2 and the buying position of f_1.

Acknowledgment. The author kindly acknowledges the support of the EGRES.

References

1. Baïou, M., Balinski, M.: The stable allocation (or ordinal transportation) problem. Math. Oper. Res. 27(3), 485–503 (2002)
2. Cechlárová, K., Fleiner, T.: On a generalization of the stable roommates problem. ACM Trans. Algorithms 1(1), 143–156 (2005)
3. Dean, B.C., Munshi, S.: Faster algorithms for stable allocation problems. In: Proceedings of the MATCH-UP (Matching Under Preferences) Workshop at ICALP 2008, Reykjavik, pp. 133–144 (2008)
4. Fleiner, T.: A fixed point approach to stable matchings and some applications. Mathematics of Operations Research 28(1), 103–126 (2003)
5. Fleiner, T.: On stable matchings and flows. Technical Report TR-2009-11, Egerváry Research Group, Budapest (2009), http://www.cs.elte.hu/egres
6. Gale, D., Shapley, L.S.: College admissions and stability of marriage. Amer. Math. Monthly 69(1), 9–15 (1962)
7. Knuth, D.E.: Stable marriage and its relation to other combinatorial problems. American Mathematical Society, Providence (1997); An introduction to the mathematical analysis of algorithms, Translated from the French by Martin Goldstein and revised by the author
8. Ostrovsky, M.: Stability in supply chain networks. American Economic Review 98(3), 897–923 (2006)
9. Roth, A.E.: On the allocation of residents to rural hospitals: a general property of two-sided matching markets. Econometrica 54(2), 425–427 (1986)
10. Roth, A.E., Oliveria Sotomayor, M.A.: Two-sided matching. Cambridge University Press, Cambridge (1990); A study in game-theoretic modeling and analysis, With a foreword by Robert Aumann
11. Tarski, A.: A lattice-theoretical fixpoint theorem and its applications. Pacific J. of Math. 5, 285–310 (1955)

Narrowing Down the Gap on the Complexity of Coloring P_k-Free Graphs

Hajo Broersma, Petr A. Golovach, Daniël Paulusma, and Jian Song*

School of Engineering and Computing Sciences, Durham University,
Science Laboratories, South Road, Durham DH1 3LE, UK
{hajo.broersma,petr.golovach,daniel.paulusma,jian.song}@durham.ac.uk

Abstract. A graph is P_k-free if it does not contain an induced subgraph isomorphic to a path on k vertices. We show that deciding whether a P_8-free graph can be colored with at most four colors is an NP-complete problem. This improves a result of Le, Randerath, and Schiermeyer, who showed that 4-coloring is NP-complete for P_9-free graphs, and a result of Woeginger and Sgall, who showed that 5-coloring is NP-complete for P_8-free graphs. Additionally, we prove that the pre-coloring extension version of 4-coloring is NP-complete for P_7-free graphs, but that the pre-coloring extension version of 3-coloring is polynomially solvable for $(P_2 + P_4)$-free graphs, a subclass of P_7-free graphs.

1 Introduction

Due to the fact that the usual ℓ-COLORING problem is NP-complete for any fixed $\ell \geq 3$, there has been a considerable interest in studying its complexity when restricted to certain graph classes, in particular graph classes that can be characterized by forbidden induced subgraphs. We refer to [14, 17] for surveys. Instead of repeating what has been written in so many papers over the years, and in order to save as much space as possible for relevant details related to our results, we also refer to these surveys for motivation and background. Here we continue the study of ℓ-COLORING for P_k-free graphs. This setting has been studied in several earlier papers by different groups of researchers (see, e.g., [3, 5, 9–13, 18]). Before we summarize their results we first introduce the necessary terminology.

Terminology. We only consider finite undirected graphs without loops and multiple edges. We refer to [2] for any undefined graph terminology. The graph P_k denotes the path on k vertices. The disjoint union of two graphs G and H is denoted $G + H$, and the disjoint union of k copies of G is denoted kG. A *linear forest* is the disjoint union of a collection of paths. Given two graphs G and H we say that G is *H-free* if G has no induced subgraph isomorphic to H.

A *(vertex) coloring* of a graph $G = (V, E)$ is a mapping $\phi : V \to \{1, 2, \ldots\}$ such that $\phi(u) \neq \phi(v)$ whenever $uv \in E$. Here $\phi(u)$ is referred to as the *color*

---

* This work has been supported by EPSRC (EP/G043434/1).

D.M. Thilikos (Ed.): WG 2010, LNCS 6410, pp. 63–74, 2010.
© Springer-Verlag Berlin Heidelberg 2010

of u. An ℓ-*coloring* of G is a coloring ϕ of G with $\phi(V) \subseteq \{1, \ldots, \ell\}$. Here we use the notation $\phi(U) = \{\phi(u) \mid u \in U\}$ for $U \subseteq V$. We let $\chi(G)$ denote the *chromatic number* of G, i.e., the smallest ℓ such that G has an ℓ-coloring. The problem ℓ-COLORING is the problem to decide whether a given graph admits an ℓ-coloring.

In *list-coloring* we assume that $V = \{v_1, v_2, \ldots, v_n\}$ and that for every vertex v_i of G there is a list L_i of *admissible* colors (a subset of the natural numbers). We say that a coloring $\phi : V \rightarrow \{1, 2, \ldots\}$ *respects* these lists if $\phi(v_i) \in L_i$ for all $i \in \{1, 2, \ldots, n\}$. We also call ϕ a *list-coloring* in this case.

In *pre-coloring extension* we assume that a (possibly empty) subset $W \subseteq V$ of G is pre-colored with $\phi_W : W \rightarrow \{1, 2, \ldots\}$ and the question is whether we can extend ϕ_W to a coloring of G. If ϕ_W is restricted to $\{1, 2, \ldots, \ell\}$ and we want to extend it to an ℓ-coloring of G, we say we deal with the *pre-coloring extension version of* ℓ-COLORING.

Known results. Results of Hoàng et al. [9] imply that the pre-coloring extension version of ℓ-COLORING is polynomially solvable on P_5-free graphs for any fixed ℓ. In contrast, determining the chromatic number is NP-hard for P_5-free graphs [10], whereas this problem is polynomially solvable for P_4-free graphs (because a P_4-free graph is perfect, and the chromatic number of a perfect graph can be determined in polynomial time [8]). Le, Randerath, and Schiermeyer [12] proved that 4-COLORING is NP-complete for P_9-free graphs. Woeginger and Sgall [18] showed that 5-COLORING is NP-complete for P_8-free graphs. In [3] we established the following three results. Firstly we proved that 6-COLORING is NP-complete for P_7-free graphs, secondly that the pre-coloring extension version of 3-COLORING is polynomially solvable for P_6-free graphs, and thirdly that the pre-coloring extension version of 5-COLORING is NP-complete for P_6-free graphs. All these results together lead to the following table that shows the current status of ℓ-COLORING and its pre-coloring extension version for P_k-free graphs. This table also shows which cases are still open.

Table 1. The complexity of ℓ-COLORING and its pre-coloring extension version (marked by *) on P_k-free graphs for combinations of fixed k and ℓ

P_k-free	$\ell \rightarrow$ 3	3*	4	4*	5	5*	≥ 6	≥ 6*
$k \leq 5$	P	P	P	P	P	P	P	P
$k = 6$	P	P	?	?	?	NP-c	?	NP-c
$k = 7$	?	?	?	?	?	NP-c	NP-c	NP-c
$k = 8$	?	?	?	?	NP-c	NP-c	NP-c	NP-c
$k \geq 9$	?	?	NP-c	NP-c	NP-c	NP-c	NP-c	NP-c

Our results and paper organization. In Section 2 we present a common improvement to results in [12] and [18] by showing that 4-COLORING is NP-complete for P_8-free graphs. In Section 3 we give a closely related result showing that the pre-coloring extension version of 4-COLORING is NP-complete for P_7-free graphs. It seems hard to extend our result from [3] on the pre-coloring extension version of 3-COLORING for P_6-free graphs to P_7-free graphs. This motivates our focus on subclasses of P_7-free graphs, namely H-free graphs, where H is a linear forest on at most 6 vertices. We show in Section 4 that the first nontrivial case is $H = P_2 + P_4$ and that the pre-coloring extension version of 3-COLORING is polynomially solvable for (P_2+P_4)-free graphs. Section 5 contains the conclusions and mentions open problems.

2   4-Coloring for P_8-Free Graphs

In this section we prove that 4-COLORING is NP-complete for P_8-free graphs. We use a reduction from 3-SATISFIABILITY (3SAT), which is an NP-complete problem [7]. We consider an arbitrary instance I of 3SAT that has variables $\{x_1, x_2, \ldots, x_n\}$ and clauses $\{C_1, C_2, \ldots, C_m\}$ and define a graph G_I. Next we show that G_I is P_8-free and that G_I is 4-colorable if and only if I has a satisfying truth assignment.

Here is the construction that defines G_I.

– For each clause C_j we introduce a 7-vertex cycle with vertex set

$$\{b_{j,1}, b_{j,2}, c_{j,1}, c_{j,2}, c_{j,3}, d_{j,1}, d_{j,2}\}$$

and edge set

$$\{b_{j,1}c_{j,1}, c_{j,1}d_{j,1}, d_{j,1}c_{j,2}, c_{j,2}d_{j,2}, d_{j,2}c_{j,3}, c_{j,3}b_{j,2}, b_{j,2}b_{j,1}\}.$$

We say that these vertices are of b-type, c-type and d-type, respectively. They induce disjoint 7-cycles (i.e., cycles on 7 vertices) in G_I which we call *clause-components* in the sequel.

– For each variable x_i we introduce a copy of a K_2, i.e., two vertices joined by an edge $x_i\overline{x}_i$. We say that both x_i and $\overline{x}_i$ are of x-type, and we call the corresponding disjoint K_2s in G_I *variable-components* in the sequel.

– For every clause C_j we fix an arbitrary order of its variables $x_{i_1}, x_{i_2}, x_{i_3}$. For $h = 1, 2, 3$ we either add the edge $c_{j,h}x_{i_h}$ or the edge $c_{j,h}\overline{x}_{i_h}$ depending on whether x_{i_h} or $\overline{x}_{i_h}$ is a literal in C_j, respectively.

– We add an edge between any x-type vertex and any b-type vertex. We also add an edge between any x-type vertex and any d-type vertex.

– We introduce one additional new vertex a which we make adjacent to all b-type, c-type and d-type vertices.

See Figure 1 for an example of a graph G_I. In this example C_1 is a clause with ordered literals $x_{i_1}, \overline{x}_{i_2}, x_{i_3}$ and C_m is a clause with ordered literals $\overline{x}_1, x_{i_3}, x_n$. The thick edges indicate the connections between the literal vertices and the

c-type vertices of the clause gadgets. We omitted the indices from the labels of the clause gadget vertices to increase the visibility.

We complete this section by proving two lemmas. Lemma 1 shows that the graph G_I is P_8-free (in fact it shows a slightly stronger statement as this will be of use for us in Section 3). In Lemma 2 we prove that G_I admits a 4-coloring if and only if I has a satisfying truth assignment.

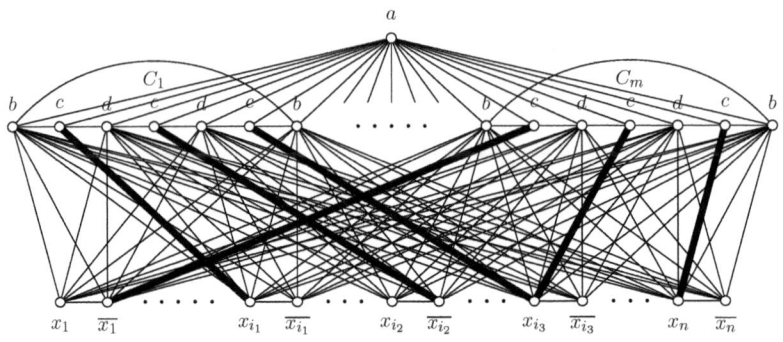

Fig. 1. The graph G_I in which clauses $C_1 = \{x_{i_1}, \overline{x}_{i_2}, x_{i_3}\}$ and $C_m = \{\overline{x}_1, x_{i_3}, x_n\}$ are illustrated

Lemma 1. *The graph G_I is P_8-free. Moreover, every induced path in G_I on seven vertices contains a.*

Proof. Let P be an induced path in G_I. We show that G_I is P_8-free by proving that P has at most seven vertices. We also show that P contains a in case P has exactly seven vertices. We distinguish a number of cases and subcases.

Case 1. $a \notin V(P)$.

Case 1a. P contains no x-type vertex.
This means that P is contained in one clause-component, which is isomorphic to an induced 7-cycle. Consequently, P has at most 6 vertices.

Case 1b. P contains exactly one x-type vertex.
Let x_i be this vertex. Then P contains vertices of at most two clause-components. Since x_i is adjacent to all b-type and d-type vertices, we then find that P contains at most two vertices of each of the clause-components. Hence P has at most 5 vertices.

Case 1c. P contains exactly two x-type vertices.
First suppose that these vertices are adjacent, say P contains x_i and $\overline{x}_i$. By the same reasoning as above we find that P has at most 4 vertices.

Now suppose the two x-type vertices of P are not adjacent. By symmetry, we may assume that P contains x_h and x_i. If P contains no b-type vertex and no d-type vertex, then there is no subpath in P from x_h to x_i, a contradiction. If P contains two or more vertices of b-type and d-type, then P contains a cycle, another contradiction. Hence P contains exactly one vertex z that is of b-type or d-type. Then $x_h z x_i$ is a subpath in P. If both x_h and x_i have a neighbor in $V(P)\backslash\{z\}$, then this neighbor must be of c-type, and consequently an end vertex of P (because a c-type vertex is adjacent to only one x-type vertex). Hence P contains at most five vertices.

Case 1d. P contains at least three x-type vertices.
Then P contains no b-type vertex and no d-type vertex, because such vertices would have degree 3 in P. However, on the other hand the three x-type vertices come from at least two different variable-components. Since any c-type vertex is adjacent to exactly one x-type vertex, P must contain a b-type or d-type vertex to connect the x-type vertices of P to one another. We conclude that this subcase is not possible.

Case 2. a ∈ V(P).
First suppose a is an end vertex of P. If $|V(P)| \geq 2$ then P contains exactly one vertex that is of b-type, c-type or d-type. Since every x-type vertex is adjacent to only one other x-type vertex, this means that P can have at most four vertices.

Now suppose a is not an end vertex of P. Then P contains exactly two vertices that are of b-type, c-type or d-type. By the same arguments as above, we then find that P has at most 7 vertices. This completes the proof of Lemma 1.   □

Lemma 2. *The graph G_I is 4-colorable if and only if I has a satisfying truth assignment.*

Proof. Suppose we have a 4-coloring of G_I with colors $\{1, 2, 3, 4\}$. We may assume without loss of generality that a has color 1, that $b_{1,1}$ has color 3 and that $b_{1,2}$ has color 4. This implies that all x-type vertices have a color from $\{1, 2\}$. Furthermore, for $i = 1, \ldots, n$, if x_i has color 1 then $\overline{x}_i$ has color 2, and vice versa. Hence we find that all b-type and d-type vertices have a color from $\{3, 4\}$. Then by symmetry we may assume that every $b_{j,1}$ has color 3 and every $b_{j,2}$ has color 4. This means that every $c_{j,1}$ has a color from $\{2, 4\}$, every $c_{j,2}$ has a color from $\{2, 3, 4\}$ and every $c_{j,3}$ has a color from $\{2, 3\}$. Now suppose there is a clause C_j with each of its three literals colored by color 2. Then $c_{j,1}$ must have color 4 and $c_{j,3}$ must have color 3. Consequently, $d_{j,1}$ has color 3 and $d_{j,2}$ has color 4. Then $c_{j,2}$ cannot have a color in a proper 4-coloring of G_I. Hence this is not possible and we find that at least one literal in every clause is colored by color 1. This means we can define a truth assignment that sets a literal to FALSE if the corresponding x-type vertex has color 2, and to TRUE otherwise. So a 4-coloring of G_I implies a satisfying truth assignment for I.

For the converse, suppose I has a satisfying truth assignment. We use color 1 to color the x-type vertices representing the true literals and color 2 for the false literals. Since each clause contains at least one true literal, we note that we can color $c_{j,1}$, $c_{j,2}$ and $c_{j,3}$ and also all other remaining vertices in a straightforward way. This implies a 4-coloring for G_I and completes the proof of Lemma 2.   □

3   Pre-coloring Extension of 4-Coloring for P_7-Free Graphs

In this section we show that the pre-coloring extension version of 4-COLORING is NP-complete for the class of P_7-free graphs. We use a reduction from NOT-ALL-EQUAL 3-SATISFIABILITY with positive literals only, which we denote as NAE 3SATPL. This NP-complete problem [15] is also known as HYPERGRAPH 2-COLORABILITY and is defined as follows. Given a set $X = \{x_1, x_2, \ldots, x_n\}$ of logical variables, and a set $C = \{C_1, C_2, \ldots, C_m\}$ of three-literal clauses over X in which all literals are positive, does there exist a truth assignment for X such that each clause contains at least one true literal and at least one false literal?

We consider an arbitrary instance I of NAE 3SATPL that has variables $\{x_1, x_2, \ldots, x_n\}$ and clauses $\{C_1, C_2, \ldots, C_m\}$, and we define a graph G_I^* with a pre-coloring on some vertices of G_I^*. Then we show that G_I^* is P_7-free and that the pre-coloring on G_I^* can be extended to a 4-coloring of G_I^* if and only if I has a satisfying truth assignment in which each clause contains at least one true literal and at least one false literal.

Here is the construction that defines G_I^* with a pre-coloring.

– For each clause C_j we introduce a gadget with vertex set

$$\{a_{j,1}, a_{j,2}, a_{j,3}, b_{j,1}, b_{j,2}, c_{j,1}, c_{j,2}, c_{j,3}, d_{j,1}, d_{j,2}\}$$

and edge set

$$\{a_{j,1}c_{j,1}, a_{j,2}c_{j,2}, a_{j,3}c_{j,3}, b_{j,1}c_{j,1}, c_{j,1}d_{j,1}, d_{j,1}c_{j,2}, c_{j,2}d_{j,2}, d_{j,2}c_{j,3}, c_{j,3}b_{j,2}, b_{j,2}b_{j,1}\},$$

and a disjoint gadget called the *copy* with vertex set

$$\{a'_{j,1}, a'_{j,2}, a'_{j,3}, b'_{j,1}, b'_{j,2}, c'_{j,1}, c'_{j,2}, c'_{j,3}, d'_{j,1}, d'_{j,2}\}$$

and edge set

$$\{a'_{j,1}c'_{j,1}, a'_{j,2}c'_{j,2}, a'_{j,3}c'_{j,3}, b'_{j,1}c'_{j,1}, c'_{j,1}d'_{j,1}, d'_{j,1}c'_{j,2}, c'_{j,2}d'_{j,2}, d'_{j,2}c'_{j,3}, c'_{j,3}b'_{j,2}, b'_{j,2}b'_{j,1}\}.$$

We say that all these vertices (so including the vertices in the copy) are of a-type, b-type, c-type and d-type, respectively. They induce $2m$ disjoint 10-vertex components in G_I^* which we will call *clause-components* in the sequel. We pre-color every $a_{j,h}$ by 1 and every $a'_{j,h}$ by 2.

– Every variable x_i will be represented by a vertex in G_I^*, and we say that these vertices are of x-type.

– For every clause C_j we fix an arbitrary order of its variables $x_{i_1}, x_{i_2}, x_{i_3}$ and add edges $c_{j,h}x_{i_h}$ and $c'_{j,h}x_{i_h}$ for $h = 1, 2, 3$.

– We add an edge between every x-type vertex and every b-type vertex. We also add an edge between every x-type vertex and every d-type vertex.

– We add an edge between every a-type vertex and every b-type vertex. We also add an edge between every a-type vertex and every d-type vertex.

In Figure 2 we illustrate an example in which C_j is a clause with ordered variables $x_{i_1}, x_{i_2}, x_{i_3}$. The thick edges indicate the connection between the variables vertices and the c-type vertices of the two copies of the clause gadget. The dashed thick edges indicate the connections between the (pre-colored) a-type and c-type vertices of the two copies of the clause gadget. We omitted the indices from the labels of the clause gadget vertices to increase the visibility.

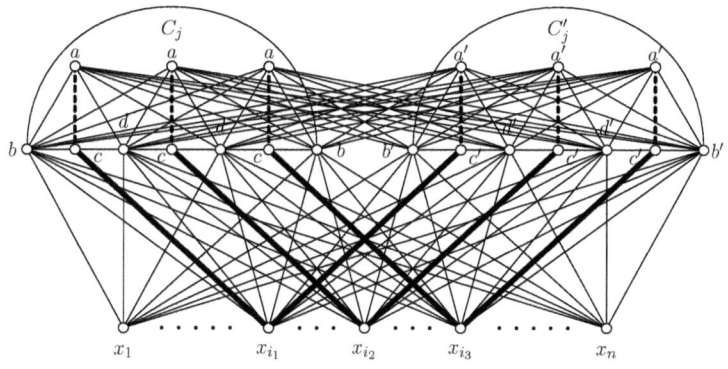

Fig. 2. The graph G_I^* for the clause $C_j = \{x_{i_1}, x_{i_2}, x_{i_3}\}$

We complete this section by proving two lemmas. Lemma 3 shows that the graph G_I^* is P_7-free. Its proof is postponed to the journal version of our paper. In Lemma 4 we prove that the pre-coloring of G_I^* can be extended to a 4-coloring of G_I^* if and only if I has a truth assignment in which each clause contains at least one true and at least one false literal.

Lemma 3. *The graph G_I^* is P_7-free.*

Lemma 4. *The pre-coloring of G_I^* can be extended to a 4-coloring of G_I^* if and only if I has a truth assignment in which each clause contains at least one true and at least one false literal.*

Proof. Suppose the pre-coloring of G_I^* can be extended to a 4-coloring of G_I^*. Since $a_{1,1}$ with color 1 and $a'_{1,1}$ with color 2 are adjacent to every b-type vertex, we may assume by symmetry that every $b_{j,1}$ and every $b'_{j,1}$ has color 3, whereas every $b_{j,2}$ and every $b'_{j,2}$ has color 4. This implies the following. Firstly, it implies that all x-type vertices have a color from $\{1, 2\}$. Consequently, all d-type vertices must have a color from $\{3, 4\}$. Secondly, it implies that every $c_{j,1}$ has a color from $\{2,4\}$, every $c_{j,2}$ has a color from $\{2,3,4\}$ and every $c_{j,3}$ has a color from $\{2,3\}$. Thirdly, it implies that every $c'_{j,1}$ has a color from $\{1,4\}$, every $c_{j,2}$ has a color from $\{1,3,4\}$ and every $c_{j,3}$ has a color from $\{1,3\}$.

Now suppose there is a clause C_j with each of its three literals colored by color 2. Then $c_{j,1}$ must have color 4 and $c_{j,3}$ must have color 3. Consequently, $d_{j,1}$ has color 3 and $d_{j,2}$ has color 4. Then $c_{j,2}$ cannot have a color in a proper 4-coloring. Hence this is not possible and we find that at least one literal in every clause is colored by color 1. By considering the copies, in a similar way we find that at least one literal in every clause is colored by color 2. Hence, we can define a truth assignment that sets a literal to FALSE if the corresponding x-type vertex has color 2, and to TRUE otherwise. So a 4-coloring of G_I^* that extends the pre-coloring on G_I^* implies a truth assignment for I in which each clause contains at least one true and at least one false literal.

For the converse, suppose I has a satisfying truth assignment in which each clause contains at least one true and at least one false literal. We use color 1 to color the x-type vertices representing the true literals and color 2 for the false literals. Since each clause contains at least one true literal, we can color $c_{j,1}$, $c_{j,2}$ and $c_{j,3}$, respecting the pre-coloring. Similarly, since each clause contains at least one false literal, we can color $c'_{j,1}$, $c'_{j,2}$ and $c'_{j,3}$, respecting the pre-coloring. We color all other remaining uncolored vertices in a straightforward way. This completes the proof of Lemma 4.   □

4   Pre-coloring Extension of 3-Coloring for Subclasses of P_7-Free Graphs

Here we consider the pre-coloring extension version of 3-COLORING for H-free graphs, where H is a subgraph of P_7 on at most 6 vertices. We can use the polynomial-time algorithm of [3] for solving this problem when H is an induced subgraph of P_6, because then any H-free graph is also P_6-free. Then the following cases remain:

$$
\begin{array}{lll}
H_1 = 6P_1 & H_6 = 3P_1 + P_3 & H_{11} = P_1 + P_2 + P_3 \\
H_2 = 5P_1 & H_7 = 2P_1 + 2P_2 & H_{12} = P_1 + P_5 \\
H_3 = 4P_1 & H_8 = 2P_1 + P_3 & H_{13} = 3P_2 \\
H_4 = 4P_1 + P_2 & H_9 = 2P_1 + P_4 & H_{14} = P_2 + P_4 \\
H_5 = 3P_1 + P_2 & H_{10} = P_1 + 2P_2 & H_{15} = 2P_3.
\end{array}
$$

We first consider H_i for $i = 1, \ldots, 12$. For these graphs we need the following observation, the proof of which follows from the fact that the decision problem in this case can be modeled and solved as a 2SAT-problem. This approach has been introduced by Edwards [6] and is folklore now, see also [9] and [13].

Observation 1 ([6]). *Let G be a graph in which every vertex has a list of admissible colors of size at most 2. Then checking whether G has a coloring respecting these lists is solvable in polynomial time.*

Proposition 1. *Let H be a graph. If the pre-coloring extension version of 3-*Coloring *is solvable in polynomial time for H-free graphs, then it is also solvable in polynomial time for $(H + P_1)$-free graphs.*

Proof. Let G be an $(H + P_1)$-free graph with pre-coloring $\phi_W : W \to \{1, 2, 3\}$ for some $W \subseteq V(G)$. If G is H-free, we are done. Otherwise, we use ϕ_W to construct a list of admissible colors for each vertex in G.

Suppose G contains an induced subgraph H' that is isomorphic to H. Because G is $(H + P_1)$-free, every vertex in $V(G) \backslash V(H')$ must be adjacent to a vertex in H'. We guess a coloring of $V(H')$ that respects the lists. Afterwards we apply Observation 1. Since H' has a fixed size, the number of guesses is polynomially bounded.   □

Using the polynomial-time algorithm of [3] that solves the pre-coloring extension version of 3-Coloring for P_6-free graphs, and (repeatedly) applying Proposition 1 yields polynomial-time results of the same problem for H_i-free graphs for $i = 1, \ldots, 12$.

The case $H_{13} = 3P_2$ follows from the more general result that 3-Coloring is polynomial-time solvable for sP_2-free graphs for any $s \geq 1$. This is known already and can be seen as follows. Balas and Yu [1] showed that for any $s \geq 1$ the number of maximal independent sets in an sP_2-free graph $G = (V, E)$ is bounded by a polynomial. These maximal independent sets can then be efficiently enumerated by applying the algorithm of Tsukiyama, Ide, Ariyoshi and Shirakawa [16]. We note that G has a 3-coloring if and only if V can be partitioned into at most 3 independent sets V_1, V_2, V_3, one of which may be assumed to be maximal. Hence, for each maximal independent set I in G we check if the subgraph induced by $V \backslash I$ is bipartite. This can be done in polynomial time.

In Section 4.1 we consider H_{14}.

4.1   Pre-coloring Extension of 3-Coloring for $(P_2 + P_4)$-Free Graphs

Below we describe how to test in polynomial time whether a given $(P_2 + P_4)$-free graph G with pre-coloring $\phi_W : W \to \{1, 2, 3\}$ for some $W \subseteq V(G)$ allows a coloring $\phi : V(G) \to \{1, 2, 3\}$ with $\phi(u) = \phi_W(u)$ for all $u \in W$.

We start by making two assumptions. Firstly, we assume that G is connected as otherwise we apply our algorithm on each component of G. Secondly, we assume that G contains an induced subgraph H isomorphic to P_6. If not, then G would be P_6-free and we could use the polynomial-time algorithm for P_6-free graphs of [3] to solve our problem.

We use ϕ_W to construct a list of admissible colors for each vertex in G. We guess a coloring of H respecting these lists and start our algorithm, which we run at most 3^6 times as this is an upper bound on the number of possible 3-colorings of H. From the description of the algorithm it will be immediately clear that its running time is polynomial in $|V(G)|$.

Our algorithm first applies the following subroutine. Let $U \subseteq V(G)$ contain all vertices that have a list consisting of exactly one color. For every vertex $u \in U$

we remove this single color $c(u)$ from the lists of its neighbors. If this results in an empty list at some vertex, then we output No. We remove u from G and repeat this process in the remaining graph as long as there exists a vertex with a list of size 1. This process is called *updating* the graph. Note that during this procedure we removed all vertices of H. We restore them back into G. We may assume that G is still connected; otherwise, due to the $(P_2 + P_4)$-freeness of G, every component not containing H is a single vertex and can be colored trivially. Let S be the set of vertices that still have a list of admissible colors of size 3. If $S = \emptyset$ then we can apply Observation 1.

Suppose $S \neq \emptyset$. Let T be the set of vertices of $V(G) \setminus V(H)$ that have at least one neighbor in H. Because we colored every vertex in H and updated G, every vertex of T has a list of exactly two admissible colors, and consequently, $S \cap (V(H) \cup T) = \emptyset$. Since G contains no induced $P_2 + P_4$, we find that $V(G) \setminus (V(H) \cup T)$, and consequently S, is an independent set in G. Since G is connected, each vertex in S has at least one neighbor in T (so $T \neq \emptyset$).

For convenience we order the vertices of H along the P_6 as $p_1, p_2, \ldots, p_6$, starting with vertex p_1 with degree 1 in H. Let $T^* \subseteq T$ consist of all vertices in T that have a neighbor in S. Let T_1 denote the subset of vertices of T^* adjacent to p_1, p_3, p_5 and not to p_2, p_4, p_6; let T_2 denote the subset of vertices of T^* adjacent to p_2, p_4, p_6 and not to p_1, p_3, p_5; let T_3 denote the subset of vertices of T^* adjacent to p_2, p_5 and not to p_1, p_3, p_4, p_6.

Because every vertex $u \in T$ has a list of two admissible colors, u is not adjacent to two adjacent vertices of H (as these vertices have different colors). By considering a vertex in T^* together with one of its neighbors in S and using the $(P_2 + P_4)$-freeness of G, we then find that $T^* = T_1 \cup T_2 \cup T_3$.

Claim 1. *Either $T_1 \cup T_2$ or T_3 is empty.*

We prove Claim 1 as follows. Assume $T_1 \cup T_2 \neq \emptyset$ and $T_3 \neq \emptyset$. Without loss of generality, assume there is a vertex $u \in T_1$ and a vertex $v \in T_3$. By definition, u is adjacent to p_1, p_3 and p_5. Since u has a list of 2 admissible colors, p_1, p_3 and p_5 are colored by the same color, say color 1. Because p_2 is adjacent to p_1, vertices p_1 and p_2 have different colors. Thus the colors of p_2 and p_5 are different. Then v has only one admissible color in its list. This contradiction proves Claim 1.

Using Claim 1 we distinguish two cases.

Case 1. $T_1 \cup T_2$ is empty and T_3 is not empty.

Since every vertex in T_3 has a list of 2 admissible colors, p_2 and p_5 are colored the same. Recall that S is an independent set. Hence we can safely color all the vertices in S by the same color as p_2 and p_5. We are left to apply Observation 1.

Case 2. T_3 is empty and $T_1 \cup T_2$ is not empty.

If one of T_1 and T_2 is empty, say $T_2 = \emptyset$, we proceed as in Case 1. We now assume that none of T_1 and T_2 is empty. As before, this means that p_1, p_3, p_5 must have the same color, say color 1, whereas p_2, p_4, p_6 also have the same color, say color 2. Recall that S is an independent set. Hence, we can safely color all vertices of

Table 2. An update of Table 1

P_k-free	$\ell \to$ 3	3*	4	4*	5	5*	≥ 6	$\geq 6^*$
$k \leq 5$	P	P	P	P	P	P	P	P
$k = 6$	P	P	?	?	?	NP-c	?	NP-c
$k = 7$	?	?	?	NP-c	?	NP-c	NP-c	NP-c
$k = 8$	?	?	NP-c	NP-c	NP-c	NP-c	NP-c	NP-c
$k \geq 9$	?	?	NP-c	NP-c	NP-c	NP-c	NP-c	NP-c

S that only have neighbors in T_1 by color 1, and all vertices of S that only have neighbors in T_2 by color 2. Afterwards we remove them from G. If no vertices of S remain we apply Observation 1. Suppose S did not become empty. Then each (remaining) vertex of S has a neighbor in T_1 and T_2. We first try the case that all vertices of T_1 receive color 2. For this coloring of T_1, all vertices in S get reduced lists of size at most 2, so we can again apply Observation 1.

We are left to consider the possibility that color 3 is used on at least one vertex of T_1. We try all possible $O(|V(G)|)$ choices in which we give one fixed vertex $x \in T_1$ color 3. Below we describe what we do for each such choice.

We first update G. If G then only contains vertices that have a list of admissible colors of size 2, we apply Observation 1. Otherwise, we restore x and all vertices of H back into G and redefine sets T_1, T_2 and S accordingly. We find that no vertex in T_2 is adjacent to x, because such vertex would have received color 1 and would have been removed when we were updating G. Furthermore, by definition of S, no vertex in S is adjacent to x, and we may again assume that each vertex in S is adjacent to a vertex in T_1 and to a vertex in T_2.

Let y be an arbitrary vertex of T_2. Suppose there exists an edge ab such that $a \in T_2$, $b \in S$ and y is not adjacent to a, b. Then G contains an induced $P_2 + P_4$ formed by bap_6y and xp_1. This is not possible. Hence, the vertex y is adjacent to at least one of the vertices of every edge ab with $a \in T_2$ and $b \in S$. We consider all possible colorings of y. This way we reduce the list of admissible colors of each vertex in S by at least one (either directly or via one of its neighbors in T_2) and we apply Observation 1. This finishes Case 2, and thus the description of our algorithm is completed.

5 Conclusions

Due to our new results we can update Table 1. This yields Table 2. Positions in this table marked by "?" are still open. We also showed that 3-COLORING is polynomial-time solvable for H-free graphs if H is any fixed linear forest on at most 6 vertices, except when $H = 2P_3$. Recently, we showed that 3-COLORING is also polynomial-time solvable for $2P_3$-free graphs [4].

References

1. Balas, E., Yu, C.S.: On graphs with polynomially solvable maximum-weight clique problem. Networks 19, 247–253 (1989)
2. Bondy, J.A., Murty, U.S.R.: Graph Theory. In: Springer Graduate Texts in Mathematics, vol. 244 (2008)
3. Broersma, H.J., Fomin, F.V., Golovach, P.A., Paulusma, D.: Three complexity results on coloring P_k-free graphs. In: Fiala, J., Kratochvíl, J., Miller, M. (eds.) IWOCA 2009. LNCS, vol. 5874, pp. 95–104. Springer, Heidelberg (2009)
4. Broersma, H.J., Golovach, P.A., Paulusma, D., Song, J.: On coloring graphs without induced forests. In: ISAAC 2010 (to appear, 2010)
5. Bruce, D., Hoàng, C.T., Sawada, J.: A certifying algorithm for 3-colorability of P_5-free graphs. In: Dong, Y., Du, D.-Z., Ibarra, O. (eds.) ISAAC 2009. LNCS, vol. 5878, pp. 594–604. Springer, Heidelberg (2009)
6. Edwards, K.: The complexity of coloring problems on dense graphs. Theoret. Comput. Sci. 43, 337–343 (1986)
7. Garey, M.R., Johnson, D.S.: Computers and Intractability: A Guide to the Theory of NP-Completeness. Freeman, San Francisco (1979)
8. Grötschel, M., Lovász, L., Schrijver, A.: The ellipsoid method and its consequences in combinatorial optimization. Combinatorica 1, 169–197 (1981)
9. Hoàng, C.T., Kamiński, M., Lozin, V., Sawada, J., Shu, X.: Deciding k-colorability of P_5-free graphs in polynomial time. Algorithmica 57, 74–81 (2010)
10. Král', D., Kratochvíl, J., Tuza, Z., Woeginger, G.J.: Complexity of coloring graphs without forbidden induced subgraphs. In: Brandstädt, A., Le, V.B. (eds.) WG 2001. LNCS, vol. 2204, pp. 254–262. Springer, Heidelberg (2001)
11. Kratochvíl, J.: Precoloring extension with fixed color bound. Acta Math. Univ. Comen. 62, 139–153 (1993)
12. Le, V.B., Randerath, B., Schiermeyer, I.: On the complexity of 4-coloring graphs without long induced paths. Theoret. Comput. Sci. 389, 330–335 (2007)
13. Randerath, B., Schiermeyer, I.: 3-Colorability $\in$ P for P_6-free graphs. Discrete Appl. Math. 136, 299–313 (2004)
14. Randerath, B., Schiermeyer, I.: Vertex colouring and forbidden subgraphs - a survey. Graphs Combin. 20, 1–40 (2004)
15. Schaefer, T.J.: The complexity of satisfiability problems. In: Conference Record of the Tenth Annual ACM Symposium on Theory of Computing, San Diego, Calif., pp. 216–226. ACM, New York (1978)
16. Tsukiyama, S., Ide, M., Ariyoshi, H., Shirakawa, I.: A new algorithm for generating all the maximal independent sets. SIAM J. Comput. 6, 505–517 (1977)
17. Tuza, Z.: Graph colorings with local restrictions - a survey. Discuss. Math. Graph Theory 17, 161–228 (1997)
18. Woeginger, G.J., Sgall, J.: The complexity of coloring graphs without long induced paths. Acta Cybernet. 15, 107–117 (2001)

Computing the Cutwidth of Bipartite Permutation Graphs in Linear Time[*]

Pinar Heggernes[1], Pim van 't Hof[2], Daniel Lokshtanov[1], and Jesper Nederlof[1]

[1] Department of Informatics, University of Bergen,
P.O. Box 7803, N-5020 Bergen, Norway
{pinar.heggernes,daniel.lokshtanov,jesper.nederlof}@ii.uib.no
[2] School of Engineering and Computing Sciences, Durham University,
Science Laboratories, South Road, Durham DH1 3LE, United Kingdom
pimvanthof@gmail.com

Abstract. The problem of determining the cutwidth of a graph is a notoriously hard problem which remains NP-complete under severe restrictions on input graphs. Until recently, non-trivial polynomial-time cutwidth algorithms were known only for subclasses of graphs of bounded treewidth. In WG 2008, Heggernes et al. initiated the study of cutwidth on graph classes containing graphs of unbounded treewidth, and showed that a greedy algorithm computes the cutwidth of threshold graphs. We continue this line of research and present the first polynomial-time algorithm for computing the cutwidth of bipartite permutation graphs. Our algorithm runs in linear time. We stress that the cutwidth problem is NP-complete on bipartite graphs and its computational complexity is open even on small subclasses of permutation graphs, such as trivially perfect graphs.

1   Introduction

A large variety of problems in many different domains can be formulated as graph layout problems [8]. A well known problem of this type is *cutwidth*. Given a graph G and a positive integer k, the cutwidth problem is to decide whether there is an ordering of the vertices of G such that any line inserted between two consecutive vertices in the ordering cuts at most k edges of the graph. The cutwidth of the input graph is the smallest integer for which the question can be answered positively. This problem was first proposed as a model to minimize the number of channels in a circuit [1,14], and later it has found applications in areas like protein engineering [3], network reliability [12], automatic graph drawing [16], and as a subroutine in the cutting plane algorithm for TSP [11].

As most graph problems of practical interest, cutwidth is NP-complete [9], even when input graphs are restricted to planar graphs of maximum degree 3 [15], split graphs [10], unit disk graphs, partial grids [7], and consequently

---

[*] This work is supported by the Research Council of Norway and by EPSRC UK grant EP/D053633/1.

D.M. Thilikos (Ed.): WG 2010, LNCS 6410, pp. 75–87, 2010.

bipartite graphs. There is a polynomial-time $O(\log^2 n)$-approximation algorithm for general graphs [13], and a polynomial-time constant factor approximation algorithm for dense graphs [2].

The knowledge on polynomial-time algorithms for the exact computation of cutwidth on restricted inputs is very limited. Cutwidth of certain trivial graph classes, like meshes or complete p-partite graphs, can be computed easily as there exist closed formulas for their cutwidth [8]. Cutwidth of proper interval graphs has a trivial solution following an interval ordering of the vertices [21]. However, there are very few graph classes whose cutwidth is non-trivially computable in polynomial time. Until recently, polynomial-time cutwidth algorithms were known only for subclasses of graphs of bounded treewidth. In particular, Yannakakis [20] gave a sophisticated and technical algorithm for trees (see also [6]). Furthermore, Thilikos et al. gave an algorithm for computing the cutwidth of bounded cutwidth graphs [18], and extended this result to graphs of bounded treewidth and maximum degree [19]. As a recent development, in a WG 2008 paper the study of cutwidth on graph classes containing graphs of unbounded treewidth was initiated, resulting in a linear-time algorithm for computing the cutwidth of threshold graphs [10].

In this paper, we continue this line of research by showing that the cutwidth of a bipartite permutation graph can be computed in linear time. As mentioned above, the cutwidth problem is NP-complete on bipartite graphs, and its computational complexity is open on permutation graphs. Thus bipartite permutation graphs are natural candidates for studying the computational complexity of the cutwidth problem. Our algorithm relies heavily on a characterization of bipartite permutation graphs by strong orderings [17]. We would like to point out that bipartite permutation graphs and threshold graphs are two unrelated subclasses of permutation graphs; the intersection of these two graph classes is restricted to stars. We would also like to point out that bipartite permutation graphs form the first graph class of unbounded clique-width [5] whose cutwidth is shown to be computable in polynomial time.

2 Preliminaries

We consider undirected finite graphs with no loops or multiple edges. For a graph $G = (V, E)$, we denote its vertex set and edge set by V and E, respectively, with $n = |V|$ and $m = |E|$. Let $S \subseteq V$. The subgraph of G induced by S is denoted by $G[S]$. We write $G - S$ to denote the graph $G[V \setminus S]$, and we simply write $G - v$ instead of $G - \{v\}$ in case $S = \{v\}$. For two vertices $u, v \in V$ with $uv \notin E$, we write $G + uv$ to denote the graph $(V, E \cup \{uv\})$. The set of *neighbors* of a vertex x of G is $N(x) = \{v \mid xv \in E\}$. The *degree* of x is $d(x) = |N(x)|$. A graph is *connected* if there is a path between any pair of its vertices. A *connected component* of a disconnected graph is a maximal connected subgraph of it.

In a bipartite graph $G = (A, B, E)$, vertex sets A and B are called *color classes*. The partition of the vertex set into color classes of a connected bipartite graph is unique, up to symmetry. Vertices of A and of B are called *A-vertices* and *B-vertices*, respectively. We say that a vertex is *bipartite universal* if it is adjacent to all the vertices of the opposite color class.

An *ordering* of a set A is a one-to-one mapping $\sigma : A \rightarrow \{1, \ldots, |A|\}$. We also use the notation $\sigma = \langle a_1, a_2, \ldots, a_{|A|} \rangle$, meaning that $\sigma(a_i) < \sigma(a_j)$ when $i < j$, where each a_i is a distinct element of A, for $1 \leq i \leq |A|$. Integers $1, 2, \ldots, |A|$ are called the *positions* of σ, and $\sigma(a)$ is the *position* of a in σ. Intuitively, we will refer to the end of the ordering with a_1 as *the left* and the end of the ordering with $a_{|A|}$ as *the right*. For two elements a and a' of A, we say that a *appears before* (or *to the left of*) a' in σ, denoted $a \prec_\sigma a'$, if $\sigma(a) < \sigma(a')$. If $\sigma(a) > \sigma(a')$, then we say that a *appears after* (or *to the right of*) a' in σ and write $a \succ_\sigma a'$. We will also use the notion of a *leftmost*, *rightmost*, and *middle* vertex or neighbor, analogously and intuitively. A subset of k elements of A are *consecutive* in σ if they occupy positions $i + 1, \ldots, i + k$, for some i between 0 and $|A| - k$. When we say that we *delete* an element a of A from σ, we get a new ordering in which all elements before a in σ keep their original positions, and the position of each element after a decreases by 1. We denote the new ordering by $\sigma - a$. For any subset of $A' \subseteq A$, we write $\sigma - A'$ to denote the ordering obtained from σ by consecutively deleting all the elements of A' from σ.

A *layout* of a graph $G = (V, E)$ is an ordering of V. We write $\Phi(G)$ to denote the set of all layouts of G. The *rank* of a vertex v with respect to a layout φ, denoted $rank_\varphi(v)$, is the number of neighbors of v appearing after v in φ minus the number of neighbors of v appearing before v in φ, i.e., $rank_\varphi(v) = |\{w \in N(v) \mid w \succ_\varphi v\}| - |\{w \in N(v) \mid w \prec_\varphi v\}|$. Note that the rank of a vertex can be negative. Given layout φ of a graph G and an integer $1 \leq i \leq n$, we define $L(i, \varphi, G) = \{u \in V \mid \varphi(u) \leq i\}$ and $R(i, \varphi, G) = \{u \in V \mid \varphi(u) > i\}$. The *ith gap* of φ is between $L(i, \varphi, G)$ and $R(i, \varphi, G)$, or equivalently, between positions i and $i + 1$ of φ. For any set $S \subseteq V$, we define the *cut* of S to be $\theta(S, G) = \{uv \in E \mid u \in S, v \notin S\}$. The *cut of G at the ith gap* of φ is defined as $\theta(i, \varphi, G) = \{uv \in E \mid u \in L(i, \varphi, G) \wedge v \in R(i, \varphi, G)\}$. Note that by definition $\theta(i, \varphi, G) = \theta(L(i, \varphi, G), G)$. We call an edge set $\theta \subseteq E$ a *cut* of φ if $\theta = \theta(i, \varphi, G)$ for some $i \in \{1, 2, \ldots n - 1\}$. The *size* of a cut θ is $|\theta|$. The *cutwidth of a layout* φ of G is $cw\varphi(G) = \max_{1 \leq i \leq n} |\theta(i, \varphi, G)|$. A cut $\theta(i, \varphi, G)$ with $|\theta(i, \varphi, G)| = cw_\varphi(G)$ is called a *worst cut* of φ. The *cutwidth* of G is $cw(G) = \min_{\varphi \in \Phi(G)} \{cw_\varphi(G)\}$, where the minimum is taken over all layouts of G. An *optimal layout* of G is a layout φ such that $cw(G) = cw_\varphi(G)$. The cutwidth of a graph G equals the maximum cutwidth over all connected components of G.

As the name already indicates, bipartite permutation graphs are permutation graphs that are bipartite. For the definition and properties of permutation graphs, we refer to [4]. The study of bipartite permutation graphs was initiated by Spinrad et al. in [17]. They present two characterizations of bipartite permutation graphs, leading to a linear-time recognition algorithm of this class as well as polynomial-time algorithms for some NP-complete problems restricted to bipartite permutation input graphs.

A *strong ordering* (σ_A, σ_B) of a bipartite permutation graph $G = (A, B, E)$ consists of an ordering σ_A of A and an ordering σ_B of B such that for all

$ab, a'b' \in E$, where $a, a' \in A$ and $b, b' \in B$, $a \prec_{\sigma_A} a'$ and $b' \prec_{\sigma_B} b$ implies that $ab' \in E$ and $a'b \in E$. An ordering σ_A of A has the *adjacency property* if, for every $b \in B$, $N(b)$ consists of vertices that are consecutive in σ_A. The ordering σ_A has the *enclosure property* if, for every pair b, b' of vertices of B with $N(b) \subseteq N(b')$, the vertices of $N(b') \setminus N(b)$ appear consecutively in σ_A, implying that b is adjacent to the leftmost or the rightmost neighbor of b' in σ_A.

Theorem 1 ([17]). *The following statements are equivalent for a bipartite graph* $G = (A, B, E)$.

1. G *is a bipartite permutation graph.*
2. G *has a strong ordering.*
3. *There exists an ordering of* A *which has the adjacency and enclosure properties.*

A strong ordering of a bipartite permutation graph can be computed in linear time [17]. If the graph G in Theorem 1 is connected, then it follows from the proof of Theorem 1 in [17] that we can combine statements 2 and 3 in Theorem 1 as follows.

Lemma 1 ([17]). *Let* (σ_A, σ_B) *be a strong ordering of a connected bipartite permutation graph* $G = (A, B, E)$. *Then both* σ_A *and* σ_B *have the adjacency and enclosure properties.*

3   Cutwidth of Bipartite Permutation Graphs

In this section we prove that the cutwidth of bipartite permutation graphs can be computed in linear time. The complete algorithm is given in the proof of Theorem 2. The main ingredient is an algorithm that we call MinCutBPG. This algorithm takes as input a connected bipartite permutation graph G and a strong ordering of G, and it outputs an optimal layout of G. We will spend most of this section describing and proving the correctness of Algorithm MinCutBPG. Before we give the algorithm, we define an operation to modify a given layout in an intuitive way. Given a layout φ of a graph, when we *move* a vertex v from position i to position j, with $i < j$, only vertices in positions from i to j are affected. We get a new layout φ' in which v gets position $\varphi'(v) = j$, the vertex x with $\varphi(x) = j$ gets position $\varphi'(x) = j - 1$, and each of the other affected vertices decrease their positions by 1, similarly. All other vertices have the same position in φ' as they had in φ. What we described is a *move toward the right*. A *move toward the left* is defined symmetrically.

3.1   Description of Algorithm MinCutBPG

We now give an outline of Algorithm MinCutBPG, which takes as input a connected bipartite permutation graph $G = (A, B, E)$ and a strong ordering (σ_A, σ_B) of G. It outputs an optimal layout φ of G. Let $A = \{a_1, \ldots, a_s\}$ where $a_1 \prec_{\sigma_A}$

$\cdots \prec_{\sigma_A} a_s$, and let $B = \{b_1, \ldots, b_t\}$ where $b_1 \prec_{\sigma_B} \cdots \prec_{\sigma_B} b_t$. The vertices of A will appear in the final layout φ in the same order as they appear in σ_A. Similarly, the order in which the vertices of B appear in φ corresponds to the order in which they appear in σ_B.

Before deciding where the vertices of A will appear in φ with respect to the vertices of B, the algorithm first assigns the vertices of B to "boxes". There are two types of boxes: a box X_i for every vertex $a_i \in A$, and a box $X_{i,i+1}$ for every pair of consecutive vertices $a_i, a_{i+1} \in A$. Recall that the neighbors of any vertex $b \in B$ appear consecutively in σ_A by Lemma 1. If b has even degree and its two middle neighbors are a_i and a_{i+1}, then b is assigned to box $X_{i,i+1}$. If b has odd degree and its middle neighbor is a_i, then b is assigned to box X_i. For convenience, we also define the boxes $X_{0,1} = \emptyset$ and $X_{s,s+1} = \emptyset$. Observe that some boxes might be empty and the collection of non-empty boxes is a partition of B. The following observation is a direct consequence of Lemma 1, the properties of a strong ordering, and the definition of boxes.

Observation 1. *Given a connected bipartite permutation graph $G = (A, B, E)$ with $|A| = s$ and a strong ordering (σ_A, σ_B), where $\sigma_A = \langle a_1, a_2, \ldots, a_s \rangle$, let boxes $X_{0,1}, X_1, X_{1,2}, \ldots, X_s, X_{s,s+1}$ be defined as above. Then we have the following:*

1. *every vertex of X_i appears before every vertex of $X_{i,i+1}$ in σ_B, and every vertex of $X_{i,i+1}$ appears before every vertex of X_{i+1} in σ_B, for $1 \leq i \leq s$;*
2. *$N(b) = N(b')$ for any two vertices b and b' appearing in the same box.*

We start with an initial layout of G in which a_1 is placed first, vertices of X_1 are placed in the immediately following positions, vertices of $X_{1,2}$ are placed in the next positions, then a_2 is placed, followed by vertices of $X_2, X_{2,3}, \{a_3\}$, $X_3, \ldots, \{a_{s-1}\}, X_{s-1}, X_{s-1,s}, \{a_s\}$, and X_s. Within each box, the vertices of B belonging to that box are ordered according to σ_B. For $1 \leq i \leq s$, a_i appears just before the vertices of box X_i. To define and obtain the final layout φ, we just need to move each a_i to its final position. This will be one of the initial positions of $\{a_i\} \cup X_i$. As a consequence, we can observe already now that, for every $b \in B$, $rank_\varphi(b) \in \{-1, 0.1\}$. The ranks of the A-vertices might have a larger range of values. Let i be any index satisfying $1 \leq i \leq s$. Recall that $rank_\varphi(a_i)$ depends on the position where a_i is placed: the further to the left a_i appears, the higher its rank. The algorithm moves a_i in such a way that $rank_\varphi(a_i)$ is as close to 0 as possible, i.e., the value of $|rank_\varphi(a_i)|$ is as small as possible, subject to the condition that the position of a_i is one of the initial positions of $\{a_i\} \cup X_i$. This is done in the following way. Note first that the set of possible positions for a_i does not intersect with the set of possible positions for any other A-vertex a_j with $i \neq j$. Furthermore, $rank_\varphi(a_i)$ is only dependent on the neighbors of a_i and no two A-vertices are adjacent. Therefore, the placement of each a_i among the positions of $\{a_i\} \cup X_i$ can be decided independently of the placements of the other A-vertices. By Lemma 1, the neighbors of a_i appear consecutively in σ_B. If a_i has odd degree then let b be the middle neighbor of a_i in σ_B. If a_i has even degree then let b be the right one of the two middle neighbors of a_i. If $b \in X_i$, then we move a_i to the position just before the position of b. If b appears in a box to the left of X_i then we do not move a_i. If b appears in a box to the right of X_i then we move a_i to the last position among the positions of X_i. Thus, if a_i is placed between two vertices of X_i then its rank is 0 or 1. If a_i is placed before or after all vertices of X_i then its rank can be higher or lower. This completes the definition and computation of φ.

We make the following observations about the layout φ generated by Algorithm MinCutBPG, which are direct consequences of Lemma 1.

Observation 2. *Let $G = (A, B, E)$ be a connected bipartite permutation graph and let (σ_A, σ_B) be a strong ordering of G, where $\sigma_A = \langle a_1, a_2, \ldots, a_s \rangle$. Let φ be the layout of G generated by Algorithm MinCutBPG on input G and (σ_A, σ_B). Then, for $1 \leq i \leq s$, we have the following:*

1. *for any $b \in X_{i,i+1}$, $rank_\varphi(b) = 0$;*
2. *for any $b \in X_i$, $rank_\varphi(b) = 1$ if $b \prec_\varphi a_i$ and $rank_\varphi(b) = -1$ if $a_i \prec_\varphi b$;*
3. *every $b \in X_{i-1,i} \cup X_i \cup X_{i,i+1}$ is adjacent to a_i;*

3.2   Correctness of Algorithm MinCutBPG

We show that Algorithm MinCutBPG produces an optimal layout when the input is a connected bipartite permutation graph and a strong ordering of that graph. We assume for contradiction that there is a connected bipartite permutation graph G for which the algorithm outputs a layout φ such that $cw_\varphi(G) > cw(G)$. Such a graph is called a *counterexample*, and we write $\mathcal{G}$ to denote the set of all counterexamples. Let $\mathcal{G}' \subseteq \mathcal{G}$ be the set of counterexamples having the minimum number of vertices among all counterexamples, and let $\mathcal{G}'' \subseteq \mathcal{G}'$ be the set of graphs in $\mathcal{G}'$ having the maximum number of edges among all graphs in $\mathcal{G}'$. A graph in $\mathcal{G}''$ is called a *tight counterexample*. If there exists a counterexample, then there also exists a tight counterexample.

For the statements and the proofs of the following lemmas, let $G = (A, B, E)$ with $E \neq \emptyset$ be a connected bipartite permutation graph that is a tight counterexample, and let (σ_A, σ_B) be a strong ordering of G such that $\sigma_A = \langle a_1, \ldots, a_s \rangle$ and $\sigma_B = \langle b_1, \ldots, b_t \rangle$. Furthermore, let $\varphi = \langle v_1, \ldots, v_n \rangle$ be the layout of G generated by Algorithm MinCutBPG on input G and (σ_A, σ_B).

Lemma 2. *Let $\theta(j, \varphi, G)$ be a worst cut of φ. Then we have the following:*

1. *a_1 is adjacent to the rightmost B-vertex of $L(j, \varphi, G)$;*
2. *b_1 is adjacent to the rightmost A-vertex of $L(j, \varphi, G)$;*
3. *a_s is adjacent to the leftmost B-vertex of $R(j, \varphi, G)$;*
4. *b_t is adjacent to the leftmost A-vertex of $R(j, \varphi, G)$.*

Proof. We only prove claim 1; the proofs of claims 2, 3, and 4 are very similar and have therefore been omitted. Let $\theta = \theta(j, \varphi, G)$ and let b be the rightmost B-vertex of $L = L(j, \varphi, G)$. If $b \prec_\varphi a_1$ then all B-vertices in L appear before a_1 in φ, and b is the vertex just before a_1 in φ, implying that $\varphi(b) = \varphi(a_1) - 1$.

Hence $b \in X_1$, and by Observation 2, $a_1 b \in E$. Now assume that $a_1 \prec_\varphi b$, and suppose for contradiction that a_1 is not adjacent to b. Note that this means that $b \notin X_1$, since every vertex in box X_1 is adjacent to a_1 by Observation 2. We claim that $G' = G - (\{a_1\} \cup X_1)$ is a counterexample, contradicting the assumption that G is a tight counterexample. Observe that G' is a connected bipartite permutation graph and $(\sigma_A - a_1, \sigma_B - X_1)$ is a strong ordering of G'. We will prove the claim by showing that θ is a cut of the layout φ' returned by Algorithm MinCutBPG on input G' and $(\sigma_A - a_1, \sigma_B - X_1)$. Since $a_1 b \notin E$, a_1 has no neighbors in $R = R(j, \varphi, G)$ as a result of the properties of a strong ordering. None of the vertices in X_1 has a neighbor in R either, because they are adjacent to a_1 only. Therefore, θ is a cut of $\varphi - (\{a_1\} \cup X_1)$. We will show that all vertices of $L \setminus (\{a_1\} \cup X_1)$ that appear to the left of b in $\varphi - (\{a_1\} \cup X_1)$ also appear to the left of b in φ'. This will imply that θ is a cut of φ' as well. Clearly, the relative orderings of the A-vertices and of the B-vertices are the same in φ' as in φ. Let us analyze how the deletion of the vertices in $\{a_1\} \cup X_1$ can affect the ranks of vertices and the boxes that they belong to. Deleting $\{a_1\} \cup X_1$ does not change the rank of any A-vertex or the rank of b, since these vertices were not adjacent to any of the vertices in $\{a_1\} \cup X_1$. Consequently, b appears in the same box after the deletion of a_1 as it did before. Let $a \neq a_1$ be the rightmost A-vertex of L; note that a might not be defined in case a_1 is the only A-vertex of L. Either a or b is the rightmost vertex of L in φ. In either case, since the ranks of a and b did not change, a and b have the same relative order to each other in φ' as in φ. The only vertices whose ranks might change by the deletion of $\{a_1\} \cup X_1$ are the B-vertices of L that were adjacent to a_1. However, these vertices cannot appear to the right of b in φ', as the algorithm respects the strong ordering $(\sigma_A - a_1, \sigma_B - X_1)$. As a result, the set of vertices that appear to the left of b is the same in φ' as in φ, which means that θ is a cut of φ'. Since $cw(G') \leq cw(G)$ and the size of the cut did not change, we conclude that G' is a counterexample with at least one fewer vertex than G, giving us the desired contradiction.

Lemma 3. *Let $\theta(j, \varphi, G)$ be a worst cut of φ. Then both $G[L(j, \varphi, G)]$ and $G[R(j, \varphi, G)]$ are complete bipartite graphs.*

Proof. Let a and b be the rightmost A-vertex and B-vertex of $L = L(j, \varphi, G)$, respectively. By Lemma 2, a_1 is adjacent to b and b_1 is adjacent to a. By the definition of a strong ordering, a_1 is adjacent to b_1 and a is adjacent to b. Since G is connected, and σ_A and σ_B have the adjacency property by Lemma 1, a and a_1 are adjacent to all B-vertices in L, and b and b_1 are adjacent to all A-vertices in L. As a result, every vertex of $A \cap L$ is adjacent to every vertex of $B \cap L$. This means that $G[L(j, \varphi, G)]$ is complete bipartite. By symmetry the same holds for $G[R(j, \varphi, G)]$. $\qed$

Lemma 4. *There is a worst cut $\theta(j, \varphi, G)$ of φ such that v_j and v_{j+1} belong to different color classes.*

Proof. Let $L = L(j, \varphi, G)$ and let $R = R(j, \varphi, G)$. Assume that either L or R, say L, contains vertices of only one color class. Since G is connected, G

contains vertices from both color classes. Let us consider the smallest index $k \geq j$ such that there is a vertex of the other color class in position $k + 1$. Then $|\theta(k, \varphi, G)| \geq |\theta(j, \varphi, G)|$ because $L \subseteq L(k, \varphi, G)$, there are no edges between the vertices of $L(k, \varphi, G)$, and each vertex of $L(k, \varphi, G)$ has a neighbor in $R(k, \varphi, G)$. Hence we can conclude that there is a worst cut at the gap between two vertices of opposite color. The case where R contains only vertices of one color class is completely symmetric. For the rest of the proof, assume that both L and R contain vertices of both color classes.

Assume first that both v_j and v_{j+1} are B-vertices. Let a_i be the rightmost A-vertex in L, which means that a_{i+1} is the leftmost A-vertex in R. Both $b = v_j$ and $b' = v_{j+1}$ are between a_i and a_{i+1}; more precisely, $a_i \prec_\varphi b \prec_\varphi b' \prec_\varphi a_{i+1}$. If $rank_\varphi(b) = 1$ then $b \in X_{i+1}$ by Observation 2. Then by Observation 1, $b' \in X_{i+1}$ as well, and consequently $rank_\varphi(b') = 1$. Thus we can conclude that b and b' have the same neighborhood and they have one more neighbor in R than in L. In this case $\theta(j, \varphi, G)$ cannot be a worst cut, because the cut just to the right of b' has larger size. Therefore, $rank_\varphi(b) \leq 0$, which means that b has at least as many neighbors to the left as it has to the right. Since b has no neighbors appearing between a_i and b, the cut just to the right of a_i is of size at least $|\theta(j, \varphi, G)|$. Hence we can take that cut as the worst cut. Consequently there is a worst cut at the gap between an A-vertex and a B-vertex.

Assume now that both v_j and v_{j+1} are A-vertices, say a_i and a_{i+1}. First we show that in this case both a_i and a_{i+1} are bipartite universal. Assume for contradiction that this is not true, and let b be the leftmost B-vertex in R which is not a neighbor of a_i. We claim that $G' = G + a_i b$ is also a counterexample, contradicting the assumption that G is a tight counterexample. Recall that $G[L]$ and $G[R]$ are complete bipartite graphs due to Lemma 3. Now observe that G' is a bipartite permutation graph and (σ_A, σ_B) is a strong ordering of G'. Let φ' be the layout computed by Algorithm MinCutBPG on input G' and (σ_A, σ_B). Let us analyze how the layout φ can change to φ' due to the addition of edge $a_i b$. Observe that $X_{i,i+1}$ is empty before the addition of edge $a_i b$, since a_i and a_{i+1} are consecutive in φ. When we add edge $a_i b$, vertex b gets one more neighbor to the left, and thus might appear in a box further to the left than the box it was in before. By Observation 1 we know that b was not in X_i or $X_{i,i+1}$ before the addition of edge $a_i b$. Now it can enter $X_{i,i+1}$ but it cannot enter X_i, since it only gained one more neighbor. This means that it can move past a_{i+1} toward the left, but it cannot move past a_i. Thus $L(j, \varphi', G') = L$ and $R(j, \varphi', G') = R$, although some vertices in R might have changed positions. Consequently, $\theta(j, \varphi', G') = \theta(j, \varphi, G) \cup \{a_i b\}$ is a cut of φ', which means that φ' has a cut whose size is 1 more than a worst cut of φ. Since $cw(G') \leq cw(G) + 1$, G' is a counterexample, contradicting the assumption that G is a tight counterexample. Thus there cannot be a B-vertex in R that a_i is not adjacent to. By Lemma 3 we know that a_i is adjacent to all B-vertices in L, and hence a_i is bipartite universal. By symmetry and with similar arguments, a_{i+1} is also bipartite universal. This means that $rank_\varphi(a_i) = rank_\varphi(a_{i+1})$. If this rank is negative, then the cut at the $(j - 1)$th gap is a larger cut than $\theta(j, \varphi, G)$ since a_i and a_{i+1} have more neighbors in L than in R. Symmetrically, if this rank is positive then the cut at the $(j+1)$th gap is a larger cut. Therefore $rank_\varphi(a_i) = rank_\varphi(a_{i+1}) = 0$, because otherwise we get a contradiction to the assumption that $\theta(j, \varphi, G)$ is a worst cut. This means that a_i and a_{i+1} have as many neighbors in L as they have in R. Since a_i and a_{i+1} are both bipartite universal and they are not adjacent to each other, the cut at the $(j-1)$th gap and the cut at the $(j+1)$th gap have the same size as $\theta(j, \varphi, G)$. Hence we can take one of these cuts as a worst cut. We can repeat this argument until we reach a B-vertex on the other side of a worst cut.

Lemma 5. *There is a worst cut $\theta(j, \varphi, G)$ of φ such that both v_j and v_{j+1} are bipartite universal.*

Proof. By Lemma 4, we know that there is a worst cut $\theta = \theta(j, \varphi, G)$ such that v_j and v_{j+1} belong to different color classes. Let us now show that both v_j and v_{j+1} are bipartite universal. Let $a = v_j \in A$ and let $b = v_{j+1} \in B$. By Lemma 3 we know that a is adjacent to every B-vertex in L and b is adjacent to every A-vertex in R. If $ab_t \in E$ then a is bipartite universal as a result of the properties of a strong ordering. If $ab_t \notin E$, we claim that $G' = G - b_t$ is also a counterexample, contradicting the assumption that G is a tight counterexample. We observe that G' is a bipartite permutation graph with strong ordering $(\sigma_A, \sigma_B - b_t)$. Since b_t has no neighbors in L as a result of the properties of a strong ordering, θ is a cut of $\varphi - b_t$. Let φ' be the layout computed by MinCutBPG on input G' and $(\sigma_A, \sigma_B - b_t)$. Since no B-vertex was adjacent to b_t, every remaining B-vertex appears in the same box after the deletion of b_t as it did before. However, an A-vertex a_i that was adjacent to b_t might move one position to the left inside the box X_i. Hence a_i can move past b toward the left, but it cannot move past a, since the algorithm respects the strong ordering. Consequently, all vertices of L to the left of a in φ appear also to the left of a in φ'. Thus θ is a cut of φ'. Since $cw(G') \leq cw(G)$ and the size of the cut did not change, G' is a counterexample, contradicting the assumption that G is a tight counterexample. Hence a is bipartite universal. To show that b is bipartite universal we use similar arguments: by symmetry, if $a_1 b \notin E$ then $G' = G - a_1$ is a counterexample as well. Finally, the case where $v_j \in B$ and $v_{j+1} \in A$ is completely symmetric. $\qquad\qquad\square$

Corollary 1. *There is a worst cut $\theta(j, \varphi, G)$ of φ such that v_j and v_{j+1} belong to different color classes and they are both bipartite universal.*

Proof. The proof of Lemma 5 takes a cut as mentioned in Lemma 4, and shows the claim of Lemma 5 using the same cut. Hence, there is a cut that satisfies both lemmas at the same time, and the corollary follows. $\qquad\qquad\square$

The proof of the following lemma has been omitted due to page restrictions.

Lemma 6. *There is a worst cut $\theta(j, \varphi, G)$ such that there are $\lfloor |A|/2 \rfloor$ A-vertices and $\lceil |B|/2 \rceil$ B-vertices on one side of the jth gap of φ, and there are $\lceil |A|/2 \rceil$ A-vertices and $\lfloor |B|/2 \rfloor$ B-vertices on the other side of the jth gap.*

We are now ready to prove the main theorem of this paper.

Theorem 2. *The cutwidth of a bipartite permutation graph can be computed in linear time.*

Proof. We describe the main algorithm for computing the cutwidth of a bipartite permutation graph G. First we compute a strong ordering of each connected component of G. Then we run MinCutBPG on each connected component with the computed strong ordering of that connected component. We concatenate the returned layouts from each of these calls into one layout φ for G. The order in which the layouts are concatenated does not matter, as the cuts at the concatenation points are empty. We check every position j with $1 \leq j < n$ to find a largest cut $\theta(j, \varphi, G)$, and we output $|\theta(j, \varphi, G)|$ as the cutwidth of G. If Algorithm MinCutBPG is correct then clearly the output of the described algorithm is equal to $cw(G)$.

Before we prove the correctness of Algorithm MinCutBPG, let us analyze the running time of the above algorithm. By the results of [17], computing a strong ordering for each connected component of G takes in total $O(n + m)$ time. The running time of Algorithm MinCutBPG is also $O(n + m)$. To see this, observe that in the first loop, when deciding the box of a B-vertex, we never need to consider boxes to the left of the most recently considered box. By Observation 1, the next B-vertex is placed in either the box in which the previous B-vertex was placed, or a box further to the right. Thus running MinCutBPG on each connected component takes $O(n + m)$ time for the whole graph. Concatenating the returned layouts and finding the largest cut takes $O(n)$ time, and the overall running time follows.

Let us prove that Algorithm MinCutBPG correctly computes the cutwidth of a connected bipartite permutation graph. Assume for contradiction that there is a tight counterexample $G = (A, B, E)$. By Lemma 6, we know that there is a worst cut $\theta = \theta(j, \varphi, G)$ of the layout φ computed by Algorithm MinCutBPG on G, such that there are $\lfloor |A|/2 \rfloor$ A-vertices and $\lceil |B|/2 \rceil$ B-vertices on one side of the jth gap of φ, and $\lceil |A|/2 \rceil$ A-vertices and $\lfloor |B|/2 \rfloor$ B-vertices on the other side. Let $F = \{ab \notin E \mid a \in A \land b \in B\}$. Then $F \cap E = \emptyset$ and $(A, B, (E \cup F))$ is a complete bipartite graph. Since by Lemma 3 vertices on either side of θ induce a complete bipartite graph, we have that for each $ab \in F$, a and b are on different sides of θ. Thus we can conclude the following about the size of θ:

$$|\theta| = \left\lfloor \frac{|A|}{2} \right\rfloor \left\lfloor \frac{|B|}{2} \right\rfloor + \left\lceil \frac{|A|}{2} \right\rceil \left\lceil \frac{|B|}{2} \right\rceil - |F| \; .$$

Let S be *any* set of $\lfloor |A|/2 \rfloor + \lceil |B|/2 \rceil$ vertices of G. We claim that $|\theta(S, G)| \geq |\theta|$, regardless of how many A-vertices and how many B-vertices there are in S. To consider all possibilities, let there be $\lfloor |A|/2 \rfloor - x$ A-vertices and $\lceil |B|/2 \rceil + x$ B-vertices in S, for an appropriate (positive, zero or negative) integer x. Consequently, there are $\lceil |A|/2 \rceil + x$ A-vertices and $\lfloor |B|/2 \rfloor - x$ B-vertices in $(A \cup B) \setminus S$. Some of the set F of missing edges might have endpoints on different sides of the cut $\theta(S,G)$ and some might not. Since $(A, B, (E \cup F))$ is a complete bipartite graph, we know the following about the size of $\theta(S, G)$:

$$|\theta(S,G)| \geq \left(\left\lfloor \frac{|A|}{2} \right\rfloor - x\right)\left(\left\lfloor \frac{|B|}{2} \right\rfloor - x\right) + \left(\left\lceil \frac{|A|}{2} \right\rceil + x\right)\left(\left\lceil \frac{|B|}{2} \right\rceil + x\right) - |F|$$

$$= \left\lfloor \frac{|A|}{2} \right\rfloor \left\lfloor \frac{|B|}{2} \right\rfloor - \left\lfloor \frac{|A|}{2} \right\rfloor x - \left\lfloor \frac{|B|}{2} \right\rfloor x + x^2 + \left\lceil \frac{|A|}{2} \right\rceil \left\lceil \frac{|B|}{2} \right\rceil + \left\lceil \frac{|A|}{2} \right\rceil x + \left\lceil \frac{|B|}{2} \right\rceil x + x^2 - |F|$$

$$= |\theta| + 2x^2 + x\left(\left\lceil \frac{|A|}{2} \right\rceil - \left\lfloor \frac{|A|}{2} \right\rfloor + \left\lceil \frac{|B|}{2} \right\rceil - \left\lfloor \frac{|B|}{2} \right\rfloor\right).$$

Note that the value of the expression in parentheses in the last line of the equation is 0, 1, or 2. Consequently, for all possible values of x, we have that $|\theta(S,G)| \geq |\theta|$.

Let φ^* be an optimal layout of G, and let $j = \lfloor |A|/2 \rfloor + \lceil |B|/2 \rceil$. Let $S^* = L(j, \varphi^*, G)$. Hence S^* contains $\lfloor |A|/2 \rfloor + \lceil |B|/2 \rceil$ vertices and $\theta(j, \varphi^*, G) = \theta(S^*, G)$. Clearly $\mathrm{cw}(G) \geq |\theta(j, \varphi^*, G)| = |\theta(S^*, G)|$. However, for any such set S^*, we have shown above that a worst cut θ of the layout computed by Algorithm MinCutBPG has the property $|\theta(S^*, G)| \geq |\theta|$. Therefore, $\mathrm{cw}(G) \geq |\theta|$, contradicting the assumption that G is a counterexample. Consequently, no counterexample exists, and the algorithm correctly computes the cutwidth of every connected bipartite permutation graph.

4   Concluding Remarks

Algorithm MinCutBPG takes as input a connected bipartite permutation graph $G = (A, B, E)$ and a strong ordering (σ_A, σ_B) of G. Before the algorithm is called, $O(n + m)$ time is spent on recognizing G as a bipartite permutation graph and computing a strong ordering of G. Within the same running time one can assign two integers $\ell(v)$ and $r(v)$ to every vertex $v \in A \cup B$ for the following purpose. If $v \in A$ then $\ell(v)$ and $r(v)$ are the positions of the leftmost and the rightmost neighbor of v in σ_B. If $v \in B$ then $\ell(v)$ and $r(v)$ are the positions of the leftmost and the rightmost neighbor of v in σ_A. Observe that with this information, $d(v)$ can be computed in constant time, and the middle neighbor of a vertex can be found in constant time. Consequently, if $\ell(v)$ and $r(v)$ are supplied to MinCutBPG as input for every $v \in A \cup B$, the running time of MinCutBPG is in fact $O(n)$.

With our results in addition to the results of [10], the cutwidth of two unrelated subclasses of permutation graphs can be computed in linear time: threshold graphs and bipartite permutation graphs. We leave as an open problem to decide the computational complexity of computing the cutwidth of permutation graphs. In fact, it would be interesting to know the computational complexity of cutwidth on other well known subclasses of permutation graphs, like cographs or even their subclass trivially perfect graphs.

References

1. Adolphson, D., Hu, T.C.: Optimal linear ordering. SIAM J. Appl. Math. 25, 403–423 (1973)
2. Arora, S., Frieze, A., Kaplan, H.: A new rounding procedure for the assignment problem with applications to dense graphs arrangements. In: Proceedings of FOCS 1996, pp. 21–30. IEEE, Los Alamitos (1996)
3. Blin, G., Fertin, G., Hermelin, D., Vialette, S.: Fixed-parameter algorithms for protein similarity search under RNA structure constraints. In: Kratsch, D. (ed.) WG 2005. LNCS, vol. 3787, pp. 271–282. Springer, Heidelberg (2005)
4. Brandstädt, A., Le, V.B., Spinrad, J.: Graph Classes: A Survey. SIAM, Philadelphia (1999)
5. Brandstädt, A., Lozin, V.V.: On the linear structure and clique-width of bipartite permutation graphs. Ars Combinatorica 67, 273–289 (2003)
6. Chung, M.J., Makedon, F., Sudborough, I.H., Turner, J.: Polynomial time algorithms for the min cut problem on degree restricted d trees. In: Proceedings of FOCS 1982, pp. 262–271. IEEE, Los Alamitos (1982)
7. Díaz, J., Penrose, M., Petit, J., Serna, M.J.: Approximating layout problems on random geometric graphs. Journal of Algorithms 39, 78–117 (2001)
8. Díaz, J., Petit, J., Serna, M.J.: A survey of graph layout problems. ACM Computing Surveys 34, 313–356 (2002)
9. Gavril, F.: Some NP-complete problems on graphs. In: 11th Conference on Information Sciences and Systems, pp. 91–95. John Hopkins University, Baltimore (1977)
10. Heggernes, P., Lokshtanov, D., Mihai, R., Papadopoulos, C.: Cutwidth of split graphs, threshold graphs, and proper interval graphs. In: Broersma, H., Erlebach, T., Friedetzky, T., Paulusma, D. (eds.) WG 2008. LNCS, vol. 5344, pp. 218–229. Springer, Heidelberg (2008)
11. Jünger, M., Reinelt, G., Rinaldi, G.: The traveling salesman problem. In: Handbook on Operations Research and Management Sciences, vol. 7, pp. 225–330. North-Holland, Amsterdam (1995)
12. Karger, D.R.: A randomized fully polynomial approximation scheme for the all terminal network reliability problem. In: Proceedings of STOC 1996, pp. 11–17. ACM, New York (1996)
13. Leighton, F.T., Rao, S.: An approximate max-flow min-cut theorem for uniform multicommodity flow problems with applications to approximation algorithms. In: Proceedings of FOCS 1988, pp. 422–431. IEEE, Los Alamitos (1988)
14. Makedon, F., Sudborough, I.H.: Minimizing width in linear layouts. In: Díaz, J. (ed.) ICALP 1983. LNCS, vol. 154, pp. 478–490. Springer, Heidelberg (1983)
15. Monien, B., Sudborough, I.H.: Min cut is NP-complete for edge weighted trees. In: Kott, L. (ed.) ICALP 1986. LNCS, vol. 226, pp. 265–274. Springer, Heidelberg (1986)
16. Mutzel, P.: A polyhedral approach to planar augmentation and related problems. In: Spirakis, P.G. (ed.) ESA 1995. LNCS, vol. 979, pp. 497–507. Springer, Heidelberg (1995)
17. Spinrad, J., Brandstädt, A., Stewart, L.: Bipartite permutation graphs. Discrete Applied Mathematics 18, 279–292 (1987)
18. Thilikos, D.M., Serna, M.J., Bodlaender, H.L.: Cutwidth I: A linear time fixed parameter algorithm. Journal of Algorithms 56, 1–24 (2005)

19. Thilikos, D.M., Serna, M.J., Bodlaender, H.L.: Cutwidth II: Algorithms for partial w-trees of bounded degree. Journal of Algorithms 56, 24–49 (2005)
20. Yannakakis, M.: A polynomial algorithm for the min cut linear arrangement of trees. Journal of ACM 32, 950–988 (1985)
21. Yuan, J., Zhou, S.: Optimal labelling of unit interval graphs. Appl. Math. J. Chinese Univ. Ser. B (English edition) 10, 337–344 (1995)

Solving Capacitated Dominating Set by Using Covering by Subsets and Maximum Matching*

Mathieu Liedloff[1], Ioan Todinca[1], and Yngve Villanger[2]

[1] LIFO, Université d'Orléans, BP 6759, F-45067 Orléans Cedex 2, France
{mathieu.liedloff,ioan.todinca}@univ-orleans.fr
[2] Department of Informatics, University of Bergen, N-5020 Bergen, Norway
yngve.villanger@uib.no

Abstract. The CAPACITATED DOMINATING SET problem is the problem of finding a dominating set of minimum cardinality where each vertex has been assigned a bound on the number of vertices it has capacity to dominate. Cygan et al. showed in 2009 that this problem can be solved in $O(n^3 m\binom{n}{n/3})$ or in $O^*(1.89^n)$ time using maximum matching algorithm. An alternative way to solve this problem is to use dynamic programming over subsets. By exploiting structural properties of instances that can not be solved fast by the maximum matching approach, and "hiding" additional cost related to considering subsets of large cardinality in the dynamic programming, an improved algorithm is obtained. We show that the CAPACITATED DOMINATING SET problem can be solved in $O^*(1.8463^n)$ time.

1 Introduction

The problem of finding a vertex subset of cardinality at most k that dominates all remaining vertices in a graph has received a considerable amount of attention over the last two decades. This problem is known as the DOMINATING SET problem and is a classical NP-complete and also $W[2]$-complete problem [5]. If each vertex is equipped with a bound on the number of neighbors that it has capacity to dominate additionally to itself, the problem is called CAPACITATED DOMINATING SET. By simply assigning the degree as the capacity to each vertex, the hardness results carry over to the CAPACITATED DOMINATING SET problem. Recently it has been proven that this problem remains $W[1]$-hard even in the planar case [2], and thus distinguishing it from DOMINATING SET which is Fixed Parameter Tractable on planar graphs [1].

For the general domination problem there has been a long sequence of moderately exponential time algorithms starting in 2004 by [8], where the currently last result is [10]. Also in the case where we ask for a connected dominating set the trivial bound was broken back in 2006 [6].

---

* This work has been supported by ANR Blanc AGAPE (ANR-09-BLAN-0159-03) and EGIDE Aurora (18809RM).

D.M. Thilikos (Ed.): WG 2010, LNCS 6410, pp. 88–99, 2010.

The Capacitated Dominating Set problem have a slightly different story. The question if the "trivial" $O^*(2^n)$ could be broken was first asked at IWPEC 2008, and repeated later that year by Johan van Rooij at Dagstuhl [7]. Cygan et al. show in [4] that the Capacitated Dominating Set problem could be solved exactly in $O^*(1.89^n)$ time. They first provide an algorithm that, given a set U of vertices, computes in polynomial time a minimum dominating set D with $U \subseteq D$ and such that only the elements of U are allowed to dominate two or more vertices outside D. This algorithm is based on a reduction to a maximum matching problem. Then the result is obtained by guessing U, which is of size no more than $n/3$.

Koivisto [9] gave a clever algorithm for the following problem: the input is a family $\mathcal{F}$ of subsets of a universe V and the output is a partition of V into a minimum number of elements of $\mathcal{F}$. The running time is $o(2^n)$ if all the sets of $\mathcal{F}$ are small (bounded by a constant). This yields an $o(2^n)$ algorithm for the Capacitated Dominating problem when all capacities are all upper bounded by a constant k. Indeed, we can put in $\mathcal{F}$ all subsets $X \subseteq N[v]$ such that $|X| \leq c(v)+1$, for all vertices v (see next section for details).

For both Cygan's et al. and Koivisto's algorithms there exist instances that forces the algorithms to spend respectively $O^*(1.89^n)$ and $O^*(2^n)$ time. There are even instances that force both algorithms to their respective maximum bound. Consider a Capacitated Dominating Set D where a constant number of vertices W dominate ρn vertices for a constant $0 < \rho < 1$, and every vertex of $D \setminus W$ uses its capacity to dominate exactly two vertices in $V \setminus D$. Such an instance will force Koivisto's algorithm to consider dominating vertices that use their capacity to dominate ρn other vertices (hence the sets of $\mathcal{F}$ are not of bounded size), and the maximum matching approach by Cygan et al. has to test all subsets U of size $|W| + (n - |W| \rho n)/3$. Thus, no significant improvement can be obtained by simply balancing the two approaches.

In this paper we adapt the algorithm by Koivisto to avoid the "constant" restriction on the capacities, meaning that the elements of the family $\mathcal{F}$ will not necessarily be of bounded size. This generalization works under the condition that there are not too many vertices in the dominating set using unbounded capacity. Maybe the most interesting contribution here is that handling elements of $\mathcal{F}$ of unbounded size does not add to the total running time of the algorithm. The reason for this comes form the fact that all unbounded sets are handled in the beginning of the dynamic programming, before the considering smaller sets. This is then balanced with the maximum matching approach of Cygan et al. to optimize the time bound. As a result of this we get that the Capacitated Dominating Set problem can be solved in $O^*(1.8463^n)$ time. When we look further into the polynomial factors hidden by the big-Oh notation, we can actually notice that there is a trade-off between the polynomial factor and the basis of the exponent, with running times ranging from $O(n^9 \cdot 1.8844^n)$ to $O(n^{40005} \cdot 1.8463^n)$.

2   Preliminaries

In this paper we consider simple and undirected graphs. Given a graph $G = (V, E)$, we denote by n the number of its vertices. If the given graph is equipped with a capacity function $c : V \to \mathbb{N}$, we say that a subset $D \subseteq V$ is a Capacitated Dominating Set if there exists a function $f : V \setminus D \to D$ such that $f(u) \in N(u) \cap D$ and $|f^{-1}(v)| \leq c(v)$ for each $v \in D$. Alternatively the CAPACITATED DOMINATING SET problem can be viewed as a partitioning problem. Let $\mathcal{F}$ be the family of subsets of V such that $X \subseteq V$ is a member of $\mathcal{F}$ if and only if there exists a vertex $v \in X$ such that $X \subseteq N[v]$ and $|X| \leq c(v) + 1$. Such a vertex v will be refereed to as the *representative* of X and is denoted $v = R(X)$. Any partitioning $S_1, S_2, \ldots, S_d$ of V where $S_i \in \mathcal{F}$ for every $i \in \{1, \ldots, d\}$ defines a Capacitated Dominating Set D of G, where $|D| = d$. The capacitated dominating set D can be retrieved from the partitioning by selecting the representative of each set, i.e. $D = \cup_{i=1}^{d} R(S_i)$. In the next section, we will view the CAPACITATED DOMINATING SET problem as the problem of finding such a partitioning where d is minimized.

3   The Two Main Ingredients

We start by briefly recalling the construction of Cygan et al. and Koivisto's algorithms.

3.1   Cygan et al.'s Algorithm

Let $G = (V, E)$ be a graph with a capacity function c. In [4], Cygan, Pilipczuk and Wojtaszczyk give a $O^*(1.89^n)$ time exact algorithm to compute a capacitated dominating set. This algorithm heavily relies on a reduction to a matching problem. Namely, they consider the Constrained CDS problem defined as follows: Given a set U of representatives of sets of size at least 3, compute a smallest CDS $D \subseteq V$ such that $U \subseteq D$, and $D \setminus U$ are the representatives of sets of size at most 2. In words, the vertices of $D \setminus U$ will dominate at most one vertex outside D. Cygan et al. give a polynomial algorithm for this problem. We refer to this algorithm as ExtendSolution(G, c, U). By enumerating all subsets U of size at most $n/3$, one can solve the CDS problem in $O^*(\binom{n}{n/3}) = O^*(1.89^n)$ time.

Theorem 1 (see [4]). *Given a set U of representatives of sets of size at least 3, algorithm ExtendSolution (G, c, U) computes a smallest CDS $D \subseteq V$ such that $U \subseteq D$, in $O(n^2 m)$ running time. Consequeltly, the CDS problem can be solved in $O^*(1.89^n)$ time.*

3.2   Koivisto's Algorithm: Partitioning into Sets

In [9], Koivisto investigates the problem of finding a partition of a universe V into k disjoints sets $S_1, S_2, \ldots, S_k$ from a family $\mathcal{F}$ of subsets of V under the restriction that the subsets of $\mathcal{F}$ are of bounded cardinality. Thanks to an (arbitrary) linear order $<$ on the elements of V (implying a lexicographic order $\prec$ on the subsets of V), Koivisto designs an algorithm with a proved worst-case running-time $O^*(c^n)$ with $c < 2$. We recall briefly the approach developped by Koivisto. In the next section, we show how to extend this approach if the familly $\mathcal{F}$ contains *some* sets of *unbounded* cardinality. Formally, the following was proved in [9]:

Theorem 2. *Given an n-element universe V, a number k, and a familly $\mathcal{F}$ of subsets of V, each of cardinality at most r, the partitions of V into k members of $\mathcal{F}$ can be counted in time $O^*(|\mathcal{F}|2^{n\lambda_r})$ where $\lambda_r = \dfrac{2r-2}{\sqrt{(2r-1)^2-2\ln 2}}$.*

To achieve this result, one idea of Koivisto approach is to define an arbitrary linear order $<$ on the universe and to count the lexicographic ordered k-partitions $(S_1, S_2, \ldots, S_k)$ of V such that $S_i \prec S_j$ whenever $1 \le i < j \le k$. It is shown in [9] that w.l.o.g the number of lexicographic ordered k-partition is equal to the number of k-partitions.

The algorithm is very simple and is based on dynamic programming: for any $W \subseteq V$ and integer j, $1 \le j \le k$, let $f_j(W)$ be the number of ordered partitions of W into j sets of $\mathcal{F}$. Clearly, we have (see [9] for details):

$$f_1(W) = [W \in \mathcal{F}] \quad \text{and} \quad f_j(W) = \sum_{X \subseteq W} f_{j-1}(W \setminus X)[X \in \mathcal{F}] \quad \text{for } j > 1.$$

Here $[W \in \mathcal{F}]$ counts the occurences of the set W in the family $\mathcal{F}$.

As observed in [9], considering only lexicographic ordered partitions leads to a reduction in the number of subsets of V needed to be considered by the algorithm. Indeed, the set S_j must contain the smallest element of V not in $S_1 \cup S_2 \cup \ldots \cup S_{j-1}$. Let $\mathcal{R}_j$ be the family of sets W that is the union of j such sets $S_1, S_2, \ldots, S_j$. This family is defined recursively by:

$$\mathcal{R}_1 = \{X \text{ s.t. } X \in \mathcal{F}, \ \min V \in X\};$$
$$\mathcal{R}_j = \{Y \cup X \text{ s.t. } Y \in \mathcal{R}_{j-1}, \ X \in \mathcal{F}, \ Y \cap X = \emptyset, \ \min V \setminus Y \in X\}.$$

It is shown in [9] that the running-time of the algorithm is proportional to $(|\mathcal{R}_1| + |\mathcal{R}_2| + \cdots + |\mathcal{R}_k|) \cdot |\mathcal{F}|$. Note that if we require the size of each sets in $\mathcal{F}$ to be bounded by a constant r, then the size of the family $\mathcal{F}$ is no more than n^r. Now we derive an upper bound on each $|\mathcal{R}_j|$ being sharper than the one of Theorem 2. However, contrary to Theorem 2 which provides an algebraic expression on the running-time, our result will involve calculus, see the next theorem and Table 1 for some values.

Theorem 3. *Given an n-element universe V, a number k, and a familly $\mathcal{F}$ of subsets of V, each of cardinality at most r, the partitions of V into k members of $\mathcal{F}$ can be counted in time $O^*\left(|\mathcal{F}| \cdot \binom{(1-\lambda)n}{(r-1)\lambda n}\right)$ where λ is the unique solution of $\dfrac{(1-\lambda r)^r}{(\lambda(r-1))^{r-1}} = 1 - \lambda$ in $[0; \frac{1}{2r-1}]$.*

Proof. Let V an n-element universe with an arbitrary linear order on its elements: $v_1, v_2, \ldots, v_n$. Let j be an integer of $\{1, 2, \ldots, k\}$. Recall that $\mathcal{R}_j$ is the family of sets W being the union of j disjoint sets $S_1, S_2, \ldots, S_j$, each of size no more than r. Also by definition of $\mathcal{R}_j$, for any set $W \in \mathcal{R}_j$, $\{v_1, v_2, \ldots, v_j\} \subseteq W$. Thus the size of $\mathcal{R}_j$ is upper bounded by the maximum number of different sets W that it can contain. This number is no more than the largest value of $\binom{n-j}{w-j}$ for $j \le w \le rj$. Indeed, the size of W, denoted here by w is at least j since W contains the j first vertices and at most rj since W is the disjoint union of j sets of size at most r. As W has to contain the j first vertices, the $|W| - j$ others vertices have to be choosen among the $n - j$ remaining vertices.

By denoting ρn, with $0 \le \rho \le 1$, the value of j and by w', with $0 \le w' \le (r-1)\rho$, the value of $w - j$, the expression $\binom{n-j}{w-j}$ can be rewritten $\binom{n(1-\rho)}{nw'}$. This latter expression is maximum for $w' = \frac{1-\rho}{2}$ whenever $\frac{1-\rho}{2} \le (r-1)\rho$ or for $w' = (r-1)\rho$ otherwise. (This can easily be seen from the well-known binomial formula.) Note that $\frac{1-\rho}{2} = (r-1)\rho$ for $\rho = \frac{1}{2r-1}$.

Also, by Stirling's approximation, $\binom{\alpha n}{\beta n}$ is asymptotically bounded by $B(\alpha, \beta)^n$ where $B : (\alpha, \beta) \mapsto \frac{\alpha^\alpha}{\beta^\beta (\alpha - \beta)^{\alpha - \beta}}$.

Thus in the next part of the proof we study the functions $f_1 : \rho \mapsto B(1 - \rho, (r-1)\rho)$ defined over $[0; \frac{1}{2r-1}]$ and $f_2 : \rho \mapsto B(1-\rho, \frac{1-\rho}{2})$ defined over $[\frac{1}{2r-1}; 1]$ (see Fig. 1 for a plot of these two functions).

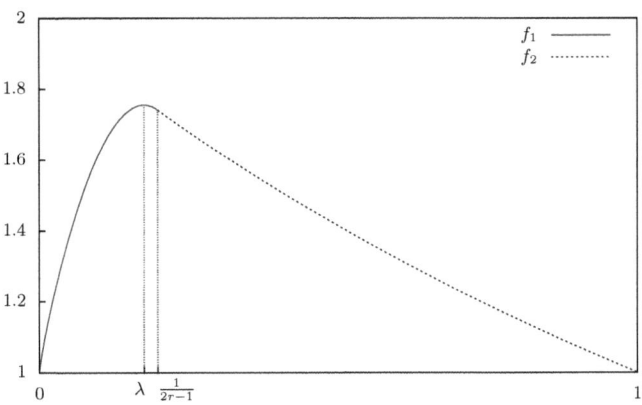

Fig. 1. The two functions f_1 and f_2 plotted for $r = 3$. The maximum of f_1 is reached for λ.

We first start by studying f_2. Its derivative $f_2' = -(1 - \rho)^{1-\rho}(\frac{1-\rho}{2})^{\rho-1} \ln(2)$ is non positive on $[\frac{1}{2r-1}; 1]$. Thus the maximum of f_2 is obtained for $\rho = \frac{1}{2r-1}$. Since $f_2(\frac{1}{2r-1}) = f_1(\frac{1}{2r-1})$, it sufficient to restrict ourselves to the analysis of function f_1 over $[0; \frac{1}{2r-1}]$.

Then we consider f_1. Its derivative is $f_1' = (1 - \rho)^{1-\rho}((r-1)\rho)^{(1-r)\rho}(1 - \rho r)^{\rho r-1}(-\ln(1 - \rho) - r \ln((r-1)\rho) + \ln((r-1)\rho) + r \ln(1 - \rho r))$ and is equal to zero over $[0; \frac{1}{2r-1}]$ if and only if ρ satisfies $(-\ln(1 - \rho) - r \ln((r-1)\rho) +$

$\ln((r-1)\rho) + r\ln(1-\rho r)) = 0$. In other words, whenever $\rho \in [0; \frac{1}{2r-1}]$ satifies $\frac{(1-\rho r)^r}{(\rho(r-1))^{r-1}} = 1 - \rho$ in $[0; \frac{1}{2r-1}]$. Let us show that such a zero exists and is unique. (The zero will correspond to λ on Fig. 1.) Consider $\tilde{f}_1' : \rho \mapsto -\ln(1-\rho) - r\ln((r-1)\rho) + \ln((r-1)\rho) + r\ln(1-\rho r)$; its derivative is $\tilde{f}_1'' : \frac{1-r}{\rho(\rho-1)(\rho r-1)}$ and is negative over $[0; \frac{1}{2r-1}]$ for any $r > 1$. Thus $\tilde{f}_1'$ is strictly decreasing over $[0; \frac{1}{2r-1}]$ and admits a unique zero since $\tilde{f}_1'(0) \geq 0$ and $\tilde{f}_1'(\frac{1}{2r-1}) \leq 0$.   □

Now we shortly explain how to compute a *good* approximation of λ (defined as in Theorem 3) and then provide an asymptotic upper bound on the running-time claimed by Theorem 3.

We have shown that f_1' has a unique zero over $[0; \frac{1}{2r-1}]$; it can easily be shown that f_1' is decreasing and monotone over $[0; \frac{1}{2r-1}]$ with f_1' positive over $[0; \lambda]$ and negative over $[\lambda; \frac{1}{2r-1}]$. Here, λ is the unique solution of $\frac{(1-\lambda r)^r}{(\lambda(r-1))^{r-1}} = 1 - \lambda$ in $[0; \frac{1}{2r-1}]$.

Let ϵ be a positive real. Let $\tilde{\lambda}$ "close enough to λ", i.e. so that $f_1'(\tilde{\lambda} - \epsilon) \geq 0$ and $f_1'(\tilde{\lambda} + \epsilon) \leq 0$. Since the zero of f_1' is unique, such a $\tilde{\lambda}$ can be found by binary search. Since f_1' is monotone, it follows that $\tilde{\lambda} - \epsilon > \lambda > \tilde{\lambda} + \epsilon$ and thus $|\tilde{\lambda} - \lambda| \leq \epsilon$. Let $\Delta = \max\left(|f_1'(\tilde{\lambda} - \epsilon)|, |f_1'(\tilde{\lambda} + \epsilon)|\right)$. By the classical *mean value* theorem, $|f(\tilde{\lambda}) - f(\lambda)| \leq \Delta \cdot |\tilde{\lambda} - \lambda| \leq \Delta \cdot \epsilon$. Consequently we have $f(\lambda) \leq f(\tilde{\lambda}) + \Delta \cdot \epsilon$. Finally we recall that $O^*(f(\lambda)^n) = \binom{(1-\lambda)n}{(r-1)\lambda n}$.

Thus it is sufficient to find a $\tilde{\lambda}$ close enough to λ by binary search to establish an upper bound on the running-time of Koivisto algorithm. Table 1 provides values of $\tilde{\lambda}$ for some values of r. We also give the corresponding running-time obtained by Theorem 3 (named as "Koivisto*") and the running-time from the original analysis by Koivisto [9] (see column "Koivisto").

Table 1. The table provides, for several values of r, the running-time of Koivisto's algorithm [9] according to Koivisto's analysis (column "Koivisto") and according to the new analysis devised by Theorem 3 (column "Koivisto*"). We also provide the value of $\tilde{\lambda}$ which act as a certificate together with $\epsilon = 10^{-5}$ for proving the correctness of values given by Koivisto*. Note that for the special case $r = 2$, the analysis provided by [9] is already sharp.

r	Koivisto	**Koivisto***	$\tilde{\lambda}$
2	1.6181	**1.6181**	0.27629
3	1.7693	**1.7549**	0.17701
4	1.8271	**1.8192**	0.13051
10	1.9308	**1.9296**	0.05081
20	1.9654	**1.9651**	0.02520
50	1.9862	**1.9861**	0.01003
100	1.9931	**1.9931**	0.00501

3.3   A New Recipe for Capacitated Dominating Set: Combining and Cooking the Ingredients

We are now ready to explain how the Capacitated Dominating Set problem can be solved in time $O^*(1.8463^n)$ by combining the approach of Cygan et al. [4] and an extended version of the result by Koivisto [9] for bounded cardinalty sets family. Namely, we will show that Koivisto's approach can be used even if *some* sets of the family are of unbounded size. To do that, instead of fixing an arbitrary ordering over the elements of a universe, we use this freedom to put some elements playing a special role at the begining of the ordering. These elements are the representatives of sets being of unbounded cardinality (i.e. not constant bounded). As we will see, in our Capacitated Dominating Set algorithm, these elements have to be guessed. It is worth to note that if these elements are given as part of the input, the running-time analysis can be improved. Then, once the unbounded sets have been guessed, the algorithm has to deal only with bounded-cardinality sets.

Description of the Algorithm. Let $G = (V, E)$ be a graph with a capacity function c defined over V. Let $\mathcal{F}$ be the family of subsets $X \subseteq V$ such that $X \in \mathcal{F}$ if and only if there exists a vertex $v \in X$ such that $X \subseteq N[v]$ and $|X| \leq c(v) + 1$. We recall here that such a $v = R(X)$ is called the representative of X. Note that it is possible that two sets X_1 and X_2 with $X_1 = X_2$, $R(X_1) = u$, $R(X_2) = v$, $u \neq v$ satisfy the required properties for belonging to $\mathcal{F}$. In that case, it is sufficient to keep only one of these sets in $\mathcal{F}$ together with its corresponding representative. In that way, given a $X \in \mathcal{F}$, its representative is unique. From now on a family $\mathcal{F}$ constructed as explained is called a *family*; we will omit to precise the universe V and the capacity function c when it is clear from the context.

Lemma 1. *Let $\mathcal{F}$ be a family. If $S_1, S_2, \ldots, S_d$, with $S_i \in \mathcal{F}$ for each $i \in \{1, \ldots, d\}$, is a partition of the universe V then their representatives are pairwise distinct.*

Proof. Since $S_1, S_2, \ldots, S_d$ is a partition of V and for each $i \in \{1, \ldots, d\}$, $R(S_i) \in S_i$, the lemma follows.   □

Let $\beta \in \mathbb{N}$ with $3 < \beta \leq n$. Let $\mathcal{F}$ be a family and let $S_1, S_2, \ldots, S_d$ be a partition of V. We enrich the notion of representatives with the notions of *big representatives*, *medium representatives* and *tiny representatives* with respect to this partition. We say that a vertex v is a

- *big representative* if there is a S_i, $1 \leq i \leq d$, with $v = R(S_i)$ and $|S_i| \geq \beta$;
- *medium representative* if there is a S_i, $1 \leq i \leq d$, with $v = R(S_i)$ and $\beta > |S_i| \geq 3$;
- *tiny representative* if there is a S_i, $1 \leq i \leq d$, with $v = R(S_i)$ and $3 > |S_i|$;

The set of all big (resp. medium and tiny) representative is denoted by (BR) (resp. (MR) and (TR)).

To simplify further the description, given a disjoint collection $S_1, S_2, \ldots, S_d$ of subsets and a representative v, we denote by $S(v)$ the set S_i such that $v = R(S_i)$.

How does the algorithm work? A complete and formal description of the algorithm is given by Algorithm minCDS. Suppose that $D \subseteq V$ is a solution of size d of the minimum Capacitated Dominating Set problem. Then, there exists a partition of V into d sets $S_1, S_2, \ldots, S_d$ such that each $v \in D$ is the representative of one set S_i with $|S_i| \leq c(v) + 1$. Thus D can also be partitioned into (BR), (MR) and (TR).

First we start in "Step 1" (see Algorithm minCDS) by computing all possible solutions (possibly not of minimum size) such that $|(\text{BR}) \cup (\text{MR})| \leq \gamma n$, for some fixed $\gamma \in [1/4; 1/3]$. This is done using Algorithm ExtendSolution. Then in all solutions computed in "Step 2" we may assume that $|(\text{BR}) \cup (\text{MR})| \geq \gamma n$. Since all sets $S(v)$ with $v \in (\text{BR}) \cup (\text{MR})$ are of cardinality at least 3, it follows that (BR) is of moderate size as shown by the next lemma. This set (BR) of big representatives is guessed by our algorithm.

Lemma 2. *Suppose that there exists a solution $S_1, S_2, \ldots, S_d$ such that $|(\text{BR}) \cup (\text{MR})| \geq \gamma n$, for $\gamma \in [0; 1]$. Then the size of (BR) is at most $\frac{n-3\gamma n}{\beta - 3}$.*

Proof. Each set S_i such that $R(S_i) \in (\text{BR}) \cup (\text{MR})$ has cardinality at least 3. Thus only $n - 3\gamma n$ vertices can be distributed over the S_i's having a big representative. Since sets with a big representative are of size at least β (and already contain 3 vertices), it follows that $|(\text{BR})| \leq \frac{n-3\gamma n}{\beta - 3}$.   □

In addition, for each S_i with a big representative, the size of S_i is not necessary bounded by a constant, but a linear bound can be established on its size:

Lemma 3. *Suppose that there exists a solution $S_1, S_2, \ldots, S_d$ such that $|(\text{BR}) \cup (\text{MR})| \geq \gamma n$, for $\gamma \in [0; 1]$. Then the size of each S_i such that $R(S_i) \in (\text{BR})$ is at most $n - 3\gamma n + 3$. Moreover, the following is satisfied $|\cup_{v \in (\text{BR})} S(v)| \leq \frac{n\beta - 3\gamma\beta n}{\beta - 3}$.*

Proof. As shown in the proof of Lemma 2, only $n - 3\gamma n$ vertices can be distributed over the S_i having a big representative, assuming that each S_j with $R(S_j) \in (\text{BR}) \cup (\text{MR})$ has cardinality at least 3. Thus, each such S_i has at most $n - 3\gamma n + 3$ vertices. Since by Lemma 2, the number of sets with a big representative is at most $\frac{n-3\gamma n}{\beta - 3}$, it follows that $|\cup_{v \in (\text{BR})} S(v)| \leq 3 \cdot \frac{n-3\gamma n}{\beta - 3} + n - 3\gamma n$.   □

In "Step 2.1", our algorithm deals with such sets S_i having a big representative. These sets can only appear during this step since we put the big representatives (guessed by the foreach-loop) at the very beginning of ordering $<$. Then we start to compute partitions of V into sets using the dynamic programming approach recalled in Section 3.2 (see also [9]). Here the size of the sets of $\tilde{\mathcal{F}}$ are not bounded by a constant; nevertheless we will show in Section 3.3 that a *good* bound on the running-time can be established. Finally, in "Step 2.2" it only remain sets of size bounded by the constant $\beta - 1$, and the dynamic programming is pursued. We combine the possible cases in order to retrieve the global optimum solution.

Remark 1. It is straightforward to adapt our Algorithm minCDS so that it returns a minimum capacitated dominating set instead of its size.

Algorithm minCDS$(G = (V, E), c : V \to \mathbb{N})$
Input: A graph $G = (V, E)$ and a capacity function c.
Output: The size of a minimum Capacitated Dominating Set of G.
```
/* γ is a constant that has to be choosen in [1/4; 1/3] and β is a
   constant so that β > 3 (see Section 3.3)                        */
```
$\gamma \leftarrow 31/100$
$\beta \leftarrow 15$
MinSol $\leftarrow \infty$

```
/* --- Step 1 : based on Cygan et al. approach ---                */
```
for $\ell = 0$ *to* γn **do**
  foreach $U \subseteq V$ *of size* ℓ **do**
    MinSol $\leftarrow \min\{\text{MinSol}, \ \ell + \text{ExtendSolution}(G, c, U)\}$

```
/* --- Step 2 : based on Koivisto approach ---                    */
```
for $\ell = 0$ *to* $\frac{n - 3\gamma n}{\beta - 3}$ **do**
  foreach (BR) $\subseteq V$ *of size* ℓ **do**
    Define an ordering $<$ by putting first the vertices of (BR) (in arbitrary order) and then the vertices of $V \setminus$ (BR) (in arbitrary order)
    forall $i \in \{0, 1, \ldots, n\}$ **do** $\mathcal{R}_i \leftarrow \emptyset$
```
    /* -- Step 2.1: dealing with sets of size ≥ β --             */
```
    $z \leftarrow n - 3\gamma n$
    Let
    $\tilde{\mathcal{F}} = \big\{ X \subseteq V, |X| \le z + 3, \exists v \in \text{(BR) s.t. } X \subseteq N[v] \text{ and } |X| \le c(v) + 1 \big\}$
    for $i = 1$ *to* ℓ **do**
      foreach $Y \in \mathcal{R}_{i-1}$ *and* $X \in \tilde{\mathcal{F}}$ *s.t.* $Y \cap X = \emptyset$ *and* $\min V \setminus Y \in X$ **do**
        Add $Y \cup X$ to $\mathcal{R}_i$

```
    /* -- Step 2.2: dealing with sets of size < β --             */
```
    Let
    $\mathcal{F} = \big\{ X \subseteq V, |X| < \beta, \exists v \in V \setminus \text{(BR) s.t. } X \subseteq N[v] \text{ and } |X| \le c(v) + 1 \big\}$
    for $i = \ell + 1$ *to* n **do**
      foreach $Y \in \mathcal{R}_{i-1}$ *and* $X \in \mathcal{F}$ *s.t.* $Y \cap X = \emptyset$ *and* $\min V \setminus Y \in X$ **do**
        Add $Y \cup X$ to $\mathcal{R}_i$

    Let i be the smallest index such that $V \in \mathcal{R}_i$
    MinSol $\leftarrow \min\{\text{MinSol}, \ i\}$

return MinSol

Running-Time Analysis. In this section we show that the worst-case running-time of Algorithm minCDS is $O^*(1.8573^n)$ (using $\gamma = 31/100$ and $\beta = 15$ as stated in the algorithm). With some appropriate values for γ and β which are used as constants by Algorithm minCDS, this worst-case running-time can be lowered to $O^*(1.8463^n)$. We already emphasys that a big polynomial is hidden in this lattest big-Oh notation. This issue will be discussed in Section 3.3 (see also Table 2).

Lemma 4. *The running-time of Step 1 is bounded by $O(n^3 m \binom{n}{\gamma n})$.*

Proof. The total number of sets U is $\sum_{\ell=1}^{\gamma n} \binom{n}{\ell} \le n\binom{n}{n/3}$ since $\gamma \le n/3$. Each call to `ExtendSolution` costs $O(n^2 m)$ time by Theorem 1. Total time for this step becomes $O(n^3 m\binom{n}{\gamma n})$. $\qquad\square$

By the same argument as above combined with Lemma 2 and Lemma 3 we get:

Lemma 5. *The number of sets* (BR) *considered by the outmost ForEach-loop in Step 2 is at most* $O\left(n\binom{n}{\frac{n-3\gamma n}{\beta-3}}\right)$.

Lemma 6. *The size of the family* $\tilde{\mathcal{F}}$ *is bounded by* $O(n\binom{n}{n-3\gamma n+3})$.

Lemma 7. *Let* $1/4 < \gamma$. *The running-time of Step 2.1 is bounded by* $O\big(n^4|\tilde{\mathcal{F}}| \cdot \binom{(1-\lambda)n}{(2\lambda+1-3\gamma)n}\big)$ *where* λ *is the (possible) unique real solution of* $27(\gamma-\lambda)^3 = (1-\lambda)(1-3\gamma+2\lambda)^2$ *over* $[0; \frac{1-3\gamma}{\beta-3}]$ *if such a solution exists, or the running-time is bounded by* $O\big(n^4|\tilde{\mathcal{F}}| \cdot \binom{(\frac{\beta-4+3\gamma}{\beta-3})n}{(\frac{1+3\gamma+\beta-3\gamma\beta}{\beta-3})n}\big)$ *otherwise.*

Proof. We first provide a bound on the size of the family $\mathcal{R}_i$, $1 \le i \le \ell$, where ℓ denotes the size of (BR). For each $W \in \mathcal{R}_i$ its size is at most $3i + n - 3\gamma n$ (recall that W is the union of at most i sets S of cardinality at least 3; furthermore at most $n-3\gamma n$ vertices can be distributed over all these sets — see Lemma 3 and its proof). However, due to the ordering $<$ and by the construction of $W \in \mathcal{R}_i$, each $W \in \mathcal{R}_i$ has to contains the first i elements. Thus there are at most $n\binom{n-i}{2i+n-3\gamma n}$ possible W in $\mathcal{R}_i$.

By Stirling's approximation, $\binom{\alpha n}{\beta n}$ is asymptotically bounded by $B(\alpha, \beta)^n$ where $B : (\alpha, \beta) \mapsto \frac{\alpha^\alpha}{\beta^\beta (\alpha-\beta)^{\alpha-\beta}}$. As done in the proof of Theorem 3, consider the function $f_1 : \rho \mapsto B(1-\rho, 2\rho+1-3\gamma)$ defined over $[0; \frac{1-3\gamma}{\beta-3}]$. Its derivative is equal to zero if and only if $3\ln(3) + 3\ln(\gamma-\rho) - \ln(1-\rho) - 2\ln(1-3\gamma+2\rho) = 0$, or in other words whenever $27(\gamma-\rho)^3 = (1-\rho)(1-3\gamma+2\rho)^2$. By standard calculation, this equation has a unique solution, if it exists, over $[0; \frac{1-3\gamma}{\beta-3}]$. Otherwise, if no solution exists, then f_1 is increasing over $[0; \frac{1-3\gamma}{\beta-3}]$ and its maximum is $f_1(\frac{1-3\gamma}{\beta-3})$. For the polynomial contribution of the running time we get n^2 for running from 0 to ℓ and then 0 to n, each set $\mathcal{R}_i$ might contain subsets of $O(n)$ different sizes, and finaly we need $O(n)$ time to find sets and check presence of edges. $\qquad\square$

Lemma 8. *Let* $\gamma \in [0.18995; 1/3]$. *The running-time of Step 2.2 is bounded by* $O\big(n^4 \cdot n^\beta \cdot \binom{(1-\lambda)n}{(1-3\gamma+2\lambda)n}\big)$ *where* λ *is the unique real root of* $(1-\lambda)(1-3\gamma+2\lambda)^2 - 27(\gamma-\lambda)^3$ *over* $[0; \frac{6\gamma-1}{5}]$.

Proof. The proof is quite similar to the previous one. Again each $W \in \mathcal{R}_i$, $\ell < i \le n$, is of size at most $3i + n - 3\gamma n$. Since it is required that each such set W contains the first i elements, it follows that the size of $\mathcal{R}_i$ is at most $\binom{n-i}{2i+n-3\gamma n}$.

Table 2. The table provides worst-case running-times of Algorithm `minCDS`, depending on the values for γ and β. The order of the hidden polynomial term in the big-Oh notation is given by the second column.

Running-Time	order of the polynomial	γ	β
$O^*(1.8844^n)$	$n^5 \cdot n^4$	0.32914	4
$O^*(1.8798^n)$	$n^5 \cdot n^5$	0.32574	5
$O^*(1.8649^n)$	$n^5 \cdot n^{10}$	0.31520	10
$O^*(1.8573^n)$	$n^5 \cdot n^{15}$	0.31000	15
$O^*(1.8486^n)$	$n^5 \cdot n^{50}$	0.30424	50
$O^*(1.8463^n)$	$n^5 \cdot n^{40000}$	0.30275	40000

Now we consider the functions $f_1 : \rho \mapsto B(1 - \rho, 1 - 3\gamma + 2\rho)$ defined over $[0; \frac{6\gamma-1}{5}]$ and $f_2 : \rho \mapsto B(1 - \rho, \frac{1-\rho}{2})$ defined over $[\frac{6\gamma-1}{5}; 1]$. (To justify this cut between f_1 and f_2, observe that $(1 - \rho)/2 \leq 1 - 3\gamma + 2\rho$ whenever $\rho \geq \frac{6\gamma-1}{5}$.) It can easily be shown that f_2 is decreasing over $[\frac{6\gamma-1}{5}; 1]$ and thus we can restrict ourself on f_1. Again, by studying its derivative, we claim that f_1 is maximum over $[0; \frac{6\gamma-1}{5}]$ for λ being the unique real root of $(1-\lambda)(1-3\gamma+2\lambda)^2 - 27(\gamma-\lambda)^3$. By the same agruments as used in the proof of Lemma 7, we get the polynomial factor to be n^4. □

By combining the previous lemmata, we establish the following bound on the worst-case running-time:

Theorem 4. *The worst-case running-time of Algorithm* **minCDS** *is the maximum over:*

- *Step 1 : $O\big(n\binom{n}{\gamma n}\big)$ (by Lemma 4);*
- *Step 2.1 : $O\big(n^6\big(\frac{n}{\frac{n-3\gamma n}{\beta-3}}\big) \cdot \big(\frac{n}{n-3\gamma n+3}\big) \cdot \big(\frac{(1-\lambda)n}{(2\lambda+1-3\gamma)n}\big)\big)$ if a solution λ of $27(\gamma - \lambda)^3 = (1-\lambda)(1-3\gamma+2\lambda)^2$ exists over $[0; \frac{1-3\gamma}{\beta-3}]$; otherwise $O^*\big(n^6\big(\frac{n}{\frac{n-3\gamma n}{\beta-3}}\big) \cdot \big(\frac{n}{n-3\gamma n+3}\big) \cdot \big(\frac{(\frac{\beta-4+3\gamma}{\beta-3})n}{(\frac{1+3\gamma+\beta-3\gamma\beta}{\beta-3})n}\big)\big)$ (by Lemma 5, Lemma 6 and Lemma 7);*
- *Step 2.2 : $O\big(n^5\big(\frac{n}{\frac{n-3\gamma n}{\beta-3}}\big) \cdot \big(\frac{(1-\lambda)n}{(1-3\gamma+2\lambda)n}\big)\big)$ where λ is the unique real root of $(1-\lambda)(1-3\gamma+2\lambda)^2 - 27(\gamma-\lambda)^3$ over $[0; \frac{6\gamma-1}{5}]$ (by Lemma 5 and Lemma 8).*

We finally derive to the following corollary:

Corollary 1. *By setting $\gamma = 31/100$ and $\beta = 15$, Algorithm* **minCDS** *runs in $O(n^{20}1.8573^n)$ and exponential space.*

A Trade-Off between Polynomial and Exponential Terms. As shown in Theorem 4, the running-time of Algorithm `minCDS` depends on two parameters: γ and β. The parameter β has a direct influence on the order of the polynomial term which appears in the running-time. As well, we recall that the size of the family $\mathcal{F}$ of subsets also contributes to the running-time of the algorithm in [9]. Thus, by adequately tuning the parameters (i.e. with $\gamma = 0.30275$ and $\beta = 40000$) Theorem 4 shows that the algorithm runs in $O^*(1.8463^n)$. However the big-Oh notation hides a *huge* polynomial term of order n^β. In Table 2 we give some possible running-times achieve by our algorithm for several values of γ and β.

References

1. Alber, J., Bodlaender, H.L., Fernau, H., Kloks, T., Niedermeier, R.: Fixed parameter algorithms for dominating set and related problems on planar graphs. Algorithmica 33(4), 461–493 (2002)
2. Bodlaender, H.L., Lokshtanov, D., Penninkx, E.: Planar capacitated dominating set is $W[1]$-Hard. In: Chen and Fomin [3], pp. 50–60
3. Chen, J., Fomin, F.V. (eds.): 4th International Workshop on Parameterized and Exact Computation, IWPEC 2009, Copenhagen, Denmark, Revised Selected Papers, September 10-11. LNCS, vol. 5917. Springer, Heidelberg (2009)
4. Cygan, M., Pilipczuk, M., Wojtaszczyk, J.O.: Capacitated domination faster than $O(2^n)$. In: Kaplan, H. (ed.) SWAT 2010. LNCS, vol. 6139, pp. 74–80. Springer, Heidelberg (2010)
5. Downey, R.G., Fellows, M.R.: Parameterized Complexity. Springer, Heidelberg (1999)
6. Fomin, F.V., Grandoni, F., Kratsch, D.: Solving connected dominating set faster than 2^n. In: Arun-Kumar, S., Garg, N. (eds.) FSTTCS 2006. LNCS, vol. 4337, pp. 152–163. Springer, Heidelberg (2006)
7. Fomin, F.V., Iwama, K., Kratsch, D., Kaski, P., Koivisto, M., Kowalik, L., Okamoto, Y., van Rooij, J., Williams, R.: 08431 open problems moderately exponential time algorithms. In: Fomin, F.V., Iwama, K., Kratsch, D. (eds.) Moderately Exponential Time Algorithms. No. 08431 in Dagstuhl Seminar Proceedings, Schloss Dagstuhl - Leibniz-Zentrum fuer Informatik, Germany, Dagstuhl, Germany (2008), http://drops.dagstuhl.de/opus/volltexte/2008/1798
8. Fomin, F.V., Kratsch, D., Woeginger, G.J.: Exact (exponential) algorithms for the dominating set problem. In: Hromkovič, J., Nagl, M., Westfechtel, B. (eds.) WG 2004. LNCS, vol. 3353, pp. 245–256. Springer, Heidelberg (2004) c
9. Koivisto, M.: Partitioning into sets of bounded cardinality. In: Chen and Fomin [3], pp. 258–263
10. van Rooij, J.M.M., Nederlof, J., van Dijk, T.C.: Inclusion/exclusion meets measure and conquer. In: Fiat, A., Sanders, P. (eds.) ESA 2009. LNCS, vol. 5757, pp. 554–565. Springer, Heidelberg (2009)

Efficient Algorithms for Eulerian Extension

Frederic Dorn[1], Hannes Moser[2,*], Rolf Niedermeier[2], and Mathias Weller[2,**]

[1] Department of Informatics, University of Bergen, Norway
frederic.dorn@ii.uib.no
[2] Institut für Informatik, Friedrich-Schiller-Universität Jena, D-07743 Jena, Germany
{hannes.moser,rolf.niedermeier,mathias.weller}@uni-jena.de

Abstract. Eulerian extension problems aim at making a given (directed) (multi-)graph Eulerian by adding a minimum-cost set of edges (arcs). These problems have natural applications in scheduling and routing and are closely related to the CHINESE POSTMAN and RURAL POSTMAN problems. Our main result is to show that the NP-hard WEIGHTED MULTIGRAPH EULERIAN EXTENSION is fixed-parameter tractable with respect to the number k of extension edges (arcs). For an n-vertex multigraph, the corresponding running time amounts to $O(4^k \cdot n^3)$. This implies a fixed-parameter tractability result for the "equivalent" RURAL POSTMAN problem. In addition, we present several polynomial-time algorithms for natural Eulerian extension problems.

1 Introduction

Edge modification problems in graphs have many applications and are well-studied in algorithmic graph theory [4,14]. The corresponding minimization problems ask to modify as few (potential) edges as possible such that an input graph is transformed into a graph with a desired property. Most studies in this context relate to undirected graphs whereas we are aware of only few studies of "arc modification" problems on directed graphs (digraphs). One example in this direction is given by the NP-hard TRANSITIVITY EDITING problem, asking to make a digraph transitive by adding and deleting as few arcs as possible [18]. In this work, as part of a larger project on Eulerian graph modification problems, we study the problem of making a (directed) (multi-)graph Eulerian by edge (arc) additions.[1]

A (directed) (multi-)graph is called *Eulerian* if it contains an oriented cycle visiting every edge (arc) exactly once. An *Eulerian extension* is a set of edges (arcs) to add to a (directed) (multi-)graph so that it becomes Eulerian.

EULERIAN EXTENSION (EE)
Input: A (directed) graph $G = (V, E)$ and $\omega_{\max} \in \mathbb{N}$.
Question: Is there an Eulerian extension $\mathcal{E}$ for G with $|\mathcal{E}| \leq \omega_{\max}$?

---

* Supported by the DFG, project AREG (NI 369/9).
** Supported by the DFG, project DARE (NI 369/11).
[1] Here, following previous work, we call this "extension" problem. In the graph modification context, this is also known as "completion" or "addition" problem.

D.M. Thilikos (Ed.): WG 2010, LNCS 6410, pp. 100–111, 2010.
© Springer-Verlag Berlin Heidelberg 2010

Variants of EE include WEIGHTED EULERIAN EXTENSION (WEE), where an additional weight function $\omega : V \times V \to \mathbb{N}$ is given[2] and the sum of the weights of the arcs in the Eulerian extension we are looking for must not exceed $\omega_{\max}$, and the multigraph variants (where parallel arcs are allowed as input and output) MULTIGRAPH EULERIAN EXTENSION (MEE) and WEIGHTED MULTIGRAPH EULERIAN EXTENSION (WMEE), respectively. This work focuses on the latter problem, which has applications in scheduling [11]. Furthermore, the various applications of RURAL POSTMAN [7] carry over to WMEE since both problems are equivalent.

Related Problems and Previous Work. Lesniak and Oellermann [13] presented an overview of undirected Eulerian graphs. The unweighted and undirected extension problems for graphs and multigraphs were already discussed by Boesch et al. [3], who developed a linear time algorithm for the multigraph case and a matching based algorithm for the graph case. Recently, Höhn et al. [11] initiated a study of Eulerian extension problems applied to sequencing problems. To the best of our knowledge, WEE has not been considered in the literature so far.

EE is closely related to the well-known CHINESE POSTMAN problem [6] and the more general RURAL POSTMAN problem [7,12]. More specifically, RURAL POSTMAN and WMEE are "equivalent" (see Section 2 for details). With this equivalence, the NP-hardness of WMEE directly follows from the known NP-hardness result for RURAL POSTMAN [12]. Moreover, the fact that RURAL POSTMAN is solvable in polynomial time if the the set of required arcs is connected [10] directly implies that WMEE is solvable in polynomial time if the input is (weakly) connected.

Our Results. Our main achievement is to show that WMEE is fixed-parameter tractable with respect to the parameter "number of extension arcs"[3] denoted by k. The running time is $O(4^k \cdot n^3)$, where n denotes the number of vertices in the input multigraph and k denotes the number of additional arcs. Using the above-mentioned equivalence, this implies a first fixed-parameter tractability result for RURAL POSTMAN. In contrast to RURAL POSTMAN, whose unweighted variant is NP-hard [12], we can show that EE and MEE are polynomial-time solvable. Altogether, our work complements and extends known results for WMEE with restricted weight function [11] and RURAL POSTMAN, for which mainly approximation, heuristic, and some polynomial-time algorithms for special cases are known [7,10].

Due to the lack of space, several technical details are deferred to a full version of the paper.

---

[2] We assume the weight function to also assign weights to so far nonexistent arcs.

[3] Replacing each weight by a shortest-path-weight, much like in the proof of Lemma 2, decreases the number of arcs needed for an optimal Eulerian extension. It seems possible to extend our results to the corresponding stronger parameter "number of extension arcs after shortest-path-preprocessing". Further considerations in the direction are deferred to a full version of this paper.

2   Preliminaries and Basic Observations

The main focus of this work is on directed (multi-)graphs and, therefore, preliminaries for undirected (multi-)graphs are omitted if they follow trivially from the directed case. In the context of directed (multi-)graphs, connectivity always means weak connectivity, that is, connectivity of the underlying undirected graph. Let $G = (V, A)$ be a directed graph or multigraph (that is, a graph with parallel arcs allowed—we also use the letter M to refer to multigraphs). The set of connected components of G that are not isolated vertices is denoted by $\mathcal{C}_G$. In this work we sometimes apply definitions for graphs to connected components or sets of connected components. For example, we use $V(G)$ to refer to the vertices of the graph G and $V(C)$ to refer to the vertices of the connected component C. For a vertex set $V' \subseteq V$, let $G[V'] := (V', A \cap (V' \times V'))$ denote the directed (multi-)graph that is *induced* by V'. For an arc set $\mathcal{E}$ and some arc a, we abbreviate $\mathcal{E} \cup \{a\}$ to $\mathcal{E} + a$. If G is not a multigraph, then the *complement* $\overline{G}$ of G is the digraph on the vertex set V that contains exactly the arcs that are not in A. An *Eulerian cycle* in a directed (multi-)graph G is a (not necessarily simple) directed cycle that visits all arcs of G exactly once. If such a cycle exists, then we call G *Eulerian*. We call a (multi-)set $\mathcal{E} \subseteq V \times V$ an *Eulerian extension* for G if $(V, A \cup \mathcal{E})$ is Eulerian. Furthermore, $\mathcal{E}$ is called *optimal* if there is no Eulerian extension of less total weight for G. A *walk* W in G is a sequence of arcs of A such that each arc starts in the end vertex of the previous arc. Walks may also be considered as multisets of arcs. For a vertex v of a directed (multi-)graph G, the *outdegree* (*indegree*) of v, denoted by $\mathrm{outdeg}(v)$ ($\mathrm{indeg}(v)$), is the number of arcs in A that are outgoing of (incoming to) v. The *balance* of a vertex v is

$$\mathrm{bal}(v) := \mathrm{indeg}(v) - \mathrm{outdeg}(v).$$

Specifically, let $\mathcal{I}_G^+$ ($\mathcal{I}_G^-$) denote the set of vertices v of G for which $\mathrm{bal}(v) > 0$ ($b(v) < 0$), that is, $\mathrm{indeg}(v) > \mathrm{outdeg}(v)$ ($\mathrm{indeg}(v) < \mathrm{outdeg}(v)$). In an undirected graph, we define the balance $\mathrm{bal}(v)$ of a vertex v to be one if the number of its neighbors is odd and zero otherwise. For both directed and undirected (multi-)graphs G, vertices v of G with $\mathrm{bal}(v) = 0$ are called *balanced*, while all other vertices of G are called *imbalanced*, with $\mathcal{I}_G$ denoting the set of imbalanced vertices of G. With the concept of vertex balance, we can state a well-known fact about Eulerian graphs and multigraphs.

Lemma 1 (Folklore). *A (directed) (multi-)graph is Eulerian if and only if all edges (arcs) are in the same connected component and all vertices are balanced.*

Eulerian extension and Related Problems. In the most general problem that we study, we have weights and allow the input and output to be multigraphs.

> Weighted Multigraph Eulerian Extension (WMEE)
> **Input**: A directed multigraph $M = (V, A)$, a weight function $\omega : V \times V \to \mathbb{N}$, and positive integers k and $\omega_{\max}$.
> **Question**: Is there an arc multiset $\mathcal{E}$ with $|\mathcal{E}| \leq k$ and total weight at most $\omega_{\max}$ such that $(V, A \cup \mathcal{E})$ is an Eulerian multigraph?

Since multigraphs allow the presence of parallel arcs, we may also add arcs that are already present in the input. If we restrict the problem to digraphs, that is, we prohibit parallel arcs in both the input and the resulting digraph, then we arrive at the WEIGHTED EULERIAN EXTENSION problem (WEE). Both WMEE and WEE are also considered in their unweighted versions, where all arcs have weight one (and, hence, the extension set $\mathcal{E}$ may contain at most $\omega_{\max}$ arcs). Since being Eulerian is defined for both directed and undirected graphs, all presented variants of EULERIAN EXTENSION also have an undirected version.

Eulerian extensions are closely related to arc routing. An important role in this relation plays the following problem:

RURAL POSTMAN (RP)
Input: A digraph $G = (V, A)$, a nonempty set $R \subseteq A$ of "required" arcs, a weight function $\omega : A \to \mathbb{N}$, and integers q and $\omega_{\max} \geq 0$.
Question: Is there a closed walk W in G such that W visits all arcs in R and contains at most $q + |R|$ arcs whose total weight is at most $\omega_{\max}$?

If $R = A$, then RP degenerates to the also well-known CHINESE POSTMAN problem.

Parameterized Complexity. Our results are in the context of parameterized complexity, which is a two-dimensional framework for studying computational complexity [5,9,15]. One dimension is the input size n, and the other one is the *parameter* (usually a positive integer). A problem is called *fixed-parameter tractable* (fpt) with respect to a parameter k if it can be solved in $f(k) \cdot n^{O(1)}$ time, where f is a computable function only depending on k. A parameterized problem P_1 is *parameterized reducible* to a parameterized problem P_2 if P_1 can be reduced to P_2 in "fpt-time" such that the new parameter exclusively depends on the old parameter. If P_1 is parameterized reducible to P_2 and vice versa, then P_1 and P_2 are *parameterized equivalent*. If used as parameterized problems, all variants of EE are parameterized by the number k of allowed arcs in a solution and RP is parameterized by the number q of allowed *additional* arcs, that is, the number of arcs outside of R that are visited by the walk W. Note that for RP, q is a "stronger" parameter than the number of arcs in W, because it is always smaller. Since all solutions guarantee to contain R, choosing q can be considered an above-guarantee parameterization of RP.

Helpful Observations. We present observations that help us prove our results and give insights into the structure of the considered problems. First, observe that, over all vertices of a graph, the balance always adds up nicely, that is, for each "missing" incoming arc, there is also a "missing" outgoing arc.

Observation 1. *Let G be a directed (multi-)graph. Then, $\sum_{v \in V(G)} \mathrm{bal}(v) = 0$.*

In undirected graphs and multigraphs we can observe that the sum over all balances is even. Observation 1 can also be applied to connected components. Next, we note the relation between RP and WMEE.

Table 1. Polynomial-time solvable Eulerian extension problems. Here, n,m, and $\overline{m}$ are defined as in Section 3. In general, weighted variants of Eulerian extension problems are NP-hard if the input (multi-)graph is not connected [11,12].

	unweighted	weighted, connected
undir. graph	$O(\overline{m}\sqrt{n})$ (Theorem 1)	$O(n^3 \log n)$ (Corollary 2)
undir. multigraph	$O(n + m)$ (Proposition 4)	$O(n^3 \log n)$ (Corollary 2)
dir. graph	$O(\overline{m}^2 + n\overline{m} \log n))$ (Prop. 3)	$O(\overline{m}^2 + n\overline{m} \log n))$ (Prop. 2)
dir. multigraph	$O(n + m)$ (Proposition 4)	$O(n^3 \log n)$ (Corollary 1)

Proposition 1. RP *is parameterized equivalent to* WMEE.

This implies that the NP-hardness of RP for disconnected arc sets R carries over to WMEE for disconnected inputs. The basic idea is to let R be the arc set in the WMEE instance and identify an Eulerian extension with the set of additional arcs in a walk that visits all arcs in R.

3   Polynomial-Time Cases of Eulerian Extension

In this section, we present polynomial-time algorithms for various variants of Eulerian extension problems and their weighted versions. All running times are given as functions in n (the number of vertices in the input), m (the number of arcs (edges) in the input), and $\overline{m}$ (the number of arcs (edges) in the complement of the input). We refer to Table 1 for an overview of the results of this section. So far, the following result was known.

Theorem 1 ([3,13]). EULERIAN EXTENSION *on undirected graphs can be solved in* $O(\overline{m}\sqrt{n})$ *time.*

In the following, we present polynomial-time algorithms for weighted variants of Eulerian extension problems if the input (multi-)graph is connected. Then, we consider the unweighted variant and allow disconnected (multi-)graphs.

Algorithms for Connected Weighted Variants. Keeping in mind that the disconnected versions of WEE and WMEE are NP-hard [11] (see also [12]), we provide polynomial-time algorithms for both problems in case of connected inputs. Most algorithms are based on computing flows or matchings. First, we present an algorithm for digraphs, which is then modified to work for directed multigraphs and undirected graphs as well.

Proposition 2. WEIGHTED EULERIAN EXTENSION *on connected digraphs can be solved in* $O(\overline{m}^2 + n\overline{m} \log n))$ *time.*

Proof. Consider an instance $(G, \omega, \omega_{\max})$ for WEE, where G is a connected digraph, and a function $b : V(G) \to \mathbb{Z}$ measuring the balance of each vertex (see Section 2). Consider the flow network $\overline{G}$ with supply determined by b (negative supply indicates demand), arc capacity one for each arc, and arc-costs determined by ω. It is easy to see that a flow of value $\frac{1}{2}\sum_{v \in V} |\operatorname{bal}(v)|$ in this network corresponds to an Eulerian extension for G and, thus, the minimum cost of such a flow is also the minimum cost of an Eulerian extension for G. Such a flow can be computed in $O(\overline{m}^2 + n\overline{m}\log n))$ time.[4] $\square$

Next, for a directed multigraph M let G_M be the complete digraph (containing all possible arcs) on the vertex set of M. Analogously to the proof of Proposition 2, we can use a min-cost flow algorithm on G_M with arc capacities ∞ and weights according to ω to solve WEE on connected directed multigraphs M. The uncapacitated version of the min-cost flow algorithm (running in $O(n^3 \log n)$ time [2]) suffices in this case.

Corollary 1. WEIGHTED MULTIGRAPH EULERIAN EXTENSION *on connected directed multigraphs can be solved in* $O(n^3 \log n)$ *time.*

To handle undirected multigraphs, we replace the min-cost flow in the auxiliary graph G_M with a min-cost perfect matching in the complete undirected graph on the vertex set $\mathcal{I}_M$ with the weight of each edge $\{u, v\}$ equal to the weight of a minimum weight path between u and v in M. These paths are computed by an all-pairs shortest path algorithm. For each edge in the perfect matching, all edges of the corresponding shortest path are added to the extension set. With some effort we can show that the same algorithm can be used for WEE (assuming connected inputs).

Corollary 2. WEIGHTED (MULTIGRAPH) EULERIAN EXTENSION *on connected undirected graphs and multigraphs can be solved in* $O(n^3 \log n)$ *time.*

Algorithms for General Unweighted Variants. Since EE is a special case of WEE, we can solve EE for connected digraphs using the algorithm from the proof of Proposition 2 with a unit-weight version of the min-cost flow algorithm running in $O(\overline{m}^2)$ time.[5]

Corollary 3. EULERIAN EXTENSION *on connected directed graphs can be solved in* $O(\overline{m}^2)$ *time.*

This algorithm cannot handle multiple components. A more general algorithm that also allows to solve the problem on disconnected digraphs (at the cost of increased running time) will be presented in the full paper.

Proposition 3. EULERIAN EXTENSION *on disconnected digraphs can be solved in* $O(\overline{m}^2 + n\overline{m}\log n))$ *time.*

This stands in contrast with RP being NP-hard for unweighted digraphs [12], which seems to be due to the possibility to prohibit arcs by choosing the input

---

[4] See Exercise 10.17 of [2], a solution to which can be found in [1].

[5] Combine the solution found in [1] for Exercise 10.17 in [2] with breadth-first search as shortest path algorithm, which is valid in case of unit weights.

digraph. In fact, the subsequent Proposition 4 implies that unweighted RP is solvable in polynomial time if the input digraph is complete. More precisely, we can solve MEE for directed inputs by a straightforward greedy strategy much like the algorithm known for undirected multigraphs [3].

Proposition 4 (See [3]). MULTIGRAPH EULERIAN EXTENSION *on directed and undirected multigraphs can be solved in $O(n + m)$ time.*

4   Weighted Eulerian Extension on Directed Multigraphs

We prove that WMEE is fpt with respect to the size k of a solution by describing a dynamic programming algorithm to solve WMEE. More precisely, our algorithm computes the solution with smallest total weight over all solutions of size at most k. First, we modify the input, to obtain an equivalent but simpler instance. This preprocessing is described in the first paragraph. Next, we transform the preprocessed instance into an instance of a modified problem called BLACK/GRAY WEIGHTED MULTIGRAPH EULERIAN EXTENSION (BGWMEE). This problem has the advantage that a corresponding Eulerian extension has a particularly simple structure to be exploited by a dynamic programming algorithm. In the last paragraph, we present such an algorithm for BGWMEE.

Preprocessing the Input. We present two preprocessing algorithms that compute an equivalent instance in which (a) the balance of each vertex v is in $\{-1, 0, 1\}$, and (b) there are no isolated vertices. To achieve (a), we repeatedly find a vertex v with $|\operatorname{bal}(v)| > 1$ and split v into two vertices: one vertex v' with $|\operatorname{bal}(v')| = 1$ and one vertex v'' with $|\operatorname{bal}(v'')| < |b(v)|$. To achieve (b), we replace the weight of a direct connection between two vertices u and v with the weight of the cheapest path of potential arcs from u to v that visits only isolated vertices.

Lemma 2. *Let $(M, \omega, \omega_{max})$ be an instance of* WEIGHTED MULTIGRAPH EULERIAN EXTENSION *and let V_I denote the set of isolated vertices in M. Then, in $O(n^3)$ time, we can compute a weight function ω' such that $(M - V_I, \omega', \omega_{max})$ is equivalent to $(M, \omega, \omega_{max})$.*

Lemma 3. *Let $(M, \omega, \omega_{max})$ be an instance of* WEIGHTED MULTIGRAPH EULERIAN EXTENSION *and let $\mathcal{C}_M$ be the set of connected components of M. One can modify $(M, \omega, \omega_{max})$ in $O(k(n + m) + k^2)$ time to obtain an equivalent instance $(M', \omega', \omega_{max})$ such that $|\operatorname{bal}(u)| \leq 1$ for each vertex u in M'.*

Lemma 2 and Lemma 3 imply a preprocessing algorithm that removes all isolated vertices and assures $|\operatorname{bal}(v)| \leq 1$ for all vertices v in $O(n^3)$ time. In the following, we assume all inputs to be preprocessed in this way.

Transformation To BGWMEE. The following observation helps to picture Eulerian extensions as collections of paths between imbalanced vertices which is fundamental for the algorithm. The observation is based on the fact that for each balanced vertex u, each Eulerian extension contains as many arcs outgoing of u as arcs incoming to u.

Observation 2. *Let M be a directed multigraph and let $\mathcal{E}$ be an Eulerian extension of M. Then, $\mathcal{E}$ can be decomposed into a collection of paths that start at a vertex in $\mathcal{I}_M^+$ and end at a vertex in $\mathcal{I}_M^-$ or start and end at a balanced vertex.*

Observation 2 implies that an Eulerian extension $\mathcal{E}$ can be decomposed into paths. Our idea to attack WMEE is to use dynamic programming to construct such paths arc by arc. There are, however, a few obstacles to this approach. Assuming that each path visits a component of the input multigraph at most once, that is, no path contains two vertices of the same component, proved helpful in overcoming these obstacles. Since this is not always the case, we modify the input multigraph in order to use it in a slightly different (unfortunately more technical) problem, for which this assumption is valid.

BLACK/GRAY WMEE (BGWMEE)
Input: A directed multigraph $M = (V, A_{\text{black}} \cup A_{\text{gray}})$ where each connected component of (V, A_{black}) has either no imbalanced vertex or exactly two imbalanced vertices (one in $\mathcal{I}_M^-$ and one in $\mathcal{I}_M^+$), a weight function $\omega : V \times V \to \mathbb{N}$, and an integer $\omega_{\max} \geq 0$.
Question: Is there an Eulerian extension $\mathcal{E}'$ of total weight at most $\omega_{\max}$ for M such that in each component C_{black} of (V, A_{black}) there is exactly one start vertex of an arc in $\mathcal{E}'$ and exactly one end vertex of an arc in $\mathcal{E}'$ (that is, $|(V(C_{\text{black}}) \times V) \cap \mathcal{E}'| = 1$ and $|(V \times V(C_{\text{black}})) \cap \mathcal{E}'| = 1$)?

Again, we can decompose a black/gray Eulerian extension into paths analogously to Observation 2. The advantage of BGWMEE is that each black component is visited exactly once by such a path. The gray arcs (arcs in A_{gray}) are used to model the connectivity constraints given by the original WMEE instance. We first describe how WMEE can be solved using an algorithm for BGWMEE and then present such an algorithm in the next paragraph. The main idea is to transform an instance $(M, \omega, \omega_{\max})$ of WMEE into an instance $(M', \omega', \omega'_{\max})$ of BGWMEE by duplicating each component C of M as many times as it is visited by paths of a solution for $(M, \omega, \omega_{\max})$. To model that the copies of C originate from one connected component of M, the copies of C are connected by adding *gray* arcs. In the following, we describe the exact transformation algorithm.

First, find pairs of imbalanced vertices sharing a component. By Observation 1 and Lemma 3, there is a bijection $m : \mathcal{I}_M^- \to \mathcal{I}_M^+$ such that for all $v \in \mathcal{I}_M^-$, the vertices v and $m(v)$ are in the same component. We use an arbitrary bijection that respects this condition. Second, for a fix solution $\mathcal{E}$ of $(M, \omega, \omega_{\max})$ and all arcs $(u, v) \in \mathcal{E}$, make a copy of the component of M that contains u. In the following, we denote the number of copies of C by $\#(C)$. Since $\#(C)$ depends on $\mathcal{E}$, we do not know it in advance. Hence, we will try all possibilities. However, not all functions $\#$ are feasible: The total number of copies cannot exceed $|\mathcal{E}|$ $(= k)$ and since each copy has at most two imbalanced vertices, each component C must have at least $|\mathcal{I}_C|/2$ copies. Thus, we need only consider functions of the form $\# : \mathcal{C}_M \to \mathbb{N}^+$ with $\sum_{C \in \mathcal{C}_M} \#(C) \leq k$ and $\#(C) \geq |\mathcal{I}_C|/2$ for all $C \in \mathcal{C}_M$. It can be shown that there are at most 2^k such functions. Third, for each component C of M, assign a copy C' of C to each pair $(v, m(v))$ of imbalanced

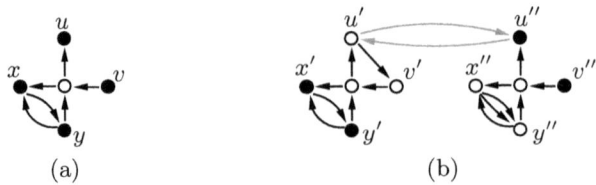

Fig. 1. (a): A component C of a directed multigraph. With $\#(C) = 2$, $m(v) = u$ and $m(y) = x$, the transformation $\mathrm{tr}_{\#}^{m}$ transforms (a) into (b). Here, white vertices are balanced, black vertices are imbalanced. Note that, (v', u') is fixed in the first copy and (y'', x'') is fixed in the second copy. Furthermore, u' and u'' are connected by gray arcs, according to the last step of the transformation.

vertices of C and "fix" C', that is, add an arc from the copy of $m(u)$ to the copy of u in C' for all $u \in \mathcal{I}_C^{-} - v$. This assures that each copy of C contains at most one pair of imbalanced vertices, and each pair of imbalanced vertices (paired by m) is represented in one copy. All copies that have not been assigned to an imbalanced pair are balanced completely in the above mentioned way. Fourth, for each component C of M, its copies are pairwisely connected by adding gray arcs. To this end, select a vertex v of each component of M and add all possible arcs between all copies of v. Note that only copies of the same component of M are connected by gray arcs. We denote the transformed instance by $(M', \omega', \omega_{\max}) := \mathrm{tr}_{\#}^{m}(M, \omega, \omega_{\max})$. See Figure 1 for an example of the described transformation.

An Algorithm for BGWMEE. Having transformed an instance of WMEE to an instance of BGWMEE using the algorithm presented in the previous paragraph, we can now exploit the simple structure of BGWMEE in a dynamic programming algorithm. The main idea in this algorithm is to construct an Eulerian extension arc by arc while maintaining a set of connected components of the input multigraph that have already been visited.

In the following, we describe a dynamic programming algorithm that solves BGWMEE. Let $(M, \omega, \omega_{\max})$ be an instance of BGWMEE and let $\mathcal{C}_M^{\text{black}}$ be the set of black connected components of M. For each subset $S \subseteq \mathcal{C}_M^{\text{black}}$ and each pair of vertices $u, v \in V(S)$, our algorithm computes an entry $\omega(S, u, v)$ with

$$\omega(S, u, v) = \begin{array}{l} \text{minimum weight } \omega(\mathcal{E}) \text{ of an arc set } \mathcal{E} \text{ such that } \mathcal{E} + (v, u) \\ \text{is a black/gray Eulerian extension for } M[V(S)]. \end{array} \quad (1)$$

If no black/gray Eulerian extension is possible with S, u, and v, then the entry $\omega(S, u, v)$ is assigned "∞". The set S represents a subgraph of M and the two vertices correspond to the endpoints of a (possibly "unfinished") path of an Eulerian extension (see Observation 2). The dynamic programming starts with computing the entries for sets S that contain exactly one component and augments S step by step, finally computing the entries for $S = \mathcal{C}_M^{\text{black}}$, which are used to derive a minimum weight black/gray Eulerian extension for M with respect to ω. In the following, we describe the update process for the entries.

For each $C \in \mathcal{C}_M^{\text{black}}$ not containing imbalanced vertices and each $u, v \in V(C)$, set

$$\omega(\{C\}, u, v) := \begin{cases} 0, & \text{if } u = v, \\ \infty, & \text{otherwise.} \end{cases}$$

This assignment is correct, that is, it satisfies (1) by setting $\mathcal{E} := \emptyset$ (which has obviously minimum weight) because adding an arc to a balanced component can only keep the component balanced if the added arc is a loop. Thus $\mathcal{E} + (v, u)$ is an Eulerian extension for $M[V(C)]$. Moreover, $\mathcal{E} + (v, u)$ is also a black/gray Eulerian extension for $M[V(C)]$ since the only connected component has exactly one incoming arc as well as one outgoing arc in $\mathcal{E}$ (in this case, the incoming and outgoing arc is (v, u)).

For each $C \in \mathcal{C}_M^{\text{black}}$ containing two imbalanced vertices $x \in \mathcal{I}_M^-$ and $y \in \mathcal{I}_M^+$, and each $u, v \in V(C)$, set

$$\omega(\{C\}, u, v) := \begin{cases} 0, & \text{if } u = x \text{ and } v = y, \\ \infty, & \text{otherwise.} \end{cases}$$

This assignment satisfies (1) since, by definition of black/gray Eulerian extension, x and y are the only imbalanced vertices of C and both are balanced adding (y, x) (that is, by using $\mathcal{E} = \emptyset$). For the same reasons as above, $\mathcal{E} + (v, u)$ is also a black/gray Eulerian extension for $M[V(C)]$.

Next, we describe the computation of the entries for larger sets S. When we compute the entry for a set S, we assume that all the entries for sets S' with $|S'| < |S|$ have already been computed. For a given $S \subseteq \mathcal{C}_M^{\text{black}}$ with $|S| > 1$, and vertices $u, v \in V(S)$, the entry $\omega(S, u, v)$ is computed as follows. Let $C \in S$ denote the black component of M that contains v and let $S' := S \setminus \{C\}$. If C is balanced, then distinguish the following three subcases:

1. If $u = v$ and there is a gray arc between C and S', then set

$$\omega(S, u, v) := \min_{u', v' \in V(S'), u' \neq v'} \omega(S', u', v') + \omega(v', u').$$

2. If $u \in V(S')$, then set

$$\omega(S, u, v) := \min_{w \in V(S')} \{\omega(S', u, w) + \omega(w, v)\}.$$

3. Otherwise, set $\omega(S, u, v) := \infty$.

If C contains two imbalanced vertices $x \in \mathcal{I}_M^-$ and $y \in \mathcal{I}_M^+$, then we distinguish the following three subcases:

1. If $u = x$ and $v = y$, and there is a gray arc between C and S', then set

$$\omega(S, u, v) := \min_{u', v' \in V(S'), u' \neq v'} \omega(S', u', v') + \omega(v', u').$$

2. If $u \in V(S')$ and $v = y$, then set

$$\omega(S, u, v) := \min_{w \in V(S')} \{\omega(S', u, w) + \omega(w, x)\}.$$

3. Otherwise, set $\omega(S, u, v) := \infty$.

Finally, the weight ω_{opt} of an optimal black/gray Eulerian extension for (M', ω) is computed as follows:

$$\omega_{\text{opt}} := \min_{u, v \in V(\mathcal{C}_M^{\text{black}}), u \neq v} \omega(\mathcal{C}_M^{\text{black}}, u, v) + \omega(v, u)$$

This follows immediately from (1). A corresponding black/gray Eulerian extension can be computed by storing each solution $\mathcal{E}$ in addition to its weight in each entry in the dynamic programming algorithm. Altogether, this algorithm takes $O(2^k \cdot n^3)$ time for solving a given instance.

Lemma 4. BLACK/GRAY WEIGHTED MULTIGRAPH EULERIAN EXTENSION *can be solved in $O(2^k \cdot n^3)$ time.*

The Complete Algorithm. The complete algorithm to solve WMEE runs in three steps. First, the input multigraph M is preprocessed in $O(n^3)$ time such that it does not contain isolated vertices or vertices with absolute balance more than one (see Lemma 2 and Lemma 3). Second, a component-respecting bijection $m : \mathcal{I}_M^- \to \mathcal{I}_M^+$ is chosen arbitrarily. Third, for all 2^k possible functions $\# : \mathcal{C}_M \to \mathbb{N}^+$, the instance is transformed and the resulting instance of BGWMEE is solved in $O(2^k \cdot n^3)$ time (see Lemma 4). The correctness of this algorithm follows directly from the correctness of the transformation algorithm and Lemma 4. The overall running time is $O(4^k \cdot n^3)$.

Theorem 2. WEIGHTED MULTIGRAPH EULERIAN EXTENSION *can be solved in $O(4^k \cdot n^3)$ time.*

Consequently, we can analogously solve RURAL POSTMAN parameterized by q (see Proposition 1).

Corollary 4. RURAL POSTMAN *can be solved in $O(4^q \cdot n^3)$ time.*

5   Conclusion

We focused on Eulerian extension problems (and due to equivalence, the RURAL POSTMAN problem), leaving yet unstudied other Eulerian graph modification problems including the editing version. Eulerian extension problems alone still offer a rich field of challenges for future research in terms of multivariate algorithmics [8,16]. More specifically, we concentrated on the parameterized complexity with respect to the parameter "number of extension arcs", but there are many natural structural parameters that make sense. For instance, it would be interesting to determine the parameterized complexity with respect to the parameter

"number of weakly connected components" in a WEIGHTED MULTIGRAPH EU-
LERIAN EXTENSION instance. In this context, Orloff [17] observed that "the
determining factor in the complexity of the problem seems to be the number
(c) of connected components in the required edge set"; Frederickson [10] noted
"the existence of an exact recursive algorithm that is exponential only in the
number of disconnected components." However, it is doubtful that this meant
fixed-parameter tractability with respect to c. In further future work, we also
want to study the undirected and non-multigraph versions of WMEE. Here, we
conjecture that similar algorithmic approaches may allow for similar results.

References

1. http://jorlin.scripts.mit.edu/Solution_Manual.html
2. Ahuja, R.K., Magnanti, T.L., Orlin, J.B.: Network Flows: Theory, Algorithms, and Applications. Prentice-Hall, Englewood Cliffs (1993)
3. Boesch, F.T., Suffel, C., Tindell, R.: The spanning subgraphs of Eulerian graphs. J. Graph Theory 1(1), 79–84 (1977)
4. Burzyn, P., Bonomo, F., Durán, G.: NP-completeness results for edge modification problems. Discrete Appl. Math. 154(13), 1824–1844 (2006)
5. Downey, R.G., Fellows, M.R.: Parameterized Complexity. Springer, Heidelberg (1999)
6. Eiselt, H.A., Gendreau, M., Laporte, G.: Arc routing problems part I: The chinese postman problem. Oper. Res. 43(2), 231–242 (1995)
7. Eiselt, H.A., Gendreau, M., Laporte, G.: Arc routing problems part II: The rural postman problem. Oper. Res. 43(3), 399–414 (1995)
8. Fellows, M.: Towards fully multivariate algorithmics: Some new results and directions in parameter ecology. In: Fiala, J., Kratochvíl, J., Miller, M. (eds.) IWOCA 2009. LNCS, vol. 5874, pp. 2–10. Springer, Heidelberg (2009)
9. Flum, J., Grohe, M.: Parameterized Complexity Theory. Springer, Heidelberg (2006)
10. Frederickson, G.N.: Approximation algorithms for some postman problems. J. ACM 26(3), 538–554 (1979)
11. Höhn, W., Jacobs, T., Megow, N.: On Eulerian extension problems and their application to sequencing problems. Technical Report 008, Combinatorial Optimization and Graph Algorithms, TU Berlin (2009)
12. Lenstra, J.K., Kan, A.H.G.R.: On general routing problems. Networks 6(3), 273–280 (1976)
13. Lesniak, L., Oellermann, O.R.: An Eulerian exposition. J. Graph Theory 10(3), 277–297 (1986)
14. Natanzon, A., Shamir, R., Sharan, R.: Complexity classification of some edge modification problems. Discrete Appl. Math. 113, 109–128 (2001)
15. Niedermeier, R.: Invitation to Fixed-Parameter Algorithms. Oxford University Press, Oxford (2006)
16. Niedermeier, R.: Reflections on multivariate algorithmics and problem parameterization. In: Proc. 27th STACS, IBFI Dagstuhl, Germany. LIPIcs, vol. 5, pp. 17–32 (2010)
17. Orloff, C.S.: On general routing problems: Comments. Networks 6(3), 281–284 (1976)
18. Weller, M., Komusiewicz, C., Niedermeier, R., Uhlmann, J.: On making directed graphs transitive. In: Dehne, F., et al. (eds.) Proc. 11th WADS. LNCS, vol. 5664, pp. 542–553. Springer, Heidelberg (2009)

On the Small Cycle Transversal of Planar Graphs

Ge Xia[1] and Yong Zhang[2]

[1] Department of Computer Science, Lafayette College, Easton, PA 18042
gexia@cs.lafayette.edu
[2] Department of Computer Science, Kutztown University, Kutztown, PA 19530
zhang@kutztown.edu

Abstract. We consider the problem of finding a k-edge transversal set that intersects all (simple) cycles of length at most s in a planar graph, where $s \geq 3$ is a constant. This problem, referred to as SMALL CYCLE TRANSVERSAL, is known to be NP-complete. We present a polynomial-time algorithm that computes a kernel of size $36s^3k$ for SMALL CYCLE TRANSVERSAL. In order to achieve this kernel, we extend the region decomposition technique of Alber et al. [*J. ACM, 2004*] by considering a *unique* region decomposition that is defined by shortest paths. Our kernel size is an exponential improvement in terms of s over the kernel size obtained under the meta-kernelization framework by Bodlaender et al. [*FOCS, 2009*].

Keywords: Parameterized Complexity, Kernelization, Planar Graphs, Cycle Transversal.

1 Introduction

Graphs without small cycles (or with large *girth*) are well studied objects in areas such as extremal graph theory [9,2] and graph coloring [18]. Finding a maximal subgraph without small cycles has applications in areas such as computational biology [12]. Raman and Saurabh [13] showed that several problems such as DOMINATING SET and t-VERTEX COVER that are hard for various parameterized complexity classes on general graphs become fixed parameter tractable (FPT) when restricted to graphs without small cycles. On planar graphs, Timmons [14] showed that every planar graph with girth at least nine can be star colored using 5 colors and every planar graph with girth at least 14 can be star colored using 4 colors.

Problem kernelization is a useful preprocessing technique in practically dealing with NP-hard problems. The *kernelization* of a parameterized problem is a reduction to a *problem kernel*, that is, to apply a polynomial-time algorithm to transform any input instance (x, k) to an equivalent reduced instance (x', k') such that $k' \leq k$ and $|x'| \leq g(k)$ for some function g solely dependent on k. We refer interested readers to [6,7] for more details on parameterized complexity and kernelization.

D.M. Thilikos (Ed.): WG 2010, LNCS 6410, pp. 112–122, 2010.
© Springer-Verlag Berlin Heidelberg 2010

In this paper we study the problem of finding a maximum subgraph without small cycles in a graph through edge deletions. Fix a constant $s \geq 3$. We call a cycle *small* if its length is at most s. A set S of edges in a graph G is called a *small cycle transversal set* if S intersects every small cycle in G. For simplicity, we refer to a small cycle transversal set of size k as a k-*transversal set*. We consider the following problem henceforth called SMALL CYCLE TRANSVERSAL: Given an undirected graph G and an integer k, is there a k-transversal set in G?

SMALL CYCLE TRANSVERSAL is known to be NP-complete on general graphs [17]. Kortsarz et al. [10] showed that the approximation ratio of 2 is likely the best possible for case $s = 3$, and they also presented $(s-1)$-approximation algorithms for the case when $s > 3$ is any odd number. Brügmann et al. [5] showed that SMALL CYCLE TRANSVERSAL remains NP-complete on planar graphs when $s = 3$. For $s = 3$ they gave data reduction rules to yield a kernel with $6k$ vertices for SMALL CYCLE TRANSVERSAL on general graphs and an $11k/3$ kernel on planar graphs. The proof by Brügmann et al. [5] for the NP-completeness of SMALL CYCLE TRANSVERSAL on planar graphs when $s = 3$ can be generalized to prove the NP-completeness of SMALL CYCLE TRANSVERSAL on planar graphs for any fixed $s \geq 3$ [16].

A multitude of problems have been shown to admit linear kernels on planar graphs using the so called *region decomposition* technique, which was first developed by Alber et al. [1] and was later generalized by Guo and Niedermeier [8]. All these previous results have recently been subsumed into a unifying meta-kernelization framework by Bodlaender et al. [4], which can be informally stated as follows: If a parameterized problem is *quasi-compact* and has *finite integer index* then it admits a linear kernel on graphs of bounded genus. Bodlaender et al. [4] proved that the problems known to have linear kernels from the previous results all satisfy *strong monotonicity* [4], which is a sufficient condition of finite integer index. Even though SMALL CYCLE TRANSVERSAL is not strongly monotone, it is not difficult to prove that it has finite integer index [15] (by an anonymous reviewer). Since SMALL CYCLE TRANSVERSAL is also quasi-compact, this implies that SMALL CYCLE TRANSVERSAL has a kernel of size linear in k on graphs of bounded genus. However, the size of the kernel could be superpolynomial in s.

The main contribution of this paper is a kernelization algorithm that computes a problem kernel of size $36s^3k$ for SMALL CYCLE TRANSVERSAL on planar graphs, which is a significant improvement in terms of s over the kernel size obtained under the meta-kernelization framework by Bodlaender et al. [4].

In order to obtain this kernel, we extend the region decomposition technique of Alber et al. [1]. We propose an *enhanced region decomposition* technique, in which the region decomposition is based on a special set of shortest paths called "witness-paths". This technique produces a *unique* region decomposition of the graph.

The rest of the paper is organized as follows. In Section 2 we give the necessary definitions and background. Section 3 contains several structural results that will be used in the design and analysis of the kernelization algorithm. Section 4 contains the kernelization algorithm and the proof of its correctness. In Section 5, we show that the size of the kernel produced by our algorithm is $36s^3k$.

Due to the lack of space, some proofs are omitted and the interested reader is referred to a technical report [15] that contains all the proofs.

2   Preliminaries

Fix an undirected simple plane graph $G = (V, E)$. A walk in G is a sequence $W = v_0v_1 \ldots v_l$ of vertices such that v_{i-1} and v_i are adjacent in G, $1 \leq i \leq l$. $\overleftarrow{W} = v_lv_{l-1} \ldots v_0$ denotes the reversal of W. We refer to the vertex set of W as $V(W) = \{v_0, \ldots, v_l\}$ and the edge set of W as $E(W) = \{(v_0, v_1), \ldots, (v_{l-1}, v_l)\}$. If $v_0 = x$ and $v_l = y$, we say that W connects x to y, and refer to W as an xy-walk, denoted by $W(xy)$. The vertices x and y are called the *ends* (or the *end points*) of the walk, x being its initial vertex and y being its terminal vertex, and the vertices $v_1, \ldots, v_{l-1}$ are its *internal vertices*. The length of W, denoted by $|W|$, is the number of edges in W. If u, v are two vertices in W and u precedes v in W, then we write $u \prec_W v$ and call the subsequence of W starting with u and ending with v the *subwalk* of W from u to v, denoted by $W(uv)$. If w is an internal vertex of $W(uv)$, we sometimes refer to $W(uv)$ as $W(uwv)$ to signify that $W(uv)$ contains w. For notational simplicity, we may also refer to $W(uv)$ as $W(uev)$ if $W(uv)$ contains an edge e. Let $W_1 = u_0 \ldots u_l$ and $W_2 = v_0 \ldots v_m$ be two walks. If $u_l = v_0$, then we can apply a *concatenation operation* $\circ$ to form a new walk $W = W_1 \circ W_2 = u_0 \ldots u_l(v_0) \ldots v_m$.

A *simple path* is a walk in which all vertices are distinct. All paths referred to in this paper are assumed to be simple. A *closed walk* is one whose initial vertex and terminal vertex are identical. A *cycle* is a closed walk that has no other repeated vertices than the initial and terminal vertices. The notations defined above on walks extend naturally to paths and cycles.

Let $\mathcal{W} = \{W_1, \ldots, W_l\}$ be a set of walks in G. The subgraph of G defined by $\mathcal{W}$ is $G_{\mathcal{W}} = (V(W_1) \cup \ldots \cup V(W_l), E(W_1) \cup \ldots \cup E(W_l))$. We say that $\mathcal{W}$ *contains a cycle* C if $G_{\mathcal{W}}$ contains C. Note that $|C| \leq |W_1| + \ldots + |W_l|$.

Let C be a cycle. Let e be an edge in C and u, v be two different vertices in C, where u precedes e and v succeeds e. We denote by $C(uev)$ the part of C between u and v that contains e and by $C(v\bar{e}u)$ the part of C between v and u that does not contain e. $C(uev)$ and $C(v\bar{e}u)$ are paths between u and v.

The following propositions are easy to verify.

Proposition 1 (proof in [15]). *Let W be a closed walk. If an edge e occurs only once in W, then W contains a cycle C and e is in C.*

Proposition 2 (proof in [15]). *If no edge occurs immediately after itself in a walk W, then either W contains a cycle, or W is a path.*

Proposition 3 (proof in [15]). *Let $P_1(uv)$ and $P_2(uv)$ be two different paths between u and v. Then the walk $W = P_1(uv) \circ \overleftarrow{P_2}(uv)$ contains a cycle.*

Let $P = u_0u_1 \ldots u_l$ and $Q = v_0v_1 \ldots v_m$ be two paths in G. We say that P and Q *cross* at a vertex w if $w = u_i = v_j$, $0 < i < l$, $0 < j < m$ and the subpaths $P(u_0w), P(wu_l), Q(v_0w)$ and $Q(wv_m)$ are all distinct. Note that our definition of two paths crossing not only includes crossing in the topological sense, i.e., the first path crosses from one side of the second path to the other side of the second path, but also includes the case where the paths merge at a vertex and diverge at a later vertex without changing sides.

Lemma 1 (proof in [15]). *Let $P(uv)$ and $Q(uv)$ be two paths between u and v. Suppose that $|P|, |Q| \le s - 1$. Then the following statements are true:*

1. *If P and Q cross at a vertex w, then $P \cup Q$ contains a small cycle.*
2. *If there are two vertices r, t such that $r \prec_P t$ and $t \prec_Q r$, then $P \cup Q$ contains a small cycle.*
3. *If there exists an edge $e = (r, t)$ such that r is in P and t is in Q, but e is neither in P nor in Q, then $P \cup Q \cup e$ contains a small cycle.*

For simplicity, we impose the condition that between any two vertices there is a unique shortest path. This condition can be easily achieved by a standard *perturbation technique* (see for example [3]): First assign a unit weight to each edge in G and then slightly perturb the edge weights such that no two paths have the same weight and that shorter paths have lower weights than longer paths. Note that the notion of *path weight* should not be confused with the previously defined notion of *path length* (the number of edges in a path). For this reason, we call a path of lower weight "lighter" instead of "shorter".

3   The Structural Results

In this section we present some structural results on *witness-paths* that will be used in both Section 4 and Section 5 that follow.

Definition 1. Let X be a set of vertices in G. A vertex $w \notin X$ is said to be *restricted* by X if w is contained in at least one small cycle and every small cycle containing w contains at least two vertices in X. Let Y be a set of vertices restricted by X. For every vertex $w \in Y$, define the *witness-path* of w with respect to X, denoted by P_w^X, to be the lightest path among all paths containing w with both ends in X. Since w is restricted by X, the witness-path P_w^X exists, is unique, and $|P_w^X| \le s - 1$. Let $\mathcal{P}_Y^X = \bigcup_{w \in Y} P_w^X$. We say that the set $\mathcal{P}_Y^X$ is "nice" if no two paths in $\mathcal{P}_Y^X$ induce a small cycle.

Lemma 2 (proof in [15]). *If $\mathcal{P}_Y^X$ is "nice", then no two paths P, Q in $\mathcal{P}_Y^X$ cross.*

Definition 2. If $\mathcal{P}_Y^X$ is "nice", then define $\mathcal{P}_Y^X(u, v)$ to be the subset of $\mathcal{P}_Y^X$ that consists of witness-paths whose ends are $\{u, v\}$, and define an auxiliary *directed* graph $\mathcal{D}_Y^X(u, v)$ to be the subgraph of G defined by $\mathcal{P}_Y^X(u, v)$, in which each edge is directed in the same direction as it appears in a path P in $\mathcal{P}_Y^X(u, v)$ with start vertex u.

Each edge in $\mathcal{D}_Y^X(u,v)$ will receive a unique direction because by Statement 2 of Lemma 1, each edge appears in the same direction in all paths in $\mathcal{P}_Y^X(u,v)$. The following lemma indicates that every directed path in $\mathcal{D}_Y^X(u,v)$ is contained in a witness-path.

Lemma 3 (proof in [15]). *Let $Q = v_0 \ldots v_l$ be a directed path in $\mathcal{D}_Y^X(u,v)$. Then there exists a path $P \in \mathcal{P}_Y^X(u,v)$ containing Q.*

Corollary 1 (proof in [15]). *$\mathcal{D}_Y^X(u,v)$ is a directed acyclic graph.*

4   A Kernelization Algorithm

In this section, we will present a kernelization algorithm for SMALL CYCLE TRANSVERSAL that runs in polynomial time. We will show in the next section that the algorithm produces a linear size kernel.

Let u, v be two vertices in G. We say that a vertex $w \notin \{u, v\}$ is *locked* by $\{u, v\}$ if w is restricted by $\{u, v\}$, and the witness-path of w with respect to $\{u, v\}$ has length greater than $s/2$, i.e., $|P_w^{\{u,v\}}| > s/2$. We say that an edge e is locked by $\{u, v\}$ if at least one of its ends is locked by $\{u, v\}$. A path $P(xy)$ between x and y is called a *locked path* of $\{u, v\}$ if $|P(xy)| \geq 2$ and every internal vertex w in $P(xy)$ is locked by $\{u, v\}$. A locked path is said to be *maximal* if x, y are not locked by $\{u, v\}$.

Let $X = \{u, v\}$ and Y be the set of vertices locked by $\{u, v\}$. Recall that by Definition 1, $\mathcal{P}_Y^{\{u,v\}} = \bigcup_{w \in Y} P_w^{\{u,v\}}$, where $P_w^{\{u,v\}}$ is the witness-path of w with respect to $\{u, v\}$. Since w is locked by $\{u, v\}$, we have $|P_w^{\{u,v\}}| > s/2$. Also recall that the length of any witness-path is at most $s - 1$, and thus $|P_w^{\{u,v\}}| \leq s - 1$. Also define the auxiliary directed graph $\mathcal{D}_Y^{\{u,v\}}$ based on $\mathcal{P}_Y^{\{u,v\}}$ as in Definition 2.

Lemma 4 (proof in [15]). *$\mathcal{P}_Y^{\{u,v\}}$ is "nice".*

Lemma 5 (proof in [15]). *Let u, v be two vertices in G. If G has a k-transversal set, then G has a k-transversal set that does not contain any edge locked by $\{u, v\}$.*

The above lemma shows that there is a k-transversal set that does not contain the locked edges and hence the locked edges can be pruned by the following kernelization algorithm, which consists of repeatedly applying the procedure **Reduce**(G) until the number of vertices in G cannot be further reduced.

Theorem 1 (proof in [15]). *The kernelization algorithm runs in $O(s^2 n^4)$ time.*

Lemma 6. *After **Reduce**(G) is applied, every remaining locked path $P(st)$ in $\mathcal{D}_Y^{\{u,v\}}$ is contained in a "selected" path.*

Proof. Proceed by induction on the length of P. If $|P| = 1$, the statement is obviously true. Let $P = v_1 \ldots v_{l-1} v_l$, and $P' = v_1 \ldots v_{l-1}$. By the inductive hypothesis, let P_1 be a "selected" path containing P', and let P_2 be a "selected"

Algorithm: **Reduce**(G)

1. Find a set B of vertices in G that are not contained in any small cycles; we call such vertices *baseless*. Remove B from G. Running a breadth-first search starting from a vertex v can determine whether v is baseless.
2. For every vertex v in G, find a set B_v of vertices that are baseless in $G - v$.
3. For every pair of vertices $\{u, v\}$, do the following:
   3.1. Let $Z_{u,v} = B_u \cap B_v$. Note that $Z_{u,v}$ is the set of vertices that are restricted by $\{u, v\}$.
   3.2. For every $w \in Z_{u,v}$, compute the witness-path $P_w^{\{u,v\}}$. If $|P_w^{\{u,v\}}| > s/2$, then w is locked by $\{u, v\}$; in this case, add w to the set Y of vertices locked by $\{u, v\}$ and add $P_w^{\{u,v\}}$ to the set $\mathcal{P}_Y^{\{u,v\}}$. For every w, the witness-path $P_w^{\{u,v\}}$ can be computed in $O(n^2)$ time using a min-cost max-flow algorithm [11, Lemma 3].
   3.3. For every path $P \in \mathcal{P}_Y^{\{u,v\}}$, if Q is a subpath of P and Q is a maximal locked path of $\{u, v\}$, then add Q to $\mathfrak{P}$, where $\mathfrak{P}$ is the set of maximal locked paths that are subpaths of paths in $\mathcal{P}_Y^{\{u,v\}}$. Group the paths in $\mathfrak{P}$ according to their end points. Mark the lightest one in each group as "selected".
   3.4. Remove all locked vertices in $\mathcal{P}_Y^{\{u,v\}}$ that are not contained in a "selected" path.

---

path containing (v_{l-1}, v_l). If P_1 contains (v_{l-1}, v_l) or P_2 contains P' then we are done. Otherwise since v_{l-1} has both incoming and outgoing edges in $\mathcal{D}_Y^{\{u,v\}}$, $v_{l-1} \notin \{u, v\}$. Therefore P_1 and P_2 cannot have v_{l-1} as an end vertex. This means that P_1 and P_2 cross at v_{l-1}. By Lemma 3, there are two paths in $\mathcal{P}_Y^{\{u,v\}}$ that contain P_1 and P_2, respectively. They will also cross, a contradiction to Lemma 2.   □

Lemma 7. *After* **Reduce***(G) is applied, there is at most one locked path between any two vertices in* $\mathcal{D}_Y^{\{u,v\}}$.

Proof. Let s, t be two vertices in $\mathcal{D}_Y^{\{u,v\}}$. Suppose that there are two locked paths P and Q between s and t. By Corollary 1, $\mathcal{D}_Y^{\{u,v\}}$ is a directed acyclic graph, P and Q must have the same direction. Without loss of generality, assume that $P(st)$ is lighter than $Q(st)$. By Lemma 6, Q is contained in a "selected" path Q'. Replacing $Q(st)$ by $P(st)$ in Q' yield a path Q'' lighter than Q' and hence Q' should not be marked as "selected", a contradiction.   □

Theorem 2. *The procedure* **Reduce***(G) is correct.*

Proof. Let G' be the subgraph of G obtained after **Reduce**(G) is applied. We will show that G has a k-transversal set if and only if G' has one. The only-if part is obvious because G' is a subgraph of G.

Now suppose that G' has a k-transversal set S'. By Lemma 5, we can assume that S' does not contain any edge locked by $\{u, v\}$. Suppose that G has a small cycle C that is not intersected by S'. C contains at least one edge e that was removed by **Reduce**(G). This means that e is locked by $\{u, v\}$ because only locked vertices are removed by **Reduce**(G) and the edges removed along with the locked vertices are locked edges. Thus C contains u and v. Let x be the last vertex preceding e in $C(uev)$ that is not locked. Let y be the first vertex succeeding e in $C(uev)$ that is not locked. Then $C(xey)$ is a maximal locked path. Since $|C(xey)| \leq s - 1$, by Statement 2 of Lemma 1, the edges in $C(xey)$ appear in the same direction as in $\mathcal{D}_Y^{\{u,v\}}$. This means that $C(xey)$ is a directed path in $\mathcal{D}_Y^{\{u,v\}}$. By Lemma 3, $C(xey)$ is a subpath of a path $P \in \mathcal{P}_Y^{\{u,v\}}$. This means that $C(xey) \in \mathfrak{P}$. There is a lightest path P' between x and y that is selected by **Reduce**(G). Thus $P' \neq C(xey)$ because e is removed by **Reduce**(G). $P' \leq |C(xey)|$ and P' is in G'.

Since P' and $C(xey)$ are directed paths in $\mathcal{D}_Y^{\{u,v\}}$, by Lemma 3, there are two paths in $\mathcal{P}_Y^{\{u,v\}}$ that contain P' and $C(xey)$, respectively. This means that P' and $C(xey)$ do not contain a small cycle because $\mathcal{P}_Y^{\{u,v\}}$ is "nice". But $C(y\bar{e}x)$ and $C(xey)$ form a small cycle. Hence $C(y\bar{e}x) \neq P'$ and $|C(y\bar{e}x)| < |P'| \leq |C(xey)|$. This means that $|C(y\bar{e}x)| < s/2$ because $|C(y\bar{e}x)| + |C(xey)| = s$. As a consequence, no vertex in $C(y\bar{e}x)$ is locked and hence $C(y\bar{e}x)$ is in G'. $P' \cup C(y\bar{e}x)$ contains a cycle and this cycle is small because $|P'| + |C(y\bar{e}x)| \leq |C(xey)| + |C(y\bar{e}x)| \leq s$. This small cycle is not intersected by S' because $C(y\bar{e}x)$ is not intersected by S' and P', being a locked path, is also not intersected by S'. Since both P' and $C(y\bar{e}x)$ are in G', we have a small cycle in G' that is not intersected by S', a contradiction to the fact that S' is a k-transversal set of G'.   □

5   A Linear Size Kernel

Let G be a plane graph in which the application of **Reduce**(G) does not further reduce its size. In this case, we call G a reduced graph. Suppose that G has a transversal set S, where $|S| \leq k$. For simplicity, we assume that S is minimal, i.e, for any edge $e \in S$, $S - e$ is not a transversal set. Let X be the set of the end points of the edges in S and let $Y = V(G) - X$. Note that Y is the set of vertices restricted by X. Recall that by Definition 1, $\mathcal{P}_Y^X = \bigcup_{w \in Y} P_w^X$, where P_w^X is the witness-path of w with respect to X, $|P_w^X| \leq s - 1$. If P_w^X is a path between two vertices $u, v \in X$, we say that w is (uniquely) *witnessed* by $\{u, v\}$. Since $\mathcal{P}_Y^X$ does not contain any edge in S, no two paths in it contain small cycles. This means that $\mathcal{P}_Y^X$ is "nice".

Definition 3. A region $R(u, v)$ between two vertices $u, v \in X$ is a closed subset of the plane whose boundary is formed by two paths $P, Q \in \mathcal{P}_Y^X(u, v)$ and whose interior is devoid of any vertex in X. A region is *maximal* if there is no region $R'(u, v) \supsetneq R(u, v)$. A *region decomposition* of G is a maximal set $\mathcal{R}$ of maximal regions between vertices in X whose interiors are pairwise disjoint.

Lemma 8. *Let w be a vertex in the interior of a region $R(u,v)$. Then any witness-path containing w is between u and v. Furthermore, w is witnessed by $\{u,v\}$.*

Proof. Let $Q(xwy)$ be a witness-path containing w, where $x, y \in X$ and $\{x,y\} \neq \{u,v\}$. Since Q connects w to a vertex outside of $R(u,v)$, Q must cross the boundary of $R(u,v)$ at a vertex $t \notin \{x,y\}$. Since Q has no vertices in X in its interior, $t \notin \{u,v\}$. This implies that Q crosses a witness-path on the boundary of $R(u,v)$, a contradiction to the fact that witness-paths in $\mathcal{P}_Y^X$ do not cross.

In particular, w's witness-path is between u and v, i.e. w is witnessed by $\{u,v\}$. $\qquad\square$

We say that two regions *cross* if their boundary paths cross.

Lemma 9. *Two regions do not cross.*

Proof. Since the boundaries of regions are witness-paths in $\mathcal{P}_Y^X$, they do not cross. $\qquad\square$

Corollary 2. *The number of maximal regions in a region decomposition is at most $6k$.*

Proof. Create an auxiliary graph $G_\mathcal{R}$ whose vertex set is X and each edge (u,v) in $G_\mathcal{R}$ corresponds to a maximal region between u and v. By [1, Lemma 5], $G_\mathcal{R}$ has at most $6k$ edges, which implies that the number of maximal regions is at most $6k$. $\qquad\square$

Let $\mathcal{P}_R$ be the set of witness-paths in the region $R(u,v)$. $\mathcal{P}_R \subseteq \mathcal{P}_Y^X(u,v)$. Let $\mathcal{D}_R$ be the subgraph of the auxiliary directed graph $\mathcal{D}_Y^X(u,v)$ defined in Definition 2, whose edges correspond to elements of $\mathcal{P}_R$. By Corollary 1, $\mathcal{D}_Y^X(u,v)$ is a directed acyclic graph and so is $\mathcal{D}_R$. By Statement 3 of Lemma 1, all edges in $R(u,v)$ are in $\mathcal{D}_R$ because otherwise, there is a small cycle that is not intersected.

Corollary 3. *Let P be an directed path in $D(u,v)$, then there is a witness-path that contains P.*

Proof. Implied by Lemma 3. $\qquad\square$

Lemma 10. *Let P be a path from u to v in $R(u,v)$. If $|P| \leq s-1$, then P is a witness-path.*

Proof. By Statement 2 of Lemma 1, each edge in P receives a direction in $D(u,v)$ that is consistent with the sequence of P. This means that P is a directed path in $D(u,v)$. By Corollary 3, P is a witness-path because the end points of P are in X. $\qquad\square$

Definition 4. *Let x,y be two vertices on the boundary of $R(u,v)$. Define a subregion $R^{sub}(x,y)$ to be a closed subset of $R(u,v)$ whose boundary is formed by two paths $P(xy), Q(xy)$, which are subpaths of $P, Q \in \mathcal{P}_R$ between u and v. A subregion is maximal if there is no subregion $R_1^{sub}(x,y) \supsetneq R^{sub}(x,y)$.*

Note that a subregion $R^{sub}(x, y)$ lies entirely in the interior of $R(u, v)$ except for x and y. Since paths in $\mathcal{P}_R$ do not cross, similar to Lemma 9 two subregions do not cross, although they can share vertices or edges on the boundaries.

Corollary 4. *Two subregions do not cross.*

The following proposition is needed for the proofs that follow.

Proposition 4. *Let H be a plane simple graph. Let $\mathcal{C}$ be a closed subset of the plane whose boundary is a cycle in H and whose interior is devoid of any vertex of H. Let E_1 be the set of edges of H in the interior of $\mathcal{C}$. Let E_2 be the set of edges on the boundary of $\mathcal{C}$. Then $|E_1| \leq |E_2| - 3$.*

Proof. Let F be the set of faces inside $\mathcal{C}$. Since each edge in E_2 appears in one face in F while each edge in E_1 appears in two faces in F, we have $3|F| \leq 2|E_1| + |E_2|$. Also observe that if $|E_1| = 0$ then $|F| = 1$ and each additional edge in E_1 increases $|F|$ by 1. Hence $|F| = |E_1| + 1$. Combining this with the above inequality, we have $|E_1| \leq |E_2| - 3$.   □

Lemma 11. *There are at most $2s - 3$ subregions in a region $R(u, v)$.*

Proof. First note that if x, y are two adjacent vertices on the boundary of $R(u, v)$, then there is no subregion between x and y because otherwise the edge (x, y) with a path of length at most $s-1$ in the subregion between x and y form a small cycle that is not intersected. There is at most one maximal subregion between a pair of non-adjacent vertices on the boundary of $R(u, v)$. If we replace every such pair of vertices on the boundary of $R(u, v)$ by an edge, then by Proposition 4, there are at most $2s - 3$ such edges. This implies that there are at most $2s - 3$ subregions in $R(u, v)$.   □

The following lemma shows that the subregions satisfy the *local property* that any small cycle involving a vertex in the interior of a subregion must pass through the two ends of the subregion.

Lemma 12 (proof in [15]). *Let $R^{sub}(x, y)$ be a subregion between x, y in a region $R(u, v)$. Then every vertex in the interior of $R^{sub}(x, y)$ is restricted by $\{x, y\}$.*

Lemma 13. *A subregion $R^{sub}(x, y)$ contains no more than $3s^2 - 5s$ vertices in its interior.*

Proof. In the interior of $R^{sub}(x, y)$, all vertices are restricted by $\{x, y\}$. Any vertex w in the interior of $R^{sub}(x, y)$ that is not locked by $\{x, y\}$ is contained in a path P between x and y of length at most $s/2$. All such vertices that are not locked by $\{x, y\}$ must appear in a single path P because otherwise there is a small cycle in $R^{sub}(x, y)$ that is not intersected. The path P, if it exists, divides $R^{sub}(x, y)$ into two smaller regions R_1^* and R_2^*, each with $3s/2$ vertices on its boundary. In the interior of each smaller region R_i^*, $i \in \{1, 2\}$, all vertices are locked by $\{x, y\}$ and they are contained in locked paths between pairs of non-adjacent vertices on the boundary of R_i^* (if such a path exists between two adjacent vertices on the boundary of R_i^*, then they form a small cycle that is not intersected). By Proposition 4, there are at most $3s/2 - 3$ pairs of vertices on the boundary of R_i^* that are connected by a locked path inside R_i^*. By Lemma 7, there is at most one locked path of length at most $s - 1$ between each of these pairs. Thus R_i^* contains at most $(3s/2 - 3)(s - 1)$ vertices in its interior, and $R^{sub}(x, y)$ contains no more than $2(3s/2 - 3)(s - 1) + s/2 \leq 3s^2 - 5s$ vertices in its interior. By a similar argument, if the path P does not exist in $R^{sub}(x, y)$, there are at most $(2s - 3)(s - 1) \leq 3s^2 - 5s$ vertices in its interior, for $s \geq 3$. □

Theorem 3. *Let G be a reduced graph. Then G has at most $36s^3 k$ vertices.*

Proof. Consider the region $R(u, v)$. By Lemma 11, there are at most $2s - 3$ subregions in $R(u, v)$, each of which has at most $3s^2 - 5s$ vertices in its interior. The boundaries of the subregions in $R(u, v)$ have at most $(2s-2)(2s-3)$ vertices. The boundary of $R(u, v)$ has at most $2s$ vertices. Hence there are at most $(2s - 3)(3s^2 - 5s) + (2s - 2)(2s - 3) + 2s \leq 6s^3 - 1$ vertices in $R(u, v)$ for $s \geq 3$. By Corollary 2, the number of maximal regions in a region decomposition is at most $6k$. Since every vertex not in X belongs to a maximal region and the set X has size $2k$, the problem kernel has size at most $(6s^3 - 1) \cdot 6k + 2k \leq 36s^3 k$, which is linear in k.   □

References

1. Alber, J., Fellows, M.R., Niedermeier, R.: Polynomial-time data reduction for dominating set. J. ACM 51(3), 363–384 (2004)
2. Alon, N., Bollobás, B., Krivelevich, M., Sudakov, B.: Maximum cuts and judicious partitions in graphs without short cycles. J. Comb. Theory Ser. B 88(2), 329–346 (2003)
3. Bley, A., Grötschel, M., Wessly, R.: Design of broadband virtual private networks: Model and heuristics for the B-WiN. In: Robust Communication Networks: Interconnection and Survivability. DIMACS Series, vol. 53, pp. 1–16. AMS, Providence (1998)
4. Bodlaender, H.L., Fomin, F.V., Lokshtanov, D., Penninkx, E., Saurabh, S., Thilikos, D.M.: (Meta) kernelization. In: FOCS CoRR, abs/0904.0727 (2009)
5. Brügmann, D., Komusiewicz, C., Moser, H.: On generating triangle-free graphs. Electronic Notes in Discrete Mathematics 32, 51–58 (2009)
6. Downey, R., Fellows, M.: Parameterized Complexity. Springer, Heidelberg (1999)
7. Guo, J., Niedermeier, R.: Invitation to data reduction and problem kernelization. SIGACT News 38(1), 31–45 (2007)
8. Guo, J., Niedermeier, R.: Linear problem kernels for NP-hard problems on planar graphs. In: Arge, L., Cachin, C., Jurdziński, T., Tarlecki, A. (eds.) ICALP 2007. LNCS, vol. 4596, pp. 375–386. Springer, Heidelberg (2007)
9. Hoory, S.: The size of bipartite graphs with a given girth. J. Comb. Theory Ser. B 86(2), 215–220 (2002)
10. Kortsarz, G., Langberg, M., Nutov, Z.: Approximating maximum subgraphs without short cycles. In: Goel, A., Jansen, K., Rolim, J.D.P., Rubinfeld, R. (eds.) APPROX and RANDOM 2008. LNCS, vol. 5171, pp. 118–131. Springer, Heidelberg (2008)

11. Krasikov, I., Noble, S.D.: Finding next-to-shortest paths in a graph. Inf. Process. Lett. 92(3), 117–119 (2004)
12. Pevzner, P., Tang, H., Tesler, G.: De novo repeat classification and fragment assembly. Genome Research 14(9), 1786–1796 (2004)
13. Raman, V., Saurabh, S.: Short cycles make W-hard problems hard: FPT algorithms for W-hard problems in graphs with no short cycles. Algorithmica 52(2), 203–225 (2008)
14. Timmons, C.: Star coloring high girth planar graphs. The Electronic Journal of Combinatorics 15(R124) (2008)
15. Xia, G., Zhang, Y.: On the small cycle transversal of planar graphs. Technical Report, http://www.cs.lafayette.edu/~gexia/research/sctrans.pdf
16. Xia, G., Zhang, Y.: Kernelization for cycle transversal problems. In: AAIM, pp. 293–303 (2010)
17. Yannakakis, M.: Node-and edge-deletion NP-complete problems. In: STOC 1978, pp. 253–264 (1978)
18. Zhu, J., Bu, Y.: Equitable list colorings of planar graphs without short cycles. Theor. Comput. Sci. 407(1-3), 21–28 (2008)

Milling a Graph with Turn Costs:
A Parameterized Complexity Perspective[*]

Mike Fellows[1],[**], Panos Giannopoulos[2],[***], Christian Knauer[3],
Christophe Paul[4],[†], Frances Rosamond[1],[‡], Sue Whitesides[5],[§], and Nathan Yu[6]

[1] PCRU, Office of DVC(Research), University of Newcastle, Australia
Michael.Fellows@newcastle.edu.au, mathgypsie@yahoo.com
[2] Institut für Informatik, Freie Universität Berlin, Berlin, Germany
panos@mi.fu-berlin.de
[3] Institut für Informatik, Universität Bayreuth, Bayreuth, Germany
christian.knauer@uni-bayreuth.de
[4] NRS - LIRMM, Montpellier, France
Christophe.Paul@lirmm.fr
[5] Department of Computer Science, University of Victoria, Canada
sue@uvic.ca
[6] nuo.nathan.yu@gmail.com

Abstract. The DISCRETE MILLING problem is a natural and quite general graph-theoretic model for geometric milling problems: Given a graph, one asks for a walk that covers all its vertices with a minimum number of *turns*, as specified in the graph model by a 0/1 turncost function f_x at each vertex x giving, for each ordered pair of edges (e, f) incident at x, the *turn cost* at x of a walk that enters the vertex on edge e and departs on edge f. We describe an initial study of the parameterized complexity of the problem.

1 Introduction

We study the parameterized complexity of the following problem:

DISCRETE MILLING

Instance: A simple graph $G = (V, E)$ and for each vertex x, a *turncost function* f_x indicating whether a *turn* is required, with $f_x : E(x) \times E(x) \to \{0, 1\}$, where $E(x)$ is the set of edges incident on x.

---

[*] Research initiated at the 6th McGill - INRIA Barbados Workshop on Computational Geometry in Computer Graphics, 2007.
[**] Research supported by the Australian Research Council through the ARC Centre of Excellence in Bioinformatics and Discovery Project DP0773331.
[***] Research supported by the German Science Foundation (DFG) under grant Kn 591/3-1.
[†] Research conducted while the author was on sabbatical at McGill University, School of Computer Science, Canada. Supported by the project ANR-06-BLAN-0148.
[‡] Research supported by the Australian Research Council.
[§] Research supported by NSERC and FQRNT.

D.M. Thilikos (Ed.): WG 2010, LNCS 6410, pp. 123–134, 2010.
© Springer-Verlag Berlin Heidelberg 2010

Question: Is there a walk making at most k turns that visits every vertex of G? The GRID MILLING problem restricts the input to *grid graphs*: rectilinearly plane-embedded graphs that are subgraphs of the integral grid, with the natural turncost function.

Related Work. DISCRETE MILLING was introduced by Arkin *et al.* [1] as a graph model for studying geometric milling problems with turn costs and other constraints. Such problems are common in manufacturing applications such as numerically controlled machining and automatic tool path generation; see Held [6] for a survey. In DISCRETE MILLING, a solution path must visit a set of vertices that are connected by edges representing the different directions ("channels") that the "cutter" can take. Arkin et al. studied a restricted version of the problem where incident edges to a vertex x are *paired* in the cost function f_x in the sense that for each incident edge e there is at most one incident edge f such that $f_x(e, f) = 0$, and symmetric: if $f_x(e, f) = 0$ then $f_x(f, e) = 0$. Here, we consider also a more general version that allows an arbitrary 0/1 turncost function at each vertex.

Arkin *et al.* [1] showed that DISCRETE MILLING is NP-hard (even for grid graphs) and described a constant-factor approximation algorithm for minimizing the number of turns in a solution walk. They also described a PTAS for the case where the cost is a linear combination of the length of the walk and the number of turns. No PTAS is known for the case of turn costs only.

Results. We start by showing that GRID MILLING is fixed-parameter tractable when parameterized by the numbers of turns. For this, two approaches are presented: one that is based on monadic second-order logic of graphs of bounded treewidth, and another, more practical one, based on dynamic programming on branch decompositions. Generalizing the former approach, we give an FPT result for DISCRETE MILLING, parameterized by (k, t, d), where k is the number of turns, t is the tree-width of the input graph G, and d is the maximum degree of G. We then explore whether this positive result can be further strengthened. However: DISCRETE MILLING, even in its restricted version, is $W[1]$-hard when parameterized by (k, p), where k is the number of turns and p is the path-width of G (and therefore also when parameterized by (k, t)). Our negative result provides one of the few problems known to be $W[1]$-hard when parameterized by pathwidth.

Definitions and Preliminaries. We will assume that the basic ideas of parameterized complexity theory and bounded tree-width algorithmics up through the basic form of Courcelle's Theorem and monadic second-order logic (MSO) are known to the reader. For background on these topics, see [3,5,7]. Details of routine deployments of MSO in the proofs of our theorems (that can be laborious in full formality) are relegated to the full version of the paper due to space limitations.

For a graph G, let tw(G) be its treewidth. We assume that all graphs G are simple (no loops or multiple edges). A *walk* $W = [x_0, \dots, x_l]$ on a graph $G = (V, E)$ is a sequence of vertices such that every pair x_i, x_{i+1} of consecutive vertices of the sequence are adjacent (we use $x_i x_{i+1}$ to refer to the edge between them). The *turn cost* of a walk W is defined as

$$\text{tc}(W) = \sum_{i=1}^{l-1} f_{x_i}(x_{i-1} x_i, x_i x_{i+1}).$$

A walk that visits every vertex of a graph is termed a *covering walk*. Note that in DISCRETE MILLING a solution covering walk may visit a vertex *many times*.

2   Grid Milling is Fixed-Parameter Tractable

We prove here that GRID MILLING is FPT for parameter k, the number of turns. We first argue that instances with large tree-width are no-instances.

Lemma 1. *Let $G = (V, E)$ be a connected grid graph with $tw(G) > 6k - 5$. Then G does not contain a $(k - 2)$-turn covering walk.*

Proof. We show that G contains k vertices that have pairwise different x- and y-coordinates. Then, any covering walk needs to take at least one turn between any two such vertices, and thus it needs at least $k - 1$ turns in total.

Since G is planar and $tw(G) > 6k - 5$, by the Excluded Grid Theorem for planar graphs (c.f. [5]), it has a $(k \times k)$-grid H as a minor. H contains $k/2$ vertex-disjoint consecutively nested cycles. Since taking minors can destroy or merge cycles but not create completely new ones, in the "pre-images" (under the operation taking minor) of these cycles there must be $k/2$ vertex-disjoint subgraphs in G, each containing a cycle. Thus, G contains a set $\mathcal{C}$ of $k/2$ nested vertex-disjoint cycles. Consider a straight line L of unit slope that intersects the innermost cycle of $\mathcal{C}$ at two vertices (grid points). L must also intersect every other cycle at at least two vertices. This produces a set of at least k vertices in G with the claimed property.

2.1   FPT *via* MSO

We give an MSO-based approach first, which also serves as a good starting point for our result on the general DISCRETE MILLING problem.

We associate to the grid graph G an annotated graph $\mathcal{M}(G)$: we simply regard the horizontal edges as being of one type, and the vertical edges as being of a second type. Equivalently, we can think of G as presented to us with a partition of the edge set: $G = (V, E_h, E_v)$. The idea is to show that the property of being a yes-instance of the problem can be expressed as an MSO property of $\mathcal{M}(G)$.

Intuitively, G has a k-covering walk if and only if there exist a start vertex v_0, turn vertices $v_1, \ldots, v_k$, and an end vertex v_{k+1}, and sets of vertices $S_0, \ldots, S_k$, such that:

(i) the graph induced by S_i, $i = 0, \ldots, k$, is a monochromatic path, i.e. a path whose edges are all either in E_h or in E_v,

(ii) the path induced by S_i starts at v_i and ends at v_{i+1}, and
(iii) $V = \cup S_i$, i.e. all vertices of G are covered.

This can be straightforwardly formalized in MSO.

Lemma 2. *Let $G = (V, E_h, E_v)$ be a grid graph. The property of having a k-covering walk on G is expressible in MSO.*

Easily, $\mathcal{M}(G)$ has treewidth bounded as a function of $\mathrm{tw}(G)$, and from Lemmata 1, 2 and Courcelle's Theorem we get:

Theorem 1. GRID MILLING *is FPT with respect to k (number of turns).*

2.2   Dynamic Programming on a Sphere Cut Decomposition

A *branch decomposition* (T, μ) of a graph G is an unrooted ternary tree T together with a bijection μ between the leaves of T and the edge set of G. For an edge e of T, let T_1, T_2 be the two subtrees obtained by removing e, and let G_1, G_2 be the subgraphs of G induced by the edges of the leaves of T_1 and T_2 respectively. The *middle set* of e is defined as $\mathrm{mid}(e) = V(G_1) \cap V(G_2)$. The *width* of a branch decomposition is the maximum size of a middle set over all edges. The *branchwidth* $\mathrm{bw}(G)$ of G is the minimum width over all possible branch decompositions.

A *noose* of a plane graph G is a simple closed curve on the plane that intersects G only at vertices and every face at most once. The noose separates the plane into two regions, which have the noose as a common boundary, and G into two subgraphs, each lying inside one of the regions; the subgraphs meet only at vertices on the noose. A *sphere cut (sc-) decomposition* (T, μ) is a branch decomposition with the property that for every edge e of T there is a noose of G such that its two corresponding subgraphs are the subgraphs G_1, G_2 associated with e. Note that the noose intersects G in $\mathrm{mid}(e)$. An sc-decomposition of a plane graph G (with no degree-1 vertices) of width $\mathrm{bw}(G)$ can be constructed in $O(n^3)$ time [2].

Let G be a grid graph (with no degree-1 vertices) and (T, μ) a sphere cut decomposition of G with minimum width. Since $\mathrm{tw}(G) = \Theta(\mathrm{bw}(G))$, from Lemma 1 we can assume that $\mathrm{bw}(G) = O(k)$. We root T as explained in [2]. For a node v of T, other than the root, let O_v be the noose corresponding to the edge between v and its parent. Let G_v be the subgraph of G that is associated with this edge and induced by the leaves of the subtree rooted at v, and let Δ_v be the region where G_v lies into. A covering walk of G induces a sequence of paths in Δ_v. Their union covers all vertices (of G_v) in the interior of Δ_v but not necessarily all its boundary vertices, i.e., the vertices on O_v. Note that any vertex or edge might be used more than once.

For a node v of T we define a table S_v of subproblems as follows: Let $C_v \subseteq O_v \cap V(G_v)$ be a set of boundary vertices and $D_v \subseteq E(G_v)$ be the set of edges with at least one endpoint in C_v. Also, let $Q_v = (e_1, e'_1, \ldots, e_l, e'_l)$ be a sequence of edges with $e_i \in D_v$ for $i = 1, \ldots, l$ and some $l \in \mathcal{N}^*$. We want to compute $S_v(C_v, Q_v)$, which is a set $\mathcal{P} = \{P_1, \ldots, P_{|Q_v|}\}$ of paths satisfying the following:

(i) The first (last) edge of P_i is e_i (e_i'),

(ii) P_i is contained in G_v,

(iii) the union of the paths in $\mathcal{P}$ covers all vertices in C_v along with all the interior vertices of G_v, and

(iv) the total number of turns of all paths in $\mathcal{P}$ is minimum.

We compute $S_v(C_v, Q_v)$ for all possible combinations of C_v and Q_v.

The dynamic program proceeds from the leaves to the root of T. Let x, y be the children of v. Observe that G_v is the union of the two subgraphs G_x, G_y drawn in the regions Δ_x, Δ_y (corresponding to the nooses O_x and O_y). Thus, a path in $S_v(C_v, Q_v)$ consists of a sequence of path segments in Δ_x and Δ_y. We compute $S_v(C_v, Q_v)$ by enumerating all possible combinations of $S_x(C_x, Q_x)$ and $S_y(C_y, Q_y)$ that form paths in Q_v. Note than when joining two paths an increase of the total turn cost by one might be needed.

Since $\mathrm{bw}(G) = O(k)$, there are $O(k)$ vertices on every noose and $2^{O(k)}$ possible subsets C_v. Since the degree of every vertex of G is at most 4, we have that $|D_v| = O(k)$. The crucial observation for bounding the size of Q_v is the following: any covering walk that uses an edge more than $k + 1$ times makes at least $k + 1$ turns. Thus, any edge of D_v cannot appear in Q_v more than $k + 1$ times, and so $|Q_v| = O(k^2)$. There are $O(2^k \cdot k^{k^2}) = O(2^{O(k^2 \log k)})$ possible sequences Q_v, thus each S_v has $O(2^k \cdot 2^{O(k^2 \log k)}) = O(2^{O(k^2 \log k)})$ entries, and since T has $O(n)$ nodes, the total time of the algorithm is $O(2^{O(k^2 \log k)})n$.

Theorem 2. GRID MILLING *can be solved in* $O(2^{O(k^2 \log k)}n + n^3)$ *time.*

3   Extending Tractability

What makes the GRID MILLING problem FPT? A few properties of grid graphs might lead us to tractable generalizations: (i) yes-instances must have bounded treewidth, (ii) vertices in grid graphs have bounded degree, and (iii) the turn-cost function is pairing and symmetric.

We are naturally led to three questions, by relaxing these conditions:

• What is the complexity of DISCRETE MILLING parameterized by (k, t, d), where k is the number of turns, t is a treewidth bound, and d is a bounded on maximum degree?

• What is the complexity of DISCRETE MILLING parameterized by (k, t)?

• What is the complexity of DISCRETE MILLING parameterized by (k, d)?

In the remainder of this paper, we answer the first two. The third question remains open.

Theorem 3. DISCRETE MILLING *is FPT for parameter* (k, t, d), *where k is the number of turns, t the tree-width of the graph G and d is the maximum degree of G.*

Proof. We describe how an instance of the DISCRETE MILLING problem, consisting of G and the turncost functions, can be represented by an annotated digraph $\mathcal{M}(G)$, that allows us to use MSO logic to express a property that corresponds to the question that the DISCRETE MILLING problem asks. The proof therefore consists of three parts: (1) a description of $\mathcal{M}(G)$, (2) necessary and sufficient criteria regarding $\mathcal{M}(G)$, for the instance of DISCRETE MILLING to be a yes-instance, and (3) the expression of these criteria in MSO logic.

Let $G = (V, E)$ be the graph of the DISCRETE MILLING instance. The vertex set of the digraph $\mathcal{M}(G)$ is $\mathcal{V} = \mathcal{V}_1 \cup \mathcal{V}_2$ where

$$\mathcal{V}_1 = \{l[v] : v \in V\} \quad \text{and} \quad \mathcal{V}_2 = \{t[e] : e \in E\} \cup \{t'[e] : e \in E\}$$

Intuitively (see Figure 1), we "keep a copy" of the vertex set V of G, mnemonically "$l[v]$" for v, as a vertex location we might be during a solution walk in G. Each edge e of G is replaced by two vertices $t[e]$ and $t'[e]$ that represent a "state" in a solution walk: traversing e in one direction or the other. In order to distinguish the directions, consider that the vertex set V of G is linearly ordered. Let $e = uv \in E$ with $u < v$ in the ordering. Our convention will be that $t[e]$ represents a traversal of e from u to v, and that $t'[e]$ represents a traversal of e in the direction from v to u. Thus each edge e of G is represented by two vertices in $\mathcal{M}(G)$.

In describing arcs of the digraph model $\mathcal{M}(G)$ we will use the notation $x \cdot y$ to denote an arc from x to y. The arc set of the digraph $\mathcal{M}(G)$ is

$$\mathcal{A} = \mathcal{A}_1 \cup \mathcal{A}_2 \cup \mathcal{A}_3 \cup \mathcal{A}_4 \cup \mathcal{A}_5$$

where

$$\mathcal{A}_1 = \{t[e] \cdot l[v] : e = \{u, v\} \in E \text{ with } u < v\}$$
$$\mathcal{A}_2 = \{l[u] \cdot t[e] : e = \{u, v\} \in E \text{ with } u < v\}$$
$$\mathcal{A}_3 = \{t'[e] \cdot l[u] : e = \{u, v\} \in E \text{ with } u < v\}$$
$$\mathcal{A}_4 = \{l[v] \cdot t'[e] : e = \{u, v\} \in E \text{ with } u < v\}$$

Let $\mathcal{A}'$ denote the union of these four sets of arcs. Intuitively, the arcs of $\mathcal{A}'$ just "attach" the vertices of the digraph that represent edges in G to the vertices of the digraph that represent the endpoints of the edge, so that the orientations of the arcs are compatible with the interpretation of a vertex of $\mathcal{V}_2$ as representing, say, a traversal of the edge uv in the direction from u to v; the vertex therefore has an arc *to it* from $l[u]$ and an arc *from it* to $l[v]$. An inspection of Figure 1 will help to clarify.

The arc set $\mathcal{A}_5$ is more complicated to write down formally. Its mission is to record the possibilities for cost-free passages through vertices of a solution walk in G. Suppose a is an arc in $\mathcal{A}'$. Then a is *to* or *from* either a $t[e]$ vertex, for some e, or a $t'[e]$ vertex for some e. Let $\epsilon(a)$ be defined to be this edge e of G. This is well-defined. We can then define

$$\mathcal{A}_5 = \{x \cdot y : x, y \in \mathcal{V}_2, \exists z = l[v] \in \mathcal{V}_1 \text{ and } \exists a, b \in \mathcal{A}'$$
$$\text{with } a = x \cdot z \text{ and } b = z \cdot y \text{ and } f_v(\epsilon(a), \epsilon(b)) = 0\}$$

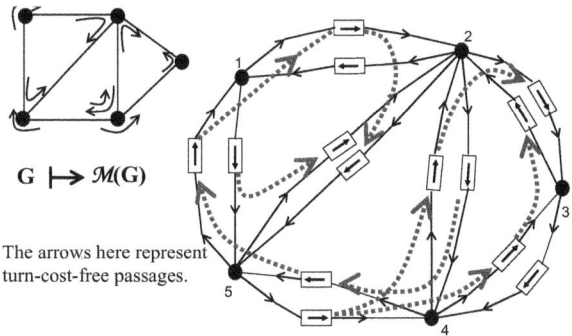

Fig. 1. The arrows drawn near the G vertices represent the turncost functions, indicating the zero-cost possibilities. These become arcs in the digraph $\mathcal{M}(G)$.

We regard $\mathcal{M}(G)$ as an annotated digraph, in the sense that there are two kinds of vertices, those of $\sqsubseteq_1$ and those of $\sqsubseteq_2$, and two kinds of arcs, those of $\mathcal{A}'$ and those of $\mathcal{A}_5$.

The rest of the proof will show that that the question of whether G admits a covering walk making at most k turns is represented by a property of the annotated digraph $\mathcal{M}(G)$ that can be expressed in MSO logic. However, before proceeding to that, it is important to verify that if the treewidth of G is bounded by t, then the treewidth of $\mathcal{M}(G)$ is bounded by a function of the parameter. This depends crucially on the fact that the maximum degree of G is part of our compound parameterization.

Suppose $\mathcal{T}(G)$ is a tree-decomposition of G of width at most t. We can describe a bounded width tree-decomposition $\mathcal{T}'$ of $\mathcal{M}(G)$ as follows. Without confusion, henceforth in this argument consider $\mathcal{M}(G)$ as an undirected graph by forgetting all arc orientations. Use the same bag-indexing tree for $\mathcal{T}'$ as for $\mathcal{T}(G)$. Suppose $B \subseteq V$ is a bag of $\mathcal{T}(G)$. Replace B with the union of the closed neighborhoods of the vertices of $\mathcal{V}_1$ corresponding to the vertices of B, in $\mathcal{M}(G)$. It is easy to check that all the axioms for a tree-decomposition hold, and that the treewidth of $\mathcal{M}(G)$ is therefore bounded by $2dt$.

In a digraph $D = (V, A)$, by a *purposeful set of arcs* (S, s, t) we refer to a set of arcs $S \subseteq A$, together with two distinguished vertices $s, t \in V$. We say that a purposeful set of arcs (S, s, t) is *walkable* if there is a directed walk W in D from s to t such that the set of arcs traversed by W (possibly repeatedly) is S.

Now consider how the information about G and its turncost functions is represented in $\mathcal{M}(G)$. A k-turn covering walk W in G that starts at a vertex s and ends at a vertex t is described by the information:

(1) a sequence of $k + 2$ vertices: $s = x_0, x_1, ..., x_{k+1} = t$, and
(2) a sequence of $k+1$ subwalks $W_0, ..., W_k$ where for $i = 0, ..., k$, W_i is a turncost-free walk from x_i to x_{i+1}, that has the property that every vertex of G is visited on at least one of the subwalks.

Let $\mathcal{D}(G)$ be the subdigraph of $\mathcal{M}(G)$ induced by the vertices of $\mathcal{V}_2$. A turncost-free walk in G corresponds to a directed walk in $\mathcal{D}(G)$, and vice versa, by the definition of $\mathcal{A}_5$.

Claim 1. G admits a k-turn covering walk if and only if:

(1) there are $k+2$ vertices $x_0, ..., x_{k+1}$ of $\mathcal{V}_1$ in $\mathcal{M}(G)$, and
(2) $k+1$ purposeful sets of arcs (S_i, s_i, t_i) in $\mathcal{D}(G)$, $0 \leq i \leq k$, such that
  (i) (S_i, s_i, t_i) is walkable in $\mathcal{D}(G)$ for $i = 0, \ldots, k$,
  (ii) there is an arc in $\mathcal{A}'$ from t_i to $x_{i+1} \in \mathcal{V}_1 = V$, and from x_i to s_i, for $i = 0, \ldots, k$,
  (iii) for every vertex $x \in \mathcal{V}_1 = V$, there is some index j, $0 \leq j \leq k$, and an arc $a = u \cdot v \in S_j$, such that there is an arc in $\mathcal{A}'$ in either direction, between x and u or v.

In one direction, the claim is easy: given a k-turn covering walk W in G, it is naturally factored into $k+1$ turncost-free subwalks W_i, and each traversal of an edge of G in a subwalk W_i corresponds in $\mathcal{M}(G)$ to a visit to a vertex of $\mathcal{V}_2$, thus the sequence of edge transversals of W_i in G corresponds 1:1 with a sequence $(y_0, ..., y_{m+1})$ of vertices of $\mathcal{V}_2$ in $\mathcal{M}(G)$. Because W_i is turncost-free, by the definition of $\mathcal{A}_5$, there is an arc in $\mathcal{D}(G)$ from y_i to y_{i+1} for $i = 0, ..., m$. We take the set of arcs to be S_i, $s_i = y_0$ and $t_i = y_{m+1}$, giving us (1) and (2) in a well-defined manner. It is straightforward to check that the conditions hold. For example, the assumption that W is a covering walk in G yields the last condition.

Conversely, suppose we have (1) and (2) in $\mathcal{M}(G)$. By the second condition, each S_i is walkable. By the definition of $\mathcal{A}_5$, a directed walk for S_i in $\mathcal{D}(G)$ corresponds to a turncost-free walk W_i in G. The third condition insures that the subwalks W_i in G can be sequenced into a k-turn walk W, where the turns occur at the vertices x_i by the first condition. W is covering in G by the fourth condition, yielding Claim 1.

Claim 2. Consider a digraph $D = (V, A)$ equipped with distinguished vertices s and t (allowing $s = t$). The property: *"there exists a directed walk from s to t that traverses (allowing repetition) every arc in A"* (that is, (A, s, t) is walkable) is expressible in MSO logic.

We first argue that (A, s, t) is walkable if and only if there is a directed path P in D from s to t, such that every arc $a \in A$ either is an arc of P, or belongs to a strongly connected subdigraph D_a that includes a vertex of P. We then argue (in Appendix A) that this property is expressible in MSO logic in a straightforward manner.

Given such a directed path $P = (s = x_0, ..., x_m = t)$ in D, we can describe a walk W that traverses every arc of A as follows. By the *arcs of P* we refer to the set of arcs

$$A[P] = \{x_0 x_1, x_1 x_2, ..., x_{m-1} x_m\}$$

The walk has m phases, one for each vertex x_i of the path P. Partition the arcs of $A - A[P]$ into m classes $A_0, ..., A_m$ where for $i = 1, ..., m$ every arc $a = uv \in A_i$

belongs to a strongly connected subdigraph D_a that includes the vertex x_i. Such a partition exists, by the supposed property of P. There is a directed path in D_a from x_i to u, and from v to x_i, by the strong connectivity of D_a, and so there is a directed cycle in D_a that includes both a and x_i. Include this cycle in W, starting from x_i and returning to x_i, for each arc $a \in A_i$. Increment i, take the arc from x_i to x_{i+1} and repeat this for $i = 0, ..., m$.

Now suppose that there is a directed walk W in D from s to t that traverses every arc in A. If there is a vertex v that is visited more than once, then we can find a shorter walk W' that, considered as a sequence of arc transversals, is a subsequence of the sequence of arc transversals of W. Therefore, by downward induction, there is a directed path P from s to t, with no repeated internal vertex visits, that considered as a sequence of arc transversals, is a subsequence of the sequence of arc transversals of W. But then, every arc a traversed in the walk W (that is, every arc $a \in A$), that is not an arc of P, must belong to a subwalk W' of W that begins and ends at a vertex of P. The vertices visited by W' therefore induce a strongly connected subdigraph containing a vertex of P.

The second part of the proof of Claim 2 is to argue the property we have identified is expressible in MSO logic. The first subtask is to describe an MSO predicate that expresses that there is a directed path P in D from s to t, quantified on the sets of vertices and arcs that form the path:

$$dipath(s, t) = \exists U (\subseteq V) \exists B (\subseteq A) : ...,$$

where the remainder of the predicate expresses that in the subdigraph $D' = (U, B)$:

- s has outdegree 1 and indegree 0
- t has indegree 1 and outdegree 0
- every vertex of U not s or t has indegree 1 and outdegree 1
- for every partition of U into U_1 and U_2 such that $s \in U_1$ and $t \in U_2$, there is a vertex $u \in U_1$ and a vertex $v \in U_2$ with an arc in B from u to v.

Being able to express that there is a directed path from s to t leads easily to an MSO predicate for strong connectivity of a subdigraph described by a set of vertices and a set of arcs. An MSO predicate for *walkability* of a set of arcs A relative to s and t is easily (but somewhat tediously) constructed on the basis of the structural characterization of *Claim 2*, using the predicates for the existence of an *s-t* path, and for strongly connected subdigraphs. An MSO formula to complete the proof of Theorem 3 is then trivial to construct by writing out *Claim 1* in the formalism.

4   Discrete Milling is Hard for Bounded Pathwidth

In this section, we see that the maximum degree restriction implicit in the parameterization for our positive result is key to tractability for this problem. In the restricted version of the DISCRETE MILLING problem the turncost functions are pairing and symmetric. This is a significant assumption, but the outcome is still negative, and the following result very much strengthens, in the parameterized setting, the NP-completeness result of Arkin *et al.* [1].

Theorem 4. DISCRETE MILLING *(with pairing and symmetric turn cost functions) $W[1]$-hard, with respect to (k, p), where k is the number of turns and p is a bound on pathwidth.*

Proof. The fpt-reduction is from MULTICOLOR CLIQUE, using an edge representation strategy, such as described, for example, in [4,8].

Suppose $G = (V, E)$ has V partitioned into color classes C_i, $i = 1, ..., r$. The MULTICOLOR CLIQUE problem asks whether G contains a r-clique consisting of one vertex from each color class C_i. We assume that each color class of G has size n [4]. The color-class partition of V induces a partition of E into $\binom{r}{2}$ classes $E_{\{i,j\}}$, for $1 \le i \ne j \le r$:

$$E_{\{i,j\}} = \{e \in E : \exists u \in C_i \text{ and } \exists v \in C_j \text{ with } e \text{ incident on } u \text{ and } v \}.$$

We can also assume that all these edge-partition classes $E_{\{i,j\}}$ have the same size m. We index the vertices and edges of G as follows:

$$C_i = \{v(i, q) : 1 \le q \le n\} \quad \text{for } i = 1, ..., r$$
$$E_{\{i,j\}} = \{e(\{i, j\}, l) : 1 \le l \le m\} \quad \text{for } 1 \le i \ne j \le r.$$

To refer to the incidence structure of G, we define functions $\pi^i_{\{i,j\}}(l)$ and $\pi^j_{\{i,j\}}(l)$ as follows:

$$\pi^i_{\{i,j\}}(l) = q : \quad \text{the edge } e(\{i, j\}, l) \text{ is incident on } v(i, q),$$
$$\pi^j_{\{i,j\}}(l) = q \quad \text{the edge } e(\{i, j\}, l) \text{ is incident on } v(j, q),$$

so the edge $e(\{i, j\}, l)$ is incident to $v(i, \pi^i_{\{i,j\}}(l))$ and $v(j, \pi^j_{\{i,j\}}(l))$.

We describe the construction of a graph G', together with the sets S_v of turn-free pairs of edges for the vertices v of G'. We first describe the vertices of G', and then specify a set of paths on these vertices. The edge set of the multi-graph G' is the (abstract) disjoint union of the sets of edges of these abstractly-defined paths, and it is understood that each path is turn-free, so that (for the most part), the sets S_v of turn-free pairs of v-incident edges for the vertices v of G' are implicit in these *generating paths* of G'.

The vertex set V' for G' is the union of the sets $V_0 \cup V_1 \cup V_2 \cup V_3 \cup V_4$,

$$V_0 = \{\sigma, \tau\}$$
$$V_1 = \{t[i, j] : 1 \le i \ne j \le r\}$$
$$V_2 = \{s[i, j] : 1 \le i \ne j \le r\}$$
$$V_3 = \{c[i, j, u] : 1 \le i \ne j \le r, \ 1 \le u \le n\}$$
$$V_4 = \{p[i, j, l] : 1 \le i \ne j \le r, \ 1 \le l \le m\}.$$

Thus $|V_1| = |V_2| = 2\binom{r}{2}$, $|V_3| = 2n\binom{r}{2}$ and $|V_4| = 2m\binom{r}{2}$.

The edge set of G' is (implicitly) described by a generating set of paths $\mathcal{P}$ (two paths for each edge of G), together with a few more edges:

$$\mathcal{P} = \{P[i, j, e(\{i, j\}, l)] : 1 \leq i \neq j \leq r, \ 1 \leq l \leq m, \},$$

where the path $P[i, j, e(\{i, j\}, l)]$ (1) starts at the vertex $p[i, j, l]$; (2) next visits $s[i, j]$; (3) then visits the vertices $c[i, j, u]$, except for $u = \pi^i_{\{i,j\}}(l)$ (the *exceptional vertex* of this block), in *consecutive order*, meaning that the vertices are visited by increasing index u, modified by skipping the exceptional vertex; (4) then visits the vertex $c[i, j^*, \pi^i_{\{i,j\}}(l)]$, where j^* is defined to be $j+1$, unless $j+1 = i$, when $j^* = j + 2$, or $j = r$ and $i \neq 1$, when $j^* = 1$, or $j = r$ and $i = 1$, when $j^* = 2$; and then (5) ends at the vertex $t[i, j]$.

Intuitively, there are two paths in $\mathcal{P}$ corresponding to each edge of G. If we fix i and consider that there are $r - 1$ blocks of vertices (each block consisting of n vertices, corresponding to the vertices of C_i), then what a path $P[i, j, e(\{i, j\}, l)]$ (corresponding to the lth edge of $E_{\{i,j\}}$) does is "hit" every vertex of its "own" $\{i, j\}^{th}$ block, except the vertex $c[i, j^*, \pi^i_{\{i,j\}}(l)]$ of the block corresponding to the vertex of C_i to which the indexing edge of G is incident, and in the "next block" in a circular ordering of the $r - 1$ blocks established by the definition of j^*, does the complementary thing: in this "next block" it hits *only* the vertex corresponding to the vertex of C_i to which the indexing edge is incident in G, and then ends at $t[i, j]$.

At this stage of the construction, the edges of G' are partitioned into (turn-free) paths that run between vertices of V_1 and vertices of V_4, where the latter have degree 1 (so far) and the vertices of V_1 have degree m (so far). We complete the construction of G' by adding a few more edges, specifying a few more turn-free pairs as we do so.

(A) Add edges between the pairs of vertices $p[i, j, l]$ and $p[j, i, l]$ for all $1 \leq i \neq j \leq r$ and $1 \leq l \leq m$. After these edges are added, we have reached a stage where all vertices in V_4 have degree 2 (and they will have degree 2 in G'). For each vertex of V_4 we make the pair of incident edges a turn-free pair.

Note that for any instance of the DISCRETE MILLING problem, the edge set is naturally and uniquely partitioned into maximal turn-free paths. At this stage of the construction, these paths all run between $t[i, j]$ and $t[j, i]$ for $1 \leq i < j \leq r$.

(B) Add some edges between the vertices of $V_0 \cup V_1$. Let $\leq_{lex}$ denote the lexicographic order on the set (of pairs of indices) $\mathcal{I} = \{[i, j] : 1 \leq i < j \leq r\}$. Let $[i, j]^*$ denote the immediate successor of $[i, j]$ in the ordering of $\mathcal{I}$ by $\leq_{lex}$. For $[i, j] \in \mathcal{I}$, let $rev[i, j] = [j, i]$. We add the edges (using the notation $u \cdot v$ for the creation of an edge between u and v):

- $t[rev[i, j]] \cdot t[[i, j]^*]$ for $1 \leq i < j \leq r$ and $[i, j] \neq [r, r - 1]$,
- $\sigma \cdot t[1, 2]$, and
- $t[r, r - 1] \cdot \tau$.

We do not specify any further turn-free pairs of vertex co-incident edges beyond (A) or implicit by being internal to the generating paths $\mathcal{P}$ of G'. That completes the description of G'.

To complete the proof, we need to show that: (1) the graph G' will admit a k-turn covering walk, where $k = 2\binom{r}{2}$, if and only if G has a multicolor r-clique; and (2) G' has path-width at most $6\binom{r}{2} + 4$. For reasons of space, the arguments will appear in the full version of the paper.

5   Open Problems

We have studied the parameterized complexity of (several versions of) the discrete milling problem with turn costs and gave an initial classification with respect to several parameterizations. We believe that there is good motivation to study "highly structured" graph problems, that is problems involving a graph together with "other information", since they are often able to engage applications better than simple graph problems. Our FPT results are impractical, but can they be improved? Our dynamic programming approach for GRID MILLING is a first step. In particular, it would be interesting to know if DISCRETE MILLING parameterized by (k, t, d) admits a polynomial kernel [7]. Our negative result provides one of the very few natural examples of a parameterized graph problem that is $W[1]$-hard, parameterized by pathwidth. Another notable open question is whether DISCRETE MILLING parameterized by (k, d), is FPT or $W[1]$-hard. Our suspicion is that it is $W[1]$-hard, but will require an even more elaborate reduction than the hardness result described here.

References

1. Arkin, E., Bender, M., Demaine, E., Fekete, S., Mitchell, J., Sethia, S.: Optimal covering tours with turn costs. SIAM J. Computing 35(3), 531–566 (2005)
2. Dorn, F., Penninkx, E., Bodlaender, H., Fomin, F.: Efficient exact algorithms on planar graphs: exploiting sphere cut branch decompositions. In: Brodal, G.S., Leonardi, S. (eds.) ESA 2005. LNCS, vol. 3669, pp. 95–106. Springer, Heidelberg (2005)
3. Downey, R.G., Fellows, M.R.: Parameterized Complexity. Springer, Heidelberg (1999)
4. Fellows, M., Fomin, F., Lokshtanov, D., Rosamond, F., Saurabh, S., Szeider, S., Thomassen, C.: On the complexity of some colorful problems parameterized by treewidth. In: Dress, A.W.M., Xu, Y., Zhu, B. (eds.) COCOA. LNCS, vol. 4616, pp. 366–377. Springer, Heidelberg (2007)
5. Flum, J., Grohe, M.: Parameterized Complexity Theory. Springer, Heidelberg (2006)
6. Held, M.: On the Computational Geometry of Pocket Machining. LNCS, vol. 500. Springer, Heidelberg (1991)
7. Niedermeier, R.: Invitation to Fixed Parameter Algorithms, vol. 31. Oxford University Press, Oxford (2006)
8. Szeider, S.: Not so easy problems for tree decomposable graphs. In: International Conference on Discrete Mathematics (ICDM), pp. 161–171 (2008) (invited talk)

Graphs that Admit
Right Angle Crossing Drawings

Karin Arikushi[1], Radoslav Fulek[2], Balázs Keszegh[2,3,*],
Filip Morić[2], and Csaba D. Tóth[1,**]

[1] University of Calgary
{karikush,cdtoth}@math.ucalgary.ca
[2] Ecole Polytechnique Fédérale de Lausanne
{radoslav.fulek,filip.moric}@epfl.ch
[3] Alfréd Rényi Institute of Mathematics
keszegh@renyi.hu

Abstract. We consider *right angle crossing (RAC) drawings* of graphs
in which the edges are represented by polygonal arcs and any two edges
can cross only at a right angle. We show that if a graph with n vertices
admits a RAC drawing with at most 1 bend or 2 bends per edge, then
the number of edges is at most $6.5n$ and $74.2n$, respectively. This is a
strengthening of a recent result of Didimo *et al.*

1 Introduction

The core problem in graph drawing is finding good and easily readable drawings
of graphs. Recent cognitive experiments [10,11] show that poly-line graph drawings with orthogonal crossings and a small number of bends per edge are just
as readable as planar drawings. Motivated by these findings, Didimo *et al.* [7]
studied the class of graphs which have a polyline drawing where crossing edges
meet at a right angle. Such a drawing is called a *right angle crossing* drawing,
or *RAC drawing*, for short.

The interior vertices of a polygonal arc are called *bends*. We say that a planar
representation of a graph is an RAC_b *drawing*, for some $b \in \mathbb{N}_0$, if the vertices
are drawn as points, the edges are drawn as polygonal arcs with at most b bends
joining the corresponding vertices, and any two polygonal arcs cross at a right
angle (and not at a bend). Let R_b, $b \in \mathbb{N}_0$, be the class of graphs that admit a
RAC_b drawing. It is clear that $R_b \subseteq R_{b+1}$ for all $b \in \mathbb{N}_0$. Didimo *et al.* [7] showed
that every graph is in R_3, hence $R_3 = R_b$ for all $b \geq 3$. They proved that every
graph with $n \geq 4$ vertices in R_0 has at most $4n - 10$ edges, and this bound is
best possible. They also showed that a graph with n vertices in the classes R_1
and R_2 has at most $O(n^{4/3})$ and $O(n^{7/4})$ edges, respectively.

Resuls. We significantly strengthen the above results, and show that every graph
with n vertices in R_1 and R_2 has at most $O(n)$ edges, and that the classes R_0,
R_1 and R_2 are pairwise distinct.

* Partially supported by grant OTKA NK 78439.
** Partially supported by NSERC grant RGPIN 35586.

D.M. Thilikos (Ed.): WG 2010, LNCS 6410, pp. 135–146, 2010.
© Springer-Verlag Berlin Heidelberg 2010

Theorem 1. *A graph G with n vertices that admits a RAC_1 drawing has at most $6.5n - 13$ edges.*

Theorem 2. *A graph G with n vertices that admits a RAC_2 drawing has less than $74.2n$ edges.*

We use two quite different methods to prove our main results. In Section 2, we use the so-called discharging method to prove Theorem 1. In Section 3, we define *block graphs* on the crossing edges, and use the Crossing Lemma to prove Theorem 2. Each method gives a linear bound for the number of edges for graphs in both R_1 and R_2, however, they would each give weaker constant coefficients for the other case (i.e., for R_2 and R_1, respectively).

We complement our upper bounds with lower bound constructions in Section 4. We construct graphs with n vertices in the classes R_1 and R_2 with $4.5n - O(\sqrt{n})$ and $7.8\dot{3}n - O(\sqrt{n})$ edges, respectively. Combined with Theorems 1 and 2, they show that $R_0 \neq R_1$ and $R_1 \neq R_2$.

Related Work. Angelini *et al.* [4] proved that every graph of maximum degree 3 admits a RAC_1 drawing, and every graph of maximum degree 6 admits a RAC_2 drawing. They also show that some planar directed graphs do not admit straight line *upward* RAC drawings.

A natural generalization of RAC drawings with straight line edges is given by Dujmović *et al.* [8]. They define α-*angle crossing* (αAC) drawings to be straight line graph drawings where every pair of crossing edges intersect at an angle at least α. In line with the results by Didimo *et al.* [7] on RAC drawings, they prove upper bounds on the number of edges for αAC graphs and give lower bound constructions. Specifically, they prove that the number of edges in an αAC graph is at most $(\pi/\alpha)(3n - 6)$ for $0 < \alpha < \pi/2$ and at most $6n - 12$ for $2\pi/5 < \alpha < \pi/2$. In addition, they give lower bound constructions based on the square and hexagonal lattices for $\alpha = \pi/k$, $k = 2, 3, 4, 6$. Di Giacomo *et al.* [6] also generalize RAC drawings in this way and call the minimum angle of any crossing the *crossing resolution*.

Preliminaries. The *crossing number* of a graph G, denoted $\mathrm{cr}(G)$, is the minimum number of edge crossings in a drawing of G in the plane. The Crossing Lemma, due to Ajtai *et al.* [3] and Leighton [12], establishes a lower bound for $\mathrm{cr}(G)$ in terms of n and m. The strongest known version is due to Pach *et al.* [13].

Lemma 1. *[13] Let G be a graph with n vertices and m edges. If $m \geq \frac{103}{6}n \approx 17.167n$, then*

$$\mathrm{cr}(G) \geq c \cdot \frac{m^3}{n^2}, \text{ where } c = \frac{1024}{31827} \approx 0.032. \tag{1}$$

Let G be a graph with n vertices and m edges, and let D be a RAC_b drawing of G. If there is no confusion, we make no distinction between the vertices (edges) of G and the corresponding points (polylines) of D.

A *plane (multi-)graph* is a (multi-)graph drawn in the plane without any edge crossings. The *faces* of a plane (multi-)graph are the connected components of the complement of G. Let D be a drawing of a graph $G = (V, E)$. A *rotation system* at a vertex $v \in V$ in drawing D is the (clockwise) circular order in which the edges leave v. A *wedge* at a vertex v in D is an ordered pair of edges (e, e') incident to v that are consecutive in its rotation system. A face f in D is *adjacent* to a wedge (e, e') if e, v, and e' are consecutive in a counterclockwise traversal of the boundary of f. Every wedge is adjacent to a unique face in D. The *size* of a face is the number of edges (counted with multiplicity) on the boundary of f.

2   RAC Drawings with One Bend per Edge

Discharging. We apply a discharging method reminiscent to that of Ackerman and Tardos [2] to prove Theorem 1. This method was apparently introduced by Wernicke [14], but it gained considerable attention only after it was extensively used in the first valid proof of the famous Four Color Theorem [5]. Since then, it was instrumental in deriving various types of results in structural graph theory, see e.g. [9]. Dujmović *et al.* [8] applied the discharging method for an alternative proof for the upper bound of $4n - 10$ on the number of edges in a graph on n vertices that admits a straight line RAC (i.e. RAC_0) drawing, originally due to Didimo *et al.* [7].

Proof (Theorem 1). Let $G = (V, E)$ be a graph in R_1. Fix a RAC_1 drawing D of G that minimizes the number of edge crossings. Partition G into two subgraphs $G_0 = (V, E_0)$ and $G_1 = (V, E_1)$, where $E_0 \subseteq E$ is the subset of crossing free edges and $E_1 \subseteq E$ is the subset of edges with at least one crossing. Since G_0 is planar, it has at most $|E_0| \leq 3n - 6$ edges.

Let C be the set of crossing points in D. We construct a plane multigraph $G' = (V', E')$ as follows: the vertices $V' = V \cup C$ are the vertices in V and all crossings in C; the edges are polygonal arcs between two consecutive vertices along the edges in E_1. That is, the edges in E' are obtained by subdividing the edges in E_1 at crossing points. Since the bends of edges in E_1 are not vertices in G', they are bends of some edges in E'. Denote by F' the set of faces of G'. A bend of an edge determines two angles: a *convex* and a *reflex* angle. We say that face $f \in F'$ is *adjacent* to a convex (resp. reflex) bend, if it has a convex (resp. reflex) interior angle at a bend point. A bounded face of size two is called a *lens*, and is adjacent to two parallel edges. A bounded face of size 3 is called a *triangle*.

Lemma 2. *Every lens $f \in F'$ is adjacent to a convex bend. If it is adjacent to exactly one convex bend, then it is incident to one vertex in C and V each, and adjacent to one convex bend and one reflex bend.*

Proof. Every lens $f \in F'$ is drawn as a simple polygon whose vertices are the incident vertices in V' and adjacent bends. Every simple polygon has at least 3 convex interior angles. A lens is incident to exactly two vertices in V', so it must have a convex interior angle at an adjacent bend.

Let $f \in F'$ be a lens adjacent to exactly one convex bend. Since every edge in E_1 crosses some other edges, no two adjacent vertices in V' are in V. At each vertex in C, the incident faces have $90°$ interior angles since D is a RAC drawing. If both vertices of lens f are in C with $90°$ interior angles, then f must have two convex bends. So, f is incident to one vertex in C and V each. If f has only one bend (see Fig. 1(a)), then we can redraw the edge $e \in E$ containing this bend in D with one fewer crossings (eliminating the crossing incident to f), which contradicts the choice of the RAC_1 drawing D. So f must be adjacent to a reflex bend (see Fig. 1(b)) as well.   □

Lemma 3. *Every triangle $f \in F'$, which is not the outerface, is adjacent to a convex bend.*

Proof. A triangle $f \in F'$ has three vertices in $V' = V \cup C$, and each of its three edges is a polygonal arc with 0 or 1 bends. Since every edge in E_1 crosses some other edges, no two adjacent vertices in V' are in V. That is, at least two vertices of f are in C, with an inner angle of $90°$. If f is adjacent to $k \in \{0, 1, 2, 3\}$ bends (at most one bend per edge), then f is a simple polygon with $k+3$ vertices, and so the sum of its interior angles is $(k+1)180°$. If all k bends are reflex, then the sum of interior angles would be more than $90° + 90° + k \cdot 180° = (k+1)180°$.   □

Lemma 4. *We have $|E_1| \leq 4n - 8$.*

Proof. Assume without loss of generality that $G' = (V', E')$ is connected. For a face $f \in F'$, let s_f be the size of f. For a vertex $v \in V' = V \cup C$, let d_v denote the degree of v in G'. We put a charge $\mathrm{ch}(v) = d_v - 4$ on each vertex $v \in V'$, and a charge $\mathrm{ch}(f) = s_f - 4$ on each face $f \in F'$. By Euler's formula the sum of all charges is

$$\sum_{v \in V'} \mathrm{ch}(v) + \sum_{f \in F'} \mathrm{ch}(f) = -8. \qquad (2)$$

Indeed, $\sum_{v \in V'}(d_v - 4) + \sum_{f \in F'}(s_f - 4) = 2|E'| - 4|V'| + 2|E'| - 4|F'| = -8$.
Since the charge at a vertex $v \in C$ is 0, we have

$$\sum_{v \in V} \mathrm{ch}(v) + \sum_{f \in F'} \mathrm{ch}(f) = -8. \qquad (3)$$

In what follows, we redistribute the charges in G' such that the total charge of all vertices and faces remains the same. The redistribution is done in two steps. In step 1, we move charges from some vertices to some faces; and in step 2 we move charges from some faces to some other faces. Our goal is to ensure that all faces have non-negative charges after the second step.

Step 1. For every edge $e \in E_1$ with one bend, we discharge $\frac{1}{2}$ unit from each of the two endpoints of e to the face adjacent to the convex bend of e. The new charge at every vertex $v \in V$ is $\mathrm{ch}'(v) \geq \frac{1}{2}d_v - 4$. Since every face in F' of size at least 4 receives a non-negative charge already at the beginning, it is enough to take care of the triangles and lenses (bounded faces of size 3 and 2), whose initial charge was -1 and -2, respectively.

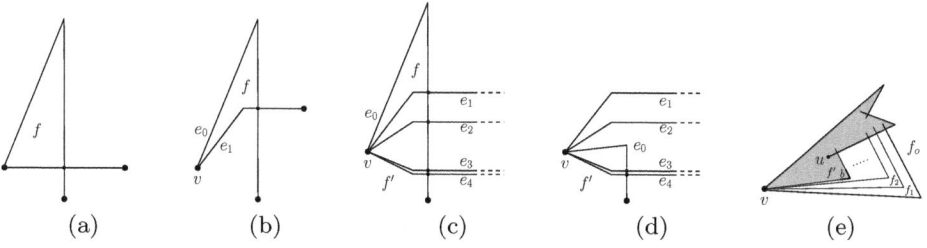

Fig. 1. (a) lens f that can be redrawn, (b) lens f having only one convex bend on its boundary, (c) situation when G could be redrawn with fewer crossings, (d) its redrawing ($i=4$), and (e) triangular outerface

By Lemma 3, each triangle $f \in F'$ except the outerface is adjacent to a convex bend, and so its charge has increased by at least 1 in step 1. Its new charge $\mathrm{ch}'(f)$ is at least 0. Similarly, if a lens $f \in F'$ is adjacent to two convex bends, then its charge after step 1 is 0. Hence, the only possible faces whose new charge is still negative are the outerface and the lenses adjacent to exactly one convex bend.

Step 2. In order to increase the charge of the outerface and lenses with exactly one convex bend from -1 to 0, we perform the second discharge step. Note that in the first step we have increased the charge of some faces of size 4 or higher (which was unnecessary), so we can now divert the "wasted" charge to faces with negative charge.

Let f be a lens with exactly one convex bend. By Lemma 2, f is incident to one vertex $v \in V$ and one in C, and it is adjacent to one convex bend and one reflex bend. Let $e_0, e_1, \ldots e_{d_v-1}$ (see Fig 1(c)) denote the edges in E_1 incident to v listed according to the rotation system at v (clockwise) such that the wedge (e_0, e_1) is adjacent to face f. We may assume without loss of generality that e_0 has a convex bend and e_1 has a reflex bend adjacent to f. Let $i \in \mathbb{N}$ be the smallest integer such that the wedge (e_i, e_{i+1}) is adjacent to the face, let us denote it by f', of size at least 4. It is easy to see that i is well-defined, since every edge in E_1 participates in a crossing.

We show that f' is adjacent to the convex bend of edge e_i. Any wedge (e_j, e_{j+1}), $1 \le j \le i - 1$, must be adjacent to a triangle bounded by parts of the edges e_j, e_{j+1}, and e_0. Since the (convex) bend of e_0 is adjacent to f, all these triangles are adjacent to a straight line portion of e_0. If any of these triangles is adjacent to the convex bend of e_j and a convex bend or no bend of e_{j+1}, then we can redraw edge e_0 to obtain a RAC_1 drawing of G with fewer crossings, eliminating the crossing incident to f (Figs. 1(c) and 1(d)). So the triangle at any wedge (e_j, e_{j+1}), $1 \le j \le i - 1$, is adjacent to the reflex bend of e_{j+1}. Hence f is adjacent to the convex bend of e_i.

Move 1 unit of charge (corresponding to the convex bend of e_i) from f' to f. This increases the charge of f to 0. Since the size of f' is at least 4, its charge remains non-negative. It is also clear that the charge corresponding to the convex bend of e_i is diverted to exactly one lens from f'.

It remains to make sure that the outerface f_o gets non-negative charge in the end as well. If f_o has a negative charge after Step 2, then it is triangle. It must have exactly one vertex v from V, otherwise three of its vertices are crossings each contributing $\frac{3}{2}\pi$ to the sum of the inner angles of the polygon which is the complement of the interior of f_o. Thus, f_o would have at least four bends, which is impossible. Moreover, f_o must be adjacent to three reflex bends, i.e. it looks like f_o on Figure 1(e). Then at least one of the inner faces adjacent to bends of f_o on edges incident to v is not a lens. Let f_1 denote such a face. If f_i is a triangle we define f_{i+1} as follows. Let f_{i+1} denote the face on the opposite side of the reflex bend of f_i. The definition of f_{i+1} is correct, since the sum of the interior angles in the grey polygon in Figure 1(e) is 4π. Eventually some $f_i = f'$ has at least four vertices and one unit of charge of the bend between f_i and f_{i-1} can be diverted to the outerface. The charge at b has not been moved in Step 2.

After the second step of redistribution, every face in D' has a non-negative charge. Let $\mathrm{ch}''(v)$ and $\mathrm{ch}''(f)$ denote the charge at each vertex $v \in V$ and $f \in F'$ after step 2. We have

$$|E_1| - 4n = \sum_{v \in V}\left(\frac{1}{2}d_v - 4\right) \leq \sum_{v \in V}\mathrm{ch}''(v) \leq \sum_{v \in V}\mathrm{ch}''(v) + \underbrace{\sum_{f \in F'}\mathrm{ch}''(f)}_{\geq 0} = -8.$$

By reordering the terms in the above inequality, we have $|E_1| \leq 4n - 8$, as required   □

At this point we have already proved that the number of edges in G is no more than $|E_0| + |E_1| \leq (3n - 6) + (4n - 8) = 7n - 14$.

We can improve this bound by applying Lemma 4 independently in each face of the plane graph $G_0 = (V, E_0)$, whose edges are the crossing-free edges in E. Notice that each edge in E_1 is fully contained in exactly one face of G_0. Let F_0 be the set of faces of G_0, and let d_f denote the number of vertices of a face $f \in F_0$. By Lemma 4, each face $f \in F_0$ contains at most $4d_f - 8$ edges of E_1, and it obviously contains no edges of E_1 if f is a triangle (i.e., $d_f = 3$). Summing this upper bound over all faces of G_0, we have

$$|E_1| \leq \sum_{f \in F_0, d_f > 3}(4d_f - 8). \tag{4}$$

Lemma 5. *If a plane graph $G_0 = (V, E_0)$ has n vertices and $3n - 6 - k$ edges, then*

$$\sum_{f \in F_0, d_f > 3}(4d_f - 8) \leq 8k. \tag{5}$$

Proof. Denote by $\tau(G_0)$ the sum on the left hand side of (5). We proceed by induction on k. For $k = 0$, the plane graph G_0 is a triangulation and $\tau(G_0) = 0$.

Assuming that the lemma holds for $k \geq 0$, we show that it holds for $k' = k+1$. Let G_0 be a plane graph with n vertices and $3n-6-k'$ edges. G'_0 can be obtained by removing an edge e from a plane graph G_0 with $3n - 6 - k$ edges, for which $\tau(G_0) \leq 8k$ by induction. If edge e is a bridge, then we have $\tau(G_0') = \tau(G_0) \leq 8k < 8k'$. Otherwise the removal of e merges two adjacent faces of G_0, say f_1 and f_2. If none of f_1 and f_2 is a triangle, then $4d_f - 8 = 4(d_{f_1} + d_{f_2} - 2) - 8 = (4d_{f_1} - 8) + (4d_{f_2} - 8)$, and so $\tau(G_0') = \tau(G_0) \leq 8k < 8k'$. If f_1 is a triangle and f_2 is a face of size more than three, then $4d_f - 8 = (4(d_{f_2} + 1) - 8) + 4$, and so $\tau(G_0') \leq \tau(G_0) + 4 \leq 8k + 4 < 8k'$. If both f_1 and f_2 are triangles, then $4d_f - 8 = 4 \cdot 4 - 8 = 8$, and $\tau(G_0') \leq \tau(G_0) + 8 \leq 8k + 8 = 8k'$. This completes the induction step, hence the proof of Lemma 5. $\qquad\square$

We have two upper bounds for m, the number of edges in G. Lemma 4 gives $m \leq |E_0| + |E_1| \leq (3n - 6 - k) + (4n - 8) = 7n - k - 14$, and Lemma 5 gives $m \leq |E_0| + |E_1| \leq (3n - 6 - k) + 8k = 3n + 7k - 6$. Therefore, we have $m \leq \max_{k \in \mathbb{N}_0} \min(7n - k - 14, 3n + 7k - 6) = 6.5n - 13$, which is attained for $k = n/2 - 1$. This completes the proof of Theorem 1. $\qquad\square$

3   RAC Drawings with Two Bends per Edge

Block graphs. The main tool in the proof of Theorem 2 is the block graph of a RAC drawing and the Crossing Lemma. Let D be a RAC_2 drawing of a graph $G = (V, E)$. Every edge is a polygonal arc that consists of line segments. Without loss of generality, we assume that every edge has two bends so that each edge has two *end segments* and one *middle segment*. A *block* of D is a connected component in the union of pairwise parallel or orthogonal segments in D. Formally, we define a binary relation on the segments in the polygonal arcs in the drawing D: two segments are related if and only if they cross. The transitive closure of this relation is an equivalence relation. We define a block of D as the union of all segments in an equivalence class. Since the union of crossing edges is connected, every block is a connected set in the plane. Furthermore, all segments in a block have at most two different (and orthogonal) orientations.

By Lemma 1, if $m \geq \frac{103}{6} n$, then the average number of crossings per segment is at least

$$\frac{2c}{3} \cdot \frac{m^2}{n^2},$$

where $c = 1024/31827 \approx 0.032$. We say a segment is *heavy* if it crosses at least $\beta c \frac{m^2}{n^2}$ other segments, where $0 < \beta < 2/3$ is the *heaviness parameter* specified later. A block is *heavy* if it contains a heavy segment.

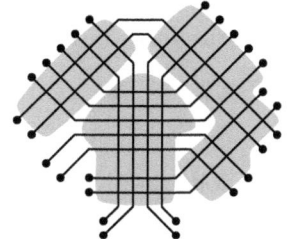

Fig. 2. A RAC_2 drawing of a graph and its heavy blocks

We define the *block graph* $B(D)$ as a bipartite multigraph whose two vertex classes are the vertices in V and the heavy blocks in D. The block graph has an edge between a vertex $v \in V$ and a heavy block for every segment incident to v and contained in the heavy block (Fig. 2). Note that if a heavy block consists entirely of middle segments, it is not adjacent to any vertex in $B(D)$.

Lemma 6. *If D is a RAC drawing of a graph, then the block graph $B(D)$ is planar.*

Proof. Recall that a heavy block u is a connected set which is incident to all vertices of G that are adjacent to u in $B(D)$. For every heavy block u, let $T_u \subseteq u$ be a spanning tree of the incident vertices of G. We can construct a planar embedding of $B(D)$. The vertices of G are represented by the same point as in D. Each heavy block u is represented by an arbitrary point r_u in the relative interior of T_u. If vertex v of G is adjacent to a heavy block u, then connect v and r_u by a Jordan arc that closely follows the shortest path between v and r_u in the tree $T_u \subseteq u$. Since shortest paths in a tree do not cross, we can draw the edges successively without crossings.   □

Denote by H the number of heavy blocks in D. The block graph $B(D)$ is bipartite and planar, with $H + n$ vertices. If it is *simple*, then it has at most $2(H + n) - 4$ edges. However, $B(D)$ is not necessarily simple: up to four segments of a heavy block may be incident to a vertex v in D.

Lemma 7. *The block graph $B(D)$ has less than $2H + 5n$ edges.*

Proof. Assume that two segments in a heavy block are incident to the same vertex v. Since the block is connected, there is a closed curve γ passing through v and the two segments such that all other blocks lie either in the interior or in the exterior of γ. Hence, multiple edges cannot interleave in the rotation order of a vertex v. Note also that segments in a block are pairwise parallel or orthogonal. It follows that $B(D)$ becomes a *simple* bipartite plane graph after removing at most 3 duplicate edges at each vertex of D. That is, after removing up to $3n$ edges, the remaining simple bipartite plane graph has at most $2(H + n) - 4$ edges.   □

Let S denote the number of segments that participate in some heavy block of D. Every heavy block contains at least one heavy segment and all other segments it crosses. That is, a heavy block contains more than $\beta cm^2/n^2$ segments. Since every segment belongs to a unique block, we have

$$H \leq \frac{S}{\beta cm^2/n^2} = \frac{Sn^2}{\beta cm^2}. \tag{6}$$

The following lemma reformulates the Crossing Lemma for heavy segments in RAC_2 drawings. We show that if a graph G has sufficiently many edges, then a constant fraction of them must contain a segment in some heavy blocks in a RAC_2 drawing of G.

Lemma 8. *Let D be a RAC_2 drawing of graph G with $m \geq \frac{103}{6}\sqrt{2/(3\beta)}n$ edges. If one can delete xm edges from D, for some $0 < x < 1$, such that every remaining edge segment crosses less than $\beta cm^2/n^2$ others, then $x > 1 - \sqrt{3\beta/2}$.*

Proof. Suppose xm edges were deleted from D to obtain D' and let G' be the graph associated with D'. The number of remaining edges is $|E(G')| = m - xm = (1-x)m$. If $(1-x)m \geq \frac{103}{6}n$, then the Crossing Lemma gives $\mathrm{cr}(G') \geq c \cdot \frac{(1-x)^3 m^3}{n^2}$, so the average number of crossings per segment in G' is at least

$$\frac{2\mathrm{cr}(G')}{3(1-x)m} \geq \frac{2c}{3} \cdot \frac{(1-x)^2 m^2}{n^2}.$$

Every segment in D' crosses less than $\beta c m^2/n^2$ others in D'. Comparing the upper and lower bounds for the average number of crossings per segment, we have

$$\frac{2c}{3} \cdot \frac{(1-x)^2 m^2}{n^2} < \beta c \frac{m^2}{n^2} \ \Rightarrow\ (1-x)^2 < 3\beta/2 \ \Rightarrow 1 < x + \sqrt{3\beta/2}.$$

If, however, $(1-x)m < \frac{103}{6}n$ but $m \geq \frac{103}{6}\sqrt{2/(3\beta)}n$, then we have again $x > 1 - \sqrt{3\beta/2}$. $\qquad\square$

Lemma 8 immediately gives a lower bound on S, the number of segments participating in heavy blocks.

Lemma 9. *Let D be a RAC_2 drawing of graph G. If $m \geq \frac{103}{6}\sqrt{2/(3\beta)}n$, then $S > (1 - \sqrt{3\beta/2})m$.*

Proof. Let E_1 be the set of edges containing a segment that participate in some heavy block in D. Clearly, we have $|E_1| \leq S$. If all edges of E_1 are deleted from D, then every remaining segment crosses less than $\beta c m^2/n^2$ others. By Lemma 8, we have $S \geq |E_1| > (1 - \sqrt{3\beta/2})m$. $\qquad\square$

Proof (Theorem 2). We set the heaviness parameter to $\beta = 0.062$. If $m \geq \frac{103}{6}\sqrt{2/(3\beta)}n > 56n$, then we can use Lemmas 8 and 9, otherwise $m \leq 56n$ and our proof is complete. Let D be a RAC_2 drawing of G. Recall that every edge has two *end segments* and one *middle segment*. Let αS be the number of end segments that participate in heavy blocks, where $0 < \alpha < 1$. The number of middle segments is m, which is a trivial upper bound on the middle segments that participate in heavy blocks. So the total number of segments in heavy blocks is at most $S \leq m + \alpha S$, which gives $S \leq \frac{1}{1-\alpha}m$.

In each heavy block, the segments can be partitioned into two sets of pairwise parallel segments. If we delete all edges that contain some segment in the smaller set of each heavy block, then the remaining segments are not heavy anymore. That is, by deleting at most $\frac{S}{2} \leq \frac{1}{2(1-\alpha)}m$ edges, we obtain a RAC_2 drawing with no heavy edge segment. By Lemma 8, we have

$$\frac{1}{2(1-\alpha)} > 1 - \sqrt{\frac{3\beta}{2}} \ \Rightarrow\ \frac{1}{1 - \sqrt{3\beta/2}} > 2(1-\alpha),$$

which implies

$$\alpha > 1 - \frac{1}{2(1 - \sqrt{3\beta/2})} \tag{7}$$

The block graph $B(D)$ has αS edges, since an edge in $B(D)$ exists if and only if a vertex of G is incident to an end segment in a heavy block. From Lemma 7, we have an upper bound on the number of edges in $B(D)$, which gives $\alpha S < 2H + 5n$. Using $S > (1 - \sqrt{3\beta/2})m$ from Lemma 9, the upper bound on H from (6), and the lower bound on α from (7), we obtain

$$\alpha S < 2H + 5n \;\Rightarrow\; \alpha S < \frac{2S}{\beta c} \cdot \frac{n^2}{m^2} + 5n$$

$$\Rightarrow\; \left(\alpha - \frac{2}{\beta c} \cdot \frac{n^2}{m^2} \right) \left(1 - \sqrt{\frac{3\beta}{2}} \right) m < 5n$$

$$\Rightarrow\; 0 < \frac{2 - 2\sqrt{3\beta/2}}{\beta c} \cdot \left(\frac{n}{m} \right)^2 + 5 \cdot \left(\frac{n}{m} \right) - \alpha \left(1 - \sqrt{\frac{3\beta}{2}} \right)$$

$$\Rightarrow\; 0 < \frac{2 - \sqrt{6\beta}}{\beta c} \cdot \left(\frac{n}{m} \right)^2 + 5 \cdot \left(\frac{n}{m} \right) - \left(\frac{1}{2} - \sqrt{\frac{3\beta}{2}} \right).$$

This is a quadratic inequality in n/m. Since $\sqrt{3\beta/2} < 1/2$, the constant term is negative, and the two roots have opposite signs. Therefore, we have

$$\frac{n}{m} > \frac{\beta c}{2(2 - \sqrt{6\beta})} \left(-5 + \sqrt{25 + \frac{4}{\beta c} \left(2 - \sqrt{6\beta} \right) \left(\frac{1}{2} - \sqrt{\frac{3\beta}{2}} \right)} \right).$$

This is maximized for $\beta = 0.062$, and gives $m < 74.2n$. $\square$

4   Lower Bound Constructions

We complement the upper bounds in Theorems 1 and 2 with lower bound constructions. We construct an infinite family of graphs which admit RAC_1 drawings and $4.5n - O(\sqrt{n})$ edges. This shows that $R_0 \neq R_1$ since every graph in R_0 has at most $4n - 10$ edges [7]. Let the vertices of G be points of the hexagonal lattice clipped in a square (Fig. 3). The edges of G are the hexagon edges and 6 diagonals with a bend in each hexagon. The diagonals connect every other vertex in the hexagon, and make a $75°$ angle with the side of the hexagon, and so they cross in right angles. The vertex degree is $3 + 3 \cdot 2 = 9$ for all but at most $O(\sqrt{n})$ lattice points around the bounding box. Hence the number of edges is $4.5n - O(\sqrt{n})$.

We also construct an infinite family of graphs which admit RAC_2 drawings and $7.8\dot{3}n - O(\sqrt{n})$ edges. This shows that $R_1 \neq R_2$ since every graph in R_1 has at most $6.5n - 13$ edges by Theorem 1. Let the vertices of G be the vertices of an Archimedean tiling $(12,12,3)$ clipped in a square. Refer to Fig. 4. In the tiling $(12,12,3)$, we can assign two triangles to each 12-gon. The edges of G are the edges of the tiling, a 6-regular graph of diagonals in each 12-gon, and two edges per 12-gon that go to vertices of the two adjacent triangles. The tiling

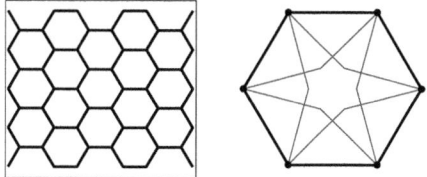

Fig. 3. Lower bound construction for a RAC_1 drawing in a hexagonal lattice

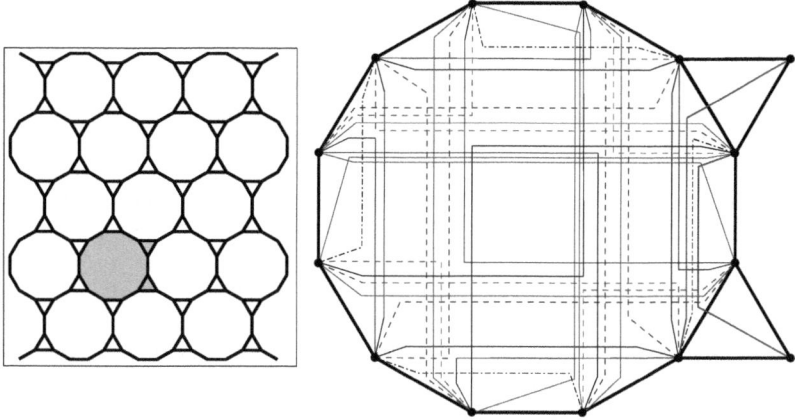

Fig. 4. Lower bound construction for a RAC_2 drawing in an Archimedean tiling (12,12,3)

and the diagonals of the 12-gons generate a vertex degree of $3 + 2 \cdot 6 = 15$ at all but at most $O(\sqrt{n})$ vertices (due to the boundary effect). The additional two edges between adjacent 12-gons and triangles increase the average degree to $15 + \frac{2}{3} - O(1/\sqrt{n})$. Hence the number of edges is $\frac{47}{6}n - O(\sqrt{n}) = 7.8\dot{3}n - O(\sqrt{n})$.

5   Concluding Remarks

It remains an open problem to determine the maximum number of edges of a graph with n vertices in the classes R_1 and R_2. Our upper bound in Theorem 1 may be slightly improved by refining the bound in Lemma 4. If we could strengthen the upper bound in Lemma 4 for small values of n, then (4) would improve. However, we did not pursue this direction as it would not lead to significant improvement without an extensive case analysis.

Let an $\alpha AC_b^=$ drawing be a polyline drawing of a graph with b bends per edge where all crossings occur at angle *exactly* α. It is easy to show that a graph with n vertices and an $\alpha AC_0^=$ drawing has at most $9n - 18$ edges. The edges in each "block" can be partitioned into 3 sets of noncrossing edges, and so the graph decomposes into 3 planar graphs. Every graph admits an $\alpha AC_3^=$ drawing, since every affine transformation deforms *all* crossing angles uniformly in the construction by Didimo *et al.* [7]. Very recently, Ackerman *et al.* [1] proved that every graph on n vertices that admit $\alpha AC_1^=$ or $\alpha AC_2^=$ drawings have $O(n)$ vertices.

References

1. Ackerman, E., Fulek, R., Tóth, C.D.: On the size of graphs that admit polyline drawings with few bends and crossing angles. In: Proc. 18th Sympos. on Graph Drawing. LNCS, Springer, Heidelberg (2010)
2. Ackerman, E., Tardos, G.: On the maximum number of edges in quasi-planar graphs. J. Combin. Theory Ser. A 114(3), 563–571 (2007)
3. Ajtai, M., Chvátal, V., Newborn, M.M., Szemerédi, E.: Crossing-free subgraphs. In: Theory and practice of combinatorics. North-Holland Math. Stud., vol. 60, pp. 9–12. North-Holland, Amsterdam (1982)
4. Angelini, P., Cittadini, L., Di Battista, G., Didimo, W., Frati, F., Kaufmann, M., Symvonis, A.: On the perspectives opened by right angle crossing drawings. In: Eppstein, D., Gansner, E.R. (eds.) GD 2009. LNCS, vol. 5849, pp. 21–32. Springer, Heidelberg (2010)
5. Appel, K., Haken, W.: The solution of the four-color-map problem. Sci. Amer. 237(4), 108–121 (1977)
6. Di Giacomo, E., Didimo, W., Liotta, G., Meijer, H.: Area, curve complexity, and crossing resolution of non-planar graph drawings. In: Eppstein, D., Gansner, E.R. (eds.) GD 2009. LNCS, vol. 5849, pp. 15–20. Springer, Heidelberg (2010) (to appear)
7. Didimo, W., Eades, P., Liotta, G.: Drawing graphs with right angle crossings. In: Proc. 11th WADS. LNCS, vol. 5664, pp. 206–217. Springer, Heidelberg (2009)
8. Dujmović, V., Gudmundsson, J., Morin, P., Wolle, T.: Notes on large angle crossing graphs. In: Proc. Computing: The Australian Theory Sympos.(CATS). CRPIT, ACS (2010) (to appear), http://arxiv.org/abs/0908.3545
9. Hliněný, P.: Discharging technique in practice, Lecture text for Spring School on Combinatorics
10. Huang, W.: Using eye tracking to investigate graph layout effects. In: 6th Asia-Pacific Sympos. Visualization (APVIS 2007), pp. 97–100 (2007)
11. Huang, W., Hong, S.-H., Eades, P.: Effects of crossing angles. In: Proc. IEEE Pacific Visualization Sympos., pp. 41–46 (2008)
12. Leighton, F.T.: Complexity issues in VLSI: optimal layouts for the shuffle-exchange graph and other networks. MIT Press, Cambridge (1983)
13. Pach, J., Radoičić, R., Tardos, G., Tóth, G.: Improving the crossing lemma by finding more crossings in sparse graphs. Discrete Comput. Geom. 36(4), 527–552 (2006)
14. Wernicke, P.: Über den kartographischen Vierfarbensatz. Math. Ann. 58(3), 413–426 (1904)

Kernelization Hardness of Connectivity Problems in d-Degenerate Graphs*

Marek Cygan, Marcin Pilipczuk,
Michał Pilipczuk, and Jakub Onufry Wojtaszczyk

Faculty of Mathematics, Computer Science and Mechanics,
University of Warsaw, Poland
{cygan@,malcin@,michal.pilipczuk@students.,onufry@}mimuw.edu.pl

Abstract. A graph is d-degenerate if its every subgraph contains a vertex of degree at most d. For instance, planar graphs are 5-degenerate. Inspired by recent work by Philip, Raman and Sikdar, who have shown the existence of a polynomial kernel for DOMINATING SET in d-degenerate graphs, we investigate kernelization hardness of problems that include connectivity requirement in this class of graphs.

Our main contribution is the proof that CONNECTED DOMINATING SET does not admit a polynomial kernel in d-degenerate graphs for $d \geq 2$ unless the polynomial hierarchy collapses up to the third level. We prove this using a problem originated from bioinformatics — COLOURFUL GRAPH MOTIF — analyzed and proved to be NP-hard by Fellows et al. This problem nicely encapsulates the hardness of the connectivity requirement in kernelization. Our technique yields also an alternative proof that, under the same complexity assumption, STEINER TREE does not admit a polynomial kernel. The original proof, via reduction from SET COVER, is due to Dom, Lokshtanov and Saurabh.

We extend our analysis by showing that, unless $PH = \Sigma_p^3$, there do not exist polynomial kernels for STEINER TREE, CONNECTED FEEDBACK VERTEX SET and CONNECTED ODD CYCLE TRANSVERSAL in d-degenerate graphs for $d \geq 2$. On the other hand, we show a polynomial kernel for CONNECTED VERTEX COVER in graphs that do not contain the biclique $K_{i,j}$ as a subgraph.

1 Introduction

In the parameterized complexity setting, an instance comes with an integer parameter k — formally, a parameterized problem Q is a subset of $\Sigma^* \times \mathbb{N}$ for some finite alphabet Σ. We say that the problem is *fixed parameter tractable* (*FPT*) if there exists an algorithm solving any instance (x, k) in time $f(k)\text{poly}(|x|)$ for some (usually exponential) computable function f. It is known that a problem is FPT iff it is kernelizable: a kernelization algorithm for a problem Q takes an instance (x, k) and in time polynomial in $|x| + k$ produces an equivalent instance (x', k') (i.e., $(x, k) \in Q$ iff $(x', k') \in Q$) such that $|x'| + k' \leq g(k)$ for some computable function g. The function g is the *size of the kernel* and if it is polynomial, we say that Q admits a polynomial kernel.

---

* The first two authors were partially supported by Polish Ministry of Science grant no. N206 355636.

D.M. Thilikos (Ed.): WG 2010, LNCS 6410, pp. 147–158, 2010.
© Springer-Verlag Berlin Heidelberg 2010

Kernelization techniques can be viewed as polynomial time preprocessing routines for tackling NP-hard problems. Parameterized complexity provides a formal framework for the analysis of such algorithms. In particular small (i.e. polynomial) kernels play an important role, and there are numerous positive results showing small kernels for various problems, including VERTEX COVER [5] and FEEDBACK VERTEX SET [19]. Recently, Bodlaender et al. [2] and Fortnow and Santhanam [12] came up with a technique that allows to prove negative results in this field: their tools provide a way to show that a parameterized problem does not admit a polynomial kernel unless the polynomial hierarchy collapses up to the third level. Up to this day, the list of non-polynomially-kernelizable (unless $PH = \Sigma_p^3$) FPT problems includes LONGEST PATH, LONGEST CYCLE [2], DIRECTED MAX LEAF OUT-BRANCHING [10], DISJOINT PATHS, DISJOINT CYCLES [4], RED-BLUE DOMINATING SET (aka SET COVER), STEINER TREE, CONNECTED VERTEX COVER [6] and CONNECTED FEEDBACK VERTEX SET [15].

On the other hand, many problems which are hard in general graphs — i.e. without a polynomial kernel or even not FPT — have small kernels in restricted graph classes, such as planar graphs, bounded genus graphs, apex-minor-free graphs or H-minor-free graphs. Recent results include linear kernels for DOMINATING SET and CONNECTED DOMINATING SET in apex-minor-free graphs and linear kernels for FEEDBACK VERTEX SET and CONNECTED VERTEX COVER in H-minor-free graphs [11].

The aforementioned results use the topological structure of the considered graph classes. However, sometimes an even weaker assumption on the graph class leads to significantly better algorithms and kernels than in the general case. One may, for instance, consider the class of d-degenerate graphs. A graph is called d-degenerate if its every induced subgraph contains a vertex of degree at most d. For instance, the class of 1-degenerate graphs is the class of forests, and all planar graphs are 5-degenerate. Moreover, every H-minor-free graph is d-degenerate, where the constant d depends on the minor [14,17,18]. Alon and Gutner [1] followed by Golovach and Villanger [13] proved that DOMINATING SET and CONNECTED DOMINATING SET, which are $W[2]$-hard in general graphs [7], become FPT when the input graph is d-degenerate. Very recently, Philip et al. [16] proved that DOMINATING SET is FPT and admits a polynomial kernel in a larger class of graphs: graphs excluding the biclique $K_{i,j}$ as a subgraph (note that a d-degenerate graph cannot contain $K_{d+1,d+1}$ as a subgraph).

A natural question arises: does the bounded degeneracy assumption help in the kernelization of other problems? In particular, the question of finding a polynomial kernel for CONNECTED DOMINATING SET in d-degenerated graphs was posted on the 1st Workshop on Kernels (WORKER'09, Bergen, Norway). In this paper we provide mostly negative answers to questions of existence of polynomial kernels for connectivity problems in graphs of bounded degeneracy. Note that this is in sharp contrast with the existence of the linear kernel for CONNECTED DOMINATING SET in apex-minor-free graphs [11].

The main contribution of this paper is the idea to use the COLOURFUL GRAPH MOTIF problem, which, intuitively, encapsulates the hardness of the connectivity requirement.

COLOURFUL GRAPH MOTIF   **Parameter:** k.
Input: A graph $G = (V, E)$, an integer k and a function $f : V \to \{1, 2, \ldots, k\}$
Question: Does there exist a connected set $S \subset V$ of cardinality k, such that $f|_S$ is bijective?

We think of the function f to be a colouring of V — each number from $\{1, 2, \ldots, k\}$ corresponds to a single colour — and we ask whether it is possible to choose a connected set containing exactly one vertex of each colour.

Fellows et al. [9] have shown that, surprisingly, this problem is NP-hard even in the class of trees of maximum degree 3. We use this fact to prove that COLOURFUL GRAPH MOTIF does not admit a polynomial kernel in 1-degenerate graphs (forests) unless $PH = \Sigma_p^3$.[1] This problem is simple enough to admit a reduction to CONNECTED DOMINATING SET in 2-degenerate graphs. As a by-product of this analysis, we obtain an alternative proof that STEINER TREE does not admit a polynomial kernel in arbitrary graphs. The original proof, via reduction from RED BLUE DOMINATING SET (aka SET COVER) is due to Dom et al. [6]. We analyze COLOURFUL GRAPH MOTIF in Section 4 and apply it to CONNECTED DOMINATING SET to show that CONNECTED DOMINATING SET does not admit a polynomial kernel in 2-degenerate graphs. In Section 4 we also show the reduction from COLOURFUL GRAPH MOTIF to STEINER TREE.

On the positive side (in Section 5) we provide a $O(k^2 + (i + j)k^{\min(i,j)})$-vertex kernel for CONNECTED VERTEX COVER in $K_{i,j}$-free graphs. In the analysis we use arguments similar to those developed by Philip et al. [16].

Preliminaries and notation are given in Section 2. As a warmup, in Section 3 by easy reductions and using already known results we show that STEINER TREE, CONNECTED FEEDBACK VERTEX SET and CONNECTED ODD CYCLE TRANSVERSAL do not admit polynomial kernels in 2-degenerate graphs. All discussed problems are parameterized by the solution size, except for STEINER TREE, which is parameterized both by the solution size and the size of the terminal set. Precise definitions of considered problems can be found in appropriate sections.

2   Preliminaries and Notation

Before we start, let us introduce some notation. All problems are considered on an undirected graph $G = (V, E)$. The set $N(v) = \{u : uv \in E\}$ is the neighbourhood of v and $N[v] = N(v) \cup \{v\}$ is the closed neighbourhood of v. We extend this notation to all subsets $A \subset V$: $N[A] = \bigcup_{v \in A} N[v]$ and $N(A) = N[A] \setminus A$. We say that a vertex v is dominated by a vertex set A if $v \in N[A]$; a vertex set A is dominating if $N[A] = V$. Whenever we speak of a parameterized problem Q, by d-deg-Q we denote the problem Q, where the class of input graphs is restricted to d-degenerate graphs.

In this section we recall all the required definitions about kernels, and ways to prove the non-existence of a polynomial kernel. In general, we follow the notation from [6]. Given a parameterized problem $Q \subset \Sigma^* \times \mathbb{N}$, its unparameterized version is a language

---

[1] In the full version of this paper we show NP-hardness and nonexistence of a polynomial kernel for COLOURFUL GRAPH MOTIF in comb graphs. A graph is called a comb graph if it is a tree, all vertices are of degree at most 3 and all the vertices of degree 3 lie on a single simple path.

$\tilde{Q} = \{x\#1^k : (x,k) \in Q\}$, i.e., we append the parameter written in unary. Let us now cite the main result of Bodlaender et al. [2] and Fortnow and Santhanam [12].

Definition 1 (Composition [2,6]). *A composition algorithm for a parameterized problem $Q \subset \Sigma^* \times \mathbb{N}$ is an algorithm that receives as input a sequence $(x_1, k), (x_2, k), \ldots, (x_t, k)$ with $(x_i, k) \in \Sigma^* \times \mathbb{N}$ for each $1 \leq i \leq t$, uses polynomial time in $\sum_{i=1}^t |x_i| + k$, and outputs $(y, k') \in \Sigma^* \times \mathbb{N}$ with $(y, k') \in Q$ iff $\exists_{1 \leq i \leq t}(x_i, k) \in Q$ and k' is polynomial in k. A parameterized problem is called* compositional *if there is a composition algorithm for it.*

Theorem 1 ([2,12]). *Let Q be a compositional parameterized problem whose unparameterized version $\tilde{Q}$ is NP-complete. Then, unless $PH = \Sigma_p^3$, there is no polynomial kernel for Q.*

To prove the non-existence of a polynomial kernel for some parameterized problem, it is not necessary to go through Theorem 1. As in the case of NP-complete problems, we can use reductions instead.

Definition 2 ([4,6]). *Let P and Q be parameterized problems. We say that P is polynomial parameter reducible to Q, written $P \leq_{Ptp} Q$, if there exists a polynomial time computable function $f : \Sigma^* \times \mathbb{N} \to \Sigma^* \times \mathbb{N}$ and a polynomial p, such that for all $(x, k) \in \Sigma^* \times \mathbb{N}$ the following holds: $(x, k) \in P$ iff $(x', k') = f(x, k) \in Q$ and $k' \leq p(k)$. The function f is called a* polynomial parameter transformation.

Theorem 2 ([4,6]). *Let P and Q be parameterized problems and $\tilde{P}$ and $\tilde{Q}$ be the unparameterized versions of P and Q respectively. Suppose that $\tilde{P}$ is NP-hard and $\tilde{Q}$ is in NP. Assume there is a polynomial parameter transformation from P to Q. Then if Q admits a polynomial kernel, so does P.*

3   Easy Cases: STEINER TREE, CONNECTED FEEDBACK VERTEX SET and CONNECTED ODD CYCLE TRANSVERSAL

We shall begin by showing that, unless $PH = \Sigma_p^3$, no polynomial kernel exists in the connected case even for 2-degenerate graphs for three problems: STEINER TREE, CONNECTED FEEDBACK VERTEX SET and CONNECTED ODD CYCLE TRANSVERSAL. We reduce them through Theorem 2 to the problems shown by other authors not to admit a polynomial kernel. We use the results of Dom et al. [6], where the authors show STEINER TREE and CONNECTED VERTEX COVER do not admit a polynomial kernel in the class of all graphs. Presented constructions are adjustments of reductions made for CONNECTED FEEDBACK VERTEX SET [15].

STEINER TREE (ST)   **Parameter:** $t := |T|$ and k.
Input: A graph $G = (V, E)$, a set of terminals $T \subset V$ and an integer k.
Question: Does there exist $S \subset V$, such that $G[S \cup T]$ is connected and $|S| \leq k$?

CONNECTED FEEDBACK VERTEX SET (CFVS)   **Parameter:** k.
Input: A graph $G = (V, E)$ and an integer k.
Question: Does there exist a set $S \subset V$ of cardinality at most k, such that $G[S]$ is connected and $G[V \setminus S]$ contains no cycles?

CONNECTED ODD CYCLE TRANSVERSAL   **Parameter:** k.
Input: A graph $G = (V, E)$ and an integer k.
Question: Does there exist a set $S \subset V$ of cardinality at most k, such that $G[S]$ is connected and $G[V \setminus S]$ is bipartite (that is, contains no cycles of odd length)?

CONNECTED VERTEX COVER   **Parameter:** k.
Input: A graph $G = (V, E)$ and an integer k.
Question: Does there exist a set $S \subset V$ of cardinality at most k, such that $G[S]$ is connected and every edge $e \in E$ has at least one endpoint in S?

Now let us note the following simple observation.

Lemma 1. *Assume that in a graph G every edge has an endpoint of degree at most* 2. *Then G is* 2-*degenerate.*

We now show reductions to each of the three aforementioned problems.

Proposition 1. CONNECTED VERTEX COVER $\leq_{Ptp}$ 2–*deg*–CONNECTED FEEDBACK VERTEX SET.

Proof. Consider any instance (G, k) of CONNECTED VERTEX COVER. We create a graph $G' = (V', E')$. We take $V' = V \cup E_1 \cup E_2$ — the vertices of G' are the vertices of G plus two new vertices e_1, e_2 for each edge e of G. For each edge $e = uv \in E$ we add four edges to E': $\{u, e_1\}, \{u, e_2\}, \{v, e_1\}$ and $\{v, e_2\}$. This means we transform each edge of G into a cycle of length 4, where the original vertices are on opposite points of the cycle. Lemma 1 implies that G' is 2-degenerate.

We now prove that the answer to CONNECTED FEEDBACK VERTEX SET for $(G', 2k-1)$ is the same as the answer to CONNECTED VERTEX COVER for (G, k). First assume we have a positive answer for (G, k). This means that there exists a connected vertex cover S of G of size at most k. As S is connected, we can create a spanning tree in $G[S]$, this consists of at most $k - 1$ edges $E_S \subset E$. Let $E'_S = \{e_1 : e \in E_S\}$ — that is, for each edge $e \in E_S$ we take into E'_S one of the two vertices in V' corresponding to e. We claim $G'[V' \setminus (S \cup E_S)]$ contains no cycles. Assume C is a cycle in $G'[V' \setminus (S \cup E_S)]$. C cannot consist only of elements of V (since V is independent in G'), thus C contains some element e_i. As $\deg e_i = 2$, C also has to contain both vertices from V which the corresponding edge $e \in E$ connects. This, however, means in particular that neither of these vertices was in S, which is a contradiction with the assumption that S was a vertex cover of G, as the edge e is not covered.

Now assume we have a connected feedback vertex set $S \subset V'$ in G' of cardinality at most $2k - 1$. Assume $|S| \geq 2$ (the case $|S| = 1$ is trivial). Notice that $|S \cap V| \leq k$ — if we have more than k vertices from V, they form at least $k + 1$ connected components, and each vertex from E' connects at most two of them — thus S would not be connected. We claim $S \cap V$ forms a connected vertex cover of G. Consider any edge $e = uv$ in E and the corresponding cycle (u, e_1, v, e_2) in G'. As S is a feedback vertex set in G', at least one of these four vertices must belong to S. As $|S| \geq 2$ and S is connected, at least one of u, v is in S — and thus e is covered in G by $S \cap V$.

Proposition 2. CONNECTED VERTEX COVER $\leq_{Ptp}$ 2–*deg*–CONNECTED ODD CYCLE TRANSVERSAL.

Proof. We proceed as above, except we transform each edge into a cycle of length five.

Corollary 1. *The problems* 2-*deg*-CONNECTED ODD CYCLE TRANSVERSAL *and* 2-*deg*-CONNECTED FEEDBACK VERTEX SET *do not admit a polynomial kernel unless* $PH = \Sigma_p^3$.

The last reduction to degenerate graphs from previously known results is for 2-*deg*-STEINER TREE. The alternative proof of the kernelization hardness of 2-*deg*-STEINER TREE, via reduction from COLOURFUL GRAPH MOTIF, can be found in Section 4.

Proposition 3. STEINER TREE $\leq_{Ptp}$ 2-*deg*-STEINER TREE *and* 2-*deg*-STEINER TREE *does not admit a polynomial kernel unless* $PH = \Sigma_p^3$.

Proof. Take a general instance (G, k, T) of STEINER TREE. Create a new graph G' by subdividing each edge — formally, let $V' = V \cup E$ and $ve \in E'$ if v is an endpoint of e in G. The graph G' is 2–degenerate by Lemma 1.

We claim that the answer for (G, k, T) is the same as the answer for $(G', 2k + |T| - 1, T)$. Assume we have a solution S of (G, k, T). Then $G[S \cup T]$ is connected. Take any spanning tree of $G[S \cup T]$, let F be the set of its edges, we have $|F| \leq k + |T| - 1$. Now $F \cup S$ is a solution in $(G', 2k + |T| - 1, T)$. In the other direction, if we have a solution S' in $(G', 2k + |T| - 1, T)$, we consider $S = S' \cap V$. Note that $S \cup T$ has cardinality at most $k + |T|$ — since $|S' \cup T| \leq 2k + 2|T| - 1$, $S \cup T$ is isolated in G', and adding a single vertex from E connects at most two components of the set. Thus S has a cardinality at most k, and $G[S \cup T]$ is connected (for otherwise $S' \cup T$ could not be connected in G').

4   From COLOURFUL GRAPH MOTIF to CONNECTED DOMINATING SET

4.1   COLOURFUL GRAPH MOTIF

The CONNECTED VERTEX COVER problem is, in a number of cases, too specific to allow easy reductions. The COLOURFUL GRAPH MOTIF problem appeared to be very handy in our case.

We show the problem has no polynomial kernel in the class of forests of maximum degree 3. Fellows et al. [9] have shown that COLOURFUL GRAPH MOTIF in this class of graphs is already NP-complete. Since trees are 1-degenerate, we use Theorem 1 and take the disjoint union of graphs and the union of functions as the composition algorithm. Note that any feasible solution is required to induce a connected subgraph and therefore it needs to be contained in one connected component of the input graph. This yields the following theorem:

Theorem 3. *The* COLOURFUL GRAPH MOTIF *problem in the class of 1-degenerate graphs (forests) of maximum degree 3 does not admit a polynomial kernel unless* $PH = \Sigma_p^3$.

4.2   Reductions

We propose COLOURFUL GRAPH MOTIF as a simpler tool to prove that various other problems do not admit a polynomial kernel unless $PH = \Sigma_p^3$. Firstly, to give some intuition on COLOURFUL GRAPH MOTIF, let us note that COLOURFUL GRAPH MOTIF is a special case of GROUP STEINER TREE.

GROUP STEINER TREE   **Parameter:** k.
Input: A graph $G = (V, E)$, sets of vertices $T_1, \ldots, T_k \subset V$ and an integer p.
Question: Does there exist $S \subset V$, such that $G[S]$ is connected, $|S| = p$ and $S \cap T_i \neq \emptyset$ for $i = 1, \ldots, k$?

Proposition 4. d–deg–COLOURFUL GRAPH MOTIF $\leq_{Ptp}$ d–deg–GROUP STEINER TREE.

Proof. Assume we have an instance (G, k, f) of d–deg–COLOURFUL GRAPH MOTIF. We create an instance of d–deg–GROUP STEINER TREE as follows: we keep the graph G, we put $p = k$ and take $T_i = f^{-1}(i)$. Now the the problem GROUP STEINER TREE asks whether there exists a connected set S of cardinality $p = k$ which has a non–empty intersection with each T_i. As $p = k$, this means that the intersection with each T_i is to contain exactly one element. This is exactly the question in COLOURFUL GRAPH MOTIF, thus the answer to COLOURFUL GRAPH MOTIF in (G, f, k) is the same as the answer to GROUP STEINER TREE in $(G, \{T_i\}, p, k)$.

Corollary 2. COLOURFUL GRAPH MOTIF *can be solved in* $2^k n^{O(1)}$ *time and polynomial space.*

Proof. We reduce COLOURFUL GRAPH MOTIF to GROUP STEINER TREE as in the proof of Proposition 4 and use $2^k n^{O(1)}$-time algorithm described in [15].

Our original motivation for analyzing COLOURFUL GRAPH MOTIF was the CONNECTED DOMINATING SET problem.

CONNECTED DOMINATING SET   **Parameter:** k.
Input: A graph $G = (V, E)$ and an integer k
Question: Does there exist a set $S \subset V$ of cardinality at most k, such that $G[S]$ is connected and every vertex $v \in V$ is adjacent or equal to some vertex $u \in S$?

Proposition 5. d–deg–COLOURFUL GRAPH MOTIF $\leq_{Ptp}$ $(d+1)$–deg–CONNECTED DOMINATING SET, *and* 2–deg–CONNECTED DOMINATING SET *admits no polynomial kernel unless* $PH = \Sigma_p^3$.

Proof. We begin with an instance (G, k, f) of d–deg–COLOURFUL GRAPH MOTIF. Due to Corollary 2, we may assume $k \geq 2$, otherwise we can solve the input instance in polynomial time. We create a graph $G' = (V', E')$ as follows:

- $V \subset V', E \subset E'$;
- for each colour $l \in \{1, 2, \ldots, k\}$ we add two vertices v_l and v_l' to V';

- for each colour $l \in \{1, 2, \ldots, k\}$ we add an edge $v_l v_l'$ to E';
- for each vertex $v \in V$ we add an edge $v v_{f(v)}$ to E'.

Firstly, we prove G' is $(d+1)$–degenerate. Consider any $S \subset V'$. Then either $S \subset V' \setminus V$ (but then every vertex in $G'[S]$ is of degree at most 1) or $S \cap V$ is non–empty. Then $G[S \cap V]$ contains a vertex v, which had degree at most d in G, so it has degree at most $d+1$ in G' (as we added one edge to each vertex of V).

Now we prove the answer to COLOURFUL GRAPH MOTIF for (G, k, f) is the same as the answer to CONNECTED DOMINATING SET for $(G', 2k)$. Assume $k > 1$. If we have a solution S of COLOURFUL GRAPH MOTIF in G, we create a solution of CONNECTED DOMINATING SET by putting $S' = S \cup \{v_1, v_2, \ldots, v_k\}$. The vertices v_l' are neighbours of v_ls, any vertex $v \in V$ is a neighbour of $v_{f(v)}$, which is in S', and S' is connected, for S was connected and each v_l is adjacent to the vertex of colour l in S. On the other hand, any solution S' to CONNECTED DOMINATING SET in G' has to contain all the vertices v_l (there are two ways to dominate v_l' — either we take v_l, or we take v_l', but in the second case we have to take v_l anyway for connectedness). To ensure connectedness we have to take at least one neighbour u_l of each v_l ($u_l \neq v_l'$). As the sets of neighbours of v_ls are disjoint and $|S'| \leq 2k$, this means exactly one neighbour of each v_l is in S'. In $G'[S']$ the vertices v_l are of degree 1 (they are not adjacent to each other, and are not adjacent to u_j for $j \neq l$), thus $G'[S' \setminus \{v_1, v_2, \ldots, v_k\}]$ is connected as $G'[S']$ is connected. This means $S' \setminus \{v_1, v_2, \ldots, v_k\}$ is a solution to COLOURFUL GRAPH MOTIF in G.

As a final example of the technique we show how to prove that the STEINER TREE problem admits no polynomial kernel in 2-degenerate graphs. The problem was studied in [6], where STEINER TREE was shown to admit no polynomial kernel in general graphs, and a simple reduction to 2-degenerate graphs was shown in Section 3. We now show a self–contained proof to demonstrate again the applicability of COLOURFUL GRAPH MOTIF.

Proposition 6. d–deg–COLOURFUL GRAPH MOTIF $\leq_{Ptp} (d+1)$–deg–STEINER TREE and 2–deg–STEINER TREE admits no polynomial kernel unless $PH = \Sigma_p^3$.

Proof. Assume we have an instance (G, k, f) of d-deg-COLOURFUL GRAPH MOTIF. We create an instance (G', T, k) of $(d+1)$-deg-STEINER TREE as follows: we keep the graph G as the set of non–terminals $V \setminus T$. Additionally for each colour $i \in \{1, 2, \ldots, k\}$ we add a vertex $t_i \in T$ and edges $v t_i$ for all $v \in f^{-1}(i)$. We ask for a Steiner tree of cardinality k in $T \cup V$ connecting all vertices from T.

First note G' is $(d+1)$-degenerate. Similarly like in the previous proof, the terminals T form an independent set, while to each non–terminal from G we added exactly one edge. Let S be the solution to STEINER TREE in G'. Note that S has to contain exactly one vertex of each colour — if some colour was excluded, the corresponding terminal could not be connected, and the number of colours is at least $|S|$. Moreover, S has to be connected in $G'[V] = G$, as there is only one vertex of each colour, each terminal is a leaf in the solution, so removing a terminal does not change the connectedness of the solution. On the other hand, it can be easily seen that any solution of COLOURFUL GRAPH MOTIF in G gives a solution of STEINER TREE in G'.

5 On the Positive Side: Polynomial Kernel for CONNECTED VERTEX COVER

As a counterpoint to the results above we show that CONNECTED VERTEX COVER in 2–degenerate graphs *does* admit a polynomial kernel. To show the problem is non–trivial we have to begin by proving CONNECTED VERTEX COVER is NP–hard in this class (otherwise finding a polynomial kernel would not be much of an achievement). This is not surprising — the CONNECTED VERTEX COVER problem was studied extensively and shown, for instance, to be NP–hard in graphs with maximum degree 4 (although it is in P for graphs of maximum degree 3, see [8]).

Proposition 7. *The unparameterized version of 2–deg–CONNECTED VERTEX COVER is NP–hard.*

Proof. We show a reduction of CNF–SAT to 2–deg–CONNECTED VERTEX COVER. Consider an instance $C_1 \wedge C_2 \wedge \ldots \wedge C_m$ with variables $x_1, \ldots, x_n$ of CNF–SAT. Let M be a total number of all literals in all clauses in this formula. We create a graph G as follows:

- we create two vertices v and v' and an edge vv';
- for each variable x we create vertices x^t and x^f, an edge $x^t x^f$ and edges vx^t and vx^f;
- for each clause C_j we create vertices C_j and C'_j and an edge $C_j C'_j$;
- for each clause C_j if x is a literal in C_j we create vertices L_{xj} and L'_{xj} and edges $L_{xj}L'_{xj}$, $L_{xj}C_j$ and $L_{xj}x^t$. If $\neg x$ is a literal in C_j we create the same vertices and edges, with the exception of the last edge being $L_{xj}x^f$;

First let us check the graph above is indeed 2–degenerate. Assume we have such a set $S \subset V$ that $G[S]$ does not contain a vertex of degree 2. The vertices v', L'_{xj} and C'_j are of degree 1 in G, so they cannot be contained in S. The vertices L_{xj} are of degree 2 in $G[V \setminus \{L'_{xj}\}]$, so they cannot be contained in S. After removing all L_{xj}s and C'_js the C_js become isolated, and the degree of x^ts and x^fs drops to 2, so S cannot contain any of them either. We are left with a single vertex v, which is isolated in $G[\{v\}]$, so S is empty.

We now claim that a solution to CNF–SAT in $C_1 \wedge \ldots \wedge C_m$ exists iff a solution to CONNECTED VERTEX COVER exists for $(G, n + m + M + 1)$. Assume we have a solution ϕ to CNF–SAT in $C_1 \wedge \ldots \wedge C_m$. We choose a set $S \subset V$ as follows:

- the vertices v, C_j for all j and L_{xj} for all x, j for which they exist are in S;
- the vertex x^t is in S if $\phi(x)$ is true, otherwise x^f is in S.

It is easy to see that the set above does indeed cover all the edges of G. It is also connected — all the x^ts and x^fs are connected with v, for each j the vertices C_j and L_{xj} are all connected, and as at least one literal of the clause C_j is set to be true by ϕ, at least one of the L_{xj}s for each j is connected to a $x^? \in S$. The solution given is of cardinality exactly $n + m + M + 1$.

On the other hand consider any solution S of CONNECTED VERTEX COVER in G. It has to contain v — to cover the edge vv' one of v, v' has to be in S, and if $v' \in S$,

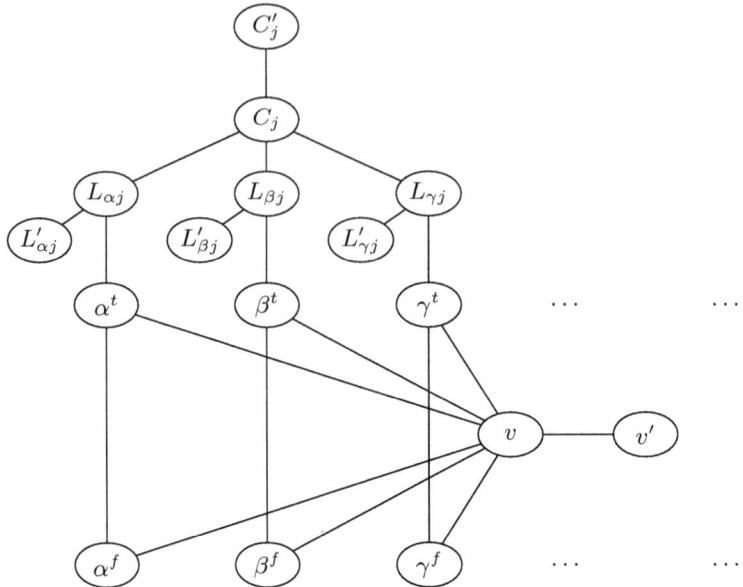

Fig. 1. Part of the constructed graph illustrating vertices added for clause $C_j = (\alpha \vee \beta \vee \gamma)$ and their interaction with vertices added for variables

then $v \in S$ to assure connectedness. For identical reasons $C_j \in S$ and $L_{xj} \in S$. We already have $m + M + 1$ vertices in S, so we can use at most n to cover the remaining edges. However the edges $x^f x^t$ form a matching of cardinality n in the remaining set, thus we have to take exactly one of $\{x^t, x^f\}$ to belong to S.

Consider a function $\phi : \{x_1, \ldots, x_n\} \rightarrow \{TRUE, FALSE\}$, setting $\phi(x) = TRUE$ iff $x^t \in S$. If for some clause C_j all of its literals were set to false by ϕ, then removing the vertices $x^?$ corresponding to these literals would split G into two connected components, one containing v, the other containing C_j, which would show S cannot be connected. As S is a solution to CONNECTED VERTEX COVER, this cannot happen, thus ϕ is a solution to our CNF–SAT instance.

Proposition 8. CONNECTED VERTEX COVER *in the class of* $K_{i,j}$*–free graphs admits a polynomial kernel for any* i, j.

Proof. Consider a graph $G = (V, E)$. We try to solve CONNECTED VERTEX COVER for (G, k). First, let $X = \{v \in V : \deg v > k\}$. Note that if there is a solution S for CONNECTED VERTEX COVER in (G, k), then $X \subset S$ — since if some $v \in X \setminus S$, then all the neighbours of v would have to be in S, but there are more than k of them. In particular if $|X| > k$, the answer is NO. Note we cannot remove X from G and analyse the remaining graph, for we could lose connectedness.

Consider the set F of edges which are not incident to X. If $|F| > k^2$, the answer is NO (for each vertex from V covers at most k such edges), let Z be the set of the endpoints of these edges, and let $Y = V \setminus (Z \cup X)$. Now for each vertex $x \in X$ add a vertex x' and an edge xx' (this is intended to assure that x is a part of any connected vertex cover not only of G, but also of the graphs we reduce G to, where the degree of x could drop). Denote the set of all vertices x' by X'.

If we do not know the answer yet, we have $|X| = |X'| \le k$ and $|Z| \le 2k^2$ (there are at most k^2 edges of which the vertices of Z are endpoints). Moreover $N(y) \subset X$ for any $y \in Y$.

Now if we have any two vertices $y_1, y_2 \in Y$ such that $N(y_1) \subset N(y_2)$, then the answer for G is the same as the answer for $G[V \setminus \{y_1\}]$. Indeed — if S is a solution in $G[V \setminus \{y_1\}]$, it is also a solution in G, since $N(y_1) \subset X \subset S$, as the edges xx' have to be covered, and thus all the edges incident to y_1 are covered by S, and of course $G[S]$ stays connected. On the other hand, if S is a solution in G, then either $y_1 \notin S$ (and then S is a solution in $G[V \setminus \{y_1\}]$) or $y_1 \in S$, and then $(S \cup \{y_2\}) \setminus \{y_1\}$ is a solution in $G[V \setminus \{y_1\}]$ (since y_2 connects everything that y_1 connected). Thus as long as a pair of vertices y_1, y_2 as above exists, we reduce G by removing y_1.

To simplify notation assume $i \le j$. Now we show that after these reductions $|Y| \le (i + j)k^i$. Consider any set $T \subset X$. There is at most one element $y \in Y$ such that $N(y) = T$ after the reductions. Moreover, if $|T| \ge i$, then there are at most $j - 1$ elements $y_1, \ldots, y_{j-1}$ of Y such that $N(y_l) \supset T$ — otherwise T and the y_ls would form a $K_{i,j}$ subgraph in G. For any element $y \in Y$ let $f(y) = N(y)$ if $|N(y)| < i$ and $f(y)$ be any $J \subset N(y), |J| = i$ if $|N(y)| \ge i$. There are at most $\sum_{l=0}^{i-1} \binom{k}{l} \le ik^i$ vertices y of the first type (as for such vertices each set appears at most once as the image of f) and at most $(j-1)\binom{k}{i} \le jk^i$ vertices of the second type (as for such vertices each set appears at most $j - 1$ times as the image of f). Thus after the reductions we have $|V| = |X| + |X'| + |Z| + |Y| \le k + k + 2k^2 + (i + j)k^i$, which is a polynomial of k.

The $K_{i,j}$–free graphs form a wider class than $(\min\{i, j\} - 1)$–degenerate graphs, thus CONNECTED VERTEX COVER admits a polynomial kernel in d–degenerate graphs for any d.

6   Conclusions and Open Problems

In this paper we investigated kernelization hardness in d-degenerate graphs for a number of problems that included the connectivity requirement. Generally, we proved that the bounded degeneracy assumption does not help much in existence of polynomial kernels. The question arises: does there exist a natural class larger than H-minor-free graphs or apex-minor-free graphs, for which CONNECTED DOMINATING SET, CONNECTED FEEDBACK VERTEX SET or CONNECTED ODD CYCLE TRANSVERSAL admit a small kernel?

Secondly, COLOURFUL GRAPH MOTIF appeared as a handy tool for proving kernelization hardness for 2–deg–CONNECTED DOMINATING SET and 2–deg–STEINER TREE. We believe this idea can inspire more negative results in the field of kernelization. In particular, such techniques may lead to a negative result for the question of existence of a polynomial kernel for PLANAR STEINER TREE, which today is a major open problem in kernelization and is not covered by meta-kernelization theorems of Bodlaender et al [3].

References

1. Alon, N., Gutner, S.: Linear time algorithms for finding a dominating set of fixed size in degenerated graphs. In: Lin, G. (ed.) COCOON 2007. LNCS, vol. 4598, pp. 394–405. Springer, Heidelberg (2007)
2. Bodlaender, H.L., Downey, R.G., Fellows, M.R., Hermelin, D.: On problems without polynomial kernels (extended abstract). In: Aceto, L., Damgård, I., Goldberg, L.A., Halldórsson, M.M., Ingólfsdóttir, A., Walukiewicz, I. (eds.) ICALP 2008, Part I. LNCS, vol. 5125, pp. 563–574. Springer, Heidelberg (2008)
3. Bodlaender, H.L., Fomin, F.V., Lokshtanov, D., Penninkx, E., Saurabh, S., Thilikos, D.M. (Meta) kernelization. In: Proc. of FOCS 2009, pp. 629–638 (2009)
4. Bodlaender, H.L., Thomasse, S., Yeo, A.: Analysis of data reduction: Transformations give evidence for non-existence of polynomial kernels, technical Report UU-CS-2008-030, Institute of Information and Computing Sciences, Utrecht University, Netherlands (2008)
5. Chen, J., Kanj, I.A., Jia, W.: Vertex cover: Further observations and further improvements. J. Algorithms 41(2), 280–301 (2001)
6. Dom, M., Lokshtanov, D., Saurabh, S.: Incompressibility through colors and IDs. In: Proc. of ICALP 2009, pp. 378–389 (2009)
7. Downey, R.G., Fellows, M.R.: Parameterized Complexity. Springer, Heidelberg (1999), http://citeseer.ist.psu.edu/downey98parameterized.html
8. Escoffier, B., Gourvès, L., Monnot, J.: Complexity and approximation results for the connected vertex cover problem. In: Brandstädt, A., Kratsch, D., Müller, H. (eds.) WG 2007. LNCS, vol. 4769, pp. 202–213. Springer, Heidelberg (2007)
9. Fellows, M.R., Fertin, G., Hermelin, D., Vialette, S.: Sharp tractability borderlines for finding connected motifs in vertex-colored graphs. In: Arge, L., Cachin, C., Jurdziński, T., Tarlecki, A. (eds.) ICALP 2007. LNCS, vol. 4596, pp. 340–351. Springer, Heidelberg (2007)
10. Fernau, H., Fomin, F.V., Lokshtanov, D., Raible, D., Saurabh, S., Villanger, Y.: Kernel(s) for problems with no kernel: On out-trees with many leaves. In: Proc. of STACS 2009, pp. 421–432 (2009)
11. Fomin, F., Lokshtanov, D., Saurabh, S., Thilikos, D.M.: Bidimensionality and kernels. In: Proc. of SODA 2010, pp. 503–510 (2010)
12. Fortnow, L., Santhanam, R.: Infeasibility of instance compression and succinct PCPs for NP. In: Proc. of STOC 2008, pp. 133–142 (2008)
13. Golovach, P.A., Villanger, Y.: Parameterized complexity for domination problems on degenerate graphs. In: Broersma, H., Erlebach, T., Friedetzky, T., Paulusma, D. (eds.) WG 2008. LNCS, vol. 5344, pp. 195–205. Springer, Heidelberg (2008)
14. Kostochka, A.V.: Lower bound of the hadwiger number of graphs by their average degree. Combinatorica 4(4), 307–316 (1984)
15. Misra, N., Philip, G., Raman, V., Saurabh, S., Sikdar, S.: FPT Algorithms for Connected Feedback Vertex Set. In: Rahman, M. S., Fujita, S. (eds.) WALCOM 2010. LNCS, vol. 5942, pp. 269–280. Springer, Heidelberg (2010)
16. Philip, G., Raman, V., Sikdar, S.: Solving dominating set in larger classes of graphs: Fpt algorithms and polynomial kernels. In: Fiat, A., Sanders, P. (eds.) ESA 2009. LNCS, vol. 5757, pp. 694–705. Springer, Heidelberg (2009)
17. Thomason, A.: An extremal function for contractions of graphs. Math. Proc. Cambridge Philos. Soc. 95(2), 261–265 (1984)
18. Thomason, A.: The extremal function for complete minors. J. Comb. Theory, Ser. B 81(2), 318–338 (2001)
19. Thomassé, S.: A quadratic kernel for feedback vertex set. In: Proc. of SODA 2009, pp. 115–119 (2009)

On the Boolean-Width of a Graph: Structure and Applications*

Isolde Adler[1], Binh-Minh Bui-Xuan[1], Yuri Rabinovich[2],
Gabriel Renault[1], Jan Arne Telle[1], and Martin Vatshelle[1]

[1] Department of Informatics, University of Bergen, Norway
[2] Department of Computer Science, Haifa University, Israel

Abstract. Boolean-width is a recently introduced graph invariant. Similar to tree-width, it measures the structural complexity of graphs. Given any graph G and a decomposition of G of boolean-width k, we give algorithms solving a large class of vertex subset and vertex partitioning problems in time $O^*(2^{O(k^2)})$. We relate the boolean-width of a graph to its branch-width and to the boolean-width of its incidence graph. For this we use a constructive proof method that also allows much simpler proofs of similar results on rank-width in [S. Oum. Rank-width is less than or equal to branch-width. *Journal of Graph Theory* 57(3):239–244, 2008]. For an n-vertex random graph, with a uniform edge distribution, we show that almost surely its boolean-width is $\Theta(\log^2 n)$ – setting boolean-width apart from other graph invariants – and it is easy to find a decomposition witnessing this. Combining our results gives algorithms that on input a random graph on n vertices will solve a large class of vertex subset and vertex partitioning problems in quasi-polynomial time $O^*(2^{O(\log^4 n)})$.

1 Introduction

Width parameters of graphs, like tree-width, branch-width, clique-width and rank-width, have many applications in the field of graph algorithms and especially in Fixed Parameter Tractable (FPT) algorithmics, see *e.g.* Downey and Fellows [7], Flum and Grohe [8], and Hliněný et al. [11]. When comparing width-parameters, we should consider the values of the parameters on various graph classes, the runtime of algorithms for finding the corresponding optimal decomposition, the classes of problems that can be solved by dynamic programming along such a decomposition, and the runtime of these algorithms. Recently, Bui-Xuan et al. [2] introduced a new width parameter of graphs called boolean-width. While rank-width is based on the number of GF(2)-sums $(1 + 1 = 0)$ of rows of adjacency matrices, boolean-width is based on the number of Boolean sums $(1 + 1 = 1)$ of these rows. Although is it open whether computing boolean-width is FPT, the number of Boolean sums of rows for a matrix is easy to compute in FPT time by an incremental approach, and surprisingly is the same for the matrix and its transpose.

---

* Supported by the Norwegian Research Council, projects PARALGO and Graph Searching.

D.M. Thilikos (Ed.): WG 2010, LNCS 6410, pp. 159–170, 2010.
© Springer-Verlag Berlin Heidelberg 2010

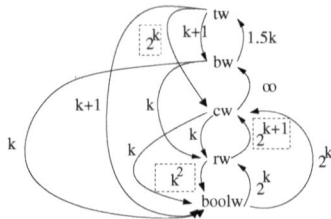

	tree-width	branch-width	clique-width	rank-width	boolean-width
MDS	$O^*(2^{1.58tw})$[22]	$O^*(2^{2bw})$ [6]	$O^*(2^{4cw})$ [17]	$O^*(2^{0.76rw^2})$ [3,9]	$O^*(2^{3boolw})$ [2]

Fig. 1. Upper bounds tying parameters tw=tree-width, bw=branch-width, cw=clique-width, rw=rank-width and $boolw$=boolean-width, and runtimes achievable for Minimum Dominating Set using various parameters. In the upper part of the figure, an arrow from P to Q labelled $f(k)$ means that any class of graphs having parameter P at most k will have parameter Q at most $f(k)$, and ∞ means that no such upper bound can be shown. Except for the labels in boxes the bounds are known to be tight, meaning that there is a class of graphs for which the bound is $\Omega(f(k))$. For the two boxes containing labels 2^k and 2^{k+1}, a $\Omega(2^{k/2})$ bound is known [4]. For the box containing label k^2 a $\Omega(k)$ bound is known [2]. The arrows $bw \rightarrow boolw$ and $tw \rightarrow boolw$ are proven in Section 4 of this paper.

This paper gives new algorithmic applications of boolean-width, and new structural properties of graphs of bounded boolean-width. It is well-known that for any class of graphs their tree-width is bounded by a constant if and only if their branch-width is bounded by a constant: we say the two parameters are equivalent. Likewise, clique-width, rank-width, and boolean-width are equivalent. For any graph class we have only three possibilities: either all five parameters are bounded (*e.g.* for trees) or none of them are bounded (*e.g.* for grids) or only clique-width, rank-width and boolean-width are bounded (*e.g.* for cliques). Capturing known results and new insights from Section 4, we show in Figure 1 information allowing for a finer comparison. Let us say that parameter P is polylog on a graph class C if the value of P for any graph G in C is polylogarithmic in the size of G. Then if P is polylog on C any algorithm with runtime[1] $O^*(2^{poly(P)})$ single exponential in parameter P runs in quasi-polynomial time on input a graph in C. From Figure 1 we see that if any of tree-width, branch-width, clique-width or rank-width is polylog on a class of graphs then so is boolean-width, while in Section 5 we show that the random graphs give an example of a class where boolean-width is polylog but none of the other parameters are. A finer comparison can be made by looking at the bounds between the parameters in combination with the runtimes achievable for a particular problem, as done for Minimum Dominating Set (MDS) in Figure 1. In this way, for MDS, and in fact all problems addressed in Section 3, boolean-width compares well to the other parameters. The paper is organized as follows.

---

[1] We use O^* notation that hides polynomial factors.

In Section 2 we define branch-width, rank-width, and boolean-width in a common framework. In Section 3 we depict algorithms that given a decomposition tree of boolean-width k of a graph, solve a large class of NP-hard vertex subset and vertex partitioning problems, namely (σ, ρ)-problems and D_q-problems [23], in time $O^*(2^{O(k^2)})$. These are monadic second order logic expressible problems related to domination, independence and homomorphism, including Max or Min Perfect Code, Max or Min Independent Dominating Set, Min k-Dominating Set, Max Induced k-Regular Subgraph, Max Induced k-Bounded Degree Subgraph, H-Coloring, H-Homomorphism, H-Covering, H-Partial Covering. From Courcelle's theorem [5] they belong to FPT when parameterized by either the tree-width, branch-width, clique-width, rank-width or boolean-width of the graph, when an appropriate decomposition is given. Although the runtime given in Courcelle's theorem contains a highly exponential factor (tower of powers), the problems behave very well for tree-width and branch-width: given a decomposition tree of tree-width tw, they can be solved in $O^*(2^{O(tw)})$ time [23]. In particular, (σ, ρ)-problems can be solved in $O^*((d(\sigma) + d(\rho) + 2)^{tw})$ time [22] for some problem specific constants $d(\sigma)$ and $d(\rho)$ (see Section 3). This is not the same situation for clique-width, where until now the best runtime contains a $O^*(2^{2^{poly(cw)}})$ double exponential factor [10]. Having small boolean-width is witnessed by a decomposition of the graph into cuts with few different unions of neighborhoods across the cut. This makes the decomposition natural to guide dynamic programming algorithms to solve problems, like Max Independent Set, where vertex sets having the same neighborhoods can be treated as equivalent [2]. Surprisingly, in this paper we extend such an observation to the much larger class of vertex subset and vertex partitioning problems. Several new techniques are introduced in order to achieve this and the runtime of these algorithms is $O^*(2^{O(boolw^2)})$, which then can also be interpreted as $O^*(2^{O(cw^2)})$ and $O^*(2^{O(rw^4)})$ by using the relationships in Figure 1, improving the $O^*(2^{2^{poly(cw)}})$ runtime in [10].

In Section 4 we relate boolean-width to branch-width. We prove for every graph G with $\mathbf{bw}(G) \neq 0$ that $\mathbf{boolw}(G) \leq \mathbf{bw}(G)$. For the proof we develop a general method of constructive manipulations of the decompositions that gives a good understanding of the connections between the graph parameters. In [20], Oum studies the relation of rank-width and branch-width using deep results from matroid theory. Our framework also allows to address this relation in a simpler and direct way. Independently, Kanté [14] gave a constructive proof showing that the rank-width of a graph is at most 4 times its tree-width plus 2. We show constructively that (except for some trivial cases) rank-width is at most branch-width, and also constructively that rank-width is at most tree-width plus one, simplifying Oum's proof and improving Kanté's construction.

In Section 5 we show for a random graph on n vertices where the edges are drawn with respect to a uniform distribution that almost surely[2] its boolean-width is $\Theta(\log^2 n)$, and it is easy to find a decomposition tree witnessing the upper bound. This contrasts sharply with a series of negative results establishing that almost

---

[2] We use term "almost surely" to denote events whose asymptotic probability is 1.

surely a random graph on n vertices has tree-width and branch-width [16], clique-width [13] and rank-width [18] all in $\Theta(n)$. The importance of this result is possibly not in the random graphs themselves, but in the indication that boolean-width is quite often much smaller than all the other parameters, and therefore potentially very useful. Our result also implies the following: any problem solvable by dynamic programming in time $O^*(2^{poly(k)})$ *given* a decomposition of boolean-width k, can be solved in quasi-polynomial time on input a random graph (where we do not need a decomposition as part of input). Such problems include Minimum Dominating Set and Maximum Independent Set which can be solved in time $O(n(n + 2^{3k}k))$ [2]. Moreover, combining our results from Sections 3 and 5 we get an algorithm that given a random graph on n vertices, solves (σ, ρ)-problems and D_q-problems in quasi-polynomial time $O^*(2^{O(\log^4 n)})$.

2   Framework

We address loopless simple undirected graphs. Let G be a graph with vertex set $V(G)$ and edge set $E(G)$. For a vertex $v \in V(G)$ let $N(v)$ be the set of all neighbours of v in G. We extend this to subsets $X \subseteq V(G)$ by letting $N(X) := \bigcup_{v \in X} N(v)$. For a tree T we denote the set of leaves by $L(T)$. A tree is *subcubic* if every vertex has degree either 1 or 3.

Let A be a finite set. For a subset $X \subseteq A$ let $\overline{X} := A \setminus X$. Let $f: 2^A \to \mathbb{R}$ be a *symmetric* set function, i.e. f satisfies $f(X) = f(\overline{X})$ for all $X \subseteq A$. A *decomposition tree* of f (on A) is a pair (T, δ), where T is a subcubic tree and $\delta: L(T) \to A$ is a bijection. Each edge $e \in E(T)$ yields a partition P_e of A, induced by the leaf labels of the two trees we get by removing e from T: if T_1 and T_2 denote the two components of $T - e$, then $P_e := \big(\delta(L(T_1) \cap L(T)), \delta(L(T_2) \cap L(T))\big)$. We extend the domain of f to edges e of T by letting $f(e) := f(X)$ for $P_e = (X, \overline{X})$. This is well-defined because f is symmetric. The *f-width* of a decomposition tree (T, δ) is $f\text{-w}(T, \delta) := \max\{f(e) \mid e \in E(T)\}$. The *width* of f is $\mathbf{width}(f) := \min\{f\text{-w}(T, \delta) \mid (T, \delta) \text{ decomposition tree of } f\}$. If $|A| \leq 1$, then f has no decomposition tree and we let $\mathbf{width}(f) := f(A)$.

We now define branch-width of a graph G. For any subset $X \subseteq E(G)$ let

$$\partial(X) := \{v \in V(G) \mid v \text{ incident to both an edge from } X \text{ and from } \overline{X}\}$$

denote the *border* of X. We define $\mathbf{cut\text{-}bw}_G := \mathbf{cut\text{-}bw}: 2^{E(G)} \to \mathbb{N}$ as $\mathbf{cut\text{-}bw}(X) := |\partial(X)|$. Clearly, $\mathbf{cut\text{-}bw}$ is symmetric. The *branch-width* of G is defined as $\mathbf{bw}(G) := \mathbf{width}(\mathbf{cut\text{-}bw})$.

For subsets $X, Y \subseteq V(G)$ let $M_{(X,Y)}$ denote the $X \times Y$-submatrix of the adjacency matrix of G. Let Δ denote the symmetric difference of sets: $A \Delta B = (A \setminus B) \cup (B \setminus A)$. We define $\mathbf{cut\text{-}rk}_G := \mathbf{cut\text{-}rk}: 2^{V(G)} \to \mathbb{N}$ as

$$\mathbf{cut\text{-}rk}(X) := \log_2 \left| \left\{ B \subseteq \overline{X} \,\middle|\, \exists A \subseteq X,\ B = \bigwedge_{v \in A} N(v) \cap \overline{X} \right\} \right| = \mathrm{rk}\big(M_{(X,\overline{X})}\big),$$

where $\mathrm{rk}(M_{(X,\overline{X})})$ denotes the GF(2)-rank of $M_{(X,\overline{X})}$. Then the *rank-width* of G is $\mathbf{rw}(G) := \mathbf{width}(\mathbf{cut\text{-}rk})$.

For boolean-width we define $\mathbf{cut\text{-}bool}_G := \mathbf{cut\text{-}bool} \colon 2^{V(G)} \to \mathbb{R}$ as

$$\mathbf{cut\text{-}bool}(X) := \log_2 \left| \{ B \subseteq \overline{X} \mid \exists A \subseteq X \text{ with } B = N(A) \cap \overline{X} \} \right|.$$

Surprisingly, the function $\mathbf{cut\text{-}bool}$ is symmetric [15, Theorem 1.2.3]. The *boolean-width* of a graph G is $\mathbf{boolw}(G) := \mathbf{width}(\mathbf{cut\text{-}bool})$. Let us give an alternative view on boolean-width. Let $R(M_{(X,Y)})$ denote the set of all vectors spanned by the rows of $M_{(X,Y)}$ by taking Boolean sums, i.e. $1 + 1 = 1$. It is easy to see that
$$\mathbf{cut\text{-}bool}(X) = \log_2 \left| R(M_{(X,\overline{X})}) \right|.$$

3 Vertex Subset and Vertex Partitioning Problems

Given a graph G together with a decomposition tree of $\mathbf{cut\text{-}bool}$ of width $boolw$, we depict algorithms with runtime $O^*(2^{O(boolw^2)})$ solving a large class of problems, the so-called (σ, ρ) vertex subset and D_q vertex partitioning problems as defined in [23].

Definition 1. Let σ and ρ be finite or co-finite subsets of natural numbers. A subset X of vertices of a graph G is a *sigma-rho set*, or simply (σ, ρ)-*set*, of G if

$$\forall v \in V(G) : |N(v) \cap X| \in \begin{cases} \sigma \text{ if } v \in X, \\ \rho \text{ if } v \in V(G) \setminus X. \end{cases}$$

The *vertex subset problems* consist of finding the size of a minimum or maximum (σ,ρ)-set in G. Several NP-hard problems are expressible in this framework, *e.g.*, Max Independent Set($\{0\}$, $\mathbb{N}$), Min Dominating Set($\mathbb{N}$, $\mathbb{N} \setminus \{0\}$), Max Strong Stable Set($\{0\}$, $\{0,1\}$), Max or Min Perfect Code($\{0\}$, $\{1\}$). Also if we let $M_k = \{0, 1, 2, \ldots k\}$ then Min k-Dominating Set($\mathbb{N}$, $\mathbb{N} \setminus M_k$), Max Induced k-Regular Subgraph($\{k\}$, $\mathbb{N}$) (see [23] for further details and a more complete list). This framework is extendible to problems asking for a partition of $V(G)$ into q classes, with each class satisfying a certain (σ, ρ)-property:

Definition 2. A *degree constraint* matrix D_q is a q by q matrix with entries being finite or co-finite subsets of natural numbers. A D_q-*partition* in a graph G is a partition $\{V_1, V_2, ..., V_q\}$ of $V(G)$ such that for $1 \leq i, j \leq q$ we have $\forall v \in V_i : |N(v) \cap V_j| \in D_q[i, j]$.

The *vertex partitioning problems* for which we give algorithms in this paper consist of deciding if G has a D_q partition, the so-called $\exists D_q$ problem. NP-hard problems fitting into this framework include *e.g.* for any fixed graph H the problems known as H-Coloring or H-Homomorphism (with q-Coloring being K_q-Coloring), H-Covering, H-Partial Covering, and in general the question of deciding if a graph has a partition into q (σ, ρ)-sets [23].

We focus on algorithms for the vertex subset problems. Let a graph G and a decomposition tree (T, δ) of $\mathbf{cut\text{-}bool}$ be given as input. Our algorithm will follow a bottom-up dynamic programming approach: subdivide an arbitrary edge of T to obtain a root r, and denote by T_r the resulting rooted tree. With each node w of T_r we associate a table data structure Tab_w, that will store optimal solutions to subproblems related to V_w, the set of vertices of G mapped to the leaves of the subtree of T_r rooted at w. Each index of the table will be associated with a certain class of equivalent subproblems that we need to define depending on the problem on which we are focusing.

Let $d(\mathbb{N}) = 0$ and let $d(\emptyset) = 0$. For every finite or co-finite set $\mu \subseteq \mathbb{N}$, let $d(\mu) = 1 + min\{max_{x \in \mathbb{N}} x : x \in \mu, max_{x \in \mathbb{N}} x : x \notin \mu\}$. We denote by $d(\sigma, \rho)$, or simply by d when it appears clearly in the context that σ and ρ are involved, the value $d = d(\sigma, \rho) = max\{d(\sigma), d(\rho)\}$. Note that when checking if a subset A of vertices is a (σ, ρ)-set, as in Definition 1, it suffices to count the number of neighbors up to d that a vertex has in A. This is the key to getting fast algorithms and motivates the following equivalence relation.

Definition 3 (d-neighbor equivalence). Let G be a graph and $A \subseteq V(G)$. Two vertex subsets $X \subseteq A$ and $X' \subseteq A$ are d-neighbor equivalent w.r.t. A, denoted by $X \equiv_A^d X'$, if

$$\forall v \in \overline{A}, \ (|N(v) \cap X| = |N(v) \cap X'|) \vee (|N(v) \cap X| \geq d \wedge |N(v) \cap X'| \geq d).$$

We now depict the entries of the table data structure Tab_w. Roughly, we aim at solving the vertex subset problems using one d-neighbor equivalence class per entry in Tab_w. For this, we first define a canonical representative for every d-neighbor equivalence class.

Lemma 1. Let G be a graph and $A \subseteq V(G)$. Then, for every $X \subseteq A$, there is $R \subseteq A$ such that $R \equiv_A^d X$ and $|R| \leq d \cdot \textbf{cut-bool}(A)$. Moreover, the number of equivalence classes of $\equiv_A^d$ is at most $2^{d \cdot \textbf{cut-bool}(A)^2}$.

We now define the canonical representative $can_{V_w}^d(X)$ of every subset $X \subseteq V_w$, and the canonical representative $can_{\overline{V_w}}^d(Y)$ of every subset $Y \subseteq \overline{V_w}$. For simplicity we define this for V_w only, but the definition can be used for $\overline{V_w}$ as well, since everything we say about $X \subseteq V_w$, $can_{V_w}^d(X)$ and $\equiv_{V_w}^d$ will hold also for $can_{\overline{V_w}}^d(Y)$, $Y \subseteq \overline{V_w}$ and $\equiv_{\overline{V_w}}^d$. Canonical representatives are to be used for indexing the table Tab_w at node w of the tree T_r. Three properties will be required. Firstly, if $X \equiv_{V_w}^d X'$, then we must have $can_{V_w}^d(X) = can_{V_w}^d(X')$. Secondly, given (X, Y), we should have a fast routine that outputs a pointer to the entry $Tab_w[can_{V_w}^d(X)][can_{\overline{V_w}}^d(Y)]$. Thirdly, we should have a list whose elements can be used as entries of the table, i.e. a list containing all canonical representatives w.r.t. $\equiv_{V_w}^d$. The following definition trivially fulfills the first requirement.

Definition 4. We assume that a total ordering of the vertices of $V(G)$ is given. For every $X \subseteq V_w$, the canonical representative $can_{V_w}^d(X)$ is defined as the lexicographically smallest set $R \subseteq V_w$ such that: $|R|$ is minimized and $R \equiv_{V_w}^d X$.

Definition 5. Let G be a graph, $A \subseteq V(G)$, and $\mu \subseteq \mathbb{N}$. For $X \subseteq V(G)$, we say that X μ-dominates A if $\forall v \in A : |N(v) \cap X| \in \mu$. For $X \subseteq A$, $Y \subseteq \overline{A}$, we say that (X, Y) σ, ρ-*dominates* A if $(X \cup Y)$ σ-dominates X and $(X \cup Y)$ ρ-dominates $A \setminus X$.

Definition 6. Let *opt* stand for either function *max* or function *min*, depending on whether we are looking for a maximum or minimum (σ, ρ)-set, respectively. For every node w of T_r, for $X \subseteq V_w$ and $Y \subseteq \overline{V_w}$, let $R_X = can_{V_w}^d(X)$ and $R_Y = can_{\overline{V_w}}^d(Y)$. We define the contents of $Tab_w[R_X][R_Y]$ as:

$$Tab_w[R_X][R_Y] \stackrel{\text{def}}{=} \begin{cases} opt_{S \subseteq V_w}\{|S| : S \equiv_{V_w}^d X \text{ and } (S, Y) \; \sigma, \rho\text{-dominates } V_w\}, \\ -\infty \text{ if no such set } S \text{ exists and } opt = max, \\ +\infty \text{ if no such set } S \text{ exists and } opt = min. \end{cases}$$

Lemma 2. *For any node w of T_r with $k = $ **cut-bool**(V_w), we can compute a list containing all canonical representatives w.r.t. $\equiv_{V_w}^d$ in time $O(m + d \cdot k \cdot 2^{2d \cdot k^2 + k})$. For any subset $X \subseteq V_w$, a pointer to $can_{V_w}^d(X)$ can be found in time $O(|X| \cdot 2^k)$.*

Note that at the root r of T_r the value of $Tab_r[X][\emptyset]$ (for all $X \subseteq V(G)$) would be exactly equal to the size of a maximum, resp. minimum, (σ, ρ)-set of G (*cf.* $\equiv_{V_r}^d$ has only one equivalence class). For initialization, the value of every entry of Tab_w will be set to $+\infty$ or $-\infty$ depending on whether we are solving a minimization or maximization problem, respectively. For a leaf l of T_r, we perform a brute-force update: let $A = \{l\}$ and $B = \overline{A}$, for every canonical representative R w.r.t. $\equiv_B^d$, we set:

- If $|N(l) \cap R| \in \sigma$ then $Tab_l[A][R] = 1$.
- If $|N(l) \cap R| \in \rho$ then $Tab_l[\emptyset][R] = 0$.

For a node w of T_r with children a and b, the algorithm proceeds as follows. For every canonical representative $R_{\overline{w}}$ w.r.t. $\equiv_{\overline{V_w}}^d$, for every canonical representative R_a w.r.t. $\equiv_{V_a}^d$, and for every canonical representative R_b w.r.t. $\equiv_{V_b}^d$, do:

- Compute $R_w = can_{V_w}^d(R_a \cup R_b)$, $R_{\overline{a}} = can_{\overline{V_a}}^d(R_b \cup R_{\overline{w}})$ and $R_{\overline{b}} = can_{\overline{V_b}}^d(R_a \cup R_{\overline{w}})$
- Update $Tab_w[R_w][R_{\overline{w}}] = opt(Tab_w[R_w][R_{\overline{w}}], Tab_a[R_a][R_{\overline{a}}] + Tab_b[R_b][R_{\overline{b}}])$.

Lemma 3. *The table at node w is updated correctly, i.e. for any canonical representatives R_w and $R_{\overline{w}}$ w.r.t. $\equiv_{V_w}^d$ and $\equiv_{\overline{V_w}}^d$, if $Tab_w[R_w][R_{\overline{w}}]$ is not $\pm\infty$ then*

$$Tab_w[R_w][R_{\overline{w}}] = opt_{S \subseteq V_w}\{|S| : S \equiv_{V_w}^d R_w \wedge (S, R_{\overline{w}}) \; \sigma, \rho\text{-dominates } V_w\}.$$

If the value of the table is $\pm\infty$ then there is no such above set S.

Theorem 1. *For every n-vertex, m-edge graph G given along with a decomposition tree (T, δ) for **cut-bool**, any (σ, ρ)-vertex subset problem on G with $d = d(\sigma, \rho)$ can be solved in time*
$$O(n(m + d \cdot \mathbf{cut\text{-}bool\text{-}w}(T, \delta)2^{3d \cdot \mathbf{cut\text{-}bool\text{-}w}(T,\delta)^2 + \mathbf{cut\text{-}bool\text{-}w}(T,\delta)})).$$

Proof. Correctness follows directly from what has been said in this section. For complexity analysis, for every node w of T_r, we basically call the first computation of Lemma 2 once, then loop through every triplet $R_{\overline{w}}$, R_a, R_b of equivalence classes, call the second computation of Lemma 2 three times, and perform the table update.   □

The algorithms for vertex partitioning problems are similar but require some graph-theoretic observations and several technical details. For space reasons this has all been omitted.

Theorem 2. *For every n-vertex, m-edge graph G given along with a decomposition tree (T, δ) of* **cut-bool***, any D_q-problem on G, with $d = \max_{i,j} d(D_q[i,j])$, can be solved in time*
$$O(n(m + qd \cdot \textbf{cut-bool-w}(T,\delta)2^{3qd\cdot\textbf{cut-bool-w}(T,\delta)^2+\textbf{cut-bool-w}(T,\delta)})).$$

4   Boolean-Width is Less than or Equal to Branch-Width

We relate boolean-width to branch-width, and show the following

Theorem 3. *Any graph G satisfies* **boolw**$(G) \leq$ **bw**(G) *(unless $E(G) \neq \emptyset$ and no two edges of G are adjacent).*

In order to clarify how the decomposition trees relate to each other, we divide our result into two steps, addressing the intermediary notion of an incidence graph (see Lemmata 4 and 5). However, we will also show how to easily derive from our method a direct proof without incidence graphs. Our framework not only applies for boolean-width, but also captures other settings including rank-width. The *incidence graph* $I(G)$ of a graph G is the graph with vertex set $V(G) \dot\cup E(G)$, where x and y are adjacent in $I(G)$ if one of x, y is a vertex of G, the other is an edge of G and x and y are incident in G.

Lemma 4. *For any graph G,* **boolw**$(I(G)) \leq$ **bw**(G) *and* **rw**$(I(G)) \leq$ **bw**(G), *unless $E(G) \neq \emptyset$ and no two edges of G are adjacent. In this case,* **bw**$(G) = 0$ *and* **rw**$(I(G)) =$ **boolw**$(I(G)) = 1$.

The proof is omitted, but let us sketch the idea. Starting with a decomposition tree (T, δ) of **cut-bw**$_G$ of width k, we modify the decomposition tree in two steps. In the first step, we replace every leaf ℓ of T by a subcubic tree with three leaves, and we label one of the three leaves with the edge $\delta(\ell)$ and we label the other two leaves with the two vertices incident with $\delta(\ell)$. In a second step, for each $v \in V(G)$ we choose one leaf with label v, we keep this leaf, and delete all other leaves that are labelled by v. In this way we obtain a decomposition tree of **cut-bool**$_{I(G)}$ (and of **cut-rk**$_{I(G)}$) of boolean-width and rank-width both at most k.

Lemma 5. *For any graph G,*
$$\max\{\textbf{boolw}(G), \textbf{rw}(G)\} \leq \min\{\textbf{boolw}(I(G)), \textbf{rw}(I(G))\}.$$

Theorem 3 now follows immediately from Lemmata 4 and 5, as well as the fact that rank-width is at most branch-width (except for the trivial cases). It is also easy to give a direct proof using the proof idea of Lemma 4. The only difference is in the first modification step. Instead of taking a subcubic tree with three leaves, we take the subcubic tree with two leaves (since we do not need to assign leaves to graph edges). Note that there is no bound in the converse direction: the class of all complete graphs has unbounded branch-width and the boolean-width is at most 1. Nevertheless, moving to incidence graphs we prove a weak converse.

Lemma 6. *For any graph G, $\mathbf{bw}(G) \leq 2 \cdot \min\{\mathbf{boolw}(I(G)), \mathbf{rw}(I(G))\}$.*

Corollary 1. *Any graph G satisfies $\mathbf{boolw}(G) \leq \mathbf{bw}(G) \leq 2 \cdot \mathbf{boolw}(I(G))$ (unless $E(G) \neq \emptyset$ and no two edges of G are adjacent).*

Corollary 2. *For any graph G,*

   1. $\mathbf{boolw}(I(G)) \leq \mathbf{bw}(I(G)) \leq 2 \cdot \mathbf{boolw}(I(G))$ and
   2. $\mathbf{boolw}(I(G)) \leq \mathbf{rw}(I(G)) + 1 \leq 2 \cdot \mathbf{boolw}(I(G)) + 1$.

Proof. Note that $\mathbf{bw}(G) = \mathbf{bw}(I(G))$, unless $E(G) \neq \emptyset$ and no two edges of G are adjacent. In this case, $\mathbf{bw}(G) = 0$ and $\mathbf{bw}(I(G)) = 1$. Then, the first statement follows from Lemmata 4 and 6. The second statement follows from the first by using a theorem from [20] stating that $\mathbf{rw}(I(G)) \in \{\mathbf{bw}(G), \mathbf{bw}(G) - 1\}$. □

5 Random Graphs

Let G_p be a random graph on n vertices where each edge is chosen randomly and independently with probability p (independent of n). There has been a series of negative results [13,16,18] establishing that almost surely G_p has rank-width, tree-width, branch-width and clique-width $\Theta(n)$. In contrast we show in this section the following.

Theorem 4. *Almost surely, $\mathbf{boolw}(G_p) = \Theta\left(\frac{\ln^2 n}{p}\right)$.*

We start with the upper bound and first prove the following lemma.

Lemma 7. *Let G_p be a graph as above, and let $k_p = \lfloor \frac{2 \ln n}{p} \rfloor$. Then, almost surely, for all subsets of vertices $S \subset V(G)$ with $|S| = k_p$ it holds that $|N(S) \backslash S| \geq |\overline{S}| - k_p$.*

Proof. In what follows, we write simply G and k. Fix a particular S with $|S| = k$. For every $v \in \overline{S}$, let X_v be 1 if $v \notin N(S)$, and 0 otherwise. Clearly, $X_v = 1$ with probability $(1 - p)^k$, and $\sum_{v \in \overline{S}} X_v = |\overline{S} \setminus N(S)|$. Observe that $E[\sum_{v \notin S} X_v] = (1 - p)^k (n - k) < (1 - p)^k n$. Call this expectation μ. By Chernoff's Bound (see e.g. [19], p.68),

$$\Pr\left[\sum_{v \in \overline{S}} X_v \geq k\right] < \left(\frac{e\mu}{k}\right)^k < \left((1 - p)^k n\right)^k = \left((1 - p)^{2 \ln n / p} n\right)^k < n^{-k},$$

the last inequality due to the fact that for $p \in (0, 1)$, $(1 - p)^{\frac{1}{p}} \leq e^{-1}$.

Applying the union bound, we conclude that the probability that there exists S of size k such that $|N(S) \setminus S| < |\overline{S}| - k$ is at most $\binom{n}{k} \cdot n^{-k} < (k!)^{-1} = o(1)$ and the statement follows. $\square$

Corollary 3. *For $G = G_p$ and $k = k_p$ as before, for all cuts $\{A, \overline{A}\}$ in G it holds almost surely that $\mathbf{cut\text{-}bool}(A) = O\left(\frac{\ln^2 n}{p}\right)$.*

Proof. The number of distinct sets $N(S) \cap \overline{A}$ contributed by the sets $S \subseteq A$ with $|S| \leq k$ is at most $\sum_{i=0}^{k} \binom{n}{i}$. By the previous lemma, for all sets $S \subseteq A$ with $|S| \geq k$, it holds almost surely that $|N(S) \cap \overline{A}| \geq |\overline{A}| - k$. Therefore, almost surely, the sets $S \subseteq A$ with $|S| \geq k$, also contribute at most $\sum_{i=0}^{k} \binom{n}{i}$ distinct sets $N(S) \cap \overline{A}$. Thus, almost surely there are at most $2 \sum_{i=0}^{k} \binom{n}{i}$ distinct sets $N(S) \cap \overline{A}$ altogether. Taking the logarithm allows to conclude. $\square$

The upper bound of Theorem 4 now follows easily: for *any* decomposition tree of **cut-bool**, all the cuts it defines will almost surely have boolean-width at most $O\left(\frac{\ln^2 n}{p}\right)$. Next, we move to the lower bound of Theorem 4. For simplicity of exposition, we restrict the discussion to the case $p = 0.5$. The lower bound for that case follows from:

Lemma 8. *Let $\{A, \overline{A}\}$ be a cut where $|A| = |\overline{A}| = m$, and the edges are chosen independently at random with probability 0.5. Then, $\Pr[\ \mathbf{cut\text{-}bool}(A) = \Omega(\log^2 m)\ ] \geq 1 - 2^{-\Omega(m^{1.3})}$. More concretely, the probability that among the neighborhoods of the subsets of A of size $k = 0.25 \cdot \log_2 m$, there are less than $2^{c \log^2 m}$ different ones (for a suitable constant c), is at most $2^{-\Omega(m^{1.3})}$.*

To prove this lemma we need some notation and preliminary results first. Let the (random) set $S_i \subseteq \overline{A}$, $i = 1, 2, \dots, m$ be the neighborhood of the vertex $i \in A$, and let $S_I = \cup_{i \in I} S_i$. We shall only be interested in the I's of size k as above. Call such I *bad* if $m - |S_I| < m^{0.5}$. Call a set I of size k *thick* if there are at least $m^{0.9}$ indices $i \in \{1, 2, \dots, m\} - I$ such that $S_i \subseteq S_I$. Lemma 8 can be proved using below Corollary 4.

Claim 1. $\Pr\left[\dfrac{\text{the number of bad } I\text{'s}}{\binom{m}{k}} \geq 0.5\right] < e^{-\Omega(m^{1.74})}.$

Claim 2. *For a fixed set I of size k, the probability that I is thick conditioned on its being good (that is, not bad), is at most $e^{-\Omega(m^{1.3})}$.*

Corollary 4. $\Pr\left[\text{the number of thick } I\text{'s} > 0.5 \cdot \binom{m}{k}\right] < e^{-\Omega(m^{1.3})}.$

Proof of Theorem 4: The upper bound has already been proved. For the lower bound we restrict for simplicity of exposition to the case $p = 0.5$. Consider a $(\frac{1}{3}, \frac{2}{3})$-balanced cut in G, that is a cut $(X, \overline{X})$ with $\frac{n}{3} \leq |X| \leq \frac{2n}{3}$. Due to the monotonicity of the **cut-bool** with respect to taking induced subcuts, Lemma 8 applies in this case with $m = n/3$. Therefore, the probability that **cut-bool** of this cut is $\Omega(\log^2 n)$ is $1 - e^{-\Omega(n^{1.3})}$. Since there at most 2^n cuts in G, we conclude that with probability $1 - e^{-\Omega(n^{1.3})}$ *all* balanced cuts have such **cut-bool**. Since any decomposition tree of G must contain a $(\frac{1}{3}, \frac{2}{3})$-balanced cut, the statement follows. □

6   Further Research

In this paper we have seen that for random graphs boolean-width is the right parameter to consider: any decomposition tree will have boolean-width polylog-arithmic in n. This also hints at the existence of large classes of graphs where boolean-width is polylogarithmic in the value of the other parameters, and raises the question of identifying these. One such class of graphs is defined by the so-called Hsu-grids [2], where boolean-width is $\Theta(\log n)$ and rank-width, branch-width, tree-width and clique-width are $\Theta(\sqrt{n})$. In contrast, we know that the boolean-width of regular graphs is $\Theta(n)$ [21], thus such an above mentioned class should exclude regular graphs.

We believe that boolean-width should be useful for *practical* applications. We have initiated research to find fast and good heuristics computing decomposi-tions of low boolean-width [12], similar to what is done for treewidth in the TreewidthLIB project [1].

A big open question is to decide if the boolean-width of a graph can be computed in FPT time. The relationship between rank-width and boolean-width is still not completely clear. Could it be that the boolean-width of any graph is linear in its rank-width? Currently the best bound is $boolw(G) \le \frac{1}{4}rw(G)^2 + \frac{5}{4}rw(G) + \log rw(G)$ [2].

The runtime of the algorithms given here for (σ, ρ)-problems and D_q-problems have the square of the boolean-width as a factor in the exponent. For problems where $d = 1$ we can in fact improve this to a factor linear in the exponent [2], but that requires a special focus on these cases. In fact, we believe that also for the other problems (with any constant value of d) we could get runtimes with an exponential factor linear in boolean-width. We must then improve the bound in Lemma 1, by showing that the number of d-neighborhood equivalence classes is no more than the number of 1-neighborhood equivalence classes raised to some function of d. This question can be formulated as a purely algebraic one as follows: First generalize the concept of Boolean sums $(1 + 1 = 1)$ to d-Boolean sums $(i + j = \min(i + j, d))$. For a Boolean matrix A let $R_d(A)$ be the set of vectors over $\{0, 1, ..., d\}$ that arise from all possible d-Boolean sums of rows of A. Is there a function f such that $|R_d(A)| \le |R_1(A)|^{f(d) \log \log |R_1(A)|}$?

References

1. Bodlaender, H., Koster, A.: Treewidth Computations I Upper Bounds. Techni-cal Report UU-CS-2008-032, Department of Information and Computing Sciences, Utrecht University (2008)
2. Bui-Xuan, B.-M., Telle, J.A., Vatshelle, M.: Boolean-width of graphs. In: 4th In-ternational Workshop on Parameterized and Exact Computation (IWPEC 2009). LNCS, vol. 5917, pp. 61–74. Springer, Heidelberg (2009)

3. Bui-Xuan, B.-M., Telle, J.A., Vatshelle, M.: H-join decomposable graphs and algorithms with runtime single exponential in rankwidth. Discrete Applied Mathematics 158(7), 809–819 (2010)
4. Corneil, D., Rotics, U.: On the relationship between clique-width and treewidth. SIAM Journal on Computing 34(4), 825–847 (2005)
5. Courcelle, B.: Graph rewriting: An algebraic and logic approach. In: Handbook of Theoretical Computer Science, Volume B: Formal Models and Sematics (B), pp. 193–242 (1990)
6. Dorn, F.: Dynamic programming and fast matrix multiplication. In: Azar, Y., Erlebach, T. (eds.) ESA 2006. LNCS, vol. 4168, pp. 280–291. Springer, Heidelberg (2006)
7. Downey, R., Fellows, M.: Parameterized Complexity. Springer, Heidelberg (1999)
8. Flum, J., Grohe, M.: Parameterized Complexity Theory. Springer, Heidelberg (2006)
9. Ganian, R., Hliněný, P.: On Parse Trees and Myhill-Nerode-type Tools for handling Graphs of Bounded Rank-width. Discrete Applied Mathematics 158(7), 851–867 (2010)
10. Gerber, M., Kobler, D.: Algorithms for vertex-partitioning problems on graphs with fixed clique-width. Theoretical Computer Science 299(1-3), 719–734 (2003)
11. Hliněný, P., Oum, S., Seese, D., Gottlob, G.: Width parameters beyond tree-width and their applications. The Computer Journal 51(3), 326–362 (2008)
12. Hvidevold, E.: Implementation of heuristics for computing boolean-width, Master thesis, University of Bergen (September 2010) (to appear)
13. Johansson, Ö.: Clique-decomposition, NLC-decomposition and modular decomposition – Relatiohships and results for random graphs. Congressus Numerantium 132, 39–60 (1998)
14. Kanté, M.: Vertex-minor reductions can simulate edge contractions. Discrete Applied Mathematics 155(17), 2328–2340 (2007)
15. Kim, K.H.: Boolean matrix theory and applications. Marcel Dekker, New York (1982)
16. Kloks, T., Bodlaender, H.: Only few graphs have bounded treewidth. Technical Report UU-CS-92-35, Department of Information and Computing Sciences, Utrecht University (1992)
17. Kobler, D., Rotics, U.: Edge dominating set and colorings on graphs with fixed clique-width. Discrete Applied Mathematics 126(2-3), 197–221 (2003); Abstract at SODA 2001
18. Lee, C., Lee, J., Oum, S.: Rank-width of Random Graphs, http://arxiv.org/pdf/1001.0461
19. Motwani, R., Raghavan, P.: Randomized Algorithms. Cambridge University Press, Cambridge (1995)
20. Oum, S.: Rank-width is less than or equal to branch-width. Journal of Graph Theory 57(3), 239–244 (2008)
21. Rabinovich, Y., Telle, J.A.: On the boolean-width of a graph: structure and applications, http://arxiv.org/pdf/0908.2765
22. Rooij, J., Bodlaender, H., Rossmanith, P.: Dynamic programming on tree decompositions using generalised fast subset convolution. In: Fiat, A., Sanders, P. (eds.) ESA 2009. LNCS, vol. 5757, pp. 566–577. Springer, Heidelberg (2009)
23. Telle, J.A., Proskurowski, A.: Algorithms for vertex partitioning problems on partial k-trees. SIAM Journal on Discrete Mathematics 10(4), 529–550 (1997)

Generalized Graph Clustering: Recognizing (p, q)-Cluster Graphs

Pinar Heggernes[1], Daniel Lokshtanov[1], Jesper Nederlof[1],
Christophe Paul[2], and Jan Arne Telle[1]

[1] Department of Informatics, University of Bergen,
P.O. Box 7803, N-5020 Bergen, Norway
{pinar.heggernes,daniel.lokshtanov,
jesper.nederlof,jan.arne.telle}@ii.uib.no
[2] CNRS, LIRMM, Université Montpellier 2, France
paul@lirrm.fr

Abstract. CLUSTER EDITING is a classical graph theoretic approach to tackle the problem of data set clustering: it consists of modifying a similarity graph into a disjoint union of cliques, i.e, clusters. As pointed out in a number of recent papers, the cluster editing model is too rigid to capture common features of real data sets. Several generalizations have thereby been proposed. In this paper, we introduce (p, q)-cluster graphs, where each cluster misses at most p edges to be a clique, and there are at most q edges between a cluster and other clusters. Our generalization is the first one that allows a large number of false positives and negatives in total, while bounding the number of these locally for each cluster by p and q. We show that recognizing (p, q)-cluster graphs is NP-complete when p and q are input. On the positive side, we show that $(0, q)$-cluster, $(p, 1)$-cluster, $(p, 2)$-cluster, and $(1, 3)$-cluster graphs can be recognized in polynomial time.

1 Introduction

Clustering is an optimization problem having applications in many fields ranging from bioinformatics [1,17] to image processing [20], with various algorithmic approaches available [16]. The general idea of clustering is to partition a set of data items into subsets, called clusters, in such a way that highly similar items belong to the same cluster and items having low similarity belong to different clusters. The input typically consists of similarity values between pairs of items and in the graph-based approach to clustering the items correspond to vertices, with two vertices being adjacent if and only if their similarity value exceeds a fixed threshold θ [13]. In a perfect setting with no noise, an appropriate threshold yields a similarity graph whose connected components (or clusters) are cliques. However, in most cases there will be noise, both false positives (presence of an edge that should not have been present) and false negatives (missing edges). Shamir et al. [19] initiated a study of clustering in terms of graph modification problems with a focus on the CLUSTER EDITING problem: modify a given graph by adding and deleting at most k edges to obtain a disjoint union of cliques.

D.M. Thilikos (Ed.): WG 2010, LNCS 6410, pp. 171–183, 2010.

CLUSTER EDITING, parameterized by the number k of false positives and negatives, is FPT [3,9,10]. Furthermore, it has a polynomial-time 4-approximation algorithm but it does not admit a PTAS unless P = NP [14]. Several drawbacks of this model have been pointed out (see e.g. [4,6]): for low values of the parameter k, it does not capture instances with a high number of false positives and negatives, nor does it allow overlap between clusters. As it has been observed that clusters do not always represent an equivalence relation (see [8,18]), overlapping clusters have been considered [5,7]. In addition, a weighted version of CLUSTER EDITING has been considered to capture the fact that the costs of fixing false positives and of false negatives can differ [2]. Other variants to tackle data sets containing a large number of false negatives have been proposed [12,11]. The p-DEFECTIVE CLIQUE EDITING problem is introduced by Guo et al. [11]: modify a given graph by adding and deleting at most k edges to obtain a disjoint union of p-defective cliques, where a p-defective clique is a graph missing at most p edges from being a clique. An FPT algorithm, parameterized by p and k, is given for this problem [11]. Note that for low values of the parameters the p-DEFECTIVE CLIQUE EDITING problem allows a high number of false negatives, as long as the noise is distributed among the clusters, but it does not allow a high number of false positives.

In this paper we present an alternative approach to graph clustering that allows a high total number of both false negatives and false positives, but little noise related to each cluster, by introducing what we call (p, q)-cluster graphs.

(p, q)-CLUSTER GRAPH RECOGNITION
Input: A graph G and two integers p, q.
Question: Can the vertex set of G be partitioned into subsets with each subset missing at most p edges from being a clique (i.e. inducing a p-defective clique) and having at most q edges going to other subsets?

Note that $(0, 0)$-cluster graphs are exactly cluster graphs. A (p, q)-cluster graph can have low values of p and q, while having a high total number of false negatives and false positives. In that case the similarity values and threshold θ satisfy the reasonable constraint that in each cluster C of similar items we find at most p pairs $u, v \in C$ with similarity less than θ, and at most q pairs $u \in C, w \notin C$ with similarity higher than θ. Observe also that tuning the p and q parameters independently is an alternative attempt of assigning different roles or importance to false positives and false negatives [2]. Moreover the transitivity constraint which has been criticized in CLUSTER EDITING [8,18] is relaxed. In this way the (p, q)-CLUSTER GRAPH RECOGNITION problem, or its editing version, answers most of the drawbacks present in the CLUSTER EDITING problem. Thus, as a first task, we want efficient algorithms for (p, q)-CLUSTER GRAPH RECOGNITION. Not surprisingly, (p, q)-CLUSTER GRAPH RECOGNITION is NP-complete, as we show in Theorem 1. However, as we summarize in Figure 1, there are various values of p and q for which (p, q)-CLUSTER GRAPH RECOGNITION can be solved in polynomial time, some trivially and some by more complicated combinatorial arguments.

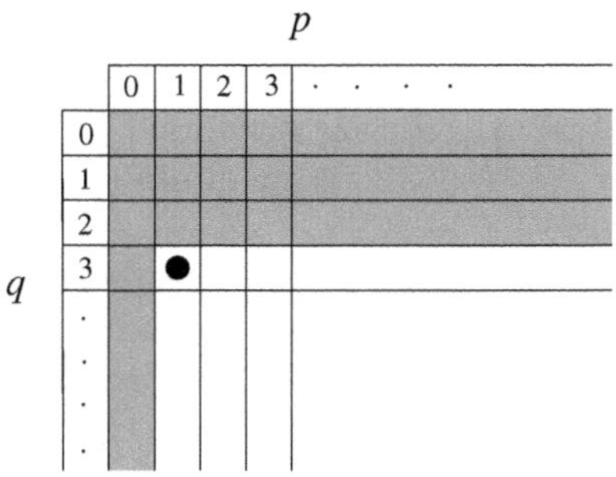

p

	0	1	2	3	·	·	·	·
0								
1								
2								
3	●							
·								
·								
·								
·								

q

Fig. 1. The shaded area and the dot indicate p and q values for which (p,q)-cluster graphs are polynomial-time recognizable

On the one hand, $(p,0)$-CLUSTER GRAPH RECOGNITION corresponds to the p-DEFECTIVE CLIQUE EDITING problem allowing zero edge modifications and is therefore trivial since the answer is yes if and only if each connected component of the input graph is a p-defective clique. On the other hand, $(0,q)$-CLUSTER GRAPH RECOGNITION is not at all simple, as there are many ways to partition the vertex set of a graph into a collection of cliques. In particular, similar problems like partitioning the vertex set of a graph into a minimum number of cliques (PARTITION INTO CLIQUES), or into subsets of bounded size each having a bounded number of edges to other subsets (MINIMUM DEGREE GRAPH PARTITION), are both NP-hard. Hence it is surprising that $(0,q)$-CLUSTER GRAPH RECOGNITION can be solved in polynomial time, as we prove in Theorem 3. We also show that $(p,1)$-cluster and $(p,2)$-cluster graphs can be recognized in polynomial time. Let us emphasize that the algorithms presented in this paper are polynomial in both p,q and the size of the graph. For example, the algorithm for $(0,q)$-CLUSTER GRAPH RECOGNITION runs in time $O(n^3)$ and by binary search one can find the smallest q such that the input graph is a $(0,q)$-cluster graph.

The polynomial-time cases mentioned so far are summarized by the shaded area of the table given in Figure 1. The first interesting case outside of the shaded area is the recognition of $(1,3)$-cluster graphs. With careful reduction rules and a computerized case analysis we are able to show that also $(1,3)$-cluster graphs can be recognized in polynomial time (Theorem 4). Answering natural follow-up complexity questions, Lokshtanov and Marx ([15]) showed recently that (p,q)-cluster graphs actually can be recognized in $2^{O(p)}nO(1)$ time and $2^{O(q)}n^{O(1)}$ randomized time.

2   Preliminaries

We consider undirected finite graphs with no loops or multiple edges. For a graph $G = (V,E)$, we denote its vertex and edge set by V and E, respectively, with $n = |V|$ and $m = |E|$. For $S \subseteq V$, the subgraph of G induced by S is denoted by $G[S]$.

The *neighborhood* of a vertex x of G is $N_G(x) = \{v \mid xv \in E\}$. The *closed neighborhood* of x is defined as $N_G[x] = N_G(x) \cup \{x\}$. If $S \subseteq V$, then the neighbors of S, denoted by $N_G(S)$, are given by $\bigcup_{x \in S} N_G(x) \setminus S$. The *degree* of a vertex x in G is $d_G(x) = |N_G(x)|$. We will omit the subscripts when there is no misunderstanding.

A *clique* is a set of vertices that are pairwise adjacent. A vertex x is called *simplicial* if $N(x)$ is a clique. If a vertex set C has exactly p pairs of non-adjacent vertices, we say that C *misses p edges*. We will call a vertex set that misses at most p edges a *p-group*. A p-group C such that there are at most q edges in G with exactly one endpoint in C is called a *(p,q)-group*.

For two non-negative integers p and q, a graph $G = (V, E)$ is a *(p,q)-cluster graph* if V can be partitioned into (p,q)-groups. Note that this condition is equivalent to the condition in the question of the (p,q)-CLUSTER GRAPH RECOGNITION problem. As deleting vertices from G cannot disturb a partition into (p,q)-groups, (p,q)-cluster graphs are hereditary, i.e., being a (p,q)-cluster graph is preserved under taking induced subgraphs.

Clearly, a graph is a (p,q)-cluster graph if and only if every connected component of it is a (p,q)-cluster graph. Hence for the rest of the paper, we will assume that the input graph is connected. If not, we can run the presented algorithms on each connected component. As a consequence, we can also restrict ourselves to identifying *connected* (p,q)-groups: a (p,q)-group C of a graph G might induce a disconnected subgraph, but then every connected component of $G[C]$ is also a (p,q)-group of G.

3 (p,q)-CLUSTER GRAPH RECOGNITION is NP-Complete

In this section we prove that, given as input a graph G and two integers p and q, it is NP-complete to decide whether G is a (p,q)-cluster graph. We use a reduction from the well known NP-complete problem CLIQUE: Given a graph G and an integer k, $0 < k < n$, does G have a clique of size at least k?

Let $G_1 = (V_1, E_1)$ and k be input to CLIQUE, where $|V_1| = n$. We show how to construct a graph G_2 and integers p and q such that G_1 has a clique of size at least k if and only if G_2 is a (p,q)-cluster graph. Let us first define

$$\alpha = nk - k^2 + 1 \qquad q = (n - k + 1)\alpha - 1 \qquad \beta = q - \alpha + 2 \qquad p = \beta k .$$

Note that $\alpha \geq n$ and $q \geq 2\alpha - 1$, since $0 < k < n$. We obtain G_2 from G_1 as follows:

1. Add a clique A of size α to G_1, and add edges between each vertex in A and each vertex in V_1. Call the resulting graph G_1'.
2. Add a clique B of size β to G_1', and add edges between each vertex in B and each vertex in A. Call the resulting graph G_2.

Lemma 1. *Let $G_1 = (V_1, E_1)$, G_2, p, and q be as described above. Then G_1 has a clique of size at least k if and only if there is a non-empty set $C \subseteq V_1$ such that $A \cup B \cup C$ is a (p,q)-group in G_2.*

Proof. Let C be a subset of V_1 and assume that $S = A \cup B \cup C$ is a (p, q)-group in G_2. Let $\ell = |C| \geq 1$. The number of edges in G_2 with exactly one endpoint in S is at least $(n - \ell)\alpha$. Since S is a (p, q)-group, $(n - \ell)\alpha \leq q = (n - k + 1)\alpha - 1$, which implies that $\ell \geq k$. Let j be the number of edges that C misses. Then S misses $\beta\ell + j$ edges. But since S is a (p, q)-group, $\beta\ell + j \leq p = \beta k$, and using $\ell \geq k$ gives $j = 0$ and $k = \ell$. Thus C is a clique of size k.

For the other direction, assume that G_1 has a clique of size at least k, and let C be a clique of size exactly k in G_1. Let $S = A \cup B \cup C$. The number edges that S misses is $\beta k = p$. The number of edges with exactly one endpoint in S is at most $(n - k)(k + \alpha)$.

$$nk - k^2 = \alpha - 1 \Leftrightarrow (n - k)\alpha + nk - k^2 \qquad = (n - k)\alpha + \alpha - 1$$
$$\Leftrightarrow (n - k)(k + \alpha) \qquad = (n - k + 1)\alpha - 1 = q$$

Lemma 2. *Let $G_1 = (V_1, E_1)$, G_2, p, and q be as described above. Then G_2 is a (p, q)-cluster graph if and only if there is a non-empty set $C \subseteq V_1$ such that $A \cup B \cup C$ is a (p, q)-group in G_2.*

Proof. Assume first that $G_2 = (V_2, E_2)$ is a (p, q)-cluster graph. We have to show that any (p, q)-group in G_2 that intersects with $A \cup B$ has to contain every vertex of $A \cup B$ and at least a vertex of V_1. Let $S \subseteq V_2$ be a (p, q)-group such that $S \cap (A \cup B) \neq \emptyset$. Observe first that S cannot be a proper subset of $A \cup B$, because any partition of $A \cup B$ into subsets results in more than q edges with an endpoint in each of the subsets, since $A \cup B$ is a clique of size $\alpha + \beta = q + 2$. Furthermore, S cannot be equal to $A \cup B$, since the number of edges between A and V_1 is αn and $q \leq \alpha n - 1$. Consequently, S must contain whole $A \cup B$ and at least a vertex of V_1.

For the other direction, assume that $S = A \cup B \cup C$ is a (p, q)-group for a non-empty set $C \subseteq V_1$. Observe that for any $v \in V_1 \setminus C$, the set $\{v\}$ is a (p, q)-group, since the degree of v is at most $n + \alpha - 1$, and $q \geq 2\alpha - 1 \geq n + \alpha - 1$. Hence S and the collection of the single vertex groups for each vertex in $V_1 \setminus C$ define a partition of V_2 into (p, q)-groups and consequently G_2 is a (p, q)-cluster graph.

Theorem 1. *Given a graph G and integers p and q, it is NP-complete to decide whether G is a (p, q)-cluster graph.*

Proof. Combining Lemmas 1 and 2, we conclude that G_1 has a clique of size at least k if and only if G_2 is a (p, q)-cluster graph. Since G_2 can be constructed from G_1 in polynomial time, the theorem follows.

4   Polynomial-Time Recognizable (p, q)-Cluster Graphs

In this section we show that for p and q values that correspond to the shaded area and the black dot in the table in Figure 1, (p, q)-cluster graphs can be recognized in polynomial time. Recall that we can assume the input graph to be connected.

As mentioned in the introduction, recognizing $(p, 0)$-cluster graphs is trivial for every integer p, as it is equivalent to checking whether the input graph misses at most p edges.

For recognizing connected $(p, 1)$-cluster graphs, note that the vertex set of such a graph is either a p-group or consists of two connected p-groups with a single edge between them. Hence we can first check whether the input graph is a $(p, 0)$-cluster graph. If not, we can check for each bridge in the graph, whether the removal of this bridge results in two connected components each of which is a p-group. This can clearly be done in polynomial time.

4.1   Polynomial-Time Recognition of $(p, 2)$-Cluster Graphs

Assume that a given connected graph $G = (V, E)$ is a $(p, 2)$-cluster graph. Then V has a partition into $(p, 2)$-groups $V_1, V_2, \ldots, V_k$. For convenience, in this subsection we call a $(p, 2)$-group simply a *group*. Let us define a graph H which has vertices $v_1, v_2, \ldots, v_k$ and edges $v_i v_j$ if G has an edge with an endpoint in V_i and an endpoint in V_j. Note that for each edge $v_i v_j$ of H, there can be at most one edge with an endpoint in V_i and an endpoint in V_j in G (except the case where the $(p, 2)$-partition consists of only two groups). Clearly H is a connected graph of maximum degree 2, which means that it is a path or a cycle. Furthermore, the removal of any two edges of H is equivalent to the removal of exactly two edges of G (except the case where H has only two vertices). We will use this property to decide whether a given graph G is a $(p, 2)$-cluster graph. For this purpose we describe a dynamic programming algorithm.

For every pair of edges $e_1 = u_1 v_1$ and $e_2 = u_2 v_2$ of G, we check whether u_1 and v_1 appear in different connected components of $G' = (V, E \setminus \{e_1, e_2\})$, and u_2 and v_2 appear in different connected components of G'. If so, then we say that $\{e_1, e_2\}$ is a *cut* of G. Let $L(e_1, e_2, u_1, u_2)$ be the disjoint union of all connected components of G' containing u_1 or u_2 considered as vertex subsets. One can think of the cut $\{e_1, e_2\}$ having two "sides", $L(e_1, e_2, u_1, u_2)$ and $L(e_1, e_2, v_1, v_2)$ respectively. We define a function $T(e_1, e_2, u_1, u_2)$ that is true if and only if $\{e_1, e_2\}$ is a cut and $L(e_1, e_2, u_1, u_2)$ can be partitioned into groups. Then the following recurrence holds for T.

- If $\{e_1, e_2\}$ is not a cut then $T(e_1, e_2, x, y)$ is False, for all x, y.
- Otherwise, if every connected component of $L(e_1, e_2, u_1, u_2)$ is a group then $T(e_1, e_2, u_1, u_2)$ is True.
- Otherwise $T(e_1, e_2, u_1, u_2)$ is True if and only if there is an edge $e = uv \in E$ with $u, v \in L(e_1, e_2, u_1, u_2)$ such that
  - every connected component of $L(e, e_1, u, u_1)$ is a group and $T(e, e_2, v, u_2)$ is true, *or*
  - every connected component of $L(e, e_2, v, u_2)$ is a group and $T(e, e_1, u, u_1)$ is true.

For all pairs of edges e_1 and e_2, we compute $T(e_1, e_2, u_1, u_2)$, $T(e_1, e_2, u_1, v_2)$, $T(e_1, e_2, v_1, u_2)$, and $T(e_1, e_2, v_1, v_2)$, using the above formula. After all this has been computed, we check for every pair of edges e_1 and e_2 whether they can be two consecutive edges of H. To do this, we simply check whether $T(e_1, e_2, x, y)$ is true and $G[V \setminus L(e_1, e_2, x, y)]$ is a connected group for some edges e_1 and e_2 and endpoints x of e_1 and y of e_2. If such edges e_1, e_2 and endpoints x, y exist we conclude that G is a $(p, 2)$-cluster graph. Otherwise it is not a $(p, 2)$-cluster graph. The necessary computations can be done in a straight forward way in time $O(m^3)$. With a few extra reduction rules and more clever dynamic programming it is possible to reduce this running time considerably.

4.2   Polynomial-Time Recognition of $(0, q)$-Cluster Graphs

Deciding whether a given graph $G = (V, E)$ is a $(0, q)$-cluster graph is deciding whether V can be partitioned into $(0, q)$-groups. This is equivalent to partitioning V into cliques, such that each of these cliques G has at most q edges with exactly one endpoint in that clique. Analogous to previous subsection, in this subsection we will call a $(0, q)$-group simply a *group*. Also, we call a vertex of v of G a *high degree vertex* if $d(v) \geq q + 1$. We start with some observations on $(0, q)$-cluster graphs, the proofs of which are given in the appendix.

Lemma 3. Let $G = (V, E)$ be a $(0, q)$-cluster graph. Then there is a partition of V into groups such that every group either consists of a single vertex or contains a high degree vertex.

Proof. Assume that G has a partition into groups $C_1, C_2, \ldots, C_k$. Let let C_i be a group containing at least two vertices, such that every vertex of C_i has degree at most q. In this case each vertex of C_i defines a group consisting only of itself. Furthermore, after dividing C_i into singletons, every other set C_j with $j \neq i$ is still a group.

Lemma 4. Let $G = (V, E)$ be a $(0, q)$-cluster graph. Then there is a partition of V into groups such that every group C with at least two vertices contains a vertex v with $N[v] = C$.

Proof. By Lemma 3 we know that there is a partition such that every group C of the partition that contains at least two vertices, contains a high degree vertex u. If u has no neighbors outside of C, since C is a clique we have that $N[u] = C$, and the proof is complete. Assume that u has at least one neighbor outside of C. Assume that every neighbor of u in C has a neighbor outside of C. Then together with the neighbors that u has outside of C, there are more than q edges in G with exactly one endpoint in C, which contradicts the assumption that C is a group. Thus u has a neighbor v in C such that v has no neighbors outside of C. Since C is a clique, $N[v] = C$.

Note that Lemma 4 is equivalent to saying that C contains a simplicial vertex. By the above observations, we can restrict our search to groups that are either singletons or contain a simplicial vertex and a high degree vertex.

Definition 1. *A group C is a good group if $C = N[v]$ for some simplicial vertex v, C contains a high degree vertex, and there are at most q edges in the graph with exactly one endpoint in C (the last condition is implicit from the definition of group).*

Lemma 5. *A graph is a $(0, q)$-cluster graph if and only if there exists a set of non-overlapping good groups whose union contains all high degree vertices.*

Proof. Let $G = (V, E)$ be a graph, and assume that G has a set of good groups $C_1, ..., C_\ell$, such that $C_i \cap C_j = \emptyset$ for every pair $i \neq j$ between 1 and ℓ, and every high degree vertex of G belongs to some C_i for $1 \leq i \leq \ell$. Since these good groups do not overlap, each high degree vertex belongs to a unique good group. Each of these groups has at most q edges leaving the group. Every vertex of G that does not appear in one of these groups, has degree at most q and is a $(0, q)$-group on its own. Let $v_1, ..., v_t$ be such vertices of G. Then $C_1, ..., C_\ell, \{v_1\}, ..., \{v_t\}$ is a partition of V into clusters, and hence G is a $(0, q)$-cluster graph.

The other direction follows from the fact that a good group containing a high degree vertex is of size at least 2 and Lemma 4.

For the next lemma, we say that a good group is *maximal* if its set of high degree vertices is not a proper subset of the set of high degree vertices of another good group.

Lemma 6. *Let G be a graph and let C_1, C_2 be two maximal good groups of G, such that C_1 has a high degree vertex not in C_2, and C_2 has a high degree vertex not in C_1. Then $C_1 \cap C_2 = \emptyset$.*

Proof. Let $C_1 \cap C_2 = X$. Let v_1 be a high degree vertex of C_1 not in C_2, and let v_2 be a high degree vertex of C_2 not in C_1. Hence $v_1, v_2 \notin X$. Observe first that $|X| \leq q$ because otherwise, since C_2 is a clique, there would be more than q edges from C_1 to v_2, contradicting that C_1 is a good group. If v_1 has a neighbor outside of C_1 then $N[v_1] \neq C_1$, hence C_1 has another vertex v_1' such that $N[v_1'] = C_1$, and $v_1' \notin X$. If v_1 has no neighbor outside of C_1, then since $d(v_1) \geq q + 1$ and $|X| \leq q$, again C_1 has a vertex $v_1' \neq v_1$ such that $v_1' \notin X$. Hence $|C_1| \geq |X| + 2$. With the same arguments, C_2 has a vertex $v_2' \neq v_2$ such that $v_2' \notin X$, and thus also $|C_2| \geq |X| + 2$.

Since $d(v_1) \geq q + 1$, v_1 has at least $q + 1 - (|C_1| - 1)$ neighbors outside of C_1. In addition, there are $|X|(|C_2| - |X|)$ edges between C_1 and $C_2 \setminus X$. Since C_1 is a good group, we must thus have: $q + 1 - (|C_1| - 1) + |X|(|C_2| - |X|) \leq q$. Symmetrically, and with the same arguments, we conclude that: $q + 1 - (|C_2| - 1) + |X|(|C_1| - |X|) \leq q$. Adding up these two inequalities and simplifying, we get:

$$4 + (|X| - 1)|C_1| + (|X| - 1)|C_2| - 2|X|^2 \leq 0$$

Recall that $|C_1| \geq |X| + 2$ and $|C_2| \geq |X| + 2$. Hence we can conclude:

$$4 + (|X| - 1)(|X| + 2) + (|X| - 1)(|X| + 2) - 2|X|^2 \leq 0$$

Doing the arithmetic, we see that the above inequality reduces to $2|X| \leq 0$, and hence we can conclude that $|X| = 0$, which completes the proof.

Consequently, maximal good groups with different sets of high degree vertices do not overlap. With this, we reach the desired characterization.

Theorem 2. *A graph is a $(0, q)$-cluster graph if and only if every high degree vertex belongs to a good group.*

Proof. Let G be a graph. If G has a high degree vertex v that does not belong to any good group, then G is clearly not a $(0, q)$-cluster graph, due to Lemma 5 and since v cannot define a good group on its own due to its high degree. For the other direction, assume that every high degree vertex of G belongs to a good group. Repeatedly take a maximal good group containing uncovered high degree vertices, and call $\mathcal{C}$ the resulting set of good groups. $\mathcal{C}$ covers all the high degree vertices of G, and the good groups of $\mathcal{C}$ pairwise have different sets of high degree vertices. Thus by Lemma 6, they are pairwise non-overlapping. Consequently, by Lemma 5, G is a $(0, q)$-cluster graph.

Theorem 3. *Given a graph G and an integer q, it can be decided in polynomial time whether G is a $(0, q)$-cluster graph.*

Proof. Note first that finding the good groups of any graph $G = (V, E)$ can be done in polynomial time, as we only need to check whether $N[v]$ is a clique, contains a high degree vertex, and G has at most q edges with exactly one endpoint in $N[v]$, for each vertex $v \in V$. Now, by Theorem 2, it simply remains to check whether every high degree vertex appears in a good group, which can be done by the procedure described in the proof of Theorem 2. A straight forward implementation gives a total running time of $O(n^3)$, which can probably be improved.

4.3 Polynomial-Time Recognition of $(1, 3)$-Cluster Graphs

Each polynomial-time algorithm that we have given so far has corresponded to a whole row or a whole column of the table given in Figure 1. In fact, with the algorithms that we have given, we have now proved that (p, q)-graphs are recognizable in polynomial time for values of p and q that correspond to the whole shaded area in that table. From here on, we see that the first natural case to study, with respect to NP-completeness versus polynomial-time computability, for a single value of p and a single value of q is the recognition of $(1, 3)$-cluster graphs. This corresponds to the dot in the table. In this subsection, we show that $(1, 3)$-cluster graphs can be recognized in polynomial time.

Analogous to previous subsections, we will refer to a $(1, 3)$-group simply as a *group* in this subsection. A *minimal group* is a group that cannot be partitioned into smaller groups. If there is a partition of the vertex set of a graph into groups then there is also a partition into minimal groups. A *high degree vertex* is now a vertex of degree at least 4. A *1-group* is a clique missing at most one edge, according to the definitions in Section 2. (Note that a 1-group is not necessarily a group.) A *maximal* 1-group is a 1-group that is not a proper subset of another

1-group. (Note the difference from the definition of maximality in the previous subsection.)

We start this subsection with some reduction rules given in Definition 2. The first steps of our algorithm will be to apply these rules until they cannot be applied anymore, to obtain a *reduced* graph. For these rules, we define the concept of *penalizing* a vertex x as follows. Let u be a vertex of G, and let $G' = G[V \setminus \{u\}]$. Let x be a vertex of $N_G(u)$. Then $d_{G'}(x) = d_G(x) - 1$. When we *penalize* x, we keep the degree of x unchanged. Hence, we let $d_{G'}(x) = d_G(x)$ and keep it artificially high.

Definition 2. *We say that a graph G is* reduced *if the following reduction rules cannot be applied to it:*

1. *If, for an edge uv, there is no group that contains both u and v, then delete edge uv and penalize u and v.*
2. *If G contains a maximal 1-group C of size at least 5, then delete C and penalize the vertices of $N_G(C)$ accordingly.*
3. *If Rule 2 cannot be applied and G contains a clique C of size at least 4, then delete C and penalize the vertices of $N_G(C)$ accordingly.*
4. *If Rules 2 and 3 cannot be applied and G contains a 1-group C of size 4, such that C has 3 vertices with neighbors outside of C, then delete C, and penalize $N_G(C)$ accordingly.*

The proofs of the following two lemmas are omitted due to space restrictions, but can be found in the full version of the paper.

Lemma 7. *Given a graph G, the reduced graph G' obtained from G by applying the reduction rules in Definition 2 can be computed in polynomial time. Moreover, G is a $(1,3)$-cluster graph if and only if G' is a $(1,3)$-cluster graph.*

Consequently, from now on we can assume that our input graph G is reduced. In particular, G has no groups of size larger than 4. We will call a group C a *leaf group* if at most one vertex of C has neighbors outside of C. Since we assume G to be connected, this is equivalent to C having exactly one vertex with neighbors outside of C.

Lemma 8. *In a reduced graph with more than 42 vertices, every high degree vertex appears in at most 2 minimal non-leaf groups.*

The above lemma enables us to show the main result of this subsection, stated in the following theorem.

Theorem 4. $(1,3)$-*cluster graphs can be recognized in polynomial time.*

Proof. Given a graph H, we can compute a reduced graph G in polynomial time by Lemma 7. If G has at most 42 vertices, we can solve the problem in constant time. Assume that G has $n > 42$ vertices.

First we show that there is a polynomial time reduction from $(1, 3)$-CLUSTER GRAPH RECOGNITION to SAT. Given a reduced graph G, we describe an instance of SAT obtained from G, as follows. For every minimal group, we make a variable x. For every high degree vertex v, we make a clause $(x_1 \vee x_2 \vee ... \vee x_t)$, where $x_1, ..., x_t$ are the variables corresponding to the t minimal groups containing v. For every pair of overlapping minimal groups with corresponding variables x and y, we make a clause $(\bar{x} \vee \bar{y})$. Clearly, G is a $(1, 3)$-cluster graph if and only if the created formula is satisfiable. By Lemma 7, the same is true for H. In G, there are at most n^4 groups, since every group is of size at most 4. Consequently, the construction of the formula from the given graph takes polynomial time.

Due to Lemma 8, in the constructed SAT formula, every clause X contains variables $x_1, ..., x_t$ corresponding to leaf groups and at most two variables a and b corresponding to minimal non-leaf groups. We can safely set $x_2, ..., x_t$ to be false, as we will let x_1 ensure the TRUE value of this clause. Every other clause that contains one of $x_2, ..., x_t$, contains it in the negated form, and hence will be true. x_1 appears in at most two other clauses: $(\bar{x_1} \vee \bar{a})$ and $(\bar{x_1} \vee \bar{b})$. Hence, according to the truth-value of a and b, we can assign TRUE or FALSE to x_1, and clause X will be true. Consequently, we can remove clauses involving all leaf groups, as they are not decisive for the satisfiability of the whole formula. The remaining clauses all have two literals, and hence we have a 2-SAT instance that can be solved in polynomial time.

Hence we have a polynomial-time reduction from $(1, 3)$-CLUSTER GRAPH RECOGNITION to 2-SAT, which means that $(1, 3)$-cluster graphs can be recognized in polynomial time.

5   Concluding Remarks

We have introduced the (p, q)-CLUSTER GRAPH RECOGNITION problem and proved that it is NP-complete. We have shown that $(p, 0)$-cluster, $(p, 1)$-cluster, $(p, 2)$-cluster, $(0, q)$-cluster, and $(1, 3)$-cluster graphs can be recognized in polynomial time. In fact, with a careful implementation we believe that $(p, 2)$-cluster and $(1, 3)$-cluster graphs can be recognized in linear time. Many interesting questions arise from these results. Some of the most obvious are: Is (p, q)-CLUSTER GRAPH RECOGNITION FPT when parameterized by either p or q, meaning that there is an algorithm with running time $f(p) \cdot poly(n)$ or $f(q) \cdot poly(n)$? Very recently, Lokshtanov and Marx ([15]) answered both these question in the affirmative, by giving $2^{O(p)} n^{O(1)}$ time and $2^{O(q)} n^{O(1)}$ randomized time algorithms.

There is also a natural extension of (p, q)-CLUSTER GRAPH RECOGNITION to (p, q, k)-CLUSTER GRAPH EDITING. Here we ask whether a (p, q)-cluster graph can be obtained by adding or removing in total at most k edges in a given graph (in fact, this was the original starting point for the authors). Hence $(p, q, 0)$-CLUSTER GRAPH EDITING is equivalent to (p, q)-CLUSTER GRAPH RECOGNITION. The editing version of the problem opens a whole range of questions of the above type involving k in addition.

References

1. Ben-Dor, Z.Y.A., Shamir, R.: Clustering gene expression patterns. J.Comput. Biol. 6(3/4), 281–292 (1999)
2. Böcker, S., Briesemeister, S., Bui, Q.A., Truß, A.: Going weighted: Parameterized algorithms for cluster editing. In: Yang, B., Du, D.-Z., Wang, C.A. (eds.) COCOA 2008. LNCS, vol. 5165, pp. 1–12. Springer, Heidelberg (2008)
3. Cai, L.: Fixed-parameter tractability of graph modification problems for hereditary properties. Information Processing Letters 58(4), 171–176 (1996)
4. Chesler, E., Lu, L., Shou, S., Qu, Y., Gu, J., Wang, J., Hsu, H., Mountz, J., Baldwin, N., Langston, M., Threadgill, D., Manly, K., Williams, R.: Complex trait analysis of gene expression uncovers polygenic and pleiotropic networks that modulate nervous system function. Nature Genetics 37(3), 233–242 (2005)
5. Damaschke, P.: Fixed-parameter enumerability of cluster editing and related problems. Theory of Computing Systems, TOCS 46(2), 261–283 (2010)
6. Dehne, F., Langston, M., Luo, X., Pitre, S., Shaw, P., Zhang, Y.: The cluster editing problem: implementations and experiments. In: Bodlaender, H.L., Langston, M.A. (eds.) IWPEC 2006. LNCS, vol. 4169, pp. 13–24. Springer, Heidelberg (2006)
7. Fellows, M., Guo, J., Komusiewicz, C., Niedermeier, R., Uhlmann, J.: Graph-based data clustering with overlaps. In: Ngo, H.Q. (ed.) COCOON 2009. LNCS, vol. 5609, pp. 516–526. Springer, Heidelberg (2009)
8. Frigui, H., Nasraoui, O.: Simultaneous clustering and dynamic key-word weighting for text documents. In: Berry, M. (ed.) Survey of Text Mining, pp. 45–70. Springer, Heidelberg (2004)
9. Gramm, J., Guo, J., Hüffner, F., Niedermeier, R.: Automated generation of search tree algorithms for hard graph modification problems. Algorithmica 39, 321–347 (2004)
10. Gramm, J., Guo, J., Hüffner, F., Niedermeier, R.: Graph-modeled data clustering: fixed-parameter algorithm for clique generation. Theory of Computing Systems 38, 373–392 (2005)
11. Guo, J., Kanj, I.A., Komusiewicz, C., Uhlmann, J.: Editing graphs into disjoint unions of dense clusters. In: Dong, Y., Du, D.-Z., Ibarra, O. (eds.) ISAAC 2009. LNCS, vol. 5878, pp. 583–593. Springer, Heidelberg (2009)
12. Guo, J., Komusiewicz, C., Niedermeier, R., Uhlmann, J.: A more relaxed model for graph-based data clustering: s-plex. In: Goldberg, A.V., Zhou, Y. (eds.) AAIM 2009. LNCS, vol. 5564, pp. 226–239. Springer, Heidelberg (2009)
13. Hartigan, J.: Clustering Algorithms. John Wiley and Sons, Chichester (1975)
14. Charikar, A.M., Guruswami, V.: Clustering with qualitative information. Journal of Computer and System Sciences 71, 360–383 (2005)
15. Lokshtanov, D., Marx, D.: Clustering with Local Restrictions. Private communication (2010)
16. Xu, D.W.R.: Survey of clustering algorithms. IEEE Transactions onNeural Networks 16(3), 645–678 (2005)
17. Sharan, R.R., Maron-Katz, A.: Click and expander: a system for clustering and visualizing gene expression data. Bioinformatics 19(14), 1787–1799 (2003)
18. Scholtens, D., Vidal, M., Gentlemand, R.:Local modeling of global interactome networks. Bioinformatics 21, 3548–3557 (2005)

19. Shamir, R., Sharan, R., Tsur, D.: Cluster graph modification problems. Discrete Applied Mathematics 144(1-2), 173–182 (2004)
20. Wu, R.L.Z.: An optimal graph theoretic approach to data clustering: theory and its application to image segmentation. IEEE Transactions on Pattern Analysis and Machine Intelligence 15(11), 1101–1113 (1993)

Colouring Vertices of Triangle-Free Graphs[*]

Konrad Dabrowski, Vadim Lozin, Rajiv Raman, and Bernard Ries

DIMAP, University of Warwick, Coventry CV4 7AL, UK

Abstract. The VERTEX COLOURING problem is known to be NP-comple-te in the class of triangle-free graphs. Moreover, it remains NP-complete even if we additionally exclude a graph F which is not a forest. We study the computational complexity of the problem in (K_3, F)-free graphs with F being a forest. From known results it follows that for any forest F on 5 vertices the VERTEX COLOURING problem is polynomial-time solvable in the class of (K_3, F)-free graphs. In the present paper, we show that the problem is also polynomial-time solvable in many classes of (K_3, F)-free graphs with F being a forest on 6 vertices.

Keywords: Vertex colouring; Triangle-free graphs; Polynomial-time algorithm; Clique-width.

1 Introduction

A vertex colouring is an assignment of colours to the vertices of a graph G in such a way that no edge connects two vertices of the same colour. The VERTEX COLOURING problem consists in finding a vertex colouring with a minimum number of colours. This number is called the chromatic number of G and is denoted by $\chi(G)$. If G admits a vertex colouring with at most k colours, we say that G is k-colourable. The k-COLOURABILITY problem consists in deciding whether a graph is k-colourable.

From a computational point of view, VERTEX COLOURING and k-COLOURABILITY ($k \geq 3$) are difficult problems, i.e. both of them are NP-complete. Moreover, the problems remain NP-complete in many restricted graph families. For instance, 3-COLOURABILITY is NP-complete for planar graphs [8], 4-COLOURABILITY is NP-complete for graphs containing no induced path on 12 vertices [28], VERTEX COLOURING is NP-complete for line graphs [11]. On the other hand, for graphs in some special classes, the problems can be solved in polynomial time. For instance, VERTEX COLOURING (and therefore, k-COLOURABILITY for any value of k) is solvable for perfect graphs, k-COLOURABILITY (for any value of k) is solvable for graphs containing no induced path on 5 vertices [10], and 3-COLOURABILITY is solvable for graphs containing no induced path on 6 vertices [26].

---

[*] Research supported by the Centre for Discrete Mathematics and Its Applications (DIMAP), University of Warwick.

Recently, much attention has been paid to the complexity of the problems in graph classes defined by forbidden induced subgraphs. Many results of this type have been mentioned before, some others can be found in [3, 5, 12, 13, 15–17, 21, 23]. In [16], the authors systematically study VERTEX COLOURING on graph classes defined by a single forbidden induced subgraph, and give a complete characterisation of those for which the problem is polynomial-time solvable and those for which the problem is NP-complete. In particular, the problem is NP-complete for K_3-free graphs, i.e. for triangle-free graphs. Moreover, the problem is NP-complete for (K_3, F)-free graphs for any graph F which is not a forest [12, 16]. Here we study the computational complexity of the problem in (K_3, F)-free graphs with F being a forest. From known results it follows that for any forest F on 5 vertices the VERTEX COLOURING problem is polynomial-time solvable in the class of (K_3, F)-free graphs. In the present paper, we show that the problem is also polynomial-time solvable in many classes of (K_3, F)-free graphs with F being a forest on 6 vertices.

2   Preliminaries

All graphs in this paper are finite, undirected, without loops or multiple edges. For any graph theoretical terms not defined here, the reader is referred to [9]. For a graph G, we denote by $V(G)$ and $E(G)$ the vertex set and the edge set of G, respectively. If v is a vertex of G, then $N(v)$ denotes the neighbourhood of v (i.e. the set of vertices adjacent to v) and $|N(v)|$ is the degree of v. The subgraph of G induced by a set of vertices $U \subseteq V(G)$ is denoted by $G[U]$. For disjoint sets $A, B \subseteq V(G)$, we say that A is complete to B if every vertex in A is adjacent to every vertex in B, and that A is anticomplete to B if every vertex in A is non-adjacent to every vertex in B.

As usual, P_n is a chordless path, C_n is a chordless cycle, and K_n is a complete graph on n vertices. Also, $K_{n,m}$ denotes a complete bipartite graph with parts of size n and m. By $S_{i,j,k}$ we denote a tree with exactly three leaves of distance i, j and k from the only vertex of degree 3. In particular, $S_{1,1,1} = K_{1,3}$ is known as a claw, and $S_{1,2,2}$ is sometimes denoted by E, since this graph can be drawn as the capital letter E. Also, by H we denote the graph that can be drawn as the capital letter H, i.e. H has vertex set $\{v_1, v_2, v_3, v_4, v_5, v_6\}$ and edge set $\{v_1v_2, v_2v_3, v_2v_4, v_4v_5, v_4v_6\}$. The graph obtained from a $K_{1,4}$ by subdividing exactly one edge exactly once is called a *cross*. Given two graphs G and G', we denote by $G + G'$ the disjoint union of G and G'. In particular, mG is the disjoint union of m copies of G.

The clique-width of a graph G is the minimum number of labels needed to construct G using the following four operations:

(i) Creation of a new vertex v with label i (denoted by $i(v)$).
(ii) Disjoint union of two labelled graphs G and H (denoted by $G \oplus H$).
(iii) Joining by an edge each vertex with label i to each vertex with label j ($i \neq j$, denoted by $\eta_{i,j}$).
(iv) Renaming label i to j (denoted by $\rho_{i \to j}$).

Every graph can be defined by an algebraic expression using these four operations. For instance, an induced path on five consecutive vertices a, b, c, d, e has clique-width equal to 3 and it can be defined as follows:

$$\eta_{3,2}(3(e) \oplus \rho_{3\to2}(\rho_{2\to1}(\eta_{3,2}(3(d) \oplus \rho_{3\to2}(\rho_{2\to1}(\eta_{3,2}(3(c) \oplus \eta_{2,1}(2(b) \oplus 1(a)))))))))$$

If a graph G does not contain induced subgraphs isomorphic to graphs from a set M, we say that G is M-free. The class of all M-free graphs is denoted by $Free(M)$, and M is called the set of forbidden induced subgraphs for this class. Many graph classes that are important from a practical or theoretical point of view can be described in terms of forbidden induced subgraphs. For instance, by definition, forests form the class of graphs without cycles, and due to König's Theorem, bipartite graphs are graphs without odd cycles. Bipartite graphs are precisely the 2-colourable graphs, and recognising 2-colourable graphs is a polynomially solvable task. However, the recognition of k-colourable graphs is an NP-complete problem for any $k \geq 3$.

In the present paper, we study the computational complexity of the VERTEX COLOURING problem in graph classes defined by two forbidden induced subgraphs one of which is a triangle K_3. The following theorem summarises known results of this type.

Theorem 1. *Let F be a graph. If F contains a cycle or $F = K_{1,5}$, then the* VERTEX COLOURING *problem is NP-complete in the class $Free(K_3, F)$. If F is isomorphic to $S_{1,2,2}, H, P_6$ or a cross, then the problem is polynomial-time solvable in the class $Free(K_3, F)$.*

Proof. If F contains a cycle, then the NP-completeness of the problem follows from the fact that it is NP-complete for graphs of girth at least $k + 1$, i.e. in the class $Free(C_3, C_4, \ldots, C_k)$, for any fixed value of k (see e.g. [12, 16]). The NP-completeness of the problem in the class of $(K_3, K_{1,5})$-free graphs was shown in [21].

In [22, 24], Randerath et al. showed that every graph in the following three classes is 3-colourable and that a 3-colouring can be found in polynomial time: $Free(K_3, H), Free(K_3, S_{1,2,2}), Free(K_3, cross)$. Therefore VERTEX COLOURING is polynomial-time solvable in these three classes.

The conclusion that the problem is solvable for (K_3, P_6)-free graphs can be derived from three facts. First, the clique-width of graphs in this class is bounded by a constant [4]. Second, the chromatic number of graphs in this class is bounded by a constant (see e.g. [27]). Third, for each fixed k, the k-colourability problem on graphs of bounded clique-width is solvable in polynomial time [6].   □

Corollary 1. *For each forest F on 5 vertices, the* VERTEX COLOURING *problem in the class $Free(K_3, F)$ is solvable in polynomial time.*

Proof. If F contains no edge, then the problem is trivial in the class of $Free(K_3, F)$, since the size of graphs in this class is bounded by a constant (by a Ramsey argument). If F contains at least one edge, then it is an induced subgraph of one of the following graphs: $H, S_{1,2,2}, cross, P_6$. Therefore $Free(K_3, F)$

is a subclass of one the classes $Free(K_3, H)$, $Free(K_3, S_{1,2,2})$, $Free(K_3, cross)$, $Free(K_3, P_6)$, and thus the result follows from Theorem 1.   □

3 (K_3, F)-Free Graphs with F Containing an Isolated Vertex

In this section, we study graph classes $Free(K_3, F)$ with F being a forest on 6 vertices at least one of which is isolated. Without loss of generality we may assume that F contains at least one edge, since otherwise there are only finitely many graphs in the class $Free(K_3, F)$ (by a Ramsey argument). Throughout the section, an isolated vertex in F is denoted by v and the rest of the graph is denoted by F_0, i.e. $F_0 = F - v$.

Lemma 1. *Let F be a forest on 6 vertices with at least one edge and at least one isolated vertex. Then the chromatic number of any graph G in the class $Free(K_3, F)$ is at most 4.*

Proof. Suppose that $F_0 \neq P_3 + P_2$. Then it is not difficult to verify that F_0 is an induced subgraph of H, $S_{1,2,2}$ or $cross$. Therefore the chromatic number of (K_3, F_0)-free graphs is at most 3 (see [22, 24]). As a result, the chromatic number of any (K_3, F)-free graph is at most 4. To see this, observe that for any vertex x, the graph $G \setminus N(x)$ is 3-colourable, while $N(x)$ is an independent set.

Now assume $F_0 = P_3 + P_2$ and let ab be an edge in a (K_3, F)-free graph G. We will show that $G_0 := G - (N(a) \cup N(b))$ is a bipartite graph. Notice that since G is K_3-free, both $N(a)$ and $N(b)$ induce an independent set. We may assume that at least one of $N(a) \setminus \{b\}, N(b) \setminus \{a\}$ is non-empty (otherwise each connected component of G has at most two vertices and thus G is trivially 4-colorable). Obviously G_0 is C_k-free for any odd $k \geq 7$, since otherwise G contains a $P_3 + P_2$. Therefore we may assume that G_0 contains a C_5 (otherwise G_0 is bipartite). Let $c \in N(b) \setminus \{a\}$. Since G is triangle-free, c either has no neighbours in the C_5, or has exactly one neighbour in the C_5, or has exactly two neighbours which are non-consecutive vertices of the C_5. Thus c is non-adjacent to at least three vertices in C_5, say d, e, f, such that $G[d, e, f]$ is isomorphic to $P_2 + K_1$. But now $G[a, b, c, d, e, f]$ is isomorphic to $P_3 + P_2 + K_1$, which is a forbidden graph for G. This contradiction shows that G_0 has no odd cycles, i.e. G_0 is a bipartite graph. If V_0^1, V_0^2 are two colour classes of G_0, then $N(b) \cup \{a\}, N(a) \cup \{b\}, V_0^1, V_0^2$ are four colour classes of G.   □

In view of Lemma 1 and the polynomial-time solvability of 2-COLOURABILITY, all we have to do to solve the problem in the classes under consideration is to develop a tool for deciding 3-colourability in polynomial time. For this, we use a result from [23]. A set $D \subseteq V(G)$ is *dominating* in G if every vertex $x \in V(G) \setminus D$ has at least one neighbour in D.

Lemma 2. *([23]) For a graph $G = (V, E)$ with a dominating set D, we can decide 3-colourability and determine a 3-colouring in time $O(3^{|D|}|E|)$.*

If a graph $G \in Free(K_3, F)$ is F_0-free, then the problem is solvable for G by Corollary 1. If G has an induced F_0, then the vertices of F_0 form a dominating set in G. Summarising the above discussion, we obtain the following result.

Theorem 2. *Let F be a forest on 6 vertices with at least one isolated vertex. Then the* VERTEX COLOURING *problem is polynomial-time solvable in the class* $Free(K_3, F)$.

4   Graphs of Bounded Clique-Width

In Section 2, we mentioned that the polynomial-time solvability of the VERTEX COLOURING problem in the class of (K_3, P_6)-free graphs follows from the facts that the clique-width and the chromatic number of graphs in this class are bounded by a constant. In the present section, we use that same idea to solve the problem in the following two classes: $Free(K_3, S_{1,1,3})$ and $Free(K_3, K_{1,3} + K_2)$.

It is known that if G is an F-free graph, where F is a subdivision of a star $K_{1,n}$, then the chromatic number of G is bounded by a function of its clique number (see e.g. [27]). Therefore the chromatic number of $(K_3, S_{1,1,3})$-free graphs and $(K_3, K_{1,3} + K_2)$-free graphs is bounded by a constant. This means that in order to prove polynomial-time solvability of the VERTEX COLOURING problem in the classes $Free(K_3, S_{1,1,3})$ and $Free(K_3, K_{1,3} + K_2)$, all we have to do is to show that the clique-width of graphs in these classes is bounded. In our proofs, we use the following helpful facts.

Fact 1: The clique-width of graphs with vertex degree at most 2 is bounded by 4 (see e.g. [7]).

Fact 2: The clique-width of $S_{1,1,3}$-free bipartite graphs [18] and $(K_{1,3} + K_2)$-free bipartite graphs [20] is bounded by a constant.

Fact 3: For a constant k and a class of graphs X, let $X_{[k]}$ denote the class of graphs obtained from graphs in X by deleting at most k vertices. Then the clique-width of graphs in X is bounded if and only if the clique-width of graphs in $X_{[k]}$ is bounded [19].

Fact 4: For a graph G, the subgraph complementation is the operation that consists in complementing the edges in an induced subgraph of G. Also, given two disjoint subsets of vertices in G, the bipartite subgraph complementation is the operation which consists in complementing the edges between the subsets. For a constant k and a class of graphs X, let $X^{(k)}$ be the class of graphs obtained from graphs in X by applying at most k subgraph complementations or bipartite subgraph complementations. Then the clique-width of graphs in $X^{(k)}$ is bounded if and only if the clique-width of graphs in X is bounded [14].

Fact 5: The clique-width of graphs in a hereditary class X is bounded if and only if it is bounded for connected graphs in X (see e.g. [7]).

Facts 2 and 5 allow us to reduce the problem to connected non-bipartite graphs in the classes $Free(K_3, S_{1,1,3})$ and $Free(K_3, K_{1,2} + K_2)$, i.e. to connected graphs in these classes that contain an odd induced cycle of length at least five.

Lemma 3. *Let G be a connected $(K_3, S_{1,1,3})$-free graph containing an odd induced cycle C of length at least 7. Then $G = C$.*

Proof. Let $C = v_1 - v_2 - \ldots - v_{2k} - v_{2k+1} - v_1$ be an induced cycle of length $2k+1$, $k \geq 3$, in G. Suppose that there exists a vertex $v \in V(G) \backslash V(C)$, which is adjacent to a vertex of C. Without loss of generality, we may assume that v is adjacent to v_1. We claim that v is non-adjacent to v_4. Otherwise, since G is K_3-free, it follows that v is non-adjacent to v_{2k+1}, v_2, v_3, v_5. But now $G[v, v_1, v_3, v_4, v_5, v_{2k+1}]$ is isomorphic to $S_{1,1,3}$, a contradiction. Thus v is non-adjacent to v_4. This implies that v is adjacent to v_3, since otherwise $G[v, v_1, v_2, v_3, v_4, v_{2k+1}]$ is isomorphic to $S_{1,1,3}$. Now repeating the same argument with v_3 playing the role of v_1, we conclude that v is adjacent to v_5. But now $G[v, v_1, v_2, v_{2k+1}, v_4, v_5]$ is isomorphic to $S_{1,1,3}$. This contradiction shows that $G = C$. $\qquad \square$

Lemma 4. *Let G be a connected $(K_3, K_{1,3} + K_2)$-free graph containing an odd induced cycle C_{2k+1}, $k \geq 3$. Then either $G = C_{2k+1}$ and $k \geq 4$, or $|V(G)| \leq 28$ and $k = 3$.*

Proof. Let $C = v_1 - v_2 - \ldots - v_{2k} - v_{2k+1} - v_1$ be an induced cycle of length $2k+1$ in G. First assume that $k \geq 4$. Suppose that there exists a vertex $v \in V(G) \backslash V(C)$ which is adjacent to some vertex of C, say v is adjacent to v_1. Since G is K_3-free, it follows that v is non-adjacent to v_{2k+1}, v_2. We claim that for every pair of vertices $\{v_i, v_{i+1}\}$, with $i = 4, 5, \ldots, 2k-2$, vertex v is adjacent to exactly one of v_i, v_{i+1}. Clearly, since G is K_3-free, v has a non-neighbour in $\{v_i, v_{i+1}\}$. If v has no neighbours in $\{v_i, v_{i+1}\}$, then $G[v_{2k+1}, v_1, v_2, v, v_i, v_{i+1}]$ is isomorphic to $K_{1,3} + K_2$, a contradiction. Now assume that v is adjacent to v_4. It follows that v is complete to $\{v_4, v_6, \ldots, v_{2k-2}\}$ and anticomplete to $\{v_5, v_7, \ldots, v_{2k-1}\}$. But now $G[v_2, v_3, v, v_{2k-3}, v_{2k-2}, v_{2k-1}]$ is isomorphic to $K_{1,3} + K_2$, a contradiction. So we may assume that v is adjacent to v_5. This implies that v is complete to $\{v_5, v_7, \ldots, v_{2k-1}\}$ and anticomplete to $\{v_4, v_6, \ldots, v_{2k-2}\}$. It follows that v is non-adjacent to v_{2k}, since G is K_3-free. But now $G[v_4, v_5, v_6, v, v_{2k}, v_{2k+1}]$ is isomorphic to $K_{1,3} + K_2$. This contradiction shows that $G = C$.

Now assume that $k = 3$ and let $v \in V(G) \backslash V(C)$ be adjacent to v_1. As before, v has exactly one neighbour in $\{v_4, v_5\}$. By symmetry we may assume that v is adjacent to v_4. Hence, v has no neighbours in $\{v_2, v_3, v_5, v_7\}$. Finally observe that v is non-adjacent to v_6, since otherwise $G[v_5, v_6, v_7, v, v_2, v_3]$ is isomorphic to $K_{1,3} + K_2$. Therefore we conclude that each vertex $v \in V(G) \backslash V(C)$ that is adjacent to some vertex $v_i \in V(C)$, is either complete to $\{v_i, v_{i+3}\}$ and anticomplete to $V(C) \backslash \{v_i, v_{i+3}\}$, or complete to $\{v_i, v_{i+4}\}$ and anticomplete to $V(C) \backslash \{v_i, v_{i+4}\}$ (here subscripts are taken modulo 7).

Denote by U_j the set of vertices at distance j from the cycle. We claim that

- $|U_1| \leq 7$. Indeed, if $|U_1| > 7$, then there exist two vertices $z, z' \in U_1$ that are complete to $\{v_i, v_{i+3}\}$ for some value of i (and thus anticomplete to $V(C) \backslash \{v_i, v_{i+3}\}$). Since G is K_3-free, z, z' are non-adjacent. But now $G[v_i, z, z', v_{i+1}, v_{i+4}, v_{i+5}]$ is isomorphic to $K_{1,3} + K_2$, a contradiction.

- *each vertex of U_1 has at most one neighbour in U_2.* Indeed, assume a vertex $x \in U_1$ has two neighbours $y, z \in U_2$, and without loss of generality let x be complete to $\{v_i, v_{i+3}\}$ (and thus anticomplete to $V(C) \backslash \{v_i, v_{i+3}\}$). Since G is K_3-free, it follows that y, z are non-adjacent. But then $G[x, y, z, v_i, v_{i+4}, v_{i+5}]$ is isomorphic to $K_{1,3} + K_2$, a contradiction.
- *each vertex of U_2 has at most one neighbour in U_3,* which can be proved by analogy with the previous claim.
- *for each $i \geq 4$, U_i is empty.* Indeed, assume without loss of generality that $U_4 \neq \emptyset$ and let u_4, u_3, u_2, u_1 be a path from U_4 to C with $u_j \in U_j$ and u_1 being adjacent to v_i. Then $G[v_i, v_{i+1}, v_{i-1}, u_1, u_3, u_4]$ is isomorphic to $K_{1,3} + K_2$, a contradiction.

From the above claims we conclude that $V(G) = V(C) \cup U_1 \cup U_2 \cup U_3$, $|U_3| \leq |U_2| \leq |U_1|$, and therefore $|V(G)| \leq 28$. □

Thus Lemmas 3 and 4 and Fact 2 further reduce the problem to graphs containing a C_5.

Lemma 5. *If G is a connected $(K_3, S_{1,1,3})$-free graph containing a C_5, then the clique-width of G is bounded by a constant.*

Proof. Let G be a connected $(K_3, S_{1,1,3})$-free graph and let $C = v_1 - v_2 - v_3 - v_4 - v_5 - v_1$ be an induced cycle of length five in G. If $G = C$ then the clique-width of G is at most 4 (Fact 1). Therefore we may assume that there exists at least one vertex $v \in V(G) \setminus V(C)$. Since G is K_3-free, v can be adjacent to at most two vertices of C, and if v has two neighbours on C, they are non-consecutive vertices of the cycle. We denote the set of vertices in $V(G) \setminus V(C)$ that have exactly i neighbours on C by N_i, $i \in \{0, 1, 2\}$. Also, for $i = 1, \ldots, 5$, we denote by V_i the set of vertices in N_2 adjacent to $v_{i-1}, v_{i+1} \in V(C)$ (throughout the proof subscripts i are taken modulo 5). We call two different sets V_i and V_j *consecutive* if v_i and v_j are consecutive vertices of C, and *opposite* otherwise. Finally, we call V_i *large* if $|V_i| \geq 2$, and *small* otherwise. The proof of the lemma will be given through a series of claims.

(1) *Each V_i is an independent set.* This immediately follows from the fact that G is K_3-free.
(2) *N_0 is an independent set.* Indeed, assume xy is an edge connecting two vertices $x, y \in N_0$, and let, without loss of generality, y be adjacent to a vertex $z \in N_1 \cup N_2$. Assume z is adjacent to $v_i \in V(C)$. Since G is K_3-free, z is non-adjacent to x, v_{i-1}, v_{i+1}. But then $G[x, y, z, v_i, v_{i+1}, v_{i-1}]$ is isomorphic to $S_{1,1,3}$, a contradiction.
(3) *Any vertex $x \in N_1 \cup N_2$ has at most one neighbour in N_0.* Suppose $x \in N_1 \cup N_2$ is adjacent to $z, z' \in N_0$, and let $v_i \in V(C)$ be a neighbour of x. Since G is K_3-free, it follows that x is non-adjacent to v_{i-1}, v_{i+1}. Furthermore x is adjacent to at most one of v_{i-2}, v_{i+2}. By symmetry we may assume that x is non-adjacent to v_{i-2}. But now $G[x, z, z', v_i, v_{i-1}, v_{i-2}]$ is isomorphic to $S_{1,1,3}$, a contradiction.

(4) $|N_1| \leq 5$. Indeed, if there are two vertices $x, x' \in N_1$ which are adjacent to the same vertex $v_i \in V(C)$, then $G[x, x', v_i, v_{i+1}, v_{i+2}, v_{i+3}]$ is isomorphic to $S_{1,1,3}$, a contradiction.

(5) *If V_i and V_j are opposite sets, then no vertex of V_i is adjacent to a vertex of V_j.* This immediately follows from the fact that G is K_3-free.

(6) *If V_i and V_j are consecutive, then every vertex x of V_i has at most one non-neighbour in V_j.* Suppose $x \in V_i$ has two non-neighbours $y, y' \in V_j$. By symmetry we may assume that $j = i+1$. But now $G[x, y, y', v_{i-1}, v_{i-2}, v_{i-3}]$ is isomorphic to $S_{1,1,3}$, a contradiction.

(7) *If V_i and V_j are two opposite large sets, then no vertex in N_0 has a neighbour in $V_i \cup V_j$.* Assume without loss of generality that $i = 1$ and $j = 4$, and suppose for a contradiction that a vertex $x \in N_0$ has a neighbour $y \in V_1$. If x is non-adjacent to some vertex $z \in V_4$, then $G[x, y, z, v_2, v_3, v_4]$ is isomorphic to $S_{1,1,3}$, a contradiction. Therefore x is complete to V_4. But now $G[x, y, v_1, v_2, z, z']$ with $z, z' \in V_4$ is isomorphic to $S_{1,1,3}$, a contradiction.

Since G is connected and N_0 is an independent set, every vertex of N_0 has a neighbour in $N_1 \cup N_2$. Let us denote by V_0 those vertices of N_0 every neighbour of which belongs to a large set V_i and by G_0 the subgraph of G induced by V_0 and the large sets. From Claims (3) and (4), it follows that at most 25 vertices of G do not belong to G_0. Therefore, by Fact 3, the clique-width of G is bounded if and only if it is bounded for G_0. We may assume that G has at least one large set, since otherwise G_0 is empty. We will show that G_0 has bounded clique-width by examining all possible combinations of large sets.

Case 1: Assume that for every large set V_i there is an opposite large set V_j. Then it follows from Claim (7) that $V_0 = \emptyset$. In order to see that G_0 has bounded clique-width, we complement the edges between every pair of consecutive large sets. By Claim (6), the resulting graph has maximum degree at most 2. From Fact 1 it follows that this graph is of bounded clique-width, and therefore, applying Fact 4, G_0 has bounded clique-width.

Case 1 allows us to assume that G contains a large set such that the opposite sets are small. Without loss of generality we let V_1 be large, and V_3 and V_4 be small. The rest of the proof is based on the analysis of the size of the sets V_2 and V_5.

Case 2: V_2 and V_5 are large. Then, by Claims (1), (2), (5) and (7), G_0 is a bipartite graph with bipartition $(V_1, V_2 \cup V_5 \cup V_0)$. Therefore by Fact 2, G_0 has bounded clique-width.

Case 3: V_2 and V_5 are small. Then G_0 is a bipartite graph with bipartition (V_1, V_0), and therefore, by Fact 2, G_0 has bounded clique-width.

Case 4: V_2 is large and V_5 is small, i.e. G_0 is induced by $V_0 \cup V_1 \cup V_2$. Consider a vertex $x \in V_0$ that has a neighbour $y \in V_1$ and a neighbour $z \in V_2$. Then y and z are non-adjacent (since G is K_3-free) and therefore, by Claim (6), y is complete to $V_2 \setminus \{z\}$ and z is complete to $V_1 \setminus \{y\}$. From the K_3-freeness of G it follows that x is anticomplete to $(V_1 \cup V_2) \setminus \{y, z\}$.

Let us denote by V_0' the vertices of V_0 that have neighbours both in V_1 and V_2, and by V_i' ($i = 1, 2$) the vertices of V_i that have neighbours in V_0'. Also, let $V_i'' = V_i - V_i'$ for $i = 0, 1, 2$, and $G_0' = G_0[V_0' \cup V_1' \cup V_2']$, $G_0'' = G_0[V_0'' \cup V_1'' \cup V_2'']$.

By Claim (3), V_0'' is anticomplete to $V_1' \cup V_2'$. Also, it follows from the above discussion that V_0' is anticomplete to $V_1'' \cup V_2''$, that V_1' is complete to V_2'', and that V_2' is complete to V_1''. Therefore by complementing the edges between V_1' and V_2'', and between V_2' and V_1'', we disconnect G_0' from G_0''. The graph G_0'' is a bipartite graph, since every vertex of V_0'' has neighbours either in V_1'' or in V_2'' but not in both. Thus it follows from Fact 2 that G_0'' has bounded clique-width. To see that G_0' has bounded clique-width, we complement the edges between V_1' and V_2'. This operation transforms G_0' into a collection of disjoint triangles. Therefore the clique-width of G_0' is bounded. Now it follows from Fact 4 that G_0 has bounded clique-width. ☐

Similarly to Lemma 5, one can prove the following result the proof of which is omitted due to space limitation.

Lemma 6. *If G is a connected $(K_3, K_{1,3} + K_2)$-free graph containing a C_5, then the clique-width of G is bounded by a constant.*

From Lemmas 3, 4, 5, and 6, we derive the main result of this section.

Theorem 3. *The clique-width of $(K_3, S_{1,1,3})$-free graphs and $(K_3, K_{1,3} + K_2)$-free graphs is bounded by a constant and therefore the* VERTEX COLOURING *problem is polynomial-time solvable in these classes of graphs.*

5   Further Results

In this section, we prove some results for graph classes $Free(K_3, F)$ with F being a forest on more than 6 vertices.

Theorem 4. *For every fixed m, the* VERTEX COLOURING *problem is polynomial-time solvable in the class $Free(K_3, mK_2)$.*

Proof. Obviously, if a graph G is k-colourable, then it admits a k-colouring in which one of the colour classes is a maximal independent set.

It is known that for every fixed m the number of maximal independent sets in the class $Free(mK_2)$ is bounded by a polynomial [1] and all of them can be found in polynomial time [29]. Therefore given a mK_2-free graph G, we can solve the 3-COLOURABILITY problem for G by generating all maximal independent sets and solving 2-COLOURABILITY for the remaining vertices of the graph. Then by induction on k, we conclude that for any fixed k the k-COLOURABILITY problem can be solved in the class $Free(mK_2)$ in polynomial time. Since the chromatic number of (K_3, mK_2)-free graphs is bounded by $2m - 2$ (see e.g. [2]), the VERTEX COLOURING problem is polynomial-time solvable in the class $Free(K_3, mK_2)$ for any fixed m. ☐

Theorem 5. *For every fixed m, the* VERTEX COLOURING *problem is polynomial-time solvable in the class $Free(K_3, P_3 + mK_1)$.*

Proof. To prove the theorem, we will show that for any fixed m, graphs in the class $Free(K_3, P_3 + mK_1)$ are either bounded in size, or they are 3-colourable and a 3-colouring can be found in polynomial time.

Let G be a $(K_3, P_3 + mK_1)$-free graph, and let S be a maximum independent set in G. Denote by R the remaining vertices of G, i.e. $R = V(G) - S$. Assume that R contains an induced odd cycle $C = v_1 - v_2 - \ldots - v_p - v_1$ with $p \geq 5$. Since S is a maximum independent set, each vertex of C has at least one neighbour in S. Let us call a vertex $v_i \in V(C)$ strong if it has at least 2 neighbours in S and weak otherwise. Since C is an odd cycle, it has either two consecutive weak vertices or two consecutive strong vertices.

If C has two consecutive weak vertices, say v_1, v_2, then jointly they are adjacent to two vertices of S, say v_1 is adjacent to s_1, and v_2 is adjacent to s_2, and therefore, they have $|S| - 2$ common non-neighbours in S. If $|S| - 2 \geq m$, then v_1, v_2, s_1 together with m vertices in $S \setminus \{s_1, s_2\}$ induce a subgraph isomorphic to $P_3 + mK_1$, a contradiction. Therefore $|S| \leq m + 1$. But then the number of vertices of G is bounded by the Ramsey number $R(3, m + 1)$.

Now assume C has two consecutive strong vertices, say v_1, v_2. Since the graph is $(P_3 + mK_1)$-free, every strong vertex has at most $m - 1$ non-neighbours in S, and since the graph is K_3-free, consecutive vertices of C cannot have common neighbours. Therefore each of v_1 and v_2 has at most $m - 1$ neighbours in S. But then $|S| \leq 2m - 2$ and hence the number of vertices of G is bounded by the Ramsey number $R(3, 2m - 2)$.

Thus, if R has an odd cycle, then the number of vertices of G is bounded by a constant. If R has no odd cycles, then $G[R]$ is bipartite, and hence G is 3-colourable. Finding a maximum independent set in a $(P_3 + mK_1)$-free graph can be done in polynomial time, so any $(K_3, P_3 + mK_1)$-free graph is either bounded in size, or can be 3-coloured in this way in polynomial time. Thus VERTEX COLOURING of $(K_3, P_3 + mK_1)$-free graphs can be solved in polynomial time. ☐

6   Concluding Remarks and Open Problem

In this paper we show that the VERTEX COLOURING problem is polynomially solvable in many classes of (K_3, F)-free graphs with F being a forest on 6 vertices. Two classes that have not been analyzed here are $(K_3, P_4 + P_2)$-free graphs and $(K_3, 2P_3)$-free graphs.

For $(K_3, P_4 + P_2)$-free graphs, we managed to prove polynomial-time solvability of the problem after the submission of the present paper and the solution will appear in the journal version of the paper. However, finding the complexity status of the VERTEX COLOURING problem in $(K_3, 2P_3)$-free graphs remains a challenging open question.

References

1. Balas, E., Yu, C.S.: On graphs with polynomially solvable maximum-weight clique problem. Networks 19, 247–253 (1989)
2. Brandt, S.: Triangle-free graphs and forbidden subgraphs. Discrete Appl. Math. 120, 25–33 (2002)
3. Brandt, S.: A 4-colour problem for dense triangle-free graphs. Discrete Math. 251, 33–46 (2002)
4. Brandstädt, A., Klembt, T., Mahfud, S.: P_6- and triangle-free graphs revisited: structure and bounded clique-width. Discrete Math. Theor. Comput. Sci. 8, 173–187 (2006)
5. Broersma, H.J., Fomin, F.V., Golovach, P.A., Paulusma, D.: Three complexity results on coloring P_k-free graphs. In: Fiala, J., Kratochvíl, J., Miller, M. (eds.) IWOCA 2009. LNCS, vol. 5874, pp. 95–104. Springer, Heidelberg (2009)
6. Courcelle, B., Makowsky, J.A., Rotics, U.: Linear time solvable optimization problems on graphs of bounded clique-width. Theory Comput. Systems 33, 125–150 (2000)
7. Courcelle, B., Olariu, S.: Upper bounds to the clique-width of a graph. Discrete Applied Math. 101, 77–114 (2000)
8. Dailey, D.P.: Uniqueness of colorability and colorability of planar 4 regular graphs are NP-complete. Discrete Math. 30, 289–293 (1980)
9. Diestel, R.: Graph theory, 3rd edn. Graduate Texts in Mathematics, vol. 173, pp. xvi+411. Springer, Berlin (2005)
10. Hoang, C., Kaminski, M., Lozin, V., Sawada, J., Shu, X.: Deciding k-colorability of P_5-free graphs in polynomial time. Algorithmica 57, 74–81 (2010)
11. Holyer, I.: The NP-completeness of edge-coloring. SIAM J. Computing 10, 718–720 (1981)
12. Kamiński, M., Lozin, V.: Coloring edges and vertices of graphs without short or long cycles. Contributions to Discrete Mathematics 2, 61–66 (2007)
13. Kamiński, M., Lozin, V.: Vertex 3-colorability of claw-free graphs. Algorithmic Operations Research 2, 15–21 (2007)
14. Kamiński, M., Lozin, V.V., Milanič, M.: Recent developments on graphs of bounded clique-width. Discrete Applied Math. 157, 2747–2761 (2009)
15. Kochol, M., Lozin, V., Randerath, B.: The 3-colorability problem on graphs with maximum degree 4. SIAM J. Computing 32, 1128–1139 (2003)
16. Král, D., Kratochvíl, J., Tuza, Z., Woeginger, G.J.: Complexity of coloring graphs without forbidden induced subgraphs. In: Brandstädt, A., van Le, B. (eds.) WG 2001. LNCS, vol. 2204, pp. 254–262. Springer, Heidelberg (2001)
17. Le, V.B., Randerath, B., Schiermeyer, I.: On the complexity of 4-coloring graphs without long induced paths. Theoret. Comput. Sci. 389, 330–335 (2007)
18. Lozin, V.V.: Bipartite graphs without a skew star. Discrete Math. 257, 83–100 (2002)
19. Lozin, V., Rautenbach, D.: On the band-, tree-, and clique-width of graphs with bounded vertex degree. SIAM J. Discrete Math. 18, 195–206 (2004)
20. Lozin, V., Volz, J.: The clique-width of bipartite graphs in monogenic classes. International Journal of Foundations of Computer Sci. 19, 477–494 (2008)
21. Maffray, F., Preissmann, M.: On the NP-completeness of the k-colorability problem for triangle-free graphs. Discrete Math. 162, 313–317 (1996)
22. Randerath, B.: 3-colorability and forbidden subgraphs. I. Characterizing pairs. Discrete Math. 276, 313–325 (2004)

23. Randerath, B., Schiermeyer, I., Tewes, M.: Three-colourability and forbidden subgraphs. II. Polynomial algorithms. Discrete Math. 251, 137–153 (2002)
24. Randerath, B., Schiermeyer, I.: A note on Brooks' theorem for triangle-free graphs. Australas. J. Combin. 26, 3–9 (2002)
25. Randerath, B., Schiermeyer, I.: Vertex colouring and forbidden subgraphs—a survey. Graphs Combin. 20, 1–40 (2004)
26. Randerath, B., Schiermeyer, I.: 3-colorability $\in$ **P** for P_6-free graphs. Discrete Appl. Math. 136, 299–313 (2004)
27. Scott, A.D.: Induced trees in graphs of large chromatic number. J. Graph Theory 24, 297–311 (1997)
28. Sgall, J., Wöginger, G.J.: The complexity of coloring graphs without long induced paths. Acta Cybernet. 15, 107–117 (2001)
29. Tsukiyama, S., Ide, M., Ariyoshi, H., Shirakawa, I.: A new algorithm for generating all the maximal independent sets. SIAM J. Computing 6, 505–517 (1977)

A Quartic Kernel for Pathwidth-One Vertex Deletion

Geevarghese Philip[1], Venkatesh Raman[1], and Yngve Villanger[2]

[1] The Institute of Mathematical Sciences, Chennai, India
{gphilip,vraman}@imsc.res.in
[2] University of Bergen, N-5020 Bergen, Norway
yngve.villanger@uib.no

Abstract. The pathwidth of a graph is a measure of how path-like the graph is. Given a graph G and an integer k, the problem of finding whether there exist at most k vertices in G whose deletion results in a graph of pathwidth at most one is NP-complete. We initiate the study of the parameterized complexity of this problem, parameterized by k. We show that the problem has a quartic vertex-kernel: We show that, given an input instance $(G = (V, E), k); |V| = n$, we can construct, in polynomial time, an instance (G', k') such that (i) (G, k) is a YES instance if and only if (G', k') is a YES instance, (ii) G' has $\mathcal{O}(k^4)$ vertices, and (iii) $k' \leq k$. We also give a fixed parameter tractable (FPT) algorithm for the problem that runs in $\mathcal{O}(7^k k \cdot n^2)$ time.

1 Introduction

The treewidth of a graph is a measure of how "tree-like" the graph is. The notion of treewidth was introduced by Robertson and Seymour in their seminal Graph Minors series [35]. It has turned out to be very important and useful, both in the theoretical study of the properties of graphs [6,27] and in designing graph algorithms [7,9]. A graph has treewidth at most one if and only if it is a forest (a collection of trees), and a set of vertices in a graph G whose removal from G results in a forest is called a *feedback vertex set* (FVS) of the graph.

Given a graph G and an integer k as input, the FEEDBACK VERTEX SET problem asks whether G has an FVS of size at most k. This is one of the first problems that Karp showed to be NP-complete [25]. The problem and its variants have extensively been investigated from the point of view of various algorithmic paradigms, including approximation and parameterized algorithms. The problem is known to have a 2-factor approximation algorithm [4], and the problem parameterized by the solution size k is fixed parameter tractable (FPT) and has a polynomial kernel[1].

The quest for fast FPT algorithms and small kernels for the parameterized FEEDBACK VERTEX SET problem presents an illuminative case study of the evolution of the field of fixed parameter tractability, and stands out among the

---

[1] See Section 2 for the terminology and notation used in this paper.

D.M. Thilikos (Ed.): WG 2010, LNCS 6410, pp. 196–207, 2010.

many success stories of this algorithmic approach towards solving hard problems. The first FPT algorithm for the problem, with a running time of $\mathcal{O}^*(k^4!)$, was developed by Bodlaender [5] and by Downey and Fellows [18]. After a series of improvements [19,24,33], a running time of the form $\mathcal{O}^*(c^k)$ was first obtained by Guo et.al [22], whose algorithm ran in $\mathcal{O}^*(37.7^k)$ time. This was improved by Dehne et.al [15] to $\mathcal{O}^*(10.6^k)$ in 2007, and to the current best $\mathcal{O}^*(3.83^k)$ by Cao et.al [13] in 2010. For classes of graphs that exclude a fixed minor H (for example, planar graphs), Dorn et.al [17] have recently obtained an FPT algorithm for the problem with a running time of the form $\mathcal{O}^*(2^{\mathcal{O}(\sqrt{k})})$.

Proving polynomial bounds on the size of the kernel for different parameterized problems has been a significant practical aspect in the study of the parameterized complexity of NP-hard problems, and many positive results are known. See [23] for a survey of kernelization results. The existence of a polynomial kernel for the FEEDBACK VERTEX SET problem was open for a long time. It was settled in the affirmative by Burrage et. al [12] as recently as 2006, when they exhibited a kernel with $\mathcal{O}(k^{11})$ vertices. This was soon improved to a cubic vertex-kernel ($\mathcal{O}(k^3)$ vertices) by Bodlaender [8,10]. The current smallest kernel, on $\mathcal{O}(k^2)$ vertices, is due to Thomassé [36].

The *pathwidth* of a graph is a notion closely related to treewidth, and was also introduced by Robertson and Seymour in the Graph Minors series [34]. The pathwidth of a graph denotes how "path-like" it is. A graph has pathwidth at most one if and only if it is a collection of *caterpillars*, where a caterpillar is a special kind of tree: it is a tree that becomes a path (called the *spine* of the caterpillar) when all its pendant vertices are removed. Graphs of pathwidth at most one are thus a very special kind of forests, and have even less structure than forests (which are themselves very "simple" graphs). As a consequence, some problems that are NP-hard even on forests can be solved in polynomial time on graphs of pathwidth at most one. Examples include (Weighted) Bandwidth [3,31,28], the Proper Interval Colored Graph problem, and the Proper Colored Layout problem [1].

In contrast to the case of forests, the corresponding vertex deletion problem for obtaining a collection of caterpillars (equivalently, a graph of pathwidth at most one) has not received much attention in the literature. In fact, to the best of our knowledge, the following problem has not yet been investigated at all: Given a graph G and an integer k as input, find whether G contains a set of at most k vertices whose removal from G results in a graph of pathwidth at most one. We call such a set of vertices a pathwidth-one deletion set (PODS), and the problem the PATHWIDTH-ONE VERTEX DELETION problem. It follows from a general NP-hardness result of Lewis and Yannakakis that this problem is NP-complete.

Our results. We study the parameterized complexity of the PATHWIDTH-ONE VERTEX DELETION problem parameterized by the solution size k, and show that (i) the problem has a vertex-kernel of size $\mathcal{O}(k^4)$, and (ii) the problem can be solved in $\mathcal{O}^*(7^k)$ time (Compare with the values $\mathcal{O}(k^2)$ and $\mathcal{O}^*(3.83^k)$ for FVS, respectively).

Note that, in general, a PODS "does more" than an FVS: It "kills" all cycles in the graph, like an FVS, and, in addition, it kills all non-caterpillar trees in the graph. In fact, the difference in the sizes of a smallest FVS and a smallest PODS of a graph can be arbitrarily large. For example, the treewidth of a binary tree is one, while for any integer c there exists a binary tree T_c of pathwidth at least $c+1$. Removing a single vertex from a graph will reduce the pathwidth by at most one, and so for T_c, the difference between the two numbers is at least c. Partly as a consequence of such differences, many of the techniques and reduction rules that have been developed for obtaining FPT algorithms and kernels for the FEEDBACK VERTEX SET problem do *not* carry over to the PATHWIDTH-ONE VERTEX DELETION problem. Instead, we use a characterization of graphs of pathwidth at most one to obtain the FPT algorithm and the polynomial kernel.

Update. After this paper was presented at WG 2010, Cygan et. al [14] improved both the results in the paper. Using the same general idea of our FPT algorithm and a clever branching strategy, they obtained an $\mathcal{O}^*(4.65^k)$ FPT algorithm for the problem. Using some of our reduction rules and a different approach based on the α-expansion Lemma of Thomassé [36], they obtained a quadratic ($\mathcal{O}(k^2)$) kernel as well.

Organization of the rest of the paper. In Section 2 we give an overview of the notation and terminology used in the rest of the paper. In Section 3 we formally define the PATHWIDTH-ONE VERTEX DELETION problem, show that the problem is NP-complete, and sketch an FPT algorithm for the problem that runs in $\mathcal{O}^*(7^k)$ time. We show in Section 4 that the problem has a vertex-kernel of size $\mathcal{O}(k^4)$. We conclude in Section 5. Due to space constraints, many proofs have been deferred to a full version [32] of the paper.

2   Preliminaries

In this section we state some definitions related to graph theory and parameterized complexity, and give an overview of the notation used in this paper; we also formally define the PATHWIDTH-ONE VERTEX DELETION problem and show that it is NP-hard. In general we follow the graph terminology of [16]. For a vertex $v \in V$ in a graph $G = (V, E)$, we call the set $N(v) = \{u \in V | (u, v) \in E\}$ the *open neighborhood* of v. The elements of $N(v)$ are said to be the *neighbors* of v, and $N[v] = N(v) \cup \{v\}$ is called the *closed neighborhood* of v. For a set of vertices $X \subseteq V$, the open and closed neighborhoods of X are defined, respectively, as $N(X) = \bigcup_{u \in X} N(u) \setminus X$ and $N[X] = N(X) \cup X$. For vertices u, v in G, u is said to be a *pendant* vertex of v if $N(u) = \{v\}$. A *caterpillar* is a tree that becomes a path (called the *spine* of the caterpillar) when all its pendant vertices are removed. A nontrivial caterpillar is one that contains at least two vertices. A T_2 is the graph on seven vertices shown in Figure 1. The *center* of a T_2 is the one vertex of degree 3, and its *leaves* are the three vertices of degree 1.

The operation of *contracting* an edge (u, v) consists of deleting vertex u, renaming vertex v to uv, and adding a new edge (x, uv) for each edge $(x, u); x \neq v$.

Multiple edges that may possibly result from this operation are preserved. Note that the operation is symmetric with respect to u and v. A graph H is said to be a *minor* of a graph G if a graph isomorphic to H can be obtained by contracting zero or more edges of some subgraph of G.

A *graph property* is a subset of the set of all graphs. Graph property Π is said to hold for graph G if $G \in \Pi$. Π is said to be nontrivial if Π and its complement are both infinite. Π is said to be *hereditary* if Π holds for every induced subgraph of graph G whenever it holds for G. The *membership testing* problem for Π is to test whether Π holds for a given input graph.

A *tree decomposition* of a graph $G = (V, E)$ is a pair (T, χ) in which $T = (V_T, E_T)$ is a tree and $\chi = \{\chi_i \mid i \in V_T\}$ is a family of subsets of V, called *bags*, such that

(i) $\bigcup_{i \in V_T} \chi_i = V$;
(ii) for each edge $(u, v) \in E$ there exists an $i \in V_T$ such that both u and v belong to χ_i; and
(iii) for all $v \in V$, the set of nodes $\{i \in V_T \mid v \in \chi_i\}$ induces a connected subgraph of T.

The maximum of $|\chi_i| - 1$, over all $i \in V_T$, is called the *width* of the tree decomposition. The *treewidth* of a graph G is the minimum width taken over all tree decompositions of G. A *path decomposition* of a graph $G = (V, E)$ is a tree decomposition of G where the underlying tree T is a path. The *pathwidth* of G is the minimum width over all possible path decompositions of G.

To describe the running times of algorithms we sometimes use the $\mathcal{O}^*$ notation. Given $f : \mathbb{N} \to \mathbb{N}$, we define $\mathcal{O}^*(f(n))$ to be $\mathcal{O}(f(n) \cdot p(n))$, where $p(\cdot)$ is some polynomial function. That is, the $\mathcal{O}^*$ notation suppresses polynomial factors in the expression for the running time.

A parameterized problem Π is a subset of $\Sigma^* \times \mathbb{N}$, where Σ is a finite alphabet. An instance of a parameterized problem is a tuple (x, k), where k is called the parameter. A central notion in parameterized complexity is *fixed-parameter tractability (FPT)* which means, for a given instance (x, k), decidability in time $f(k) \cdot p(|x|)$, where f is an arbitrary function of k and p is a polynomial. The notion of *kernelization* is formally defined as follows.

Definition 1. [Kernelization, Kernel] [21,30] *A kernelization algorithm for a parameterized problem $\Pi \subseteq \Sigma^* \times \mathbb{N}$ is an algorithm that, given $(x, k) \in \Sigma^* \times \mathbb{N}$, outputs, in time polynomial in $|x| + k$, a pair $(x', k') \in \Sigma^* \times \mathbb{N}$ such that (a) $(x, k) \in \Pi$ if and only if $(x', k') \in \Pi$ and (b) $|x'|, k' \leq g(k)$, where g is some computable function. The output instance x' is called the kernel, and the function g is referred to as the size of the kernel. If $g(k) = k^{\mathcal{O}(1)}$ then we say that Π admits a polynomial kernel.*

When a kernelization algorithm outputs a graph on $h(k)$ vertices, we sometimes say that the output is an $h(k)$ vertex-kernel.

3   The PATHWIDTH-ONE VERTEX DELETION Problem

In this section we formally define the PATHWIDTH-ONE VERTEX DELETION problem, show that it is NP-complete, and briefly sketch an $\mathcal{O}^*(7^k)$ FPT algorithm for the problem. We begin with the observation that caterpillars are the quintessential graphs of pathwidth at most one:

Fact 1. [2] *A graph G has pathwidth at most one if and only if it is a collection of vertex-disjoint caterpillars.*

A vertex set $S \subseteq V$ of a graph G is said to be a *pathwidth-one deletion set* (PODS) if $G[V \setminus S]$ has pathwidth at most one. In this paper we investigate the parameterized complexity of the following problem:

PATHWIDTH-ONE VERTEX DELETION (POVD)
Input: An undirected graph $G = (V, E)$, and a positive integer k.
Parameter: k
Question: Does there exist a set $S \subseteq V$ of at most k vertices of G such that $G[V \setminus S]$ has pathwidth at most one (i.e., S is a PODS of G)?

The following general NP-completeness result is due to Lewis and Yannakakis:

Fact 2. [29] *The following problem is NP-complete for any nontrivial hereditary graph property Π for which the membership testing problem can be solved in polynomial time:*

Input: *Graph $G = (V, E)$, positive integer k.*
Question: *Is there a subset $S \subseteq V, |S| \leq k$ such that $G[V \setminus S] \in \Pi$?*

The NP-completeness of the PATHWIDTH-ONE VERTEX DELETION problem follows directly from this result:

Theorem 1. $[\star]^2$ *The PATHWIDTH-ONE VERTEX DELETION problem is NP-complete.*

In the rest of the paper we focus on the parameterized complexity of the PATHWIDTH-ONE VERTEX DELETION problem. We now sketch an $\mathcal{O}^*(7^k)$ time FPT algorithm, and in the next section we describe an $\mathcal{O}(k^4)$ vertex-kernel for the problem. Let $(G = (V, E), k)$ be the input instance, where $|V| = n$. Let $S \subseteq V$ be a PODS of G of size at most k. Observe that if (G, k) is a YES instance, then the number of edges in G is at most $k(n-1) + (n-1) = (k+1)(n-1)$. The first term on the left is an upper bound on the number of edges that are incident the vertices in S; the second term is a loose upper bound on the number of edges in $G \setminus S$. So, if G has more than $(k+1)(n-1)$ edges, then we can immediately

---

2 Due to space constraints, proofs of results labeled with a [$\star$] have been moved to a full version [32] of the paper.

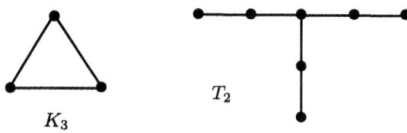

Fig. 1. The set of excluded minors for graphs of pathwidth at most one

reject the input. Since each reduction rule in the sequel is sound, and no rule increases the number of vertices or edges, from now on we assume, without loss of generality, that the graph has at most $(k + 1)(n - 1)$ edges.

The kernel arguments are based on Fact 1, while our starting point for the FPT algorithm is the following characterization, in terms of excluded minors, of graphs of pathwidth at most one:

Fact 3. [11,20] *A graph G has pathwidth at most one if and only if it does not contain K_3 or T_2 as a minor, where K_3 and T_2 are as in Figure 1.*

Fact 3 is not very helpful in the given form in checking for a small PODS. Instead, we derive and use the following alternate characterization and the two succeeding lemmas:

Lemma 1. [⋆] *A graph G has pathwidth at most one if and only if it does not contain a cycle or a T_2 as a subgraph.*

Lemma 2. [⋆] *Let $S = \{T_2, K_3, C_4\}$, where C_4 is a cycle of length 4. Given a graph $G = (V, E)$; $|V| = n$, we can find whether G contains a subgraph H that is isomorphic to one of the graphs in S, and also locate such an H if it exists, in $\mathcal{O}(kn^2)$ time.*

Lemma 3. [⋆] *Let $S = \{T_2, K_3, C_4\}$, where C_4 is a cycle of length 4. If G is a graph that does not contain any element of S as a subgraph, then each connected component of G is either a tree, or a cycle with zero or more pendant vertices ("hairs") attached to it.*

3.1   An FPT Algorithm for POVD

Let $(G = (V, E), k)$ be the input instance, where $|V| = n$. We use a branching strategy inspired by Lemmas 1 and 3. First we locate a (not necessarily induced) subgraph T of G that is isomorphic to one of $S = \{T_2, K_3, C_4\}$. From Lemma 2, this can be done in $\mathcal{O}(kn^2)$ time. At least one of the (at most seven) vertices of T must be in any PODS of G. So we branch on the vertices of T: We pick each one, in turn, into the minimal PODS that we are constructing, delete the picked vertex and all its adjacent edges, and recurse on the remaining graph after decrementing the parameter by one.

The leaves of this recursion tree correspond to graphs which do not have a subgraph isomorphic to any graph in S. By Lemma 3, each connected component of such a graph is a tree, or a cycle with zero or more pendant vertices ("hairs") attached to it. The trees can be ignored — they do not have a T_2 as a subgraph — and each cycle (with or without hairs) forces exactly one vertex into any minimal solution. Thus the base case of the recursion can be solved in linear time.

This is a 7-way branching, where the depth of the recursion is at most k, and where the algorithm spends $\mathcal{O}(kn^2)$ time at each node. Hence we have

Theorem 2. *The* PATHWIDTH-ONE VERTEX DELETION *problem parameterized by the solution size k has an FPT algorithm that runs in $\mathcal{O}(n^2 \cdot 7^k k)$ time.*

By a folklore result of parameterized complexity, it follows immediately from Theorem 2 that the PATHWIDTH-ONE VERTEX DELETION problem parameterized by the solution size k has a kernel of size $\mathcal{O}(7^k)$ (See, for example, [21]). We now show that the kernel size can be brought down significantly from this trivial bound.

4   A Polynomial Kernel for POVD

We turn to the main result of this paper. We describe a polynomial-time algorithm (the *kernelization algorithm*) that, given an instance (G, k) of POVD, returns an instance (G', k') (the *kernel*) of POVD such that (i) (G, k) is a YES instance if and only if (G', k') is a YES instance, (ii) G' has $\mathcal{O}(k^4)$ vertices, and (iii) $k' \leq k$. The kernelization algorithm (Algorithm 1) exhaustively applies the reduction rules of Section 4.1 to the input instance. The resulting instance, to which no rule applies, is said to be *reduced* with respect to the reduction rules. To demonstrate a quartic vertex-kernel for the problem, it suffices to show that

1. The rules can be exhaustively applied in polynomial time;
2. Each rule is *sound*: the output of a rule is a YES instance if and only if its input is a YES instance; and
3. If the input instance (G, k) is a YES instance, then the reduced instance (G', k') has $\mathcal{O}(k^4)$ vertices.

The reduction rules are based on the following idea: Suppose $(G = (V, E), k)$ is a YES instance of the problem that is reduced with respect to the reduction rules. Then there is a set $S \subseteq V, |S| \leq k$ such that $G[V \setminus S]$ is a collection of caterpillars, and it suffices to show that $|V \setminus S| = \mathcal{O}(k^4)$. We express $V \setminus S$ as the union of different kinds of vertices, and devise reduction rules that help us bound the total number of vertices of each kind. To be more specific, we set $V \setminus S = V_1 \cup V_2 \cup V_3 \cup V_4 \cup V_5$ where

1. $V_1 = \{v \in (V \setminus S); N(v) \cap (V \setminus S) = \emptyset \text{ and } |N(v) \cap S| \leq 1\}$
2. $V_2 = \{v \in (V \setminus S); N(v) \cap (V \setminus S) = \emptyset \text{ and } |N(v) \cap S| \geq 2\}$
3. $V_3 = \{v \in ((V \setminus S) \setminus V_1); v \text{ is on the spine of a nontrivial caterpillar in } G[V \setminus S]\}$
4. $V_4 = \{v \in (V \setminus S); |N(v) \cap S| = 0 \text{ and } v \text{ is a pendant vertex in } G[V \setminus S]\}$
5. $V_5 = \{v \in (V \setminus S); |N(v) \cap S| \geq 1 \text{ and } v \text{ is a pendant vertex in } G[V \setminus S]\}$

Algorithm 1. The kernelization algorithm

1: **procedure** KERNELIZE(G, k)
2:     $CurrentInstance \leftarrow (G, k)$
3:     **repeat**
4:         Apply Rules 1 to 6 and set $CurrentInstance$ to be the output.
5:     **until** None of the rules cause any change to $CurrentInstance$.
6: **end procedure**

It is easy to verify that these sets together exhaust $V \setminus S$. We state the reduction rules and describe their consequences in the next section; the claims of soundness of the rules and a more formal bound on the running time and kernel size are deferred to Section 4.2.

4.1   Reduction Rules

For each rule below, let $(H = (V_H, E_H), k)$ be the instance on which the rule is applied, and (H', k') the resulting instance. Let $G = (V, E)$ be a YES instance of the problem that is reduced with respect to all the reduction rules, and let $S, V_1, \ldots, V_5$ be as described above. To bound the sizes of various subsets of $V \setminus S$, we use the fact that no reduction rule applies to G.

Rule 1. *If a connected component $H[X]; X \subseteq V_H$ of H has pathwidth at most 1, then remove X from H. The resulting instance is $(H' = H[V_H \setminus X], k' = k)$.*

Rule 2. *If a vertex u in H has two or more pendant neighbors, then delete all but one of these pendant neighbors to obtain H'. The resulting instance is $(H', k' = k)$.*

Rules 1 and 2 together ensure that every caterpillar in $G[V \setminus S]$ has at least one neighbor in S, and that $|V_1| \le k$: See Lemma 5.

Rule 3. *Let u be a vertex of H with at least two neighbors. If for every two vertices $\{v, w\} \subseteq N(u)$ there exist $k + 2$ vertices excluding u that are adjacent to both v and w, then delete u from H. The resulting instance is $(H' = H[V_H \setminus \{u\}], k' = k)$.*

Rule 3 ensures that $|V_2| \le \binom{k}{2}(k+2)$: Set $A = V_2$ and $X = S$ in Lemma 6.

Rule 4. *For a vertex u of H, if there is a matching M of size $k + 3$ in H where (i) each edge in M has at least one end vertex in $N(u)$, and, (ii) u is not incident with any edge in M, then delete u and decrement k by one. The resulting instance is $(H' = H[V \setminus \{u\}], k' = k - 1)$.*

Rule 5. *Let x, y be the end vertices of the spine $x, v_1, v_2, v_3 \ldots, v_p, y$ of an induced caterpillar C in H such that (1) no $v_i; 1 \le i \le p$ is adjacent in H to any vertex outside C, and (2) every pendant vertex of C is a pendant vertex in H. If $p \ge 5$, then contract the edge (v_2, v_3) in H to obtain the graph H'. The resulting instance is $(H', k = k')$.*

From Rules 1 to 5 it follows that $|V_3| \leq 17k(k+2)$ (Lemma 7), and that $|V_5| \leq 17(k+2)^2 k(2k-1)$ (Lemma 8). Each vertex in G can have at most one pendant neighbor, or else Rule 2 would apply. From this we get $|V_4| \leq |V_3| = 17k(k+2)$. Putting all the bounds together, $|V| \leq 34k^4 + 120k^3 + 103k^2 + k$, and so we have:

Rule 6. *If none of the Rules 1 to 5 can be applied to the instance (H, k), and $|V_H| > 34k^4 + 120k^3 + 103k^2 + k$, then set the resulting instance to be the trivial NO instance (H', k') where H' is a cycle of length 3 and $k' = 0$.*

In the next section we prove that these rules are sound, and that they can all be applied exhaustively in polynomial time. Hence we get

Theorem 3. *The* PATHWIDTH-ONE VERTEX DELETION *problem parameterized by solution size k has a polynomial vertex-kernel on $\mathcal{O}(k^4)$ vertices.*

4.2   Correctness and Running time

We now show that the reduction rules are sound. That is, we show that for each rule, (using the notation of the previous section) (H, k) is a YES instance if and only if (H', k') is a YES instance. We also show that each rule can be implemented in polynomial time. For discussing the rules, we reuse the notation from the respective rule statement in Section 4.1. In each case, n is the number of vertices in the input to the kernelization algorithm.

Claim. [⋆] Rule 1 is sound, and can be applied in $\mathcal{O}(kn)$ time.

Claim. [⋆] Rule 2 is sound, and can be applied in $\mathcal{O}(kn)$ time.

Claim. [⋆] Rule 3 is sound, and can be applied in $\mathcal{O}(n^3)$ time.

Claim. [⋆] Rule 4 is sound, and can be applied in $\mathcal{O}(kn^{1.5})$ time.

Claim. [⋆] Rule 5 is sound, and can be applied in $\mathcal{O}(kn)$ time.

From these claims, we get

Lemma 4. *On an input instance $(G = (V, E), k); |V| = n$ of* PATHWIDTH-ONE VERTEX DELETION*, the kernelization algorithm (Algorithm 1) runs in $\mathcal{O}(n^4)$ time and outputs a kernel on $\mathcal{O}(k^4)$ vertices.*

Proof. From the above claims it follows that Rules 1 to 5 are sound, and that each can be applied in $\mathcal{O}(n^3)$ time. From the discussion in Section 4.1 (using Lemmas 5 to 8 below) it follows that Rule 6 is sound, and it is easy to see that this rule can be applied in $\mathcal{O}(n)$ time. Each time a rule is applied, the number of vertices in the graph reduces by at least one (contracting an edge also reduces the vertex count by one). Hence the loop in lines 3 to 5 of Algorithm 1 will run at most $|V| + 1 = n + 1$ times. The algorithm produces its output either at a step where Rule 6 applies, or when none of the rules applies and the remaining instance has $\mathcal{O}(k^4)$ vertices. Thus the algorithm runs in $\mathcal{O}(n^4)$ time and outputs a kernel on $\mathcal{O}(k^4)$ vertices.   □

We now list the lemmas used in Section 4.1 to bound the sizes of $V_1, \ldots, V_5$.

Lemma 5. [⋆] *Let $(G = (V, E), k)$ be a YES instance of the problem that is reduced with respect to Rules 1 and 2, and let S be a PODS of G of size at most k. Let $V_1 = \{v \in (V \setminus S); (N(v) \cap (V \setminus S)) = \emptyset \text{ and } |N(v) \cap S| \leq 1\}$. Then every caterpillar in $G[V \setminus S]$ has at least one neighbor in S, and $|V_1| \leq k$.*

Lemma 6. [⋆] *Let (G, k) be a YES instance of the problem that is reduced with respect to Rule 3. For a set $X \subseteq V$, if $A \subseteq V \setminus X$ is such that every $v \in A$ has (i) at least two neighbors in X, and (ii) no neighbors outside X, then $|A| \leq \binom{|X|}{2}(k + 2)$.*

Lemma 7. [⋆] *Let $(G = (V, E), k)$ be an instance of the problem that is reduced with respect to Rules 1 to 5, and let $S \subseteq V$ be such that $G[V \setminus S]$ has pathwidth at most one. Let $X \subseteq (V \setminus S)$ be the set of vertices in $(V \setminus S)$ that lie on the spines of nontrivial caterpillars in $G[V \setminus S]$. Then $|X| \leq 17k(k + 2)$.*

Lemma 8. [⋆] *Let $(G = (V, E), k)$ be a YES instance of the problem that is reduced with respect to Rules 1 to 5, and let $S \subseteq V; |S| \leq k$ be such that $G[V \setminus S]$ has pathwidth at most one. Let $P \subseteq (V \setminus S)$ be the set of pendant vertices in $G[V \setminus S]$ that have at least one neighbor in S. Then $|P| \leq 17(k + 2)^2 k(2k - 1)$.*

5   Conclusion

We defined the PATHWIDTH-ONE VERTEX DELETION problem as a natural variant of the iconic FEEDBACK VERTEX SET problem, and initiated the study of its algorithmic complexity. We established that the problem is NP-complete, and showed that the problem parameterized by the solution size k is fixed-parameter tractable. We gave an FPT algorithm for the problem that runs in $\mathcal{O}^*(7^k)$ time, and showed that the problem has a polynomial kernel on $\mathcal{O}(k^4)$ vertices.

An immediate question is whether these bounds can be improved upon[3]. A more challenging problem is to try to solve the analogous problem for larger values of pathwidth. That is, we know that for any positive integer c, the Pathwidth c Vertex Deletion problem, defined analogously to PATHWIDTH-ONE VERTEX DELETION, is FPT parameterized by the solution size. This follows from the Graph Minor Theorem of Robertson and Seymour because, for each fixed c, the set of YES instances for this problem form a minor-closed class. However, for $c = 2$, the number of graphs in the obstruction set is already a hundred and ten [26], and so our approach would probably be of limited use for $c \geq 2$. Thus the interesting open problems for $c \geq 2$ are: (i) Can we get an $\mathcal{O}^*(d^k)$ FPT algorithm for the problem for some small constant d, and (ii) Does the problem have a polynomial kernel?

Acknowledgements. We thank our anonymous reviewers for a number of useful comments for improving the paper.

---

[3] After we presented our work at WG 2010, Cygan et al. [14] improved both these bounds, to $\mathcal{O}^*(4.65^k)$ and $\mathcal{O}(k^2)$, respectively. See Introduction for more details.

References

1. Álvarez, C., Serna, M.: The Proper Interval Colored Graph problem for caterpillar trees. Electronic Notes in Discrete Mathematics 17, 23–28 (2004)
2. Arnborg, S., Proskurowski, A., Seese, D.: Monadic Second Order Logic, Tree Automata and Forbidden Minors. In: Schönfeld, W., Börger, E., Kleine Büning, H., Richter, M.M. (eds.) CSL 1990. LNCS, vol. 533, pp. 1–16. Springer, Heidelberg (1991)
3. Assmann, S.F., Peck, G.W., Sysło, M.M., Zak, J.: The bandwidth of caterpillars with hairs of length 1 and 2. SIAM Journal on Algebraic and Discrete Methods 2(4), 387–393 (1981)
4. Bafna, V., Berman, P., Fujito, T.: A 2-Approximation Algorithm for the Undirected Feedback Vertex Set problem. SIAM Journal of Discrete Mathematics 12(3), 289–297 (1999)
5. Bodlaender, H.L.: On disjoint cycles. International Journal of Foundations of Computer Science 5(1), 59–68 (1994)
6. Bodlaender, H.L.: A partial k-arboretum of graphs with bounded treewidth. Theoretical Computer Science 209(1–2), 1–45 (1998)
7. Bodlaender, H.L.: Treewidth: Characterizations, Applications, and Computations. In: Fomin, F.V. (ed.) WG 2006. LNCS, vol. 4271, pp. 1–14. Springer, Heidelberg (2006)
8. Bodlaender, H.L.: A Cubic Kernel for Feedback Vertex Set. In: Thomas, W., Weil, P. (eds.) STACS 2007. LNCS, vol. 4393, pp. 320–331. Springer, Heidelberg (2007)
9. Bodlaender, H.L., Koster, A.M.C.A.: Combinatorial Optimization on Graphs of Bounded Treewidth. The Computer Journal 51(3), 255–269 (2008)
10. Bodlaender, H.L., van Dijk, T.C.: A cubic kernel for feedback vertex set and loop cutset. Theory of Computing Systems 46(3), 566–597 (2010)
11. Bryant, R.L., Kinnersley, N.G., Fellows, M.R., Langston, M.A.: On Finding Obstruction Sets and Polynomial-Time Algorithms for Gate Matrix Layout. In: Proceedings of the 25th Allerton Conference on Communication, Control and Computing, pp. 397–398 (1987)
12. Burrage, K., Estivill Castro, V., Fellows, M.R., Langston, M.A., Mac, S., Rosamond, F.A.: The Undirected Feedback Vertex Set Problem Has a Poly(k) Kernel. In: Bodlaender, H.L., Langston, M.A. (eds.) IWPEC 2006. LNCS, vol. 4169, pp. 192–202. Springer, Heidelberg (2006)
13. Cao, Y., Chen, J., Liu, Y.: On Feedback Vertex Set: New Measure and New Structures. In: Kaplan, H. (ed.) Algorithm Theory - SWAT 2010. LNCS, vol. 6139, pp. 93–104. Springer, Heidelberg (2010)
14. Cygan, M., Pilipczuk, M., Pilipczuk, M., Wojtaszczyk, J.O.: An improved fpt algorithm and quadratic kernel for pathwidth one vertex deletion. Accepted at IPEC 2010 (2010)
15. Dehne, F., Fellows, M.R., Langston, M.A., Rosamond, F.A., Stevens, K.: An $O(2^{O(k)}n^3)$ FPT-Algorithm for the Undirected Feedback Vertex Set problem. Theory of Computing Systems 41(3), 479–492 (2007)
16. Diestel, R.: Graph Theory, 3rd edn. Springer, Heidelberg (2005)
17. Dorn, F., Fomin, F.V., Thilikos, D.M.: Subexponential parameterized algorithms. Computer Science Review 2(1), 29–39 (2008)
18. Downey, R.G., Fellows, M.R.: Fixed Parameter Tractability and Completeness. In: Complexity Theory: Current Research, pp. 191–225. Cambridge University Press, Cambridge (1992)

19. Downey, R.G., Fellows, M.R.: Parameterized Complexity. Springer, Heidelberg (1999)
20. Fellows, M.R., Langston, M.A.: On Search, Decision and the Efficiency of Polynomial-time Algorithms. In: Proceedings of STOC 1989, pp. 501–512. ACM Press, New York (1989)
21. Flum, J., Grohe, M.: Parameterized Complexity Theory. Springer, Heidelberg (2006)
22. Guo, J., Gramm, J., Hüffner, F., Niedermeier, R., Wernicke, S.: Compression-based fixed-parameter algorithms for feedback vertex set and edge bipartization. Journal of Computer and System Sciences 72(8), 1386–1396 (2006)
23. Guo, J., Niedermeier, R.: Invitation to data reduction and problem kernelization. SIGACT News 38(1), 31–45 (2007)
24. Kanj, I.A., Pelsmajer, M.J., Schaefer, M.: Parameterized Algorithms for Feedback Vertex Set. In: Downey, R.G., Fellows, M.R., Dehne, F. (eds.) IWPEC 2004. LNCS, vol. 3162, pp. 235–247. Springer, Heidelberg (2004)
25. Karp, R.M.: Reducibility among combinatorial problems. Complexity of Computer Communications, 85–103 (1972)
26. Kinnersley, N.G., Langston, M.A.: Obstruction Set Isolation for the Gate Matrix Layout problem. Discrete Applied Mathematics 54(2-3), 169–213 (1994)
27. Kloks, T.: Treewidth – computations and approximations. LNCS, vol. 842. Springer, Heidelberg (1994)
28. Lin, M., Lin, Z., Xu, J.: Graph bandwidth of weighted caterpillars. Theoretical Computer Science 363(3), 266–277 (2006)
29. Lewis, J.M., Yannakakis, M.: The node-deletion problem for hereditary properties is NP-complete. Journal of Computer and System Sciences 20(2), 219–230 (1980)
30. Niedermeier, R.: Invitation to Fixed-Parameter Algorithms. Oxford University Press, Oxford (2006)
31. Papadimitriou, C.H.: The NP-Completeness of the bandwidth minimization problem. Computing 16(3), 263–270 (1976)
32. Philip, G., Raman, V., Villanger, Y.: A quartic kernel for pathwidth-one vertex deletion. A full version of the current paper,
http://www.imsc.res.in/~gphilip/publications/pwone.pdf
33. Raman, V., Saurabh, S., Subramanian, C.: Faster fixed parameter tractable algorithms for finding feedback vertex sets. ACM Transactions on Algorithms 2(3), 403–415 (2006)
34. Robertson, N., Seymour, P.D.: Graph minors I. Excluding a forest. Journal of Combinatorial Theory, Series B 35(1), 39–61 (1983)
35. Robertson, N., Seymour, P.D.: Graph Minors. II. Algorithmic Aspects of Tree-Width. Journal of Algorithms 7(3), 309–322 (1986)
36. Thomassé, S.: A quadratic kernel for feedback vertex set. In: Proceedings of SODA 2009, Society for Industrial and Applied Mathematics, pp. 115–119 (2009)

Network Exploration
by Silent and Oblivious Robots*

Jérémie Chalopin[1], Paola Flocchini[2], Bernard Mans[3], and Nicola Santoro[4]

[1] Laboratoire d'Informatique Fondamentale de Marseille
CNRS & Aix-Marseille Université, Marseille, France
jeremie.chalopin@lif.univ-mrs.fr
[2] SITE, University of Ottawa, Ottawa, Canada
flocchin@site.uottawa.ca
[3] Macquarie University, Sydney, Australia
bernard.mans@mq.edu.au
[4] School of Computer Science, Carleton University, Ottawa, Canada
santoro@scs.carleton.ca

Abstract. In this paper we investigate the basic problem of *Exploration* of a graph by a group of identical mobile computational entities, called robots, operating autonomously and asynchronously. In particular we are concerned with what graphs can be explored, and how, if the robots do not remember the past and have no explicit means of communication. This model of robots is used when the spatial universe in which the robots operate is *continuous* (e.g., a curve, a polygonal region, a plane, etc.). The case when the spatial universe is *discrete* (i.e., a graph) has been also studied but only for the classes of acyclic graphs and of simple cycles. In this paper we consider networks of arbitrary topology modeled as connected graphs with local orientation (locally distinct edge labels). We concentrate on class $\mathcal{H}_k$ of asymmetric configurations with k robots. Our results indicate that the explorability of graphs in this class depends on the number k of robots participating in the exploration. In particular, exploration is impossible for $k < 3$ robots. When there are only $k = 3$ robots, only a subset of $\mathcal{H}_3$ can be explored; we provide a complete characterization of the networks that can be explored. When there are $k = 4$ robots, we prove that all networks in $\mathcal{H}_4$ can be explored. Finally, we prove that for any odd $k > 4$ all networks in $\mathcal{H}_k$ can be explored by presenting a general algorithm. The determination of which networks can be explored when $k > 4$ is even, is still open but can be reduced to the existence of a gathering algorithm for $\mathcal{H}_k$.

1 Introduction

Consider a team (or swarm) of identical mobile robots located in a spatial universe. Each robot operates autonomously by cyclically executing three operations: **Look** - it observes the position of the other robots; **Compute** - based on

---

* This work was partially supported by ANR Project SHAMAN, by COST Action 295 DYNAMO, by NSERC, and by Dr. Flocchini's University Research Chair.

D.M. Thilikos (Ed.): WG 2010, LNCS 6410, pp. 208–219, 2010.

this input, it computes a destination (a neighbouring node) or it decides not to move, according to a predefined algorithm (the same for all robots); Move- it then moves to its computed destination. The robots are *silent*: they have no direct means of communication; *anonymous*: externally identical with no distinct identifiers that can be used in the execution of the algorithm; *asynchronous*: the time between each operation in the Look-Compute-Move cycle as well as between successive cycles is finite but arbitrary; *oblivious*: the robots have no memory of past actions and computations, and the computation is based solely on what determined in the current cycle.

This model of robots is used when the spatial universe in which the robots operate is *continuous*, e.g., a curve, a polygonal region, a plane, etc. In this setting, the computational power of such robots has been investigated with respect to a variety of problems, such as *Pattern Formation, Gathering, Flocking*, etc. (e.g. see [1, 3–5, 9, 12, 13]).

Recently investigations have also considered the case when the spatial universe is *discrete*, e.g., when the robots operate in a network [6–8, 10, 11]. The research has focused on two fundamental problems (extensively studied in the past in a variety of other models): *Gathering* (or *Rendezvous*), which requires all robots to move to the same node (whose location is a priori undetermined); and *Exploration*, which requires every node of the network to be visited by at least one robot and all robots reach a quiescent state within finite time. The results have however been limited to two classes of graphs: *rings* [6, 7, 10, 11] and *trees* [8] and assuming that the edges incident on a vertex are indistinguishable for the robots. No results exist to date for the exploration of arbitrary graphs by oblivious robots in the Look-Compute-Move model. Note that the latter assumption (unlabeled edges) is atypical; indeed in standard models of networks (anonymous or not) in distributed computing, the links incident on a node x have distinct labels, called *port numbers*.

The computational study of such weak robots is a difficult task due to the simultaneous presence of asynchrony, obliviousness, and lack of explicit communication. Lack of explicit communication means that synchronization, interaction, and communication of information among the robots can be achieved solely by means of observing the position of the other robots. However, because of asynchrony a robot r may observe the position of the robots at some time t; based on that observation, compute the destination at time $t' > t$, and move at an even later time $t'' > t'$; thus it might be possible that at time t'' some robots are in different positions from those previously perceived by r at time t, because in the meantime they performed their Move operations (possibly several times). In other words, robots may compute destinations and move based on significantly outdated perceptions, which adds to the difficulty of exploration. Moreover, since robots are oblivious, the task of deciding the global status of the exploration process, in particular termination detection, has to be performed without memory and without communication.

In this paper we continue the investigation of the computational power of such weak robots in the discrete setting, and consider the *exploration* problem in anonymous networks, that is edge-labeled graphs (G, λ) where $G = (V, E)$ is a connected simple graph and $\lambda = \{\lambda_v : v \in V\}$ is the set of local port-numbering functions λ_v. Let Ψ be the placement function describing the position of the robots in the network, and let (G, λ, Ψ) denote the network with the placement of the robots. In particular, we are interested in determining which networks with what initial placements of the robots exploration is possible; that is, which (G, λ, Ψ) can be explored and how. The study of the specific classes of graphs investigated in absence of edge-labels, of trees and rings, becomes much simpler with edge-labels, and a complete characterization can be found in [2].

When considering networks with at most two robots, it is easy to see that exploration cannot be solved if the networks has at least three vertices. When the number of robots is larger and (G, λ, Ψ) is symmetric (i.e., there exists a non-trivial automorphism of (G, λ, Ψ) that preserves the labels and the placement of the robots), exploration cannot be generally solved: solvability depends on the number of equivalence classes of agents and the task is impossible when the number of equivalence classes is less than 2.

In this paper, we consider only initial configurations (G, λ, Ψ) that are asymmetric. We denote by $\mathcal{H}_k$ the class of configurations (G, λ, Ψ) with k robots such that there is no non-trivial automorphism that preserves the labels and the placement of the robots. Our goal is to determine which networks in this class can be explored and how.

Our results indicate that the explorability of networks in $\mathcal{H}_k$ depends on the number k of robots participating in the exploration. In particular, with one robot (resp. two robots), only a network with one vertex (resp. two vertices) can be explored. When there are only $k = 3$ robots, we prove that not all networks in $\mathcal{H}_3$ can be explored. More precisely, we do provide a complete characterization of the networks that can be explored with $k = 3$ robots, and present an algorithm that performs the exploration of those networks with three robots (Section 4). When there are $k = 4$ robots, we prove that all networks in $\mathcal{H}_4$ can be explored; the proof is constructive, and the algorithm is (unfortunately) quite involved (Section 5). Finally, we prove that for any odd $k > 4$ all networks in $\mathcal{H}_k$ can be explored by presenting a general algorithm. The determination of which networks can be explored when $k > 4$ is even, is still open, but it can be reduced to the existence of a gathering algorithm for $\mathcal{H}_k$.

Due to space limitations most proofs are omitted in this extended abstract and will be presented in the journal version of our paper.

2   Model and Basics

Let $G = (V, E)$ be an undirected connected simple graph where V is the set of vertices and E the set of edges, with $|V| = n$. The vertices of G are unlabeled. We denote by $N(v)$ the set of neighbors of v, and by $d(v)$, the degree of v. For each node v, there is a bijective function $\lambda_v : N(v) \to \{1, 2, \ldots d(v)\}$ which assigns unique labels to the edges incident to v. Each edge $uv \in E$ has two distinct labels $\lambda_u(v)$ and $\lambda_v(u)$. Let $\lambda = \{\lambda_v : v \in V\}$ be the global labeling function and let (G, λ) denote the resulting edge-labeled graph. The label $\Lambda(\pi)$ of a path

$\pi = (u_0, u_1, \ldots, u_k)$ is obtained by extending λ from edges to paths as follows: $\Lambda(\pi) = ((\lambda_{u_0}(u_1), \lambda_{u_1}(u_0)), \ldots, (\lambda_{u_{k-1}}(u_k), \lambda_{u_k}(u_{k-1})))$.

The shortest path from u to v is the minimal element we get when sorting the paths from u to v first by length, and then lexicographically using their labels. We say that a node u is closer from a node w than a node v if either $\mathsf{dist}(u, w) < \mathsf{dist}(v, w)$, or $\mathsf{dist}(u, w) = \mathsf{dist}(v, w)$ and the label $\Lambda(\pi_{uw})$ of a shortest path from u to w is lexicographically smaller (or "weaker") than the label $\Lambda(\pi_{vw})$ of any shortest path from v to w.

Operating in (G, λ) is a set $\mathcal{R}$ of k identical robots. Each robot operates in Look-Compute-Move cycles, which are performed asynchronously for each robots. When *Looking*, a robot perceives a snapshot of the labeled graph with the current position of the robots (called a *view* of the graph); when *Computing* it decides where to move on the basis of the snapshot; when *Moving* it actually moves to the chosen neighboring node. The time between Look, Compute, and Move operations is finite but unbounded, and is decided by the adversary for each action of each robot. The only constraint is that moves are instantaneous, as in [6–8, 10, 11], and hence any robot performing a Look operation sees all other robots at nodes and not on edges. We say that there is a *tower* on a node if the node is occupied by more than one robot. We call a robot *free* if it does not belong to a tower. Initially all robots are free; that is there is at most one robot in each node. During the Look operation, the robots can perceive if there is one or more robots in a given location; this ability, called *multiplicity detection* is a standard assumption in the continuous model.

Let $\Psi : \mathcal{R} \to V$ be the placement function returning the initial position of a given robot. Let (G, λ, Ψ) denote the edge-labeled graph with the initial placement of the robots. The vertices of (G, λ) (resp., (G, λ, Ψ)) can be partitioned according to the equivalence classes they belong to, where an equivalence class $[v_0]$ of vertices of G is such that for each $v \in [v_0]$, there exists an automorphism σ of G that preserves the labels (resp., and placement) such that $\sigma(v) = v_0$. We say that (G, λ) (resp., (G, λ, Ψ)) is *symmetric* if there exists a non-trivial automorphism of (G, λ) (resp., (G, λ, Ψ)) that preserves the edge-labeling λ (resp., and the placement), *asymmetric* otherwise. Note that, in (G, λ) and (G, λ, Ψ) the equivalence classes can be ordered using the fact that each identifies a different "view" of the graph. We say that a robot a is closer from a node w than a robot b if $\Psi(a)$ is closer from w than $\Psi(b)$.

Given an initial placement Ψ of k robots in (G, λ), we say that a protocol $\mathcal{A}$ solves the exploration problem in (G, λ, Ψ) if in all possible executions of $\mathcal{A}$ by the k robots, every node of the graph is visited by at least one robot and all robots enter a quiescent state within finite time. We say that exploration of (G, λ, Ψ) with k robots is *impossible* if no protocol solves the exploration problem in (G, λ, Ψ). A network (G, λ) is explorable with k robots if there exists protocol $\mathcal{A}$ that solves the exploration problem in (G, λ, Ψ) for any placement of Ψ of k robots in (G, λ).

Lemma 1. *For any network (G, λ), it is impossible to solve exploration with one (resp. two) agents if $|V| > 1$ (resp. $|V| > 2$).*

In the rest of the paper we focus on *asymmetric configurations*, i.e. the set $\mathcal{H}_k$ of all (G, λ, Ψ) with k robots where (G, λ, Ψ) is asymmetric; and we assume multiplicity detection. Since exploration is impossible for networks in $\mathcal{H}_1$ and $\mathcal{H}_2$ we consider $k \geq 3$.

3   About Gathering and Asymmetry

3.1   Gathering Algorithms

In [2], we have studied the gathering problem: we want to gather all the robots at a single node. In the exploration algorithms we present below, we use the gathering algorithm of [2] as a subroutine. We describe it here briefly.

Consider a node $v \in V$; let $PATHS(v) = \{\Lambda(\pi) \mid \pi$ is the shortest path from v to $\Psi(a) \mid a \in \mathcal{R}\}$. Let $SORT(PATHS(v))$ be the set of labeled paths sorted first by the length and then by lexicographic order of their associated sequences of labels. Given two vertices v_1 and v_2, we define the following order: $v_1 < v_2$ if $SORT(PATHS(v_1)) <_{lex} SORT(PATHS(v_2))$, where $<_{lex}$ denotes lexicographic order. Let $SD(v) = \sum_{a \in \mathcal{A}} \text{dist}(v, \Psi(a))$ denote the sum of the distances from a node v to all the robots and let $\text{code}(v) = (SD(v), SORT(PATHS(v)))$.

Consider an equivalence class $[u]$ of (G, λ), a *Minimal Weber node* of $[u]$ is a node $u' \in [u]$ such that $\text{code}(u') = \min\{\text{code}(u'') \mid u'' \in [u]\}$. Note that if (G, λ, Ψ) is asymmetric, then each class $[u]$ of (G, λ) has a unique Minimal Weber node, denoted by $MWN[u]$.

Given an equivalence class $[u]$ of (G, λ) (this class can be seen as a parameter of the algorithm), and starting from any asymmetric configuration (G, λ, Ψ) with k robots, the algorithm from [2] has the following properties.

- at each moment, at most one robot can move,
- if $k \geq 3$ is odd, then the only robot that is allowed to move is the closest from $u' = MWN[u]$ that is not on u'. After this move, we still have $MWN[u] = u'$, and $\text{code}(u')$ has strictly decreased.
- if $k = 4$, the robot that is allowed to move is either the closest one from $u' = MWN[u]$ that is not on u', or the robot that is on u'. After the move, $MWN[u]$ can be modified, but $\text{code}(MWN[u])$ has strictly decreased.

Thus, we have the following theorem.

Theorem 1 ([2]). *From any asymmetric configuration (G, λ, Ψ) with k agents, and for any equivalence class $[u]$ of (G, λ), there exists a gathering algorithm where the gathering node belongs to $[u]$ when $k \geq 3$ is odd, or when $k = 4$.*

3.2   Asymmetry

For Gathering, we know that if the initial configuration (G, λ, Ψ) is asymmetric, then it is impossible to solve gathering [2]. When considering exploration, we don't have such strong results. However, we want the robots to be able to detect that they have explored the network. Since they are oblivious, they should be able to detect from the snapshot they compute if the configuration can be an initial configuration or not. To do so, the robots will create towers of two (or more) robots that cannot be part of the initial configuration. If in the initial configuration, all robots are in the same equivalence class, we know that no tower can be created [2] and thus, it implies that exploration is impossible. It means that when the degree of symmetry is too high, exploration is impossible. To avoid this problem, in this paper, we only consider initial configurations that are asymmetric. Note that, even if we do not ask the robots to stop once they have explored all the nodes (i.e. we consider perpetual exploration), there are still some symmetric networks that cannot be explored.

4   Exploration with $k = 3$ Robots

We prove that explorability of $(G, \lambda, \Psi) \in \mathcal{H}_3$ requires the presence of a node (or an edge) in G whose removal creates either a graph with a Hamiltonian path, or a graph that is spanned by two intersecting elementary paths satisfying several conditions. This class of graphs is denoted by $\mathcal{E}_3$.

Definition 1. *A graph $G \in \mathcal{E}_3$ if at least one of the two following conditions hold:*

(Case 1) *There exists an elementary path $P = (x_1, \ldots, x_m)$ and two neighbors u_0, v_0 both different from x_1 such that:*

(A1) u_0 does not appear in P,

(A2) any vertex $v \notin \{u_0, v_0\}$ appears in P,

(A3) (G, λ, Ψ_0) is asymmetric where Ψ_0 maps a robot to x_1, one to u_0 and one to v_0.

(Case 2) *There exists two elementary paths $P_1 = (x_1, \ldots, x_m)$ and $P_2 = (y_1, \ldots, y_\ell)$ and two neighbors u_0, v_0 both different from x_1 such that:*

(B1) $x_m = y_1$,

(B2) either $u_0 = x_m$ or $u_0 \in N(x_m)$,

(B3) $u_0 \neq x_i$, for all $i \in [1, m-1]$,

(B4) any vertex $v \notin \{u_0, v_0\}$ appears either in P_1 or in P_2,

(B5) (G, λ, Ψ_0) is asymmetric where Ψ_0 maps a robot to x_1, one to u_0 and one to v_0,

(B6) for any consecutive vertices y, y' of P_2, and for any vertex x of P_1 adjacent to u_0, there is no automorphism of (G, λ) mapping x to y and u_0 to y'.

(B7) for any distinct vertices $y, y' \in P_2$, there is no automorphism of (G, λ) mapping y to y'.

Examples of graphs in $\mathcal{E}_3$ can be seen in Figure 1.

Remark 1. A graph satisfies the conditions of (Case 1) in Definition 1 if and only if there exists $u_0, v_0 \in V(G)$ such that either $G \setminus \{u_0\}$ or $G \setminus \{u_0, v_0\}$ has a Hamiltonian path.

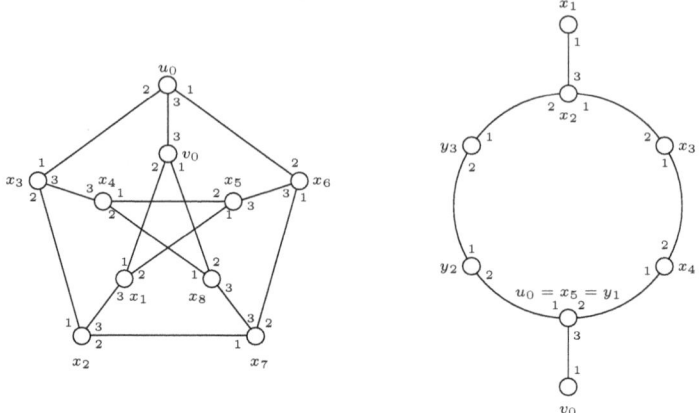

Fig. 1. On the left, a graph satisfying conditions of (Case 1) of Definition 1; on the right, a graph satisfying conditions of (Case 2) of Definition 1

Lemma 2. *If $(G, \lambda, \Psi) \in \mathcal{H}_3 \setminus \mathcal{E}_3$ exploration of (G, λ, Ψ) is impossible.*

We now briefly describe the exploration algorithm for $(G, \lambda, \Psi) \in \mathcal{E}_3$.

The algorithm is different depending on the characteristics of the graph. Suppose that the graph satisfies the conditions of (Case 1) in Definition 1 for some vertices u_0, v_0 and some elementary path $P = (x_1, \ldots, x_m)$. The idea is to first have the three robots place themselves on u_0, v_0, and x_1 (this procedure is not trivial; the details will appear in the complete version of this paper). Then the two robots that are on u_0, v_0 (up to isomorphism) create a small tower on u_0, and the third robot performs the exploration following path P until it reaches the last node x_k of this path. Suppose now that (Case 1) does not apply but (Case 2) does for some vertices u_0, v_0 and two elementary paths $P_1 = (x_1, \ldots, x_m)$ and $P_2 = (y_1, \ldots, y_\ell)$. Again one can show that we can have the three robots placed on u_0, v_0 and x_1. Then the two robots that are on u_0, v_0 (up to isomorphism) create a small tower on node u_0, and the third robot moves along the path P_1. Then, a big tower is created on y_1 and this big tower moves along P_2 until it reaches the last node y_ℓ of this path. Note that the tower might break on the way due to asynchrony, it will however recompose itself along the path. From Lemma 2 and the Algorithm above we have:

Theorem 2. *$(G, \lambda, \Psi) \in \mathcal{H}_3$ can be explored if and only if $G \in \mathcal{E}_3$.*

5   Exploration with $k = 4$ Robots

We now turn to the case of $k = 4$ robots and we show that any $(G, \lambda, \Psi) \in \mathcal{H}_4$ can be explored. The algorithm is however slightly involved. We actually need to describe two versions depending on the existence of structures called *pseudo-neck* and *neck*. If there exist in G two vertices v, v' of degree 1 with a common neighbor u, we say that (v, v', u) is a *pseudo-neck*. A *neck* (n_1, n_2) in G consists of two neighbors $n_1, n_2 \in V(G)$ such that after removing them and their adjacent edges from G the resulting graph is connected. It is easy to see that:

Lemma 3. *If a graph does not contain any pseudo-neck, it contains at least a neck.*

The general idea of the algorithms is the following. Two robots uniquely identify a pseudo-neck (or a neck if the pseudo-neck does not exist) by moving to the leaves of a pseudo-neck (or to the vertices of the neck). Once the pseudo-neck (or the neck) is occupied, the robots identify a unique spanning tree rooted in the pseudo-neck (or in the neck) and an ordering $f_1, \ldots f_m$ of its leaves. One of the robot that is not on the pseudo-neck (or the neck) moves to the first leaf f_1 of this spanning tree. Then, the two robots on the pseudo-neck (or the neck) create a tower. The two free robots occupy the first two leaves of the spanning tree, without creating symmetries and proceed with the exploration by placing on consecutive leaves f_i and f_{i+1}, and having the robot on the smaller leaf f_i move to f_{i+2} until the last leaf is reached (a technique similar to the one used in [8] for asymmetric trees).

The algorithm will be discussed in details and analyzed in the rest of this section. Note that the algorithm will employ in some cases a preliminary step of "partial gathering". For such a step, we will adapt the gathering algorithm for asymmetric (G, λ, Ψ) described in [2].

5.1 Graphs with a Pseudo-neck

We now show how to create a tower on one of the nodes of the pseudo-neck and how to have a robot move on the first leaf of a uniquely identified spanning tree. If there is more than one pseudo-neck in the graph, ties are broken using the edge labels and we assume they are ordered from "weakest" to "strongest".

Given a pseudo-neck (v, v', u), let $T(v, v', u)$ be a spanning tree obtained by a depth-first traversal starting in u. Let $f_1(T(v, v', u)), f_2(T(v, v', u)), \ldots,$ $f_m(T(v, v', u))$ be an ordering of its leaves different from v, v'. We construct $T(v, v', u)$ in such a way that $f_1(T(v, v', u)) = f_1(T(v', v, u))$ (it is easy to see that this can always be done).

Algorithm FORM-TOWER-WITH-PSEUDO-NECK is described by listing six possible cases. A robot checks the cases in this order and follows the first that applies to the observed configuration.

Algorithm. FORM-TOWER-WITH-PSEUDO-NECK
Case 1. *There is a pseudo-neck (v, v', u) with a robot in u, a robot in v and there is a robot in f_1.*
If I am the robot in u: move to v creating a tower in v.
Case 2. *There is a pseudo-neck (v, v', u) with a robot in v, a robot in v' and there is a robot in f_1.*
If I am the robot in v': move to u.

Case 3. *There is a pseudo-neck (v, v', u) with a robot in v, a robot in v' and there are no robots in $f_1(T(v, v', u)) = f_1(T(v', v, u))$.*
Let a be the closest robot to $f_1(T(v, v', u))$ in $G \setminus \{v, v'\}$. If I am a: move toward $f_1(T(v, v', u))$.
Case 4. *There is a pseudo-neck (v, v', u) with a robot in v, and robot a in u.*
If I am a: move to v'.
Case 5. *There is a pseudo-neck (v, v', u) with a robot in v, no robot in u nor in v'.*
Among all the necks like that, consider the weakest such that $\min\{\text{dist}(a, u) \mid \Psi(a) \neq v\}$ is minimal. Let a be the closest robot to u in G that is not in v. If I am a: move toward u.
Case 6. *Any other situation.*
Apply the gathering Algorithm of [2] until the tower is formed, by considering as candidate gathering points nodes of degree 1 which have another node of degree 1 at distance 2.

Lemma 4. *Algorithm* FORM-TOWER-WITH-PSEUDO-NECK *terminates and no movement can create a symmetry during the tower formation on a pseudo-neck with Algorithm* FORM-TOWER-WITH-PSEUDO-NECK.

Algorithm EXPLORE-WITH-PSEUDO-NECK proceeds by letting the other two robots move sequentially. They move to $f_1(T(v, v', u))$ and $f_2(T(v, v', u))$; they then move on the spanning tree (which is invariant to their movement) placing on consecutive leaves f_i and f_{i+1}, and having the robot on the smaller leaf f_i move to f_{i+2} until the last leaf is reached. We can conclude that:

Lemma 5. *Let G have a pseudo-neck. Then any $(G, \lambda, \Psi) \in \mathcal{H}_4$ can be explored.*

5.2   Graphs with a Neck

We consider the case when a pseudo-neck does not exist; thus a neck exists. In this case the algorithm is more complex and we describe the process of tower creation and the one of exploration separately.

Placement on the Neck. We first introduce some notation. A *block* in a graph is an inclusion-maximal 2-vertex-connected component (possibly reduced to one edge). Two blocks of G are either disjoint or share a single vertex, that is an articulation point. Any graph G admits a block-decomposition in the form of a rooted tree T: each vertex of T is a block of G, pick any block B_1 as a root of T, label it, and make it adjacent in T to all blocks intersecting it, then label that blocks and make them adjacent to all unlabeled blocks which intersect them, etc.

We describe how to have two robots move to occupy the neck. We identify four situations and for each we sketch the algorithm followed by the robots.

Algorithm. OCCUPY-NECK

Case 1. *G is 2-connected.*
In this case any edge is a neck. Consider a class $[u_0]$ of vertices of G and execute the gathering algorithm of [2] until there is a robot a in one vertex $u \in [u_0]$ and a robot b in a vertex v incident to u. Then a and b are on a neck.

Case 2. *There exists a leaf B in the block-decomposition of G is of size 2.*
Note that, since there does not exist any pseudo-neck, it implies that there exists a vertex u_0 of degree 1 adjacent to a vertex v_0 of degree 2. In that case, execute the gathering algorithm of [2] until there is a robot a in one vertex $u \in [u_0]$ and a robot b in a vertex v incident to u. Then a and b are on a neck.

Case 3. *There exists a leaf B_0 in the block-decomposition of G, whose articulation point is w_0 such that there exists a vertex $u_0 \in B_0$ at distance 2 from w_0 in G.*
Note that any edge $(u_0, v) \in E(G)$ is a neck. In this case, execute the gathering algorithm of [2] until there is a robot a in one vertex $u \in [u_0]$ and a robot b in a vertex v incident to u. Then a and b are on a neck.

Case 4. *In any leaf B of the block-decomposition, the articulation point is adjacent to all vertices of B.*

In this case, consider a leaf B_0 of the block-decomposition of size at least 3. Let w_0 be its articulation point and let u_0 be a vertex of B_0 distinct from u_0. Execute the gathering algorithm of [2] until there is a robot a in a vertex $u \in [u_0]$ and a robot b in a vertex v incident to u. If $v \in [w_0]$, b can safely move to a neighbor $t \notin [w_0]$ of u.

Lemma 6. *In Algorithm* OCCUPY-NECK *two robots move on the two nodes of a neck without creating symmetries.*

Exploration. Given a vertex u, let $T(u)$ be a spanning tree of G obtained by doing a depth-first traversal (DFT) starting from u. If (u, v) is a neck, let $T(u, v)$ be a spanning tree obtained by a DFT starting from u and using (u, v) as a first edge, i.e., $T(u, v) \setminus \{u\}$ is a DFT of $G \setminus \{u\}$ starting in v. Given a spanning tree $T(u, v)$ of G, consider an ordering of its leafs $f_1, \ldots, f_m$ such that f_1 is as far as possible from u (using the labels on the paths from u to f to break the symmetry) in $G \setminus \{u, v\}$.

We say that a neck (n_1, n_2) is symmetric if there is an automorphism σ of G such that $\sigma(n_1) = n_2$, asymmetric otherwise. Exploration proceeds differently whether the neck is symmetric or not. (Note that a pseudo-neck is by definition never symmetric because of the edge labels). In case of symmetry, once two robots are on a neck, we have to carefully proceed to have a robot move to the first leaf. We now describe the two cases.

Draft of Algorithm EXPLORE-WITH-SYMMETRIC-NECK. Assume that the neck is symmetric. First of all note that, due to the port numbers, there is at most one automorphism σ of G such that $\sigma(n_1) = n_2$. Let $T = T(n_1, n_2)$ and $T' = T(n_2, n_1)$. Let a and b be two robots that are not on the neck. Without loss of generality, assume that $\mathsf{dist}_{G \setminus \{n_1, n_2\}}(\Psi(a), f_1) \leq \min \{\mathsf{dist}_{G \setminus \{n_1, n_2\}}(\Psi(a), f'_1),$

$\text{dist}_{G\setminus\{n_1,n_2\}}(\Psi(b), f'_1)$, $\text{dist}_{G\setminus\{n_1,n_2\}}(\Psi(b), f_1)\}$ and that if there is an equality, the label of the shortest path between $\Psi(a)$ and f_1 is "weaker" than the label of the other shortest paths (since the situation is asymmetric, even if σ is an automorphism). In this case, let a move towards f_1 (on a node denoted $\Psi'(a)$) in $G \setminus \{n_1, n_2\}$ if it does not creates a symmetry. Otherwise, let b move to $\Psi'(a)$ (if a symmetry appears when a move to $\Psi'(a)$, then one can show that $\Psi(b)$ is adjacent to $\Psi'(a)$). Once the neck and the first leaf are occupied, the exploration proceeds as in Section 5.1.

Lemma 7. *With Algorithm* EXPLORE-WITH-SYMMETRIC-NECK, *a robot reaches* f_1 *without creating any symmetry.*

Draft of Algorithm EXPLORE-WITH-ASYMMETRIC-NECK. Assume now that the neck is asymmetric, i.e., there is no automorphism σ of G such that $\sigma(n_1) = n_2$. Using an arbitrary predefined order, we assume without loss of generality that n_1 is "weaker" than n_2. Let $T = T(n_1, n_2)$ and let a and b be two robots that are not on the neck. Again without loss of generality, assume that $\text{dist}_{G\setminus\{n_1,n_2\}}(\Psi(a), f_1) \le \text{dist}_{G\setminus\{n_1,n_2\}}(\Psi(b), f_1)$ (using labels to break the symmetry). In this case, let a move towards f_1 (on a node denoted $\Psi'(a)$) in $G \setminus \{n_1, n_2\}$ if it does not creates a symmetry. Otherwise, let b move to $\Psi'(a)$ (if a symmetry appears when a move to $\Psi'(a)$, then one can show that $\Psi(b)$ is adjacent to $\Psi'(a)$). Once the neck and the first leaf are occupied, the exploration proceeds as in Section 5.1.

Lemma 8. *During the execution of Algorithm* EXPLORE-WITH-ASYMMETRIC-NECK, *a robot reaches* f_1 *without creating any symmetry.*

From Lemmas 6, 7 and 8, we have that any $(G, \lambda, \Psi) \in \mathcal{H}_4$ can be explored if G has a neck. Consequently, with Lemma 5, we have the following theorem.

Theorem 3. *Any* $(G, \lambda, \Psi) \in \mathcal{H}_4$ *can be explored.*

6 Exploration with $k > 4$ Odd

Let $k > 4$ be odd. In this case, the exploration algorithm is rather simple, and is drafted below.

The robots start by applying the gathering algorithm of [2] where the gathering point is chosen within an equivalence class $[v]$ of vertices where v is not an articulation point (it is easy to see that such a class can always be selected). The gathering algorithm is executed until $k - 3$ robots have gathered in some vertex r creating a tower. At this point, the robots can select a spanning tree T - for example a Depth-First Spanning Tree - rooted in r and they can agree on one where r has a single neighbor in T. Let $f_1, \ldots, f_m$ be the ordered leaves of T encountered in some predefined traversal of T (for example depth-first). We let the robots move sequentially with the following idea: the closest robot to f_1 (among the three remaining free robots) moves to f_1. The closest robot (not on f_1 nor on r) now moves to r. Once the robot reaches r, we have a tower and two free robots one of which is in f_1. At this point the two free robots recognize the situation that signals the beginning of the exploration and collaboratively explore the graph moving from leaf to leaf as in Section 5. Based on the asymmetry of the initial placement, on the uniqueness of the chosen spanning tree and on its preservation, and on the sequentiality of the moves, is not difficult to show that the algorithm terminates correctly and we have:

Theorem 4. *Let $k > 4$ be odd; then any $(G, \lambda, \Psi) \in \mathcal{H}_k$ can be explored.*

Note that if we are given an algorithm that solves gathering for even $k > 4$ robots in asymmetric configurations, then the previous algorithm can be used to explore all $(G, \lambda, \Psi) \in \mathcal{H}_k$.

References

1. Asahiro, Y., Fujita, S., Suzuki, I., Yamashita, M.: A self-stabilizing marching algorithm for a group of oblivious robots. In: Baker, T.P., Bui, A., Tixeuil, S. (eds.) OPODIS 2008. LNCS, vol. 5401, pp. 125–144. Springer, Heidelberg (2008)
2. Chalopin, J., Flocchini, P., Mans, B., Santoro, N.: Gathering and rendezvous by oblivious robots in arbitrary graphs, rings, and trees, Technical Report, University of Ottawa (2009)
3. Cieliebak, M., Flocchini, P., Prencipe, G., Santoro, N.: Solving the robots gathering problem. In: Baeten, J.C.M., Lenstra, J.K., Parrow, J., Woeginger, G.J. (eds.) ICALP 2003. LNCS, vol. 2719, pp. 1181–1196. Springer, Heidelberg (2003)
4. Cohen, R., Peleg, D.: Convergence properties of the gravitational algorithm in asynchronous robot systems. SIAM J. Computing 34, 1516–1528 (2005)
5. Défago, X., Souissi, S.: Non-uniform circle formation algorithm for oblivious mobile robots with convergence toward uniformity. Theoretical Computer Science 396(1-3), 97–112 (2008)
6. Devismes, S., Petit, F., Tixeuil, S.: Optimal probabilistic ring exploration by semi-synchronous oblivious robots. In: Kutten, S., Žerovnik, J. (eds.) SIROCCO 2009. LNCS, vol. 5869, pp. 203–217. Springer, Heidelberg (2010)
7. Flocchini, P., Ilcinkas, D., Pelc, A., Santoro, N.: Computing without communicating: ring exploration by asynchronous oblivious robots. In: Tovar, E., Tsigas, P., Fouchal, H. (eds.) OPODIS 2007. LNCS, vol. 4878, pp. 105–118. Springer, Heidelberg (2007)
8. Flocchini, P., Ilcinkas, D., Pelc, A., Santoro, N.: Remembering without memory: Tree exploration by asynchronous oblivious robots. Theoretical Computer Science (2010) (to appear)
9. Flocchini, P., Prencipe, G., Santoro, N., Widmayer, P.: Arbitrary pattern formation by asynchronous anonymous oblivious robots. Theoretical Computer Science 407(1-3), 412–447 (2008)
10. Klasing, R., Kosowski, A., Navarra, A.: Taking advantage of symmetries: gathering of asynchronous oblivious robots on a ring. In: Baker, T.P., Bui, A., Tixeuil, S. (eds.) OPODIS 2008. LNCS, vol. 5401, pp. 446–462. Springer, Heidelberg (2008)
11. Klasing, R., Markou, E., Pelc, A.: Gathering asynchronous oblivious mobile robots in a ring. Theoretical Computer Science 390(1), 27–39 (2008)
12. Suzuki, I., Yamashita, M.: Distributed anonymous mobile robots: formation of geometric patterns. SIAM J. Comput. 28, 1347–1363 (1999)
13. Yamashita, M., Suzuki, I.: Characterizing geometric patterns formable by oblivious anonymous mobile robots. Theoretical Computer Science (2010) (to appear)

Uniform Sampling of Digraphs with a Fixed Degree Sequence⋆

Annabell Berger and Matthias Müller-Hannemann

Department of Computer Science, Martin-Luther-University Halle-Wittenberg,
Von-Seckendorff-Platz 1, 06120 Halle, Germany
{berger,muellerh}@informatik.uni-halle.de

Abstract. Many applications in network analysis require algorithms to sample uniformly at random from the set of all digraphs with a prescribed degree sequence. We present a Markov chain based approach which converges to the uniform distribution of all realizations. It remains an open challenge whether the Markov chain is rapidly mixing.

We also explain in this paper that a popular switching algorithm fails in general to sample uniformly at random because the state graph of the Markov chain decomposes into different isomorphic components. We call degree sequences for which the state graph is strongly connected *arc swap sequences*. To handle arbitrary degree sequences, we develop two different solutions. The first uses an additional operation (a reorientation of induced directed 3-cycles) which makes the state graph strongly connected, the second selects randomly one of the isomorphic components and samples inside it. Our main contribution is a precise characterization of arc swap sequences, leading to an efficient recognition algorithm. Finally, we point out some interesting consequences for network analysis.

1 Introduction

We consider the problem of sampling uniformly at random from the set of all realizations of a prescribed degree sequence as simple, labeled digraphs without loops.

Motivation. In complex network analysis, one is interested in studying certain network properties of some observed real graph in comparison with an ensemble of graphs with the same degree sequence to detect deviations from randomness [1]. For example, this is used to study the motif content of classes of networks [2]. To perform such an analysis, a uniform sampling from the set of all realizations is required. A general method to sample random elements from some set of objects is via rapidly mixing Markov chains [3,4]. Every Markov chain can be viewed as a random walk on a directed graph, the so-called *state graph*. In

---

⋆ This work was partially supported by the DFG Focus Program Algorithm Engineering, grant Mu 1482/4-2, and by a VolkswagenStiftung grant for the project "Impact on motif content on dynamic function of complex networks".

D.M. Thilikos (Ed.): WG 2010, LNCS 6410, pp. 220–231, 2010.
© Springer-Verlag Berlin Heidelberg 2010

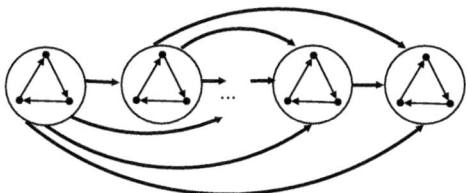

Fig. 1. Example of digraphs where no 2-swap operation can be applied

our context, its vertices (the states) correspond one-to-one to the set of all real-izations of prescribed degree sequences. For a survey on random walks, we refer to Lovász [5].

A popular variant of the Markov chain approach to sample among such realiza-tions is the so-called *switching-algorithm*. It starts with a given realization, and then performs a sequence of 2-swaps. In the undirected case, a *2-swap* replaces two non-adjacent edges $\{a, b\}, \{c, d\}$ either by $\{a, c\}, \{b, d\}$ or by $\{a, d\}, \{b, c\}$, provided that both new edges have not been contained in the graph before the swap operation. Likewise, in the directed case, given two arcs $(a, b), (c, d)$ with all vertices a, b, c, d being distinct, a *2-swap* replaces these two arcs by $(a, d), (c, b)$ which are currently not included in the realization (the latter is crucial to avoid parallel arcs). The switching algorithm is usually stopped heuristically after a certain number of iterations, and then outputs the resulting realization as a "random element". For undirected graphs, one can prove that this switching al-gorithm converges to a random stage. The directed case, however, turns out to be much more difficult. The following example demonstrates that the switching algorithm does not even converge to a random stage.

Example 1. *Consider the following class of digraphs $D = (V, A)$ with $3n$ ver-tices $V = \{v_0, v_1, \ldots, v_{3n-1}\}$, see Figure 1. Roughly speaking, this class consists of induced directed 3-cycles C_i formed by triples $V_i = \{v_{3i}, v_{3i+1}, v_{3i+2}\}$ of ver-tices, and arcs $A_i = \{(v_{3i}, v_{3i+1}), (v_{3i+1}, v_{3i+2}), (v_{3i+2}, v_{3i})\}$ for $i \in \{0, \ldots, n - 1\}$. All vertices of cycle C_i are connected to all other vertices of cycles with larger index than i. More formally, let $A' := \{(v, w) | v \in V_i, w \in V_j, i < j\}$. We set $A := A' \cup (\cup_{i=0}^{n-1} A_i)$.*

It is easy to check that no 2-swap can be applied to this digraph. However, we can independently reorient each of the n induced 3-cycles, leading to $2^{n/3}$ many (isomorphic) realizations of the same degree sequence. Thus, if we use a random walk on the state graph of all realizations of this degree sequence and use only 2-swaps to define the possible transitions between realizations, this state graph consists exactly of $2^{n/3}$ many singleton components. Hence, a "random walk" on this graph will be stuck in a single realization although exponentially many realizations exist.

It is interesting to note that slightly generalized 2-swap operations (the two replaced arcs are not restricted to be non-adjacent) suffice to sample directed graphs *with* loops. This has been proven by Ryser [6] in the context of square matrices with $\{0, 1\}$–entries which can be interpreted as node-node adjacency matrices of digraphs with loops.

Realizability of degree sequences. In order to use a Markov chain approach one needs at least one feasible realization. In applications from complex network analysis, one can usually take the degree sequence of some observed real world graph. Otherwise, one has to construct a realization.

The realization problem, i.e., characterizing the existence and finding at least one realization, has quite a long history. In 1966, Chen [7] presented necessary and sufficient conditions for the realizability of degree sequences as digraphs which can be checked in linear time. The construction of a concrete realization is equivalent to an f-factor problem on a corresponding undirected bipartite graph. Given a simple graph $G = (V, E)$ and a function $f : V(G) \mapsto \mathbb{N}_0$, an *f-factor* is a subgraph H of G such that every vertex $v \in V$ in this subgraph H has exactly degree $d_H(v) = f(v)$. Tutte did seminal work on the f-factor problem [8] and gave a polynomial time transformation to the perfect matching problem. Kleitman and Wang [9] found a greedy-type algorithm generalizing previous work by Havel [10] and Hakimi [11,12]. This approach has recently been rediscovered by Erdős et al. [13].

Related work. Kannan et al. [14] showed how to sample bipartite undirected graphs via Markov chains. They proved polynomial mixing time for regular and near-regular graphs. Cooper et al. [15] extended this work to non-bipartite undirected, d-regular graphs and proved a polynomial mixing time for the switching algorithm. More precisely, they upper bounded the mixing time in these cases by $d^{15}n^8(dn \log(dn) + \log(\varepsilon^{-1}))$, for graphs with $|V| = n$. In a break-through paper, Jerrum, Sinclair, and Vigoda [16] presented a polynomial-time almost uniform sampling algorithm for perfect matchings in bipartite graphs. Their approach can be used to sample arbitrary bipartite graphs and therefore also arbitrary digraphs with a specified degree sequence in $O(n^{14} \log^4 n)$ via the above-mentioned reduction due to Tutte. In the context of sampling binary contingency tables, Bezáková et al. [17] managed to improve the running time for these sampling problems to $O(n^{11} \log^5 n)$, which is still far from practical.

McKay and Wormald [18,19] use a configuration model and generate random undirected graphs with degrees bounded by $o(n^{1/2})$ with uniform distribution in $O(m^2 d_{max})$ time, where d_{max} denotes the maximum degree, and m the number of edges. Steger and Wormald [20] introduced a modification of the configuration model that leads to a fast algorithm and samples asymptotically uniform for degrees up to $o(n^{1/28})$. Kim and Vu [21] improved the analysis of Steger and Wormald's algorithm, proving that the output is asymptotically uniform for degrees up to $O(n^{1/3-\varepsilon})$, for any $\varepsilon > 0$. Bayati et al. [22] recently presented a nearly-linear time algorithm for counting and randomly generating almost uniformly simple undirected graphs with a given degree sequence where the maximum degree is restricted to $d_{max} = O(m^{1/4-\tau})$, and τ is any positive constant.

Random walks and Markov chains. Let us briefly review the basic notions of random walks and their relation to Markov chains. See [4,5,23] for more details. A random walk (Markov chain) on a digraph $D = (V, A)$ is a sequence of vertices $v_0, v_1, \ldots, v_t, \ldots$ where $(v_i, v_{i+1}) \in A$. Vertex v_0 represents the initial state. Denote by $d_D^+(v)$ the out-degree of vertex $v \in V$. At the tth step we move to an arbitrary neighbor of v_t with probability $1/d_D^+(v_t)$ or stay at v_t with probability $(1 - \nu(v_t))/d_D^+(v_t)$, where $\nu(v_t)$ denotes the number of neighbors of v_t. Furthermore, we define the distribution of V at time $t \in \mathbb{Z}^+$ as the function $P_t \in [0,1]^{|V|}$ with $P_t(i) := Prob(v_t = i)$. A well-known result [5] is that P_t tends to the uniform stationary distribution for $t \to \infty$, if the digraph is (1) non-bipartite (that means aperiodic), (2) strongly connected (i.e., irreducible), (3) symmetric, and (4) regular. A digraph D is d_D-*regular* if all vertices have the same in- and out-degrees d_D.

In this paper, we will view all Markov chains as random walks on symmetric d_D-regular digraphs $D = (V, A)$ whose vertices correspond to the state space V. The transition probability on each arc $(v, w) \in A$ will be the constant $1/d_D$.

Our contribution. Carefully looking at our Example 1, we observe that the state graph becomes strongly connected if we add a second type of operation to transform one realization into another: Simply reorient the arcs of an induced directed 3-cycle. We call this operation *3-cycle reorientation*. We give a graph-theoretical proof that 2-swaps and 3-cycle reorientations suffice not only here, but also in general for arbitrary prescribed degree sequences. These observations allow us to define a Markov chain, very similar to the undirected case. The difference is that two realizations are mutually connected by arcs if and only if their symmetric difference is either an alternating 4-cycle or a 6-cycle on exactly three different vertices. This digraph becomes regular by adding additional loops, see Section 2. The transition probabilities are of order $O(1/m^2)$, and the diameter can be bounded by $O(m)$, where m denotes the number of arcs in the prescribed degree sequence.

In the context of $(0, 1)$-matrices with given marginals (i.e., prescribed degree sequences in our terminology), Rao et al. [24] similarly observed that switching operations on so-called "compact alternating hexagons" are necessary. A compact alternating hexagon is a 3×3-submatrix, which can be interpreted as the adjacency matrix of a directed 3-cycle subgraph. They define a random walk on a series of state graphs, starting with a non-regular one which is iteratively updated towards regularity, i.e. their Markov chain converges asymptotically to the uniform distribution. However, it is unclear how fast this process converges and whether this is more efficient than working directly with a single regular state graph. Since Rao et al. work directly on matrices, their transition probabilities are of order $O(1/n^6)$, i.e., by several orders smaller than in our version. Very recently, Erdős et al. [13] proposed a similar Markov chain approach using 2-swaps and 3-swaps. The latter type of operation exchanges a simple directed 3-path or 3-cycle $(v_1, v_2), (v_2, v_3), (v_3, v_4)$ (the first and last vertex may be identical) by $(v_1, v_3), (v_3, v_2), (v_2, v_4)$, but is a much larger set of operations than ours.

Although in directed graphs 2-swaps alone do not suffice to sample uniformly in general, the corresponding approach is still frequently used in network analysis. One reason for the popularity of this approach — in addition to its simplicity — might be that it empirically worked in many cases quite well [1]. In this paper, we study under which conditions this approach can be applied and provably leads to correct uniform sampling. We call such degree sequences *arc-swap sequences*, and give a graph-theoretical characterization which can be checked in polynomial time. More specifically, we can recognize arc-swap sequences in $O(m^2)$ time using matching techniques. Using a parallel Havel-Hakimi algorithm by LaMar [25], originally developed to realize Euler sequences with an odd number of arcs, the recognition problem can even be solved in linear time. This algorithm also allows us to determine the number of induced directed 3-cycles which appear in *every* realization. However, the simpler approach comes with a price: our bound on the diameter of the state graph becomes mn and so is by one order of n worse in comparison with using 2-swaps and 3-cycle reorientations. Since half of the diameter is a trivial lower bound on the mixing time and the diameter also appears as a factor in known upper bounds, we conjecture that the classical switching algorithm requires a mixing time τ_ε with an order of n more steps as the variant with 3-cycle reorientation.

In those cases where 2-swaps do not suffice to sample uniformly, the state graph decomposes into 2^k strongly connected components, where k is the number of induced directed 3-cycles which appear in every realization. We can also efficiently determine the number of strongly connected components of the state graph (of course, without explicitly constructing this exponentially sized graph). However, all these components are isomorphic. This can be exploited as follows: For a non-arc-swap sequence, we first determine all those induced directed 3-cycles which appear in every realization. By reducing the in- and out-degrees for all vertices of these 3-cycles by one, we then obtain a new sequence, now guaranteed to be an arc-swap sequence. On the latter sequence we can either use the switching algorithm or our variant with additional 3-cycle reorientations. Since the bound on the diameter reduces from $n(m-3k)$ to $m-3k$ in the second variant, we conjecture the following.

Conjecture 1. *a) Sampling with a randomly chosen component of the state graph by (i) 2-swaps or by (ii) 2-swaps and 3-cycle reorientations has a polynomially bounded mixing time.*
b) Variant (ii) samples by at least one order of magnitude faster than (i).

Our results give a theoretical foundation to compute certain network characteristics on unlabeled digraphs in a single component using 2-swaps only. For example, this includes the analysis of the motif content [26]. Likewise we can still compute the average diameter among all realizations if we work in a single component. However, for other network characteristics, for example betweenness centrality on edges [27], this leads in general to incorrect estimations.

Overview. The remainder of the paper is structured as follows. In Section 2, we introduce appropriately defined state graphs underlying our Markov chains, and show for these graphs crucial properties like regularity and strong connectivity. We also upper bound their diameter. Afterwards, in Section 3, we characterize those degree sequences for which a simpler Markov chain based on 2-swaps provably leads to uniform sampling of digraphs. We also describe a few consequences and applications. Due to space restrictions, we omit the presentation of related results for the undirected case. These results and complete proofs can be obtained in the full version of the paper available from http://arxiv.org/abs/0912.0685.

2 Sampling Digraphs

Let us start with the formal problem definition and some additional notation. Afterwards, we introduce our Markov chain and analyze its properties.

Formal problem definition. In the directed case, we define a degree sequence S as a sequence of 2-tuples $\left(\binom{a_1}{b_1}, \binom{a_2}{b_2}, \ldots, \binom{a_n}{b_n} \right)$ with $a_i, b_i \in \mathbb{Z}_0^+, i \in \{1, \ldots, n\}$ where $a_i > 0$ or $b_i > 0$.

Let $G = (V, A)$ be a directed labeled graph $G = (V, A)$ without loops and parallel arcs and $|V| = n$. We define the *in-degree-function* $d_G^+ : V \to \mathbb{Z}_0^+$ which assigns to each vertex $v_i \in V$ the number of incoming arcs and the *out-degree-function* $d_G^- : V \to \mathbb{Z}_0^+$ which assigns to each vertex $v_i \in V$ the number of outgoing arcs. We denote S as *graphical sequence* if and only if there exists at least one directed labeled graph $G = (V, A)$ without any loops or parallel arcs which satisfies $d_G^+(v_i) = b_i$ and $d_G^-(v_i) = a_i$ for all $v_i \in V$ and $i \in \{1, \ldots, n\}$. Any such graph G is called *realization* of S. Let H be a subdigraph of G. We say that $H = (V_H, A_H)$ is an *induced subdigraph* of G if every arc of A with both end vertices in V_H is also in A_H. We write $H = G \langle V_H \rangle$. We define an *alternating walk* P for a directed graph $G = (V, A)$ as a sequence $P := (v_1, v_2, \ldots, v_l)$ of vertices $v_i \in V$ where either $(v_i, v_{i+1}) \in A(G)$ and $(v_i, v_{i-1}) \notin A(G)$ or $(v_i, v_{i+1}) \notin A(G)$ and $(v_i, v_{i-1}) \in A(G)$ for $i \mod 2 = 1$. Alternating here means that any two subsequent arcs alternate with respect to their orientation and to being present or not. We call an alternating walk C of even length *alternating cycle* if $v_1 = v_l$ is fulfilled. For two realizations G, G', the symmetric difference of their arc sets $A(G)$ and $A(G')$ is denoted by $G \Delta G' := (A(G) \backslash A(G')) \cup (A(G') \backslash A(G))$. Consider for example the realizations G and G' with $A(G) := \{(v_1, v_2), (v_3, v_4)\}$ and $A(G') := \{(v_1, v_4), (v_3, v_2)\}$ consisting of exactly two arcs. Then the symmetric difference is the alternating 4-cycle $C := (v_1, v_2, v_3, v_4, v_1)$ where $(v_i, v_{i+1}) \in A(G)$ for $i \in \{1, 3\}$ and $(v_{i+1}, v_i) \in A(G')$ taking indices $i \mod 4$. In general, the symmetric difference of two realizations always decomposes into a number of alternating cycles. Note that even a weakly connected symmetric difference can decompose into several alternating cycles, in contrast to undirected symmetric differences. See Figs. 2 and 3 for an example.

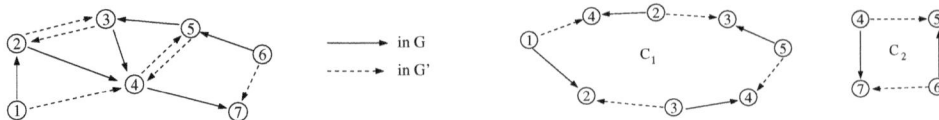

Fig. 2. Example: Two realizations G and G'

Fig. 3. Decomposition of a weakly connected symmetric difference $G \Delta G'$ of Fig. 2 into a minimum number of alternating cycles

The Markov chain. We denote the state graph for our random walk by $\Phi = (V_\phi, A_\phi)$. Its underlying vertex set V_ϕ is the set of all realizations of a given degree sequence S. For a realization G, we denote by V_G the corresponding vertex in V_ϕ. The arc set A_ϕ is defined as follows.

a) We connect two vertices $V_G, V_{G'} \in V_\phi, G \neq G'$ with arcs $(V_G, V_{G'})$ and $(V_{G'}, V_G)$ if and only if one of the two following constraints is fulfilled
    1. $|G \Delta G'| = 4$
    2. $|G \Delta G'| = 6$ and $G \Delta G'$ contains exactly three different vertices.
b) We set a directed loop (V_G, V_G)
    1. for each pair of non-adjacent arcs $(v_{i_1}, v_{i_2}), (v_{i_3}, v_{i_4}) \in A(G)$ with $i_j \in \{1, \dots, n\}$ if and only if $(v_{i_1}, v_{i_4}) \in A(G) \vee (v_{i_3}, v_{i_2}) \in A(G)$ in a realization G,
    2. for each directed 2-path $(v_{i_1}, v_{i_2}), (v_{i_2}, v_{i_3}) \in A(G)$ if and only if one of the following constraints is true for a realization G,
        i) $(v_{i_2}, v_{i_1}) \in A(G) \vee (v_{i_3}, v_{i_2}) \in A(G) \vee (v_{i_1}, v_{i_3}) \in A(G)$,
        ii) $(v_{i_3}, v_{i_1}) \notin A(G)$,
        iii) $i_3 < i_1 \vee i_3 < i_2$.
    3. if G contains no directed 2-path.

Lemma 1. *The state graph* $\Phi = (V_\phi, A_\phi)$ *is non-bipartite, symmetric, and regular.*

The next step is to show that our constructed state graph Φ is strongly connected. This is sufficient to prove the reachability of each realization independent of the starting realization.

Proposition 1. *Let S be a graphical sequence and G and G' be two different realizations. If $G \Delta G'$ is exactly one weak component and $|G \Delta G'| \neq 6$ then there exists in $G \Delta G'$ a vertex-disjoint alternating 3-walk of type P or Q, where $P = (v_1, v_2, v_3, v_4)$ with $(v_1, v_2), (v_3, v_4) \in A(G)$ and $(v_3, v_2) \in A(G')$ and $Q = (w_1, w_2, w_3, w_4)$ with $(w_1, w_2), (w_3, w_4) \in A(G')$ and $(w_3, w_2) \in A(G)$.*

Note that the above proposition does not assert that the symmetric difference contains P and Q. The smallest counter-example are the realizations $G = (V, A)$ and $G' = (V, A')$ with $V = \{v_1, v_2, v_3, v_4\}$ and $A = \{(v_1, v_3), (v_3, v_2), (v_2, v_4), (v_4, v_1)\}$ and $A' = \{(v_1, v_2), (v_2, v_1), (v_3, v_4), (v_4, v_3)\}$.

Proposition 2. *Let S be a graphical sequence and G and G' be two different realizations. If $|G\Delta G'| = 6$, then there exist*

a) *realizations G_0, G_1, G_2 with $G_0 := G$, $G_2 := G'$ and $|G_i\Delta G_{i+1}| = 4$ for $i \in \{0, 1\}$ or*
b) *G and G' are different in the orientation of exactly one directed 3-cycle.*

Lemma 2. *Let S be a graphical sequence and G and G' be two different realizations. There exist realizations $G_0, G_1, \ldots, G_k$ with $G_0 := G$, $G_k := G'$ and*

1. *$|G_i\Delta G_{i+1}| = 4$ or*
2. *$|G_i\Delta G_{i+1}| = 6$*

where $k \leq \frac{1}{2}|G\Delta G'| - 1$. In case (2), $G_i\Delta G_{i+1}$ consists of a directed 3-cycle and its opposite orientation.

Corollary 1. *State graph Φ is a strongly connected digraph.*

The properties of the state graph imply a straightforward random walk based sampling algorithm.

Theorem 1. *A random walk on state graph Φ samples uniformly at random a directed graph $G' = (V, A)$ as a realization of sequence S for $t \to \infty$.*

3   Arc-Swap Sequences

In this section, we study under which conditions the simple switching algorithm works correctly for digraphs. The Markov chain used in the switching algorithm works on the following simpler state graph $\overline{\Phi} = (V_{\overline{\phi}}, A_{\overline{\phi}})$. We define $A_{\overline{\phi}}$ as follows.

a) We connect two vertices $V_G, V_{G'} \in V_{\overline{\phi}}, G \neq G'$ with arcs $(V_G, V_{G'})$ and $(V_{G'}, V_G)$ if and only if $|G\Delta G'| = 4$ is fulfilled.
b) We set for each pair of non-adjacent arcs $(v_{i_1}, v_{i_2}), (v_{i_3}, v_{i_4}) \in A(G), i_j \in \{1, \ldots, n\}$ a directed loop (V_G, V_G) if and only if $(v_{i_1}, v_{i_4}) \in A(G) \vee (v_{i_3}, v_{i_2}) \in A(G)$.
c) We set one directed loop (V_G, V_G) for all $V_G \in V_{\overline{\phi}}$.

Lemma 3. *The state graph $\overline{\Phi} = (V_{\overline{\phi}}, A_{\overline{\phi}})$ is non-bipartite, symmetric, and regular.*

3.1   Characterization of Arc-Swap Sequences

As shown in Example 1 in the Introduction, $\overline{\Phi}$ decomposes into several components, but we are able to characterize sequences S for which strong connectivity is fulfilled in $\overline{\Phi}$. In fact, we will show that there are numerous sequences which only require switching by 2-swaps. In the following we give necessary and sufficient conditions allowing to identify such sequences in polynomial running time.

Definition 1. *Let S be a graphical sequence and let $G = (V, A)$ be an arbitrary realization. We denote a vertex subset $V' \subseteq V$ with $|V'| = 3$ as an* induced cycle set *V' if and only if for each realization $G^* = (V, A^*)$ the induced subdigraph $G^* \langle V' \rangle$ is a directed 3-cycle.*

Definition 2. *Let S be a graphical sequence and $G = (V, A)$ an arbitrary realization. We call S an* arc-swap-sequence *if and only if each subset $V' \subseteq V$ of vertices with $|V'| = 3$ is not an induced cycle set.*

This definition enables us to use our simpler state graph $\overline{\Phi}$ for sampling a realization G for arc-swap-sequences. In Lemma 5, we will show that in these cases we have only to switch the ends of two non-adjacent arcs.

Before, we study how to recognize arc-swap sequences efficiently. Clearly, we may not determine all realizations to identify a sequence as an arc-swap-sequence. Fortunately, we are able to give a characterization allowing us to identify an arc-swap-sequence in only considering one realized digraph. We need a further definition for a special case of symmetric differences.

Definition 3. *Let S be a graphical sequence and $G = (V, A)$ and $G^* = (V, A^*)$ arbitrary realizations. We call $G \Delta G^*$* simple symmetric cycle *if and only if each vertex $v \in V(G \Delta G^*)$ has vertex in-degree $d^-_{G \Delta G^*}(v) \leq 2$ and vertex out-degree $d^+_{G \Delta G^*}(v) \leq 2$, and if $G \Delta G^*$ is exactly one alternating cycle.*

Note that the alternating cycle C_1 in Fig. 3 is not a simple symmetric cycle, because $d^+_{C_1}(4) = 4$. Cycle C_1 decomposes into two simple symmetric cycles $C'_1 = \{v_1, v_2, v_3, v_4, v_1\}$ and $C''_1 = \{v_2, v_3, v_5, v_4, v_2\}$.

Theorem 2. *A graphical sequence S is an arc-swap-sequence if and only if for any realization $G = (V, A)$ the following property is true: For each induced, directed 3-cycle $G \langle V' \rangle$ of G there exists a realization $G^* = (V, A^*)$ so that $G \Delta G^*$ is a simple symmetric cycle and that the induced subdigraph $G^* \langle V' \rangle$ is not a directed 3-cycle.*

This characterization allows us to give a simple polynomial-time algorithm to recognize arc-swap-sequences. All we have to do is to check for each induced 3-cycle of the given realization, if it forms an induced cycle set. Therefore, we check for each arc (v, w) in an induced 3-cycle whether there is an alternating walk from v to w (not using arc (v, w)) which does not include all five remaining arcs of the 3-cycle and its reorientation. Moreover, each node on this walk has at most in-degree 2 and at most out-degree 2. Such an alternating walk can be found in linear time by using a reduction to an f-factor problem in a bipartite graph. In this graph we search for an undirected alternating path by growing alternating trees (similar to matching algorithms in bipartite graphs, no complications with blossoms will occur), see for example [28]. The trick to ensure that not all five arcs will appear in the alternating cycle is to iterate over these five arcs and exclude exactly one of them from the alternating path search between v and w. Of course, this loop stops as soon as one alternating path is found. Otherwise, no such alternating path exists. As mentioned in the Introduction, a linear-time recognition is possible with a parallel Havel-Hakimi algorithm of LaMar [25].

Next, we are going to prove that $\overline{\Phi}$ is strongly connected for arc-swap-sequences.

Lemma 4. *Let S be a graphical arc-swap-sequence and G and G^* be two different realizations. Assume that $V' := \{v_1, v_2, v_3\} \subseteq V$ such that $G\langle V'\rangle$ is an induced directed 3-cycle but $G^*\langle V'\rangle$ is not an induced directed 3-cycle. Moreover, assume that $G\Delta G^*$ is a simple symmetric cycle. Then there are realizations $G_0, G_1, \ldots, G_k$ with $G_0 := G, G_k := G^*, |G_i\Delta G_{i+1}| = 4$ and $k \leq \frac{1}{2}|G\Delta G^*|$.*

Proposition 3. *Let S be a graphical arc-swap-sequence and G and G' be two different realizations. If $|G\Delta G'| = 6$ and $G\Delta G'$ consists of exactly three vertices $V' := \{v_1, v_2, v_3\}$, then there exist realizations $G_0, G_1, \ldots, G_k$ with $G_0 := G, G_k := G', |G_i\Delta G_{i+1}| = 4$ and $k \leq 2n + 2$.*

Lemma 5. *Let S be a graphical arc-swap-sequence, and G and G' be two different realizations. Then there exist realizations $G_0, G_1, \ldots, G_k$ with $G_0 := G$, $G_k := G'$ and $|G_i\Delta G_{i+1}| = 4$, where $k \leq \left(\frac{1}{2}|G\Delta G'| - 1\right) \cdot (n + 1)$.*

Corollary 2. *State graph $\overline{\Phi}$ is a strongly connected directed graph if and only if a given sequence S is an arc-swap-sequence.*

An arc-swap-sequence implies the connectedness of the simple realization graph $\overline{\Phi}$. Therefore, for such sequences we are able to make random walks on the simple state graph $\overline{\Phi}$ which can be implemented easily.

Theorem 3. *A random walk on the state graph $\overline{\Phi}$ uniformly samples a digraph $G' = (V, A)$ as a realization of an arc-swap-sequence S for $t \to \infty$.*

3.2   Sampling within Randomly Chosen Components

As mentioned in the Introduction, many "practitioners" use the switching algorithm for the purpose of network analysis, regardless whether the corresponding degree sequence is an arc-swap-sequence or not. In this section we would like to discuss under which circumstances this common practice can be well justified and when it may lead to wrong conclusions. What would happen if we sample using the state graph $\overline{\Phi}$ for a sequence S which is not an arc-swap-sequence? Clearly, we get the insight that $\overline{\Phi}$ has several connected components, but as we will see $\overline{\Phi}$ consists of at most $2^{\lfloor \frac{|V|}{3} \rfloor}$ isomorphic components containing exactly the same realizations up to the orientation of directed 3-cycles each consisting of an induced cycle set V'. Fortunately, we can identify all induced cycle sets using our results in Theorem 2 by only considering an arbitrary realization G.

Proposition 4. *Let S be a graphical sequence which is not an arc-swap-sequence and has at least two different induced cycle sets V' and V''. Then it follows $V' \cap V'' = \emptyset$.*

Theorem 4. *Let S be a sequence. Then the state graph $\overline{\Phi}$ consists of at most $2^{\lfloor \frac{|V|}{3} \rfloor}$ isomorphic components.*

We propose the following new two-stage sampling algorithm. (1) For a given degree sequence S, first determine all induced cycle sets. For each of them independently flip a coin to decide upon the cycle orientation. Fixing all these orientations thus means to select uniformly at random some connected component of the state graph. (2) Reduce the in- and out-degrees of all vertices in these cycles by one (the induced cycle sets are vertex-disjoint), and obtain a new degree sequence S' which must be an arc-swap sequence. In the second stage, do a random walk on S' using 2-swaps and 3-cycle reorientations. Finally, reinsert into the obtained realization the arcs of the induced cycle sets determined in step (1) with the chosen orientation. The second stage could be done without 3-cycle reorientations, but we conjecture (Conjecture 1) that the mixing time would be at least one order of magnitude larger.

Applications in Network Analysis. Our results shed some light on claims made in complex network analysis in recent years. In particular, the traditional switching algorithm has been considered to be correct (see, for example [1]). We proved the claim to be in general false. This might have been overlooked for two reasons: On the one hand, researchers may have worked by chance on arc-swap sequences in their experiments. On the other hand, certain network statistics lead to correct results if one samples in a single component of the state graph. Examples where the traditional approach is feasible are network statistics like the average diameter or the motif content over all realizations.

For other network statistics, however, the random walk on $V_{\overline{\Phi}}$ systematically over- and under-samples the probability that an arc is present.

References

1. Milo, R., Kashtan, N., Itzkovitz, S., Newman, M., Alon, U.: On the uniform generation of random graphs with arbitrary degree sequences, arXiv:cond-mat/0312028v2 (May 30, 2004)
2. Milo, R., Itzkovitz, S., Kashtan, N., Levitt, R., Shen-Orr, S., Ayzenshtat, I., Sheffer, M., Alon, U.: Superfamilies of evolved and designed networks. Science 303, 1538–1542 (2004)
3. Sinclair, A.: Improved bounds for mixing rates of Markov chains and multicommodity flow. Combinatorics, Probability & Computing 1, 351–370 (1992)
4. Sinclair, A.: Algorithms for Random Generation and Counting: A Markov Chain Approach. Birkhäuser, Basel (1993)
5. Lovász, L.: Random walks on graphs: A survey. In: D.M., et al. (eds.) Combinatorics, Paul Erdős is Eighty, vol. 2, pp. 353–397. János Bolyai Mathematical Society (1996)
6. Ryser, H.J.: Combinatorial properties of matrices of zeroes and ones. Canadian J. of Mathematics 9, 371–377 (1957)
7. Chen, W.: On the realization of a (p,s)-digraph with prescribed degrees. J. Franklin Institute 281, 406–422 (1966)
8. Tutte, W.: The factors of graphs. Canadian J. of Mathematics 4, 314–328 (1952)
9. Kleitman, D., Wang, D.: Algorithm for constructing graphs and digraphs with given valences and factors. Discrete Math. 6, 79–88 (1973)

10. Havel, V.: A remark on the existence of finite graphs. Časopis Pěst. Mat. 80, 477–480 (1955)
11. Hakimi, S.: On the realizability of a set of integers as degrees of the vertices of a simple graph. SIAM J. Appl. Math. 10, 496–506 (1962)
12. Hakimi, S.: On the degrees of the vertices of a directed graph. J. Franklin Institute 279, 290–308 (1965)
13. Erdős, P.L., Miklós, I., Toroczkai, Z.: A simple Havel-Hakimi type algorithm to realize graphical degree sequences of directed graphs. The Electronic Journal of Combinatorics 17, #R66 (2010)
14. Kannan, R., Tetali, P., Vempala, S.: Simple Markov-chain algorithms for generating bipartite graphs and tournaments. Random Structures and Algorithms 14, 293–308 (1999)
15. Cooper, C., Dyer, M., Greenhill, C.: Sampling regular graphs and a peer-to-peer network. Combinatorics, Probability and Computing 16, 557–593 (2007)
16. Jerrum, M., Sinclair, A., Vigoda, E.: A polynomial-time approximation algorithm for the permanent of a matrix with nonnegative entries. Journal of the ACM 51, 671–697 (2004)
17. Bezáková, I., Bhatnagar, N., Vigoda, E.: Sampling binary contingency tables with a greedy start. Random Structures and Algorithms 30, 168–205 (2007)
18. McKay, B., Wormald, N.: Uniform generation of random regular graphs of moderate degree. J. Algorithms 11, 52–67 (1990)
19. McKay, B., Wormald, N.: Asymptotic enumeration by degree sequence of graphs with degrees $o(n^{1/2})$. Combinatorica 11, 369–382 (1991)
20. Steger, A., Wormald, N.: Generating random regular graphs quickly. Combinatorics, Probability, and Computing 8, 377–396 (1999)
21. Kim, J., Vu, V.: Generating random regular graphs. In: STOC, pp. 213–222 (2003)
22. Bayati, M., Kim, J.H., Saberi, A.: A sequential algorithm for generating random graphs. Algorithmica (2009) doi 10.1007/s00453-009-9340-1
23. Jerrum, M., Sinclair, A.: The Markov chain Monte Carlo method: An approach to approximate counting and integration. In: Hochbaum, D. (ed.) Approximation Algorithms for NP-hard Problems, pp. 482–520. PWS Publishing, Boston (1996)
24. Rao, A., Jana, R., Bandyopadhyay, S.: A Markov chain Monte Carlo method for generating random (0,1)–matrices with given marginals. Sankhya: The Indian Journal of Statistics 58, 225–242 (1996)
25. LaMar, M.D.: Algorithms for realizing degree sequences for directed graphs, arXiv.org:0906.0343v1 (2009)
26. Milo, R., Shen-Orr, S., Itzkovitz, S., Kashtan, N., Chklovskii, D., Alon, U.: Network motifs: simple building blocks of complex networks. Science 298, 824–827 (2002)
27. Koschützki, D., Lehmann, K.A., Peeters, L., Richter, S., Tenfelde-Podehl, D., Zlotowski, O.: Centrality indices. In: Brandes, U., Erlebach, T. (eds.) Network Analysis. LNCS, vol. 3418, pp. 16–61. Springer, Heidelberg (2005)
28. Schrijver, A.: Combinatorial Optimization: Polyhedra and Efficiency. Springer, Heidelberg (2003)

Measuring Indifference:
Unit Interval Vertex Deletion

René van Bevern*, Christian Komusiewicz**,
Hannes Moser*, and Rolf Niedermeier

Institut für Informatik, Friedrich-Schiller-Universität Jena,
Ernst-Abbe-Platz 2, D-07743 Jena, Germany
{rene.bevern,c.komus,hannes.moser,rolf.niedermeier}@uni-jena.de

Abstract. Making a graph unit interval by a minimum number of vertex deletions is NP-hard. The problem is motivated by applications in seriation and measuring indifference between data items. We present a fixed-parameter algorithm based on the iterative compression technique that finds in $O((14k + 14)^{k+1}kn^6)$ time a set of k vertices whose deletion from an n-vertex graph makes it unit interval. Additionally, we show that making a graph chordal by at most k vertex deletions is NP-complete even on {claw,net,tent}-free graphs.

1 Introduction

Being indifferent between two objects means to prefer neither of them. The indifference relation defines an undirected graph with an edge between two vertices if and only if they are judged indifferent. In this paper, we study measuring indifference in the context of seriation. The specific task is to put objects in a serial order, respecting the given indifference relation as much as possible.

Indifference corresponds to "closeness" between data items [1,2]. Accordingly, an undirected graph $G = (V, E)$ whose vertices represent data items is called an *indifference graph* if there exists a function $r\colon V \to \mathbb{R}$ such that for all $u, v \in V$

$$\{u, v\} \in E \Leftrightarrow |r(u) - r(v)| \leq \delta,$$

where δ is a positive number (the "threshold") measuring closeness. The function r induces a serial order. Informally, the above equivalence expresses that we distinguish between u and v only if, according to r, there is sufficiently high difference between them. Empirical indifference judgments (with correspondingly defined undirected graphs) usually do not permit such an assignment r satisfying the above equivalence. One possibility, however, is that "almost all" data items induce an indifference graph. In other words, given a graph based on empirical indifference judgments (which contain some "errors"), the task then is to

---

* Supported by the DFG, project AREG, NI 369/9.
** Supported by a PhD fellowship of the Carl-Zeiss-Stiftung and the DFG, project PABI, NI 369/7.

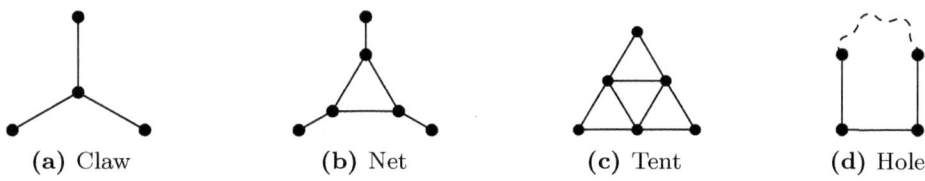

(a) Claw   (b) Net   (c) Tent   (d) Hole

Fig. 1. The (infinitely many) forbidden induced subgraphs for unit interval graphs. Holes are induced cycles of length at least four.

spot as few "outlier vertices" as possible such that after the removal of these vertices the graph becomes an indifference graph. Thus, the minimum number of vertex deletions measures how close "empirical indifference data" is to mathematically defined indifference. Since indifference graphs are precisely the unit interval graphs [3,4][1], we arrive at the central problem in this work:

UNIT INTERVAL VERTEX DELETION
Input: An undirected graph $G = (V, E)$.
Question: Is there a set $S \subseteq V$ with $|S| \leq k$ and $G[V \setminus S]$ being unit interval?

A graph is unit interval if and only if it contains no induced claw, net, tent, or hole (an induced cycle of length at least four) [5]. These infinitely many forbidden induced subgraphs are illustrated in Figure 1. A general result on vertex deletion problems implies that UNIT INTERVAL VERTEX DELETION is NP-complete [6].

Related Work. Roberts [3,4] discusses indifference and seriation and explains applications in fields such as archaeology and developmental psychology. Seriation by transforming graphs into indifference graphs belongs to the field of "seriation in the presence of errors" and can be understood as a variant of fitting Robinson structures to distances [7,8,9]. More specifically, our setting is related to the special case where distances are specified by symmetric 0/1-matrices; here, the Robinson property becomes the consecutive-ones property [4,8]. Thus, the UNIT INTERVAL VERTEX DELETION problem is equivalent to making a symmetric 0/1-matrix fulfill the consecutive-ones property by simultaneous[2] column and row deletions. We remark that making a 0/1-matrix fulfill the consecutive-ones property by means of non-simultaneous column or row deletions has recently been studied in terms of approximability and fixed-parameter tractability [10].

Our work is closely related to a result by Marx [11], who shows that CHORDAL VERTEX DELETION (asking whether a graph can be made chordal by k vertex deletions) is fixed-parameter tractable parameterized by k. One can observe that the result implies fixed-parameter tractability for UNIT INTERVAL VERTEX DELETION. However, the running time of the CHORDAL VERTEX DELETION algorithm [11] is not specified and it relies on solving CHORDAL VERTEX DELETION

---

[1] Unit interval graphs are also equivalent to proper interval graphs [3].
[2] For an $i \in \mathbb{N}$, column i is deleted from the matrix if and only if row i is deleted.

on tree decompositions of worst-case-width $\Omega(k^4)$. This renders the algorithm unimplementable. Subsequently to our work, Villanger [12] presented a search-tree based algorithm for UNIT INTERVAL VERTEX DELETION, improving the running time to $O(6^k k n^6)$.

Our Results. We present a fixed-parameter algorithm for UNIT INTERVAL VERTEX DELETION running in $O((14k + 14)^{k+1} \cdot k n^6)$ time, where k denotes the number of allowed vertex deletions and n is the number of graph vertices. Like Marx [11], we employ the iterative compression technique by Reed et al. [13,14]. However, we do not employ bounded-treewidth techniques and circumvent huge hidden constants. Before that, we show that UNIT INTERVAL VERTEX DELETION remains NP-hard when restricted to {claw, net, tent}-free graphs, where it is equivalent to CHORDAL VERTEX DELETION. Due to lack of space, some proofs are omitted [9].

Preliminaries. We only consider simple *undirected* graphs $G = (V, E)$, where $V(G) := V$ is the set of vertices and $E(G) := E$ is the set of edges. Throughout this work, let $n := |V|$ and $m := |E|$. The *neighborhood* $N(v)$ of a vertex $v \in V$ is the set of vertices adjacent to v. A *clique* is a graph in which every two distinct vertices are adjacent. For a set $V' \subseteq V$, the *induced subgraph* $G[V']$ is the graph with vertex set V' and edge set $\{\{v, w\} \in E \mid v, w \in V'\}$. We use $G - V'$ as an abbreviation for $G[V \setminus V']$. A *path* P from v_i to v_ℓ is a sequence $(v_1, v_2, \ldots, v_\ell) \in V^\ell$ with $\{v_i, v_{i+1}\} \in E$ for $i \in \{1, \ldots, \ell-1\}$; it *visits* the vertices $v_1, \ldots, v_\ell$. If $i \neq j$ implies $v_i \neq v_j$, then P is *simple*. If $\{v_i, v_j\} \notin E$ for $|i - j| > 1$, then P is *induced*. An *(induced) cycle* is an (induced) path with $\{v_1, v_\ell\} \in E$, called (induced) C_ℓ. Two vertices $v, w \in V$ are *connected* in G if there is a path from v to w in G. A *vertex-cut* between v and w is a set $C \subseteq V$ such that v and w are not connected in $G - C$. The graph G is *F-free* if G does not contain an induced subgraph isomorphic to the graph F. A *segment* of a total order $\preceq$ on V is a set $[u, w] := \{v \in V \mid u \preceq v \preceq w\}$.

2   NP-Hardness on a Restricted Graph Class

We show that UNIT INTERVAL VERTEX DELETION is NP-complete even on {claw, net, tent}-free graphs, where it is equivalent to CHORDAL VERTEX DELETION.

Theorem 1. CHORDAL VERTEX DELETION *on {claw, net, tent}-free graphs is NP-complete.*

Proof. We only show NP-hardness and employ a reduction from the NP-complete VERTEX COVER on triangle-free graphs [15]. First, we describe the reduction, then we prove its correctness. Let (G, k) be a VERTEX COVER instance, where G is triangle-free and we ask whether there is a set of k vertices whose deletion makes G edgeless. We construct an instance (G', k) for CHORDAL VERTEX DELETION as follows: let $G' := \overline{G}$, where the complement graph $\overline{G}$ of G has the same vertices as G and $\overline{G}$ has an edge $\{v, w\}$ if and only if G has not. Add to G' two disjoint cliques A and B, each containing $k + 1$ vertices, and make every vertex in A and every vertex in B adjacent to every vertex in $V(G)$. Because claw, net, and tent each contain a size-three independent set, the constructed graph G' is {claw, net, tent}-free: since G is triangle-free, $\overline{G}$ does not contain a size-three independent set. Moreover, there is no size-three independent set in G', since the vertices in the cliques A and B are adjacent to every vertex in $V(G)$. It remains to show that (G, k) is a yes-instance for VERTEX COVER if and only if (G', k) is a yes-instance for CHORDAL VERTEX DELETION.

Let S be a vertex cover of size at most k for G. Since S is a vertex cover, $G - S$ is an edgeless graph. As a consequence, the complement of $G - S$ is a clique C and thus $G' - S$ contains three cliques A, B, and C, where every vertex of A and B is adjacent to every vertex in C by construction of G'. The graph $G' - S$ is obviously chordal and S is a chordal vertex deletion set of size at most k for G'.

Let S be a chordal vertex deletion set for G' with $|S| \leq k$. Assume that $S \cap V(G)$ is not a vertex cover for G. Then, there is an edge $\{u, v\}$ in $G - S$, and, therefore, there is no edge $\{u, v\}$ in G' by construction of G'. Because A and B each contain $k + 1$ vertices and $|S| \leq k$, there are vertices $a \in A \setminus S$ and $b \in B \setminus S$. The vertex set $\{a, u, v, b\}$ induces a hole in $G' - S$, a contradiction to S being a chordal vertex deletion set for G'. Hence, $S \cap V(G)$ is a vertex cover for G. □

3   An Outline of the Algorithm

Our algorithm employs the iterative compression technique by Reed et al. [13,14]. The rough idea of this technique is to iteratively build up the input graph by adding vertices one by one and to compute in each iteration an optimal solution for the current subgraph, using the solution computed for the previous subgraph. More precisely, given an arbitrary order of the vertices from 1 to n, we start with the empty graph and an empty solution $S_0 := \emptyset$. The task of iteration i is to compute a solution for the graph $G[\{v_1, \ldots, v_i\}]$. Assume that the previously computed set S_{i-1} is a solution of size at most k for $G[\{v_1, \ldots, v_{i-1}\}]$. Then $S_{i-1} \cup \{v_i\}$ is a solution of size at most $k + 1$ for the graph $G[\{v_1, \ldots, v_i\}]$. We apply a *compression routine* that either computes a size-k solution S_i using $S_{i-1} \cup \{v_i\}$ or proves that no such solution exists. The pseudo-code of this main loop is given in Algorithm 1.

Algorithm 1. Iterative Compression

Input: A graph G and its vertices $v_1, \ldots, v_n$ in an arbitrary order, $k \in \mathbb{N}$.
Output: A unit interval vertex deletion set S for G with $|S| \leq k$ or "no".

1 $S_0 \leftarrow \emptyset$;
2 **for** $i := 1$ **to** n **do**
3      $S_i \leftarrow compress(G[\{v_1, \ldots, v_i\}], S_{i-1} \cup \{v_i\}, k)$;
4   **if** $S_i =$ "no" **then return** "no"

5 **return** S_n

The central part of the algorithm is the routine *compress* described below. Given an input graph G, a natural number k, and a unit interval vertex deletion set S' with $|S'| \leq k + 1$, the routine can return S' unchanged if $|S'| \leq k$. Thus, we assume that $|S'| = k + 1$. We now try all possible 2^{k+1} partitions of S' into two sets X and Y, where Y is a subset of the new solution S and $X \cap S = \emptyset$. For each partition, we delete the vertex set Y from G (since the vertices of Y are assumed to belong to the new solution). Then, the remaining task is to find a unit interval vertex deletion set disjoint from X and smaller than X. The crucial observation is that deleting X from $G - Y$ results in a unit interval graph. We say that X is a *unit interval vertex deletion set* for $G - Y$. Summarizing, we arrive at:

DISJOINT UNIT INTERVAL VERTEX DELETION
Input: A graph G and a unit interval vertex deletion set X for G.
Output: A unit interval vertex deletion set S with $|S| < |X|$ and $S \cap X = \emptyset$, or "no" if no such set exists.

DISJOINT UNIT INTERVAL VERTEX DELETION is NP-hard [16]. The advantage of working with DISJOINT UNIT INTERVAL VERTEX DELETION is that we can exploit $G - X$ being a unit interval graph. In the next section, we prove:

Theorem 2. DISJOINT UNIT INTERVAL VERTEX DELETION *can be solved in* $O((14|X| - 1)^{|X|-1} \cdot |X| n^5)$ *time.*

Exploiting this in the routine *compress* of Algorithm 1 leads to the main theorem of this work. The running time follows from the fact that *compress* is invoked $O(n)$ times and that each invocation solves DISJOINT UNIT INTERVAL VERTEX DELETION for all partitions of the solution from the previous iteration.

Theorem 3. UNIT INTERVAL VERTEX DELETION *can be solved in* $O((14k + 14)^{k+1} \cdot kn^6)$ *time.*

4   Finding Disjoint Unit Interval Vertex Deletion Sets

For DISJOINT UNIT INTERVAL VERTEX DELETION, given a unit interval vertex deletion set X for G, we search for a unit interval vertex deletion set S for G with $|S| < |X|$ and $S \cap X = \emptyset$. Roughly, the algorithm works as follows: first, enumerate *all* minimal vertex sets of size at most $|X| - 1$ whose deletion transforms G into a {claw,net,tent,C_4,C_5,C_6}-free graph, henceforth called *almost unit interval graph*. For each of these graphs, it remains to find a minimum-cardinality vertex set S' whose removal destroys all holes of length greater than six to make the graph unit interval. We call such a set *optimal*. If $|S'|$ and the number of vertex deletions needed to transform a graph into an almost unit interval graph add up to at most $|X| - 1$, then we have found a solution. Since we try *all* minimal vertex sets of size at most $|X| - 1$ whose deletion transforms a graph into an almost unit interval graph, we always find a size-$(|X| - 1)$ unit interval vertex deletion set if it exists.

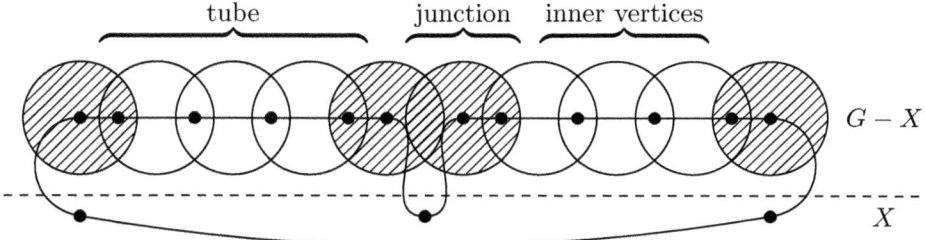

Fig. 2. A hole visiting the maximal cliques of $G - X$ (indicated by circles). Hatched circles show junctions. The vertices of $G - X$ are shown from left to right in a bicompatible elimination order. A maximal set of consecutive white cliques forms a tube. Tubes may contain vertices of junctions. For the right tube, we indicate the inner vertices.

To destroy all holes in an almost unit interval graph G, we show that each hole can be destroyed by deleting any of at most $14|X| - 1$ vertex sets, of which at least one vertex set can be assumed to be in an optimal unit interval vertex deletion set. This allows us to use a bounded search tree algorithm that, for each hole H in G, branches into $14|X| - 1$ possibilities to destroy H. Because at most $|X| - 1$ vertices may be deleted from G to transform it into a unit interval graph, the height of the corresponding search tree is bounded by $|X| - 1$. To find these $14|X| - 1$ vertex sets for each hole in G, we exploit that $G - X$ is a unit interval graph and thus allows for a linear-time computable *bicompatible elimination order* of its vertices [17]:

Definition 1. *Let $G = (V, E)$ be a graph. A total order $\preceq$ on V is a bicompatible elimination order for G if for each vertex $v \in V$, the sets $\{w \in N(v) \mid w \preceq v\}$ and $\{w \in N(v) \mid v \preceq w\}$ induce cliques in G.*

Without loss of generality, we assume that the vertices of a connected component of G form a segment of $\preceq$. We will see in Proposition 2 that, with respect to a bicompatible elimination order $\preceq$, $G - X$ forms a sequence of maximal cliques such that the vertices of each maximal clique are a segment of $\preceq$. Figure 2 illustrates this together with the following classification of the maximal cliques of $G-X$: a *junction* is a maximal clique in $G-X$ containing neighbors of vertices in X; a *tube* is a maximal set of maximal cliques of $G - X$ that are not junctions and whose vertices form a segment of $\preceq$. We say that a vertex is *contained in a tube* T if it is contained in a maximal clique of T. A hole *visits* a junction (or tube) if it contains a vertex of a junction (or tube). Vertices of a tube that are not in junctions are *inner vertices*.

Now, assume that there is a hole H in an almost unit interval graph G as illustrated in Figure 2. We show $14|X|-1$ possibilities to destroy H of which one is optimal. Each vertex of H in $G-X$ is contained in a junction or tube (or both). First, we show that H contains at most $12|X|$ vertices in junctions and that H contains inner vertices of at most $2|X| - 1$ tubes. Additionally, we show that there is an optimal unit interval vertex deletion set that contains a vertex of H in junctions or a polynomial-time computable vertex subset of one of the $2|X|-1$

tubes whose inner vertices are visited by H. Then, we solve UNIT INTERVAL VERTEX DELETION by repeatedly searching for a hole H in G in polynomial time and branching into the following $14|X| - 1$ possibilities to destroy H: delete one of the $12|X|$ vertices of H in junctions, or delete an optimal, polynomial-time determinable vertex subset of one of the $2|X| - 1$ tubes whose inner vertices are visited by H. Using this branching, the overall search tree size is $O((14|X| - 1)^{|X|-1})$, which results in the running time of Theorem 2. In the following, we show in detail the $14|X| - 1$ possibilities to destroy H of which one is optimal.

Bounding the Number of Vertices in Junctions. We now prove the following:

Lemma 1. *Let X be a unit interval vertex deletion set for an almost unit interval graph G. A hole in G contains at most $12|X|$ vertices from junctions in $G - X$.*

First, observe that a hole contains at most two vertices of a clique. We now exploit that G is an almost unit interval graph. In the following, we say that a vertex set can be covered by two cliques if it is the union of two vertex sets that induce cliques.

Proposition 1. *If a connected almost unit interval graph G contains a hole, then the neighborhood of each vertex in G can be covered by two cliques.*

Proof. If G contains a hole, then it must contain a hole with more than six vertices, since G is {claw, net, tent, C_4, C_5, C_6}-free. Thus, G contains an independent set of size three. We now exploit a result due to Fouquet [18]:

> In a connected claw-free graph containing an independent set of size three, every vertex v satisfies exactly one of the following properties:
> (i) $N(v)$ can be covered by two cliques or
> (ii) $N(v)$ contains an induced C_5.

Because G contains no induced C_5, the proposition follows immediately.   □

From Proposition 1, one can conclude that if an almost unit interval graph G contains a hole, then the neighborhood in $V \setminus X$ of a unit interval vertex deletion set X can be covered by $2|X|$ cliques.

We now prove that the maximal cliques of a unit interval graph form segments of a bicompatible elimination order and that vertices on induced paths occur in the same (or reverse) order as in a bicompatible elimination order.

Proposition 2. *Let $v_1 \preceq v_2 \preceq \ldots \preceq v_n$ be a bicompatible elimination order for a connected unit interval graph G.*
(1) If there is an induced path $P = (v_i, \ldots, v_k)$ with $v_i \preceq v_k$, then each vertex v_j on P satisfies $v_i \preceq v_j \preceq v_k$.
(2) If $v_i \preceq v_k$ and there is an edge between v_i and v_k, then the segment $[v_i, v_k]$ induces a clique in G. In particular, maximal cliques of G form segments.

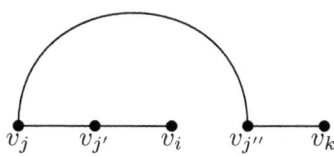

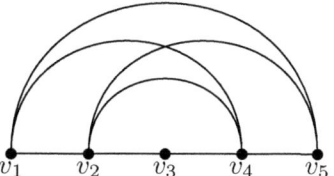

(a) The vertices $v_j \preceq v_{j'} \preceq v_{j''}$ form the induced path $(v_{j'}, v_j, v_{j''})$, in wrong order.

(b) A maximal clique induced by the non-consecutive vertices v_1, v_2, v_4, and v_5.

Fig. 3. The vertex orderings from left to right are not bicompatible elimination orders, as they violate Proposition 2

Proof. We prove the two statements independently. To show (1), for the purpose of contradiction, assume that there is an induced path $P = (v_i, \ldots, v_j, \ldots, v_k)$ with $v_j \preceq v_i \preceq v_k$ (the case $v_i \preceq v_k \preceq v_j$ can be proven analogously) such that v_j is the minimum vertex with respect to $\preceq$ that appears between v_i and v_k on P; this arrangement is illustrated in Figure 3a. Because there are induced subpaths of P from v_j to both v_i and v_k, the vertex v_j has two distinct neighbors $v_{j'}$ and $v_{j''}$ on P. Because v_j is the minimum vertex with respect to $\preceq$ that appears between v_i and v_k on P, it holds that $v_j \preceq v_{j'}$ and $v_j \preceq v_{j''}$. The vertices $v_{j'}$ and $v_{j''}$ are adjacent by Definition 1, because both are succeeding neighbors of v_j. This contradicts P being an *induced* path.

Before showing (2), we show that there is an edge between a vertex v_i and its direct successor v_{i+1} for $i \in \{1, \ldots, n-1\}$. Recall that G is connected by assumption. This implies that there is a shortest (and hence, induced) path from v_i to v_{i+1}. It follows from (1) that this path can neither contain a predecessor of v_i nor a successor of v_{i+1}. Because v_{i+1} directly succeeds v_i in $\preceq$, v_i and v_{i+1} are the only vertices on the shortest path from v_i to v_{i+1}, implying that they are adjacent.

We now show (2). Let $v_i \preceq v_k$ and assume that there is an edge between v_i and v_k. We have shown that v_i is also adjacent to its direct successor v_{i+1}. Because v_{i+1} and v_k are succeeding neighbors of v_i, the vertices v_i, v_{i+1}, and v_k form a clique by Definition 1. Inductively, it follows that all vertices v_j with $v_i \preceq v_j \preceq v_k$ are adjacent to v_k. Because all vertices v_j with $v_i \preceq v_j \preceq v_k$ are preceding neighbors of v_k in the bicompatible elimination order $\preceq$, these vertices must form a clique together with v_k. A maximal clique in G forms a segment because, with respect to $\preceq$, it contains an edge from its minimum vertex to its maximum vertex. □

To prove Lemma 1, we finally need the following definition (illustrated in Figure 4).

Definition 2. *Let G be a unit interval graph with a bicompatible elimination order $\preceq$ and let C be a clique of G. We define $\mathcal{S}(C)$ to be the set of vertices of all maximal cliques in G that contain vertices of C.*

Let c_{min} (and c_{max}) denote the minimum (or maximum, respectively) elements of $\mathcal{S}(C)$ with respect to $\preceq$. We define $\mathcal{S}_{min}(C)$ (and $\mathcal{S}_{max}(C)$) to be the vertex set

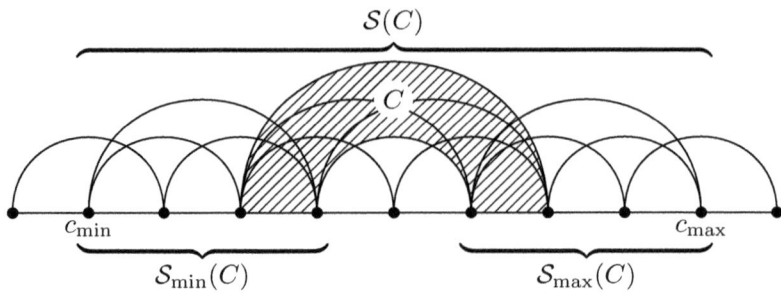

Fig. 4. Illustration for Definition 2. The vertices are shown from left to right in a bicompatible elimination order. The hatched clique is C.

of the (uniquely determined) maximal clique in G that contains c_{min} (or c_{max}, respectively) and some vertex from C.

Using Proposition 2, one can show that $\mathcal{S}(C)$ is the union of $\mathcal{S}_{min}(C)$, $\mathcal{S}_{max}(C)$, and the vertex set of any maximal clique containing C. Therefore, $\mathcal{S}(C) = [c_{min}, c_{max}]$. We have now collected the necessary observations to prove Lemma 1.

Proof (Proof of Lemma 1). Let X be a unit interval vertex deletion set for an almost unit interval graph G. Assume that G contains a hole. By Proposition 1, the neighborhood of X in $V \setminus X$ can be covered by a set $\mathcal{C}$ of at most $2|X|$ cliques. By Definition 2, the vertices of all junctions containing vertices of a clique $C \in \mathcal{C}$ are in $\mathcal{S}(C)$. We show that a hole contains at most six vertices from $\mathcal{S}(C)$ and, thus, in all junctions containing vertices of C. A hole contains at most two vertices in the vertex set C' of any maximal clique containing C, at most two vertices of the clique induced by $\mathcal{S}_{min}(C)$, and at most two vertices in the clique induced by $\mathcal{S}_{max}(C)$. Because $\mathcal{S}(C) = \mathcal{S}_{min}(C) \cup \mathcal{S}_{max}(C) \cup C'$, a hole contains at most six vertices from $\mathcal{S}(C)$. Finally, since $|\mathcal{C}| \leq 2|X|$ and for each $C \in \mathcal{C}$, a hole contains at most six vertices in junctions containing vertices of C, it follows that a hole contains at most $12|X|$ vertices in junctions of $G - X$. □

Finding Optimal Solutions in Tubes. We now show how to find optimal solutions in tubes. To this end, one can prove that a hole H visits inner vertices of at most $2|X| - 1$ tubes in $G - X$. Moreover, one can show that there is an optimal unit interval vertex deletion set containing at least one vertex of H in junctions or a polynomial-time computable vertex subset of one of the $2|X| - 1$ tubes.

Lemma 2. *Let X be a unit interval vertex deletion set for an almost unit interval graph G. A hole in G contains inner vertices of at most $2|X| - 1$ tubes with respect to a bicompatible elimination order $\preceq$ for $G - X$.*

To state the second result (Lemma 3), we need the following concepts.

Definition 3. *Let X be a unit interval vertex deletion set for a graph G and let T be a tube in $G - X$ with respect to a bicompatible elimination order $\preceq$ such that T contains inner vertices visited by a hole H.*

(1) *For two vertices v_i and v_k of H, we call (v_i, v_k) the T-boundary of H if, with respect to $\preceq$, v_i is in H the preceding neighbor of H's minimum inner vertex in T and if v_k is in H the succeeding neighbor of H's maximum inner vertex of T.*

(2) *For the T-boundary (v_i, v_k) of a hole H, we call a (minimum-cardinality) vertex-cut between v_i and v_k in $G - X$ a (minimum) H-T-cut.*

Lemma 3. *Let X be a unit interval vertex deletion set for a graph G. Let T be the set of tubes in $G - X$ with respect to a bicompatible elimination order $\preceq$ that contain inner vertices visited by a hole H.*

If a unit interval vertex deletion set S for G with $S \cap X = \emptyset$ does not contain vertices of H in junctions of $G - X$, then there is a unit interval vertex deletion set S' with $|S'| \le |S|$ and a tube $T \in T$ for which S' contains a minimum H-T-cut.

Lemma 3 can be shown exploiting Proposition 2(1), which implies that if a hole enters a tube at one side, it must leave the tube at the opposite side. Using this fact, one can show the following two claims, which together imply Lemma 3.

Claim. A unit interval vertex deletion set S for G with $S \cap X = \emptyset$ contains vertices of H in junctions or contains an H-T-cut for some tube $T \in T$.

Claim. For a vertex set S containing a H-T-cut for a tube $T \in T$, no hole in $G - S$ contains vertices of the segment $[v_i, v_k]$, where (v_i, v_k) is the T-boundary of H.

The Algorithm

Combining Lemmas 1, 2 and 3, we finally present the algorithm for DISJOINT UNIT INTERVAL VERTEX DELETION, thus proving Theorem 2. The algorithm employs the following branching rule:

Branching Rule 1. If $G - S$ contains a forbidden induced subgraph induced by a vertex set F with $|F| \le 6$, then branch into all possibilities of adding a vertex $v \in F \setminus X$ to S.

Proof (Proof of Theorem 2). Given a graph G and a unit interval vertex deletion set X for G, we search for a unit interval vertex deletion set S for G with $|S| < |X|$ and $S \cap X = \emptyset$. We start with $S := \emptyset$ and apply Branching Rule 1 as long as $|S| < |X|$ to destroy forbidden induced subgraphs with at most six vertices. Because each such forbidden induced subgraph contains one vertex in the unit interval vertex deletion set X, we find such a forbidden induced subgraph in $O(|X| n^5)$ time and branch into at most five cases to add one of its vertices to S.

If $|S| \ge |X|$ and Branching Rule 1 is still applicable, return "no" because S is not extensible to a unit interval vertex deletion set for G that is smaller than X and disjoint from X. Otherwise, proceed as follows: compute a bicompatible elimination order $\preceq$ for $G - (S \cup X)$. This works in linear time [17] because $G - (S \cup X)$ is a unit interval graph. From this bicompatible elimination order $\preceq$, a set $\mathcal{C}$ of all maximal cliques of $G-(S \cup X)$ can easily be computed in $O(n^2)$ time by finding for each vertex v in $G-(S \cup X)$ its last neighbor with respect to $\preceq$.

Now, we find junctions and tubes. For each clique $C \in \mathcal{C}$, check whether C has neighbors in X. If this is the case, which can be checked in $O(kn^2)$ time for all $C \in \mathcal{C}$, then C is a junction. To find tubes, sort the set $\mathcal{C}$ such that C_1 occurs before C_2 if, in $\preceq$, the minimum vertex of C_1 occurs before the minimum vertex of C_2. Because a unit interval graph has at most n maximal cliques, this is possible in $O(n \log n)$ time. From the sorted set $\mathcal{C}$, compute a set $\mathcal{T}$ of all tubes in $G-(S \cup X)$ in $O(n)$ time: repeatedly find the first clique C in $\mathcal{C}$ that is not a junction and not yet part of a tube and add C and all succeeding cliques in $\mathcal{C}$ to a new tube T until a junction is encountered.

Next, as long as $|S| < |X|$, repeatedly find a hole H in $G-S$ and add at least one vertex of H to S as follows: because $G-S$ is an almost unit interval graph and $G-(S \cup X)$ is a unit interval graph, recursively branch into at most $12|X|$ possibilities to choose a vertex of H from a junction for inclusion in S (Lemma 1) and into at most $2|X| - 1$ possibilities to choose tube $T \in \mathcal{T}$ for which a H-T-cut shall be included in S (Lemma 2, Lemma 3).

In each search tree node, we branch into at most $14|X| - 1$ cases (at most five cases for Branching Rule 1 and at most $14|X| - 1$ for a hole in $G-S$). In each case, at least one vertex is added to S. As a result, the corresponding search tree has depth at most $|X| - 1$ and thus at most $(14|X| - 1)^{|X|-1}$ nodes.

To analyze the running time for processing each node, it remains to analyze the running time for finding holes and minimum H-T-cuts. A hole in $G-S$ can be found in $O(|X|(n+m))$ time by breadth-first search starting at each vertex in X. A minimum H-T-cut is computable in $O(\sqrt{n}m)$ time [19, Theorem 9.8]. ☐

5 Conclusion

It remains open to study the existence of a polynomial-size problem kernel [20,21] for UNIT INTERVAL VERTEX DELETION. Another task for future study is to search for polynomial-time algorithms with low-degree polynomials in case of constant k. Villanger's algorithm [12] runs in $O(6^k kn^6)$ time and thus also is far from this goal. We remark that already Marx [11] asked for the study of the parameterized complexity of the related INTERVAL VERTEX DELETION problem, remaining a challenge for future research. In general, interval graphs do not allow for bicompatible elimination orders; as our algorithm heavily relies on them, it is not straightforward to extend it to INTERVAL VERTEX DELETION.

Acknowledgments. We thank anonymous referees for their constructive feedback.

References

1. Luce, R.D.: Semiorders and a theory of utility discrimination. Econometrica 24, 178–191 (1956)
2. Aleskerov, F., Bouyssou, D., Monjardet, B.: Utility Maximization, Choice and Preference. Studies in Economic Theory, vol. 16. Springer, Heidelberg (2007)

3. Roberts, F.S.: Indifference graphs. In: Proof Techniques in Graph Theory, pp. 139–146. Academic Press, New York (1969)
4. Roberts, F.S.: Indifference and seriation. Annals of the New York Academy of Sciences 328, 173–182 (1979)
5. Brandstädt, A., Le, V.B., Spinrad, J.P.: Graph classes: a survey. SIAM, Philadelphia (1999)
6. Lewis, J.M., Yannakakis, M.: The node-deletion problem for hereditary properties is NP-complete. J. Comput. System Sci. 20, 219–230 (1980)
7. Chepoi, V., Fichet, B., Seston, M.: Seriation in the presence of errors: NP-hardness of l_∞-fitting Robinson structures to dissimilarity matrices. J. Classification 26, 279–296 (2010)
8. Chepoi, V., Seston, M.: Seriation in the presence of errors: A factor 16 approximation algorithm for l_∞-fitting Robinson structures to distances. Algorithmica (2009); Available electronically
9. Van Bevern, R.: The Computational Hardness and Tractability of Restricted Seriation Problems on Inaccurate Data. Diplomarbeit. Institut für Informatik, Friedrich-Schiller-Universität, Jena, Germany (2010)
10. Dom, M., Guo, J., Niedermeier, R.: Approximation and fixed-parameter algorithms for consecutive ones submatrix problems. J.Comput. System Sci. 76, 204–221 (2010)
11. Marx, D.: Chordal deletion is fixed-parameter tractable. Algorithmica 57, 747–768 (2010)
12. Villanger, Y.: Proper interval vertex deletion. In: Proc. 5th IPEC. LNCS, Springer, Heidelberg (December 2010)
13. Reed, B., Smith, K., Vetta, A.: Finding odd cycle transversals. Oper. Res. Lett. 32, 299–301 (2004)
14. Guo, J., Moser, H., Niedermeier, R.: Iterative compression for exactly solving NP-hard minimization problems. In: Lerner, J., Wagner, D., Zweig, K.A. (eds.) Algorithmics of Large and Complex Networks. LNCS, vol. 5515, pp. 65–80. Springer, Heidelberg (2009)
15. Garey, M.R., Johnson, D.S., Stockmeyer, L.J.: Some simplified NP-complete graph problems. Theor. Comp. Sci. 1, 237–267 (1976)
16. Fellows, M.R., Guo, J., Moser, H., Niedermeier, R.: A complexity dichotomy for finding disjoint solutions of vertex deletion problems. In: Královič, R., Niwiński, D. (eds.) MFCS 2009. LNCS, vol. 5734, pp. 319–330. Springer, Heidelberg (2009)
17. Panda, B.S., Das, S.K.: A linear time recognition algorithm for proper interval graphs. Inf. Process. Lett. 87, 153–161 (2003)
18. Fouquet, J.L.: A strengthening of Ben Rebea's lemma. J. Combin. Theory Ser. B 59, 35–40 (1993)
19. Schrijver, A.: Combinatorial Optimization: Polyhedra and Efficiency, vol. A. Springer, Heidelberg (2003)
20. Bodlaender, H.L.: Kernelization: New upper and lower bound techniques. In: Chen, J., Fomin, F.V. (eds.) IWPEC 2008. LNCS, vol. 5917, pp. 17–37. Springer, Heidelberg (2009)
21. Guo, J., Niedermeier, R.: Invitation to data reduction and problem kernelization. ACM SIGACT News 38, 31–45 (2007)

Parameterized Complexity of the Arc-Preserving Subsequence Problem⋆

Dániel Marx[1] and Ildikó Schlotter[2]

[1] Tel Aviv University, Israel
[2] Budapest University of Technology and Economics, Hungary
{dmarx,ildi}@cs.bme.hu

Abstract. We study the ARC-PRESERVING SUBSEQUENCE (APS) problem with unlimited annotations. Given two arc-annotated sequences P and T, this problem asks if it is possible to delete characters from T to obtain P. Since even the unary version of APS is NP-hard, we used the framework of parameterized complexity, focusing on a parameterization of this problem where the parameter is the number of deletions we can make. We present a linear-time FPT algorithm for a generalization of APS, applying techniques originally designed to give an FPT algorithm for INDUCED SUBGRAPH ISOMORPHISM on interval graphs [12].

1 Introduction

Many important problems in computational biology are related to pattern matching in strings, since DNA, RNA, or protein molecules can be viewed as sequences of nucleotides or amino acids. To gain information about such molecules, we often need to compare two sequences and measure their similarity.

Given two sequences S_1 and S_2 over some alphabet, the task of the LONGEST COMMON SUBSEQUENCE (LCS) problem is to find the longest possible sequence that is the subsequence of both S_1 and S_2. In other words, we are looking for a sequence C that can be obtained both from S_1 and from S_2 by deleting characters. This problem arises in many applications, like deciding if two species are biologically related, or whether two proteins are likely to exhibit similar functionalities related to three-dimensional structure (protein folding). Another classical problem, SUBSEQUENCE, asks if a sequence is the subsequence of another.

If we only want to deal with character sequences, LCS can be solved efficiently using dynamic programming. However, recent biological research suggests that we might loose relevant information if we model DNA, RNA, or protein molecules simply as sequences. The reason for this is that in such molecules, the shape and hence the functionality is greatly affected by chemical bonds between elements that might be far apart from each other in the sequence. *Arc-annotated sequences* are widely used to represent such bonds. In this model, any two elements (or *bases*) of a sequence can be connected to each other through an *arc*.

⋆ Supported by ERC Advanced Grant DMMCA and by the Hungarian National Research Fund OTKA 67651.

For two arc-annotated sequences S_1 and S_2, the LONGEST ARC-PRESERVING COMMON SUBSEQUENCE or LAPCS asks for an arc-annotated sequence C of maximum length that can be obtained both from S_1 and from S_2 by deleting bases together with all arcs incident to them. Since LAPCS is NP-complete even if the arc structures are highly restricted [5,6,10], researchers focused on polynomial-time solvable cases and approximation algorithms [5,10,9,11].

Another direction of research is to use the parameterized complexity framework [4,7]. This area deals with NP-hard problems by giving algorithms that have an acceptable running time on many relevant instances. An algorithm is *fixed-parameter tractable* (FPT) if its running time is bounded by $f(k)n^{O(1)}$ for some function f, where n is the input size and k is the *parameter* associated with the input. The idea behind this definition is that the running time of an FPT algorithm remains tractable provided that the parameter has small value.

Parameterized complexity of LAPCS has already been studied, and FPT algorithms were presented for various parameterizations [1,6]. An interesting parameterization is where the parameter is the number of deletions we are allowed to make in order to construct the common subsequence. This models a situation where we compare two sequences which are similar. An FPT algorithm was given in [1] with this parameter, but it only applies for a restricted case.

Unlike most previous results, we considered unlimited annotations where any two bases of a sequence can be connected by arcs. Instead of concentrating on LAPCS, we dealt with the more simple ARC-PRESERVING SUBSEQUENCE problem (APS), the annotated analog of SUBSEQUENCE. Given two arc-annotated sequences P and T, the task of APS is to find out whether the pattern sequence P can be obtained by deleting some bases of the target sequence T, together with all the arcs incident to them. We remark that APS on its own is an interesting problem in computation biology, and has been widely studied in the literature. Its NP-hardness has been proved for numerous restricted cases [2], and polynomial-time algorithms have been presented [8,3] for limited arc structures.

Here, we present an FPT algorithm for the unlimited APS, where the parameter is the number k of deletions allowed. Our algorithm runs in $f(k)n$ time for some function f depending only on k, where n is the input size. In fact, we solve a generalization of APS where a few arcs can be deleted additionally. We mention that APS is W[1]-hard if the parameter is the length of the pattern [5].

The ideas and techniques applied here originate from an FPT algorithm solving a seemingly unrelated problem on interval graphs [12]. This algorithm answers the INDUCED SUBGRAPH ISOMORPHISM in FPT time: given two interval graphs G and H and a parameter k, is it possible to delete k vertices from G to obtains a graph isomorphic to H? Our work shows that research connected to interval graphs can be useful for arc-annotated sequences as well.

2   Problem Definition and Notation

We denote $\{1, \ldots, n\}$ by $[n]$. We refer to the elements of a sequence S over an alphabet Σ as *bases*. The i-th base of S is $S[i]$, and the length of S is $|S|$.

Let S_P and S_T be two sequences over Σ. Let $|S_P| = n_P$ and $|S_T| = n_T$, assume $n_P \leq n_T$. We say that S_P is a *subsequence* of S_T if S_P can be obtained by deleting bases from S_T, or equivalently, if there is a bijective mapping φ from $[n_P]$ into a subset of $[n_T]$ such that $\varphi(i_1) < \varphi(i_2)$ for each $1 \leq i_1 < i_2 \leq n_P$, and $S_P[i] = S_T[\varphi(i)]$ for each $i \in [n_P]$. We call such a φ an *alignment* of $(S_P; S_T)$. We write $S^{\mathrm{del}}(\varphi)$ to denote the set of bases that have to be deleted from S_T according to φ, i.e. $S^{\mathrm{del}}(\varphi) = [n_T] \setminus \bigcup_{i \in [n_P]} \varphi(i)$.

An *arc-annotation* A of a sequence S of length n is a multiset of pairs of integers from $[n]$, where each pair $(i_1, i_2) \in A$ satisfies $i_1 < i_2$. An *arc-annotated sequence* (S, A) is a sequence S together with an arc-annotation A for S. We say that an *arc* (i_1, i_2) *starts* at i_1, *ends* at i_2, and *connects* the positions i_1 and i_2 *incident* to it. We write $A(i_1, i_2)$ for the multiplicity of the pair (i_1, i_2) in A, and we write $A^+(i)$ and $A^-(i)$ for the set of arcs starting or ending at i, respectively. Also, we let a^{start} and a^{end} to denote the starting and ending position of an arc a. We use $|(S, A)|$ to denote the *size* of (S, A) in binary encoding.

Given two arc-annotated sequences (S_P, A_P) and (S_T, A_T), we say that (S_P, A_P) is an *arc-preserving subsequence* of (S_T, A_T) if it can be obtained from (S_T, A_T) by deleting bases from it, i.e. there is an alignment φ of $(S_P; S_T)$ such that $A_P(i, j) = A_T(\varphi(i), \varphi(j))$ for any $1 \leq i < j \leq |S_P|$. Such an alignment is an *arc-preserving alignment* of $(S_P, A_P; S_T, A_T)$. Note that by deleting a base, we also mean the deletion of the arcs incident to it. Given two arc-annotated sequence P and T, the ARC-PRESERVING SUBSEQUENCE problem (APS) asks whether P is an arc-preserving subsequence of T.

We will deal with the following generalization of APS, which we call ALMOST APS or AAPS: given two arc-annotated sequences (S_P, A_P) and (S_T, A_T) and some $k_a \in \mathbb{Z}$, we ask if we can delete some bases from S_T (together with their incident arcs) and at most k_a arcs *in addition* to obtain (S_P, A_P). Formally, we have to decide if there is a set A^{del} of at most k_a arcs in A_T such that (S_P, A_P) is an arc-preserving subsequence of $(S_T, A_T \setminus A^{\mathrm{del}})$. We call φ a k_a-*alignment* for $(S_P, A_P; S_T, A_T)$ if φ is an arc-preserving alignment of $(S_P, A_P; S_T, A_T \setminus A^*)$ for some set A^* with $|A^*| \leq k_a$. Also, we let $A^{\mathrm{del}}(\varphi)$ to denote such an A^*.

Given a sequence S, let S^{rev} denote the reverse of S. For a position i of S, we will use i^{rev} to denote the position $|S| - i + 1$ of S^{rev} corresponding to i. If A is an arc-annotation of S, then let A^{rev} denote the corresponding arc-annotation of S^{rev}, meaning $A^{\mathrm{rev}}(i_1, i_2) = A(i_2^{\mathrm{rev}}, i_1^{\mathrm{rev}})$. We also let $X^{\mathrm{rev}} = \{i^{\mathrm{rev}} \mid i \in X\}$ for any set X of positions in S.

If φ is a k_a-alignment for $(S_P, A_P; S_T, A_T)$, then φ^{rev} is the corresponding k_a-alignment for $(S_P^{\mathrm{rev}}, A_P^{\mathrm{rev}}; S_T^{\mathrm{rev}}, A_T^{\mathrm{rev}})$, i.e. $\varphi^{\mathrm{rev}}(i) = (\varphi(i^{\mathrm{rev}}))^{\mathrm{rev}}$ for each i.

Due to lack of space, we omit several proofs, see the full paper for them.

3   Fixed-Parameter Tractability of APS

In this section we present an FPT algorithm for AAPS, a generalization of APS, with the parameterization where the parameters are the number of bases to delete and the number of arcs that can be deleted additionally.

ALMOST ARC-PRESERVING SUBSEQUENCE

Input: Two arc-annotated sequences (S_P, A_P) and (S_T, A_T), and $k_a \in \mathbb{Z}$.

Parameters: k_a and $k_b = |S_T| - |S_P|$.

Task: decide whether (S_P, A_P) can be obtained from (S_T, A_T) by deleting k_b bases (together with their incident arcs) and k_a arcs in addition, i.e. whether there is a k_a-alignment φ for $(S_P, A_P; S_T, A_T)$.

Our aim is to prove the main result of the paper stated by Theorem 1.

Theorem 1. *There is an algorithm that solves any instance $(S_P, A_P; S_T, A_T; k_a)$ of the* ALMOST ARC-PRESERVING SUBSEQUENCE *problem and runs in time $k_b^{O(k_b^3 + k_b k_a)}|(S_T, A_T)|$ where $k_b = |S_T| - |S_P|$.*

3.1 Outline of the Algorithm

To prove Theorem 1, we present an algorithm that uses a bounded search tree technique in order to construct a k_a-alignment step by step. In certain situations, the algorithm might branch on a bounded number of possibilities to proceed with. Since both the number of such branchings and the possible directions of a branching will be bounded in terms of k_a and k_b, the size of the resulting search tree will be bounded by a function of k_a and k_b.

Actually, the algorithm described here has the following behavior: given an instance of APS, consisting of the arc-annotated sequences (S_P, A_P) and (S_T, A_T), and an integer k_a, it tries to construct a k_a-alignment φ for $(S_P, A_P; S_T, A_T)$. To do so, it fixes such a hypothetical solution φ, and looks for bases in $S^{\mathrm{del}}(\varphi)$ and arcs in $A^{\mathrm{del}}(\varphi)$, which we will call *removable bases* and *removable arcs* of φ, resp. More precisely, our algorithm does one of the followings in linear time:

- it produces an **arc-preserving alignment** ψ for $(S_P, A_P; S_T, A_T)$ (note that ψ is a k_a-alignment for $(S_P, A_P; S_T, A_T)$ as well),
- it correctly **rejects** the instance, or
- it produces a **removable base** or a **removable arc** of φ.

In the last case, we can delete the given base or arc, and apply the algorithm to the obtained instance. Notice that one of the parameters k_a and $k_b = |S_T| - |S_P|$ is decreased in the new instance. The presented algorithm will be shown to run in $f(k_a, k_b)|(S_T, A_T)|$ time for some functions f, which therefore implies Theorem 1 by proving that AAPS can be solved in $(k_a + k_b)f(k_a, k_b)|(S_T, A_T)|$ time.

Our algorithm might branch several times before producing an output as described above. Each such branch will be caused by guessing the answer to a question of the following form: given some position p in S_P, what is the value of the position $\varphi(p)$?[1] We interpret these branchings in the usual framework of bounded search trees: a branching happens when we do not know the exact value of a certain variable (such as the value of $\varphi(p)$ in the above example), and thus we have to investigate every possible value. A certain branch examines

---

[1] In a few cases we will also need some additional branchings, described later on.

one possible value of the variable, and it produces a correct output *if* the given variable indeed has the value associated with this branch. Since the examined cases always cover every possibilities, this implies that the output will be correct in at least one of the branches.

Although our algorithm seems to be a straightforward application of the bounded search tree methodology used frequently in parameterized algorithms, we had to overcome many difficulties to avoid any possibility of using an unbounded number of such guesses. The presented algorithm will apply considerably sophisticated methods to keep the search tree bounded.

3.2 Fragmentations and Related Concepts

Fragmentation. To describe our knowledge of the partially constructed k_a-alignment we have, we introduce a data structure called *fragmentation*. By iteratively refining the fragmentation, we can get closer and closer to actually determine a k_a-alignment. We write $|S_P| = n_P$ and $|S_T| = n_T$.

Recall that φ is a fixed k_a-alignment for $(S_P, A_P; S_T, A_T)$. For some $1 \leq i_1 \leq i_2 \leq n_P$, we define the *block* $[i_1, i_2]$ in S_P to be the set of positions $i_1, i_1+1, \ldots, i_2$, and we define blocks in S_T similarly. Given a set of f disjoint blocks $\{[p_1^h, p_2^h] \mid h \in [f]\}$ in S_P and a set of f disjoint blocks $\{[t_1^h, t_2^h] \mid h \in [f]\}$ in S_T, we let $F_h = ([p_1^h, p_2^h], [t_1^h, t_2^h])$. We say that $\{F_h \mid h \in [f]\}$ is a *fragmentation* for φ, if

- $t_1^h \leq \varphi(p_1^h)$ and $\varphi(p_2^h) \leq t_2^h$ for each $h \in [f]$, and
- $p_1^{h+1} = p_2^h + 1$ and $t_1^{h+1} = t_2^h + 1$ for each $h \in [f-1]$.

We will call the element F_h for some $h \in [f]$ a *fragment*. We define $\sigma(F_h) = (t_2^h - t_1^h) - (p_2^h - p_1^h)$ and $\delta(F_h) = t_1^h - p_1^h$, which are both clearly non-negative integers. Note that $\delta(F_{h+1}) = \delta(F_h) + \sigma(F_h)$ holds for each $h \in [f-1]$. We say that a position $i \in [n_P]$ of S_P is *contained* in the fragment F_h, if $p_1^h \leq i \leq p_2^h$.

We will say that a fragment F is *trivial* if $\sigma(F)$ is zero, and *non-trivial* otherwise. We also call a position of S_P trivial (or non-trivial) in a fragmentation, if the fragment containing it is trivial (or non-trivial, resp). Given fragmentation for φ and a position i in S_P, we will use the notation $i_{\text{left}} = i + \delta(F)$ and $i_{\text{right}} = i + \delta(F) + \sigma(F)$, where F is the fragment containing i. Observe that

$$i_{\text{left}} \leq \varphi(i) \leq i_{\text{right}}$$

always holds. We will classify a position i of S_P as follows:

- If $\varphi(i) = i_{\text{left}}$, then i is *left-aligned*.
- If $\varphi(i) = i_{\text{right}}$, then i is *right-aligned*.
- If $\varphi(i) = j$ such that $i_{\text{left}} < j < i_{\text{right}}$, then i is *skew*.

If i is trivial, then only $\varphi(i) = i_{\text{left}} = i_{\text{right}}$ is possible. Thus, each trivial position must be both left- and right-aligned.

Notice that each fragment F must contain exactly $\sigma(F)$ positions that are contained in $S^{\text{del}}(\varphi)$. This implies the following bounds.

Proposition 2. *If $\mathcal{F}$ is a fragmentation for φ, then $\sum_{F \in \mathcal{F}} \sigma(F) = k_b$. In particular, $\mathcal{F}$ can have at most k_b non-trivial fragments.*

A *marked fragmentation* for φ is a pair $(\mathcal{F}, M)$ formed by a fragmentation $\mathcal{F}$ for φ and a set M of positions in S_P such that each $m \in M$ is a trivial position in $\mathcal{F}$. We say that the trivial positions contained in M are *marked*.

For a fragment $F = ([p_1, p_2], [t_1, t_2])$ we let $F^{\mathrm{rev}} = ([p_2^{\mathrm{rev}}, p_1^{\mathrm{rev}}], [t_2^{\mathrm{rev}}, t_1^{\mathrm{rev}}])$, hence a fragmentation $\mathcal{F}$ for φ clearly yields a fragmentation $\mathcal{F}^{\mathrm{rev}} = \{F^{\mathrm{rev}} | F \in \mathcal{F}\}$ for φ^{rev} as well. Note that if a position i of S_P is left-aligned (right-aligned) in $\mathcal{F}$, then the position i^{rev} is right-aligned (left-aligned, resp.) in $\mathcal{F}^{\mathrm{rev}}$.

Pairing arcs. Given a position i in S_P, let us order the arcs c in $A_P^+(i)$ increasingly according to their right endpoint c^{end}. Similarly, we order the arcs in $A_P^-(i)$ increasingly according their left endpoint. In both cases, we break ties arbitrarily. Also, we order the arcs in $A_T^+(j)$ and $A_T^-(j)$ in the same way for each position j in S_T. Now, we "pair" arcs in $A_P^+(i)$ with arcs in $A_T^+(i_{\mathrm{left}})$, and also arcs in $A_P^-(i)$ with arcs in $A_T^-(i_{\mathrm{left}})$ according to their ranking in this ordering. To this end, we construct the sets $R_{\mathrm{left}}^+(i) \subseteq A_P^+(i) \times A_T^+(i_{\mathrm{left}})$ and $R_{\mathrm{left}}^-(i) \subseteq A_P^-(i) \times A_T^-(i_{\mathrm{left}})$ in the following way. We put a pair (c, d) into $R_{\mathrm{left}}^+(i)$, if $c \in A_P^+(i)$, $d \in A_T^+(i_{\mathrm{left}})$, and c has the same rank (according to the above ordering) in $A_P^+(i)$ as the rank of d in $A_T^+(i_{\mathrm{left}})$. Similarly, we put a pair (c, d) into $R_{\mathrm{left}}^-(i)$, if $c \in A_P^-(i)$, $d \in A_T^-(i_{\mathrm{left}})$, and c has the same rank in $A_P^-(i)$ as the rank of d in $A_T^-(i_{\mathrm{left}})$. In addition, we define the sets $R_{\mathrm{right}}^+(i)$ and $R_{\mathrm{right}}^-(i)$ analogously, by substituting i_{right} for i_{left} in the above definitions. The key properties of these sets are summarized below.

Lemma 3. *We know $\varphi(c^{\mathrm{end}}) = d^{\mathrm{end}}$ and $\varphi(c^{\mathrm{start}}) = d^{\mathrm{start}}$ in the following cases:*

(1) If $(c, d) \in R_{\mathrm{left}}^+(i)$ and $|A_P^+(i)| = |A_T^+(i_{\mathrm{left}})|$ for some left-aligned i.
(2) If $(c, d) \in R_{\mathrm{left}}^-(i)$ and $|A_P^-(i)| = |A_T^-(i_{\mathrm{left}})|$ for some left-aligned i.
(3) If $(c, d) \in R_{\mathrm{right}}^+(i)$ and $|A_P^+(i)| = |A_T^+(i_{\mathrm{right}})|$ for some right-aligned i.
(4) If $(c, d) \in R_{\mathrm{right}}^-(i)$ and $|A_P^-(i)| = |A_T^-(i_{\mathrm{right}})|$ for some right-aligned i.

Arcs connecting two non-trivial fragments. Given two non-trivial fragments F and H of a fragmentation with F preceding H, we define three disjoint subsets of those arcs of A_P that start in a position of F and end in a position of H. These sets will be denoted by $\mathcal{L}(F, H)$, $\mathcal{R}(F, H)$, and $\mathcal{X}(F, H)$, and we construct them as follows. Suppose that $c = (f, h) \in A_P$ for some f and h contained in F and H, respectively. We put c in exactly one of these three sets, if $(c, d) \in R_{\mathrm{left}}^-(h)$ for some arc $d \in A_T$ such that $f_{\mathrm{left}} \leq d^{\mathrm{start}} \leq f_{\mathrm{right}}$. If $d^{\mathrm{start}} = f_{\mathrm{left}}$ then we put c into $\mathcal{L}(F, H)$, if $d^{\mathrm{start}} = f_{\mathrm{right}}$ then we put c into $\mathcal{R}(F, H)$, and if $f_{\mathrm{left}} < d^{\mathrm{start}} < f_{\mathrm{right}}$ then we put c into $\mathcal{X}(F, H)$.

By Lemma 3, if the positions in H are left-aligned, then the left endpoints of the arcs in $\mathcal{R}(F, H)$ must be right-aligned. Similarly, the left endpoints of the arcs in $\mathcal{X}(F, H)$ must be skew in such a case. Proposition 4 states these observations in a precise manner. Since we would like to ensure each position to be left-aligned, we will try to get rid of the arcs in $\mathcal{R}(F, H)$ and $\mathcal{X}(F, H)$.

Proposition 4. *Let i be left-aligned, $|A_P^-(i)| = |A_T^-(i_{left})|$, and $c \in A_P^-(i)$.*
(1) If $c \in \mathcal{L}(F, H)$, then c^{start} is left-aligned.
(2) If $c \in \mathcal{R}(F, H)$, then c^{start} is right-aligned.
(3) If $c \in \mathcal{X}(F, H)$, then c^{start} is skew.

We say that two positions $f_1, f_2 \in [n_P]$ are *conflicting* for (F, H), if $f_1 \leq f_2$, $A_P^+(f_1) \cap \mathcal{R}(F, H) \neq \emptyset$ and $A_P^+(f_2) \cap \mathcal{L}(F, H) \neq \emptyset$. In such a case, we say that any $h \geq \max\{h_1, h_2\}$ in H is *conflict-inducing* for (F, H) (and for the conflicting pair (f_1, f_2)), where h_1 denotes the minimal position for which $(f_1, h_1) \in \mathcal{R}(F, H)$, and h_2 denotes the minimal position for which $(f_2, h_2) \in \mathcal{L}(F, H)$. Notice that if such a conflict-inducing h is left-aligned, then both h_1 and h_2 are left-aligned. By Proposition 4, this implies that f_1 is right-aligned and f_2 is left-aligned. But since f_1 precedes f_2, this cannot happen. This implies the following observation.

Proposition 5. *If a position h is conflict-inducing for (F, H) in a given fragmentation, then h cannot be left-aligned.*

In addition, if $\mathcal{L}(F, H) \neq \emptyset$, then let $L^{max}(F, H)$ denote the largest position f in F for which $A_P^+(f) \cap \mathcal{L}(F, H) \neq \emptyset$. Let the *L-critical position* for (F, H) be the smallest position h contained in H for which $(L^{max}(F, H), h) \in \mathcal{L}(F, H)$. Similarly, if $\mathcal{R}(F, H) \neq \emptyset$, then let $R^{min}(F, H)$ denote the smallest position f in F for which $A_P^+(f) \cap \mathcal{R}(F, H) \neq \emptyset$. Also, let the *R-critical position* for (F, H) be the smallest position h in H for which $(R^{min}(F, H), h) \in \mathcal{R}(F, H)$.

Now, a position h in H is *LR-critical* for (F, H), if either h is the R-critical position for (F, H) and $\mathcal{L}(F, H) = \emptyset$, or $h = \max\{h_L, h_R\}$ where h_L is the L-critical and h_R is the R-critical position for (F, H). Note that both cases require $\mathcal{R}(F, H) \neq \emptyset$. Moreover, H contains an LR-critical position for (F, H), if and only if $\mathcal{R}(F, H) \neq \emptyset$. Intuitively, if an LR-critical position in H is left-aligned, then this implies that some position in F is right-aligned.

Note that the definitions of the sets $\mathcal{L}(F, H), \mathcal{R}(F, H)$, and $\mathcal{X}(F, H)$ together with the definitions connected to them as described above depend on the given fragmentation, so whenever the fragmentation changes, these must be adjusted appropriately as well. (In particular, arcs in $\mathcal{L}(F, H), \mathcal{R}(F, H)$, and $\mathcal{X}(F, H)$ must start and end in two different non-trivial fragments.)

Properties 1-9. Let $(\mathcal{F}, M)$ be a marked fragmentation for φ. Our aim is to ensure that the properties given below hold for each position in S_P. Intuitively, these properties mirror the expectation that every position should be left-aligned. Note that although we cannot decide whether $(\mathcal{F}, M)$ is a correct marked fragmentation without knowing the k_a-alignment φ, we are able to check whether these properties hold for some position i in $(\mathcal{F}, M)$.

Property 1: $S_P[i] = S_T[i_{left}]$.
Property 2: If i is non-trivial, then $|A_P^+(i)| = |A_T^+(i_{left})|$ and $|A_P^-(i)| = |A_T^-(i_{left})|$.
Property 3: If i is non-trivial, then $A_P(y, i) = A_T(y_{left}, i_{left})$ for any $y < i$ contained in the same fragment as i.
Property 4: If i is non-trivial, then for every $(c, d) \in R_{left}^+(i)$ such that $c^{end} = y$ is non-trivial, $y_{left} \leq d^{end} \leq y_{right}$ holds. Also, for every $(c, d) \in R_{left}^-(i)$ such that $c^{start} = y$ is non-trivial, $y_{left} \leq d^{start} \leq y_{right}$ holds.

Property 5: No arc in $\mathcal{X}(F, H)$ for some (F, H) ends at i.

Property 6: i is not conflict-inducing for any (F, H).

Property 7: i is not LR-critical for any (F, H).

Property 8: If i is non-trivial, then for every $(c, d) \in R_{\text{left}}^+(i)$ such that $c^{\text{end}} = y$ is non-trivial, $d^{\text{end}} = y_{\text{left}}$ holds. Also, for every $(c, d) \in R_{\text{left}}^-(i)$ such that $c^{\text{start}} = y$ is non-trivial, $d^{\text{start}} = y_{\text{left}}$ holds.

Property 9: If i is non-trivial, then for each marked position $m \in M$, $A_P(i, m) = A_T(i_{\text{left}}, m_{\text{left}})$ holds if $m > i$, and $A_P(m, i) = A_T(m_{\text{left}}, i_{\text{left}})$ holds if $m < i$.

Observe that each of these properties depend on the fragmentation $\mathcal{F}$, and Property 9 depends on the set of marked positions M as well. Also, if some property holds for a position i in $(\mathcal{F}, M)$, then this does not imply that the property holds for i^{rev} in $(\mathcal{F}^{\text{rev}}, M^{\text{rev}})$, as most of these properties are not symmetric. For example, i_{left} and i_{right} both have a different meaning in the fragmentation $\mathcal{F}$ and in $\mathcal{F}^{\text{rev}}$. We say that a position $i \in [n_P]$ *violates* Property ℓ ($1 \le \ell \le 9$) in a marked fragmentation $(\mathcal{F}, M)$, if Property ℓ does not hold for i in $(\mathcal{F}, M)$.

If the first eight properties hold for each position both in $(\mathcal{F}, M)$ and in $(\mathcal{F}^{\text{rev}}, M^{\text{rev}})$, then we say that $(\mathcal{F}, M)$ is *8-proper*. We say that $(\mathcal{F}, M)$ is *proper*, if it is 8-proper and Property 9 holds hold for each position of S_P in $(\mathcal{F}, M)$. Note that we do not care whether Property 9 holds for the positions in the reversed instance, so $(\mathcal{F}, M)$ is proper even if Property 9 does not hold in $(\mathcal{F}^{\text{rev}}, M^{\text{rev}})$.

3.3   Description of the Algorithm

We start with a marked fragmentation where $M = \emptyset$ and the fragmentation contains only the unique fragment $([1, n_P], [1, n_T])$, which is non-trivial if $k_b > 0$. Given a marked fragmentation $(\mathcal{F}, M)$, we do the following: if one of Properties $1, 2, \ldots, 9$ does not hold for some position i in $(\mathcal{F}, M)$ or one of the first eight properties does not hold for some i in the reversed marked fragmentation $(\mathcal{F}^{\text{rev}}, M^{\text{rev}})$, then we will either **reject** the instance, output a **removable base** of φ, or modify the given marked fragmentation. If the given marked fragmentation is proper, the algorithm returns an output using Lemmas 9 and 10.

To do this, in each step we choose the first property violated by a position either in $(\mathcal{F}, M)$ or in $(\mathcal{F}^{\text{rev}}, M^{\text{rev}})$. Observe that we can assume w.l.o.g. that there is an ℓ ($1 \le \ell \le 9$) such that Properties $1, \ldots, \ell - 1$ hold for each position both in $(\mathcal{F}, M)$ and in $(\mathcal{F}^{\text{rev}}, M^{\text{rev}})$, but Property ℓ is violated by a position in S_P in $(\mathcal{F}, M)$, otherwise we simply reverse the instance. (We only reverse it if this condition is not true.)

Given ℓ, the algorithm takes the first position i violating Property ℓ, and branches on choosing $\varphi(i)$ according to $i_{\text{left}} \le \varphi(i) \le i_{\text{right}}$. By Proposition 2, this results in at most $k_b + 1$ directions. Next, the algorithm handles each of the cases in a different manner, according to whether i turns out to be left-aligned, right-aligned, or skew. We consider these cases in a general way that is essentially independent from ℓ, and mainly relies on the type of i. We suppose that i is contained in a fragment $F^i = ([p_1, p_2], [t_1, t_2])$.

Extremal cases. Assume that $i = p_1$ and i is skew or right-aligned, or $i = p_2$ and i is skew or left-aligned. In these cases, we can find at least one **removable base** of φ. First, if $i = p_1$ and i is skew or right-aligned, then each base $S_T[j]$ must be deleted for each j where $t_1 \leq j < \varphi(i)$. Second, if $i = p_2$ and i is skew or left-aligned, then $S_T[j]$ must be deleted for each j where $\varphi(i) < j \leq t_2$.

Skew position. Suppose that $i > p_1$ and j is skew, meaning that $\varphi(i) = j$ for some j with $i_{\text{left}} < j < i_{\text{right}}$. In this case, we can divide the fragment F^i, or more precisely, we can delete F^i from the fragmentation $\mathcal{F}$ and add the new fragments $([p_1, i{-}1], [t_1, j{-}1])$ and $([i, p_2], [j, t_2])$. Note that the newly introduced fragments are non-trivial by the bounds on j. We also modify M by declaring every trivial position of the fragmentation to be marked (no matter whether it was marked or not before). Observe that the number of non-trivial fragments increases in this step. By Proposition 2, this can happen at most $k_b - 1$ times.

Left-aligned position. Lemma 6 summarizes our results that show how to deal with the case when i is left-aligned and $i < p_2$. The proof of this lemma is essential in the correctness of our algorithm.

Lemma 6. *Suppose that Property ℓ ($1 \leq \ell \leq 9$) does not hold for some $i \in [n_P]$ in the marked fragmentation $(\mathcal{F}, M)$, but all the previous properties hold for each position both in $(\mathcal{F}, M)$ and in $(\mathcal{F}^{\text{rev}}, M^{\text{rev}})$. If i is left-aligned, then depending on ℓ, we can do one of the followings in linear time (without any branchings):*

A) reject correctly,
B) output a removable arc of φ,
C) find that i is incident to a removable arc of φ (this only happens if $\ell = 2$),
D) produce a skew position i', or
E) produce a set N of at most $2k_b - 1$ positions in S_T such that $N \cap S^{\text{del}}(\varphi) \neq \emptyset$.

In Case A or B, we **reject** or output a **removable arc** of φ.

In Case C, we put the non-trivial position i in a set W, which will only store positions in S_T that are incident to a removable arc of φ. (We set $W = \emptyset$ initially.) Whenever Case C happens, we examine whether $|W| \leq 2k_a$. If not, then we **reject** the input. This is correct, since there can be at most k_a removable arcs of φ, and each such arc is incident to two bases.

If $|W| \leq 2k_a$ holds, then we modify the given fragmentation, replacing F^i by new fragments $F_1 = ([p_1, i], [t_1, i_{\text{left}}])$ and $F_2 = ([i+1, p_2], [i_{\text{left}} + 1, t_2])$. By $\varphi(i) = i_{\text{left}}$, this yields a fragmentation for φ. Note that F_1 is trivial and F_2 is non-trivial. We mark each position of F_2, putting them into M. We refer to this operation as a *left split* at i. Since i becomes trivial in F_1, each position can be placed into W at most once. Thus, Case C can happen at most $2k_a$ times without rejecting.

In Cases D and E, we might branch into a bounded number of additional branches. In Case D, we branch on those choices of $\varphi(i')$ where i' is indeed skew, which means $\sigma(F^i) - 1 \leq k_b - 1$ directions, and we handle each branch according to the way described above (dividing one fragment at the skew position i'). In Case E, we branch into at most $2k_b - 1$ directions on choosing a **removable base** of φ from N and outputting it.

Note that Case D or E can happen at most k_b times, by our observation that a skew position can only be found at most $k_b - 1$ times.

We remark that if i is trivial, then we treat it as left-aligned.

Right-aligned position. Suppose that $i > p_1$ and i is right-aligned. In this case, we replace F^i by new fragments $F_1 = ([p_1, i - 1], [t_1, i_{\text{right}} - 1])$ and $F_2 = ([i, p_2], [i_{\text{right}}, t_2])$. This yields a fragmentation where F_1 is non-trivial and F_2 is trivial. We refer to this operation as performing a *right split* at j. If this happens because i violated Property ℓ for some $\ell \leq 8$, then we mark every trivial position (including those contained in F_2), by putting them into M. If $\ell = 9$, then we do not modify M, so the trivial positions of F_2 will not be marked.

The above process either produces a **removable base** of φ, **rejects** correctly, or ends by providing a marked fragmentation that is proper. In the remaining steps of the algorithm, the set M will never be modified, and the only possible modification of the actual fragmentation will be to perform a right split.

Given a proper marked fragmentation $(\mathcal{F}, M)$, we make use of Lemma 9 below. This lemma gives sufficient conditions to do one of the followings.

- Find out that some non-trivial position i is right-aligned. In this case, we perform a right split at i in the actual fragmentation.
- Find a **removable arc** of φ.
- **Reject** correctly.

Our algorithm applies Lemma 9 repeatedly, until it either stops (by rejecting or outputting a removable arc of φ), or finds that none of the conditions of Lemma 9 apply. Before stating this lemma, we need two more important observations. First, Lemma 7 shows that the repeated application of Lemma 9 results in a proper fragmentation. Second, Lemma 8 states some useful invariants that hold for each fragmentation obtained by us after a proper fragmentation is achieved.

Lemma 7. *If $(\mathcal{F}, M)$ is proper and $\mathcal{F}'$ is obtained by applying an arbitrary number of right splits to $\mathcal{F}$, then $(\mathcal{F}', M)$ is proper as well.*

Lemma 8. *Let $(\mathcal{F}, M)$ be a 8-proper marked fragmentation whose trivial positions are all marked. Suppose that $\mathcal{F}'$ is obtained by applying an arbitrary number of right splits to the fragmentation $\mathcal{F}$.*

(1) For each i that is not marked ($i \in [n_P] \setminus M$), both $A_P^+(i) = A_T^+(i_{\text{right}})$ and $A_P^-(i) = A_T^-(i_{\text{right}})$ hold in $(\mathcal{F}', M)$.

(2) Suppose that neither i nor j is marked ($i, j \in [n_P] \setminus M$) and $c = (i, j) \in A_P$. If $(c, d) \in R_{\text{right}}^+(i)$ for some $d \in A_T^+(i_{\text{right}})$, then $d^{\text{end}} = j_{\text{right}}$. Similarly, if $(c, d) \in R_{\text{right}}^-(j)$ for some $d \in A_T^-(j_{\text{right}})$, then $d^{\text{start}} = i_{\text{right}}$.

Now, we can state Lemma 9.

Lemma 9. *Let $(\mathcal{F}, M)$ be a proper marked fragmentation for φ obtained by our algorithm, and let $a, b \in [n_P]$.*

(i) Suppose that a is trivial but not marked and b is non-trivial. If $(a, b) \in A_P$ or $(b, a) \in A_P$, then b is right-aligned.

(ii) If a and b are trivial, $a < b$ and $A_P(a, b) \neq A_T(a_{\text{left}}, b_{\text{left}})$, then we can either **reject** *or output a* **removable arc** *of φ.*

After applying Lemma 9 repeatedly, the algorithm either stops by rejecting or outputting a removable arc of φ, or it finds that neither of the conditions (i) and (ii) of Lemma 9 holds. Let $(\mathcal{F}, M)$ be the final marked fragmentation obtained. Note that the algorithm does not modify the set M of marked trivial positions when applying Lemma 9, and it can only modify the actual fragmentation by performing a right split. Hence, Lemma 7 yields that $(\mathcal{F}, M)$ is proper.

Using $(\mathcal{F}, M)$, Lemma 10 claims that we can find an arc-preserving alignment for $(S_P, A_P; S_T, A_T)$ in linear time. Hence, the final step of our algorithm, finishing its description, is to output this **arc-preserving alignment**.

Lemma 10. *Let $(\mathcal{F}, M)$ be a proper marked fragmentation for φ obtained by the algorithm. If none of the conditions of Lemma 9 holds, then we can produce an* **arc-preserving alignment** *ψ for $(S_P, A_P; S_T, A_T)$ in linear time.*

Proof. We show that defining $\psi(i) = i_{\text{left}}$ for each position $i \in [n_P]$ fulfills the requirements. For this, we have to prove $S_P[i] = S_T[i_{\text{left}}]$ for each position $i \in [n_P]$, and $A_P(i, j) = A_T(i_{\text{left}}, j_{\text{left}})$ for each two positions $i \neq j \in [n_P]$.

First, as Property 1 holds for each position in $\mathcal{F}$, we know $S_P[i] = S_T[i_{\text{left}}]$ for each $i \in [n_P]$. It remains to show $A_P(i, j) = A_T(i_{\text{left}}, j_{\text{left}})$ for each $i \neq j \in [n_P]$. If both i and j are trivial positions, then this is true because the conditions of (ii) in Lemma 9 do not apply. If both i and j are non-trivial, then $A_P(i, j) = A_T(i_{\text{left}}, j_{\text{left}})$ again holds, by Properties 2 and 8 for j. Now, if i is non-trivial but j is trivial and marked (or vice versa), then Property 9 implies the required equality. Finally, if one of i and j is non-trivial and the other one is trivial but not marked, then $A_P(i, j) = 0$ holds, since (i) of Lemma 9 is not applicable. $\square$

3.4   Analysis of the Algorithm

In this section, we give some hints how to analyse the running time of the presented algorithm. The following lemma, stating the key properties of the our algorithm, proves Theorem 1.

Lemma 11. *Let $(S_P, A_P, S_T, A_T, k_a)$ be the given instance of APS. The presented algorithm branches into at most $f(k_a, k_b)$ directions in total for some function f such that in each branch it does one of the followings (supposing that the conditions of the given branch do hold):*

- *it gives an* **arc-preserving alignment** *ψ of $(S_P, A_P; S_T, A_T)$,*
- *it correctly* **rejects** *the instance, or*
- *it outputs a* **removable base** *or a* **removable arc** *of φ.*

Moreover, each branch takes linear time in the size of the input.

Although we do not prove Lemma 11 due to lack of space, we give the most important definitions used in the proof.

Given a fragmentation $\mathcal{F}$ for φ, a fragment $F \in \mathcal{F}$, and some ℓ ($1 \leq \ell \leq 8$), let $\pi(\mathcal{F}, F, \ell)$ be 1 if Property ℓ holds for each position i in F, and 0 otherwise. Let $N(\mathcal{F})$ denote the set of non-trivial fragments in $\mathcal{F}$. We define the *measure* $\mu(\mathcal{F})$ of a given fragmentation $\mathcal{F}$ for φ as follows:

$$\mu(\mathcal{F}) = \sum_{1 \leq \ell \leq 8} \left(\sum_{F \in N(\mathcal{F})} \pi(\mathcal{F}, F, \ell) + \sum_{F \in N(\mathcal{F}^{\mathrm{rev}})} \pi(\mathcal{F}^{\mathrm{rev}}, F, \ell) \right).$$

Note that $\mu(\mathcal{F}) = \mu(\mathcal{F}^{\mathrm{rev}})$ is trivial, so reversing a fragmentation does not change its measure. The importance of this definition is shown by Lemma 12.

Lemma 12. *Let* $\mathcal{F}_1, \ldots, \mathcal{F}_t, \mathcal{F}_{t+1}$ *be a series a fragmentations such that for each* $i \in [t]$ *the algorithm obtains* $\mathcal{F}_{i+1}$ *from* $\mathcal{F}_i$ *by applying a left or a right split at a position* j_i *violating Property* ℓ_i *in* $\mathcal{F}_i$. *Then (1)* $\mu(\mathcal{F}_{i+1}) \geq \mu(\mathcal{F}_i)$ *for each* $i \in [t]$, *and (2) if* $\mu(\mathcal{F}_1) = \mu(\mathcal{F}_t)$, *then* $t \leq k_b$ *holds.*

References

1. Alber, J., Gramm, J., Guo, J., Niedermeier, R.: Computing the similarity of two sequences with nested arc annotations. Theor. Comput. Sci. 312(2-3), 337–358 (2004)
2. Blin, G., Fertin, G., Rizzi, R., Vialette, S.: What makes the Arc-Preserving Subsequence problem hard? In: Sunderam, V.S., van Albada, G.D., Sloot, P.M.A., Dongarra, J. (eds.) ICCS 2005. LNCS, vol. 3515, pp. 860–868. Springer, Heidelberg (2005)
3. Damaschke, P.: A remark on the subsequence problem for arc-annotated sequences with pairwise nested arcs. Inf. Process. Lett. 100(2), 64–68 (2006)
4. Downey, R.G., Fellows, M.R.: Parameterized Complexity. Monographs in Computer Science. Springer, New York (1999)
5. Evans, P.A.: Algorithms and complexity for annotated sequence analysis. PhD thesis, University of Victoria, Canada (1999)
6. Evans, P.A.: Finding common subsequences with arcs and pseudoknots. In: Crochemore, M., Paterson, M. (eds.) CPM 1999. LNCS, vol. 1645, pp. 270–280. Springer, Heidelberg (1999)
7. Flum, J., Grohe, M.: Parameterized Complexity Theory. In: Texts in Theoretical Computer Science. An EATCS Series, p. 493. Springer, Heidelberg (2006)
8. Gramm, J., Guo, J., Niedermeier, R.: Pattern matching for arc-annotated sequences. ACM Trans. Algorithms 2(1), 44–65 (2006)
9. Jiang, T., Lin, G., Ma, B., Zhang, K.: The longest common subsequence problem for arc-annotated sequences. J. Discrete Algorithms 2(2), 257–270 (2004)
10. Lin, G., Chen, Z.-Z., Jiang, T., Wen, J.: The longest common subsequence problem for sequences with nested arc annotations. J. Comput. Syst. Sci. 65(3), 465–480 (2002)
11. Ma, B., Wang, L., Zhang, K.: Computing similarity between RNA structures. Theor. Comput. Sci. 276(1-2), 111–132 (2002)
12. Marx, D., Schlotter, I.: Cleaning interval graphs. CoRR abs/1003.1260 (2010) arXiv:1003.1260 [cs.DS]

From Path Graphs to Directed Path Graphs

Steven Chaplick[1], Marisa Gutierrez[2],
Benjamin Lévêque[3], and Silvia B. Tondato[4]

[1] University of Toronto, Canada
`chaplick@cs.toronto.edu`
[2] CONICET, Universidad Nacional de La Plata, Argentina
`marisa@mate.unlp.edu.ar`
[3] CNRS, LIRMM, Montpellier, France
`benjamin.leveque@lirmm.fr`
[4] Universidad Nacional de La Plata, Argentina
`tondato@mate.unlp.edu.ar`

Abstract. We present a linear time algorithm to greedily orient the edges of a path graph model to obtain a directed path graph model (when possible). Moreover we extend this algorithm to find an odd sun when the method fails. This algorithm has several interesting consequences concerning the relationship between path graphs and directed path graphs. One is that for a directed path graph, path graph models and directed path graph models are the same. Another consequence concerns the difference between path graphs and directed path graphs in terms of forbidden induced subgraphs. This can be used to deduce the forbidden induced subgraph characterization of directed path graphs from the forbidden induced subgraph characterization of path graphs. The last consequence is algorithmic and shows that the recognition of directed path graphs is not more difficult than the recognition of path graphs.

1 Introduction

A *hole* is a chordless cycle of length at least four. A graph is a *chordal graph* if it contains no hole as an induced subgraph. Gavril [3] proved that a graph is chordal if and only if it is the intersection graph of a family of subtrees of a tree. In this paper, whenever we talk about the intersection of subgraphs of a graph we mean that the *vertex sets* of the subgraphs intersect. A graph is an *interval graph* if it is the intersection graph of a family of intervals on the real line; or equivalently, the intersection graph of a family of subpaths of a path. The class of path graphs lies between interval graphs and chordal graphs. A graph is a *path graph* if it is the intersection graph of a family of subpaths of a tree. Two variants of path graphs have been defined when the tree is a directed graph. A *directed tree* is a directed graph whose underlying undirected graph is a tree. A *directed subpath* of a directed tree is a subpath whose edges are all oriented in the same way. A graph is a *directed path graph* if it is the intersection graph of a family of directed subpaths of a directed tree. A *rooted tree* is a directed tree in which the

D.M. Thilikos (Ed.): WG 2010, LNCS 6410, pp. 256–265, 2010.

path from a particular vertex r to every other vertex is a directed path; vertex r is called the *root*. A graph is a *rooted path graph* if it is the intersection graph of a family of directed subpaths of a rooted tree.

The following inclusions hold by definition:

$$\text{interval} \subset \text{rooted path} \subset \text{directed path} \subset \text{path} \subset \text{chordal}$$

and these inclusions are strict.

In Section 4, we present a method to greedily orient the edges of a tree T that is a path graph model to obtain a directed path graph model (when possible). The idea is very simple: Pick any non oriented edge e of T and orient it arbitrarily. Orient every edge of T that is forced by e. Repeat the process until all edges of T are oriented. In fact, to ensure that the algorithm runs in linear time the formal description of the algorithm is more complex and uses a particular order obtained by an algorithm presented in Section 3. Moreover, we extend this method to find an odd sun when the greedy path forcing fails. A *sun* is a graph with vertices $C = \{c_0, \ldots, c_r\}$, and $S = \{s_0, \ldots, s_r\}$, $r \geq 2$, where C is a clique, S is a stable set and for $0 \leq i \leq r$, $N(s_i) \cap C = \{c_{i-1}, c_i\}$ (subscripts are modulo $r + 1$). An *odd sun* is a sun where $|S|$ is odd. Finding an odd sun is interesting as it certifies that the input graph is not a path graph.

This algorithm has several interesting consequences concerning the relationship between path graphs and directed path graphs presented in Section 5. One is that for a directed path graph, every path graph model has a corresponding directed path graph model. Another consequence concerns the difference between path graphs and directed path graphs in terms of forbidden induced subgraphs. This can be used to deduce the forbidden induced subgraph characterization of directed path graphs from the forbidden induced subgraph characterization of path graphs. The last consequence is algorithmic and shows that the recognition of directed path graph is not more difficult than the recognition of path graphs.

2   Definitions and Background

In a graph G, a *clique* is a set of pairwise adjacent vertices. Let $\mathcal{C}(G)$ be the set of all (inclusionwise) maximal cliques of G. For any vertex $v \in V$, let $\mathcal{C}_v(G) = \{C \in \mathcal{C}(G) : v \in C\}$. When there is no ambiguity we write $\mathcal{C}$ and $\mathcal{C}_v$ instead of $\mathcal{C}(G)$ and $\mathcal{C}_v(G)$. Given a set X of vertices, let $G[X]$ denote the subgraph of G induced by the vertices of X.

A *clique tree* T of a graph G is a tree whose vertices are the members of $\mathcal{C}$ and, for each vertex v of G, the induced subgraph $T[\mathcal{C}_v]$ is a tree. A classical result of Gavril [3] states that a graph is chordal if and only if it has a clique tree. A *clique path tree* T of G is a clique tree of G such that, for each vertex v of G, the subtree $T[\mathcal{C}_v]$ is a path. Gavril [4] proved that a graph is a path graph if and only if it has a clique path tree. A *clique directed path tree* T of G is a clique path tree of G such that edges of the tree T are directed and for each vertex v of G, the subpath $T[\mathcal{C}_v]$ is a directed path. A *clique rooted path tree* T of G is a clique directed path tree of G such that T is a rooted tree. Monma and Wei [9] proved that a graph is a directed path graph if and only if it has a clique directed path tree, and that a graph is a rooted path graph if and only if it has a clique rooted path tree. These results allow us to restrict our attention to intersection models that are clique trees when studying the properties of these graph classes.

For more information about clique trees and chordal graphs, see [5,8].

3   Maximum Cardinality Clique Search

First, we need an algorithm (see Algorithm 1) that provides the vertex order used to accomplish the greedy path forcing algorithm (see Algorithm 2). In particular, we order the vertices of a given graph G starting from an arbitrary vertex and selecting the next vertex v_i such that the number of maximal cliques v_i shares with $v_0, ..., v_{i-1}$ is as large as possible.

Algorithm 1. *Maximum Cardinality Clique Search*

   Input: A graph G and the sets $\mathcal{C}_v$ for every vertex v.
   Output: An ordering σ on the vertices of G such that for every vertex
      v the cardinality of $\mathcal{C}_v \cap (\cup_{\sigma(u) < \sigma(v)} \mathcal{C}_u)$ is maximum.

1   All vertices and maximal cliques of G are non-marked.
2   Let $label(v) = 0$ for every vertex v.
3   **for** $i = 1$ to n **do**
4 | Choose a non-marked vertex v with maximum label.
5 | Mark v and let $\sigma(v) = i$.
6 | **foreach** non-marked clique $C \in \mathcal{C}_v$ **do**
7 | | Mark C.
8 | | **foreach** $u \in C \setminus \{v\}$ **do**
9 | | | $label(u) = label(u) + 1$.
10 **return** σ.

Notice that the above algorithm runs in linear time with respect to its input, i.e. $O(\Sigma_{v \in V} |\mathcal{C}_v|)$. In particular, this is the same as the number of ones in the vertex to maximal clique incidence matrix of the input graph G. Furthermore, we have the following result by Fulkerson and Gross [6]:

Theorem 1 ([6]). *For a chordal graph, the number of non-zero entries in the vertex to maximal clique incidence matrix is $O(n + m)$.*

Therefore Algorithm 1 runs in time $O(n+m)$ for a chordal graph with n vertices and m edges.

4   Greedy Path Forcing

We now provide a linear time algorithm that for any path graph G and any clique path tree T of G, returns either an orientation of T that is a clique directed path tree of G or an induced subgraph of G that is an odd-sun.

Algorithm 2 considers all the vertices of G one by one, using the order obtained by Algorithm 1, and orients their corresponding subpath in T to form a directed path without modifying already oriented edges. If the method fails, there is a subpath that cannot be oriented. This subpath cannot be oriented because it has two consecutive edges e, f which are already oriented in opposite directions. By following the sequence of vertices that leads from the orientation of e to f, it is possible to find an odd sun whose central clique corresponds to the common extremity of e and f.

The following theorem shows the correctness of Algorithm 2:

Theorem 2. *For any path graph G and any clique path tree T of G, Algorithm 2 returns in time $\mathcal{O}(n + m)$, either an orientation of T that is a clique directed path tree of G or an induced subgraph of G that is an odd-sun.*

Algorithm 2. *Greedy Path Forcing*

Input: A path graph G and a clique path tree T of G
Output: An orientation of T that is a clique directed path tree of G or an induced subgraph of G that is an odd-sun.

1   Extract the sets $\mathcal{C}_v$ from the clique path tree T.
2   Let $\sigma(v_i) = i$, $1 \leq i \leq n$, obtained by applying Algorithm 1 on G and sets $\mathcal{C}_v$.
3   **for** $i = 1$ to n **do**
4   **if** $T[\mathcal{C}_{v_i} \cap (\cup_{\sigma(u)<i}\mathcal{C}_u)]$ is a directed path (maybe empty) **then**
5   Orient the edges of $T[\mathcal{C}_{v_i}]$ that are not already oriented such that $T[\mathcal{C}_{v_i}]$ forms a directed path.
6   **else**   /* find an odd sun */
7   Let $C_{start}, C_{centre}, C_{stop}$ be three consecutive cliques of $T[\mathcal{C}_{v_i} \cap (\cup_{\sigma(u)<i}\mathcal{C}_u)]$ such that $T[C_{start}, C_{centre}, C_{stop}]$ is not a directed path.
8   Mark all vertices v with $\sigma(v) \geq i$ (all other vertices are non-marked).
9   Mark C_{start} (all other cliques are non-marked).
10   Let $f(C_{start}) = v_i$ and $g(v_i)$ be a vertex of $C_{start} \setminus C_{centre}$.
11   **while** C_{stop} is not marked **do**
12   Choose a non marked vertex v of a marked clique C_{parent}.
13   Mark v.
14   Let $h(v) = f(C_{parent})$.
15   **if** there exists a non marked clique $C \in \mathcal{C}_v \cap N(C_{centre})$ **then**
16   Mark C.
17   Let $f(C) = v$ and $g(v)$ be a vertex of $C \setminus C_{centre}$.
18   Let $u_0 = f(C_{stop})$ and $u_j = h^j(u_0)$, $1 \leq j \leq r$, with $u_r = v_i$.
19   **return** $G[u_0, \ldots, u_r, g(u_0), \ldots, g(u_r)]$.
20   Orient all not already oriented edges of T with an arbitrary direction.
21   **return** T with its orientation.

Proof. To distinguish between marked elements of Algorithm 1 and 2, we say that a clique or a vertex is marked 1 if it corresponds to the marking of Algorithm 1 and marked 2 if it corresponds to the marking of Algorithm 2. Similarly we distinguish between lines of the two algorithms by using 1.x and 2.x.

Let G be any path graph and T be any clique path tree of G. We prove that Algorithm 2 applied on G and T, returns a clique directed path tree when every test of line 2.4 is satisfied, and returns an odd sun when one such test is false (i.e., the algorithm executes lines 2.6 to 2.18 and returns an odd sun at line 2.19).

Case 1: Every test of line 2.4 is true. Every vertex is considered one by one and its corresponding subpath in T is oriented to form a directed path without modifying already oriented edges. At the end, there may still be some non-oriented edges (when G is disconnected), that are oriented arbitrarily at line 2.20. Clearly, at line 2.21, the algorithm returns an orientation of T that is a clique directed path tree of G.

In this case, the complexity of the algorithm is $O(\Sigma_{v \in V} |C_v|)$ and thus $O(n+m)$ by Theorem 1.

Case 2: At least one test of line 2.4 is false. Let i be the first time such that the test of line 2.4 is false for v_i. Also, consider the point when Algorithm 2 enters the else due to vertex v_i. In fact, there is only one time that this test can be false since the algorithm will return (at line 2.19) before leaving the scope of this else block. Let $U = \{v \in V$ such that $\sigma(v) < i\}$. The subgraph $T[\mathcal{C}_{v_i} \cap (\cup_{\sigma(u)<i} \mathcal{C}_u)]$ is connected by the choice of σ and thus the three cliques $C_{start}, C_{centre}, C_{stop}$ exists at line 2.7. The set $C_{start} \cap C_{stop} \cap U = \emptyset$ as $T[C_{start}, C_{centre}, C_{stop}]$ is not a directed path (line 2.7). First, we prove that the clique C_{stop} will be marked 2 during the while loop at line 2.11 (i.e., the algorithm ends).

Claim. While C_{stop} is not marked 2, there exists a non marked 2 vertex v of an already marked 2 clique and a non marked 2 clique $C \in \mathcal{C}_v \cap N(C_{centre})$ (corresponding to lines 2.12 and 2.15).

Proof. Suppose on the contrary that at one point of the loop of line 2.11 these v and C do not exist. Let $\mathcal{M}$ be the set of already marked 2 cliques. Note that $\mathcal{M} \subseteq N(C_{centre})$ as all marked 2 cliques (at line 2.9 or 2.16) are adjacent to C_{centre}. Let $\mathcal{N} = N(C_{centre}) \setminus \mathcal{M}$. Note that $C_{start} \in \mathcal{M}$ and $C_{stop} \in \mathcal{N}$. Let $A = U \cap C_{centre} \cap (\cup_{C \in \mathcal{M}} C)$ and $B = U \cap C_{centre} \cap (\cup_{C \in \mathcal{N}} C)$. The set $A \cap B$ is empty, otherwise there exists $v \in A \cap B$ and $C \in \mathcal{C}_v \cap \mathcal{N}$ that can play the role of v and C as in the claim. Edges $C_{start} C_{centre}$ and $C_{stop} C_{centre}$ are already oriented, so the sets $C_{start} \cap C_{centre} \cap U$ and $C_{stop} \cap C_{centre} \cap U$ are non empty. So A and B are non empty. Let x (resp. y) be the minimum vertex for σ in A (resp. in B). We have $x \notin B$ and $y \notin A$. Let C_x be a clique of $\mathcal{C}_x \cap \mathcal{M}$. Let $t_0 = f(C_x)$ and $t_j = h^j(t_0)$, $1 \leq j \leq s$, with $t_s = v_i$. We distinguish two cases corresponding to the relation between x and y for the order σ obtained by applying Algorithm 1 at line 2.2.

Case $\sigma(x) < \sigma(y)$. Consider the point of Algorithm 1 when y is chosen at line 1.4. Note that x is already marked 1, so C_{centre} and C_x are already marked 1. Vertex y is the first vertex of B chosen by Algorithm 1. So, when y is chosen, the only marked 1 clique in $\mathcal{C}_y$ is C_{centre} and so $label(y) = 1$ (otherwise, $y \in A \cap B$).

We claim that $\sigma(t_j) < \sigma(y)$ for $0 \leq j \leq s$. Suppose the contrary and let k be minimal such that $\sigma(t_k) > \sigma(y)$. If $k = 0$, then C_x and C_{centre} are already marked 1 cliques of C_{t_0}, so $label(t_0) \geq 2$, a contradiction to the choice of y. If $1 \leq k \leq s$, then t_{k-1} is already marked 1, so $f^{-1}(t_k)$ is already marked 1. Then $f^{-1}(t_k)$ and C_{centre} are already marked 1 cliques of C_{t_k}, so $label(t_k) \geq 2$, a contradiction to the choice of y. Thus $\sigma(v_i) = \sigma(t_s) < \sigma(y)$, contradicting $y \in U$.

Case $\sigma(y) < \sigma(x)$. Consider the point of Algorithm 1 when x is chosen at line 1.4. Note that y is already marked 1, so C_{centre} is already marked 1. Vertex x is the first chosen vertex of A in Algorithm 1. So when it is chosen, the only marked 1 clique in C_x is C_{centre} and so $label(x) = 1$ (otherwise, $x \in A \cap B$).

Suppose there exists $z \in B$ with $\sigma(x) < \sigma(z)$ and let z be minimal with this property. Clearly $label(z) \leq label(x)$ when x is chosen, so $label(z) = 1$ as $C_{centre} \in C_z$. Vertex z is the first vertex of B chosen after x, so its label remains the same until it is marked. Consider (temporarily) the point of Algorithm 1 when z is chosen at line 1.4. Note that x is already marked 1, so C_x is already marked 1. We claim that all t_j, $0 \leq j \leq s$, satisfy $\sigma(t_j) < \sigma(z)$. Suppose the contrary and let k minimal such that $\sigma(t_k) > \sigma(z)$. If $k = 0$, then C_x and C_{centre} are already marked 1 cliques of C_{t_0}, so $label(t_0) \geq 2$, a contradiction to the choice of z. If $1 \leq k \leq s$, then t_{k-1} is already marked 1, so $f^{-1}(t_k)$ is already marked 1. Then $f^{-1}(t_k)$ and C_{centre} are already marked 1 cliques of C_{t_k}, so $label(t_k) \geq 2$, a contradiction to the choice of z. Thus $\sigma(v_i) = \sigma(t_s) < \sigma(z)$, contradicting $z \in U$. So there are no vertices in B with $\sigma(x) < \sigma(z)$.

Let z be a vertex of $C_{stop} \cap C_{centre} \cap U$, thus $z \in B$. By the preceding paragraph $\sigma(z) < \sigma(x)$. We consider again the point of Algorithm 1 when x is chosen at line 1.4 with $label(x) = 1$. Cliques C_{stop} and C_{centre} are already marked 1 cliques of C_{v_i}, so $label(v_i) \geq 2$, a contradiction to the choice of x.   □

By the claim, a new clique will always be marked 2 at line 2.15 until C_{stop} is marked. So the while loop of line 2.11 ends and so the algorithm ends. Let u_j, $0 \leq j \leq r$, be as defined at line 2.18. For $0 \leq j \leq r$, let $z_j = g(u_j)$. At line 2.19, the graph G' induced by $u_0, \ldots, u_r, z_0, \ldots, z_r$ is returned. We now prove that G' is an odd sun.

For $0 \leq j \leq r$, let $C_j = f^{-1}(u_j)$. Note that $C_0 = C_{stop}$ and $C_r = C_{start}$. All of the cliques C_i are distinct and adjacent to C_{centre} so the tree $T[C_{centre}, C_0, \ldots, C_r]$ is a star centred at C_{centre}. Thus $u_0, \ldots, u_r$ is a clique Q and $z_0, \ldots, z_r$ is a stable set with $N(z_j) \cap Q = \{u_{j-1}, u_j\}$, for $0 \leq j \leq r$ and subscripts modulo $r+1$. So G' is a sun. The tree $T[C_{centre}, C_0, \ldots, C_r]$ is already oriented and $T[C_0, C_{centre}, C_r]$ is not directed by the choice of $C_{start}, C_{centre}, C_{stop}$ of line 2.7. So $C_0 C_{centre}$ and $C_{centre} C_r$ are not directed in the same way. Suppose, by symmetry, that $C_0 \rightarrow C_{centre}$ and $C_r \rightarrow C_{centre}$ (where $a \rightarrow b$ means there is a edge oriented from a to b). Vertices C_0, C_{centre}, C_1 appear in this order along $T[C_{u_0}]$ and $C_0 \rightarrow C_{centre}$, so $C_{centre} \rightarrow C_1$. Vertices C_1, C_{centre}, C_2 appear in this order along $T[C_{u_1}]$ and $C_{centre} \rightarrow C_1$, so $C_2 \rightarrow C_{centre}$. Propagating this forward, for $2 \leq j < r$, vertices C_j, C_{centre}, C_{j+1} appear in this order along $T[C_{u_j}]$, so $C_{j+1} \rightarrow C_{centre}$ when j is odd and $C_{centre} \rightarrow C_{j+1}$ when j is even. Thus r is even as $C_r \rightarrow C_{centre}$ and G' is an odd sun.

Vertices and maximal cliques are marked at most once in the while loop of line 2.11, so the complexity of the else part is $O(|V| + |\mathcal{C}|)$. Thus the total complexity of the algorithm is $O(n + m)$.   □

5 Consequences

Theorem 2 has several consequences concerning the relationship between path graphs and directed path graphs.

First, we need the following lemma that is part of the work of Panda [10]. We give a short proof of this lemma here. One consequence of this lemma, Theorem 2, and [7] is a new proof of the main result of [10] (i.e., the forbidden induced subgraph characterization of directed path graphs).

Lemma 1 ([10]). *Odd suns are minimally not directed path graphs.*

Proof. Let G be an odd sun. Let $C = \{c_0, \ldots, c_{2k}\}$, $S = \{s_0, \ldots, s_{2k}\}$, $k \geq 1$, be the vertices of G where C is a clique, S is a stable set and for $0 \leq i \leq 2k$, $N(s_i) \cap C = \{c_{i-1}, c_i\}$ (subscripts are modulo $2k + 1$).

Suppose that G is a directed path graph and let T be a clique directed path tree of G. The maximal cliques of G are C and $C_i = \{s_i, c_{i-1}, c_i\}$ for $0 \leq i \leq 2k$. For $0 \leq i \leq 2k$, the clique C is between C_i and C_{i+1} in T as otherwise c_{i-1} is adjacent to s_{i+1} or c_{i+1} is adjacent to s_i. Thus the subtree $T[\mathcal{C}_{c_i}]$ is the path C_i, C, C_{i+1}. Suppose, by symmetry, that $C_0 \to C$. Vertices C_0, C, C_1 appear in this order along $T[\mathcal{C}_{c_0}]$ and $C_0 \to C$, so $C \to C_1$. Vertices C_1, C, C_2 appear in this order along $T[\mathcal{C}_{c_1}]$ and $C \to C_1$, so $C_2 \to C$. Propagating this forward, for $2 \leq i \leq 2k$, vertices C_i, C, C_{i+1} appear in this order along $T[\mathcal{C}_{c_i}]$, where $C_{i+1} \to C$ when i is odd and $C \to C_{i+1}$ when i is even. So for $i = 2k$, we have $C \to C_{2k+1}$ and $C_{2k+1} = C_0$, contradicting $C_0 \to C$. So G is not a directed path graph.

We now prove that G is minimally not directed path graph, that is for any vertex w of G, the graph $G \setminus w$ is a directed path graph. If $w \in S$, then we can assume that $w = s_0$. Thus the tree on vertices C, C_i, $1 \leq i \leq 2k$ and edges $C_i \to C$ when i is odd and $C \to C_i$ when i is even is a clique directed path tree of $G \setminus w$. If $w \in C$, then we can assume that $w = c_{2k}$. Thus the tree on vertices $C' = C \setminus \{w\}$, $C_0' = C_0 \setminus \{w\}$, $C_{2k}' = C_{2k} \setminus \{w\}$, $C_i' = C_i$, $1 \leq i \leq 2k - 1$ and edges $C_i' \to C'$ when i is odd and $C' \to C_i'$ when i is even is a clique directed path tree of $G \setminus w$.   □

A consequence of Theorem 2 and Lemma 1 is that, for a directed path graph, clique path trees and clique directed path trees are the same. More precisely:

Theorem 3. *For any directed path graph G and any clique path tree T of G, the edges of T can be oriented to obtain a clique directed path tree of G.*

Note that there is no analogous of Theorem 3 for rooted path graph. In fact, there exists rooted path graphs having clique (directed) path tree that cannot be oriented to obtain a clique rooted path tree (see Figure 1 for an example).

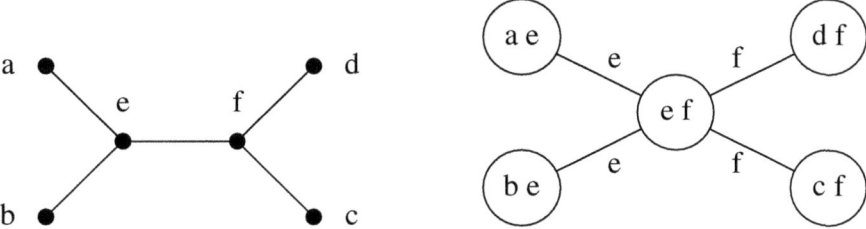

Fig. 1. A rooted path graph with a clique path tree that cannot be rooted to obtain a clique rooted path tree

Thus, there is no algorithm that can return a clique rooted path tree of a rooted path graph G by simply orienting the edges of any clique path tree of G.

Another consequence of Theorem 2 and Lemma 1 concerns the difference between path graphs and directed path graphs in terms of forbidden induced subgraphs:

Theorem 4. *A path graph is a directed path graph if and only if it does not contain an odd sun as an induced subgraph.*

Theorem 4 can be used to deduce the forbidden induced subgraph characterization of directed path graphs from the forbidden induced subgraph characterization of path graphs. The forbidden induced subgraph characterization of path graphs was obtained by Lévêque, Maffray and Preissmann [7] (see Figure 2) (an independent proof has been obtained by Tondato [12]):

Theorem 5 ([7]). *A graph is a path graph if and only if it does not contain any of $F_0(n)_{n \geq 4}$, F_1, F_2, F_3, F_4, $F_5(n)_{n \geq 7}$, F_6, F_7, F_8, F_9, $F_{10}(n)_{n \geq 8}$, $F_{11}(4k)_{k \geq 2}$, $F_{12}(4k)_{k \geq 2}$, $F_{13}(4k+1)_{k \geq 2}$, $F_{14}(4k+1)_{k \geq 2}$, $F_{15}(4k+2)_{k \geq 2}$, $F_{16}(4k+3)_{k \geq 2}$ as an induced subgraph.*

Form the list of minimal forbidden induced subgraphs of path graphs of Theorem 5, if we remove every graph that contains an odd sun, namely F_2, F_8, $F_{11}(4k)_{k \geq 2}$, $F_{12}(4k)_{k \geq 2}$, $F_{14}(4k+1)_{k \geq 2}$, and add the odd suns to the list, we obtain the list of minimal forbidden induced subgraphs of directed path graphs (see Figure 3):

Theorem 6 ([10]). *A graph is a directed path graph if and only if it contains no $F_0(n)_{n \geq 4}$, F_1, F_3, F_4, $F_5(n)_{n \geq 7}$, F_6, F_7, F_9, $F_{10}(n)_{n \geq 8}$, $F_{13}(4k+1)_{k \geq 2}$, $F_{15}(4k+2)_{k \geq 2}$, $F_{16}(4k+3)_{k \geq 2}$, $F_{17}(4k+2)_{k \geq 1}$ as an induced subgraph.*

Theorem 6 has already been proven by Panda [10] with the use of the Separator Theorem of Monma and Wei [9] and a technical case analysis. Here we obtain an independent proof using [7]. This new proof does not rely on the Separator Theorem as it is not used in the proof of Theorem 5 in [7].

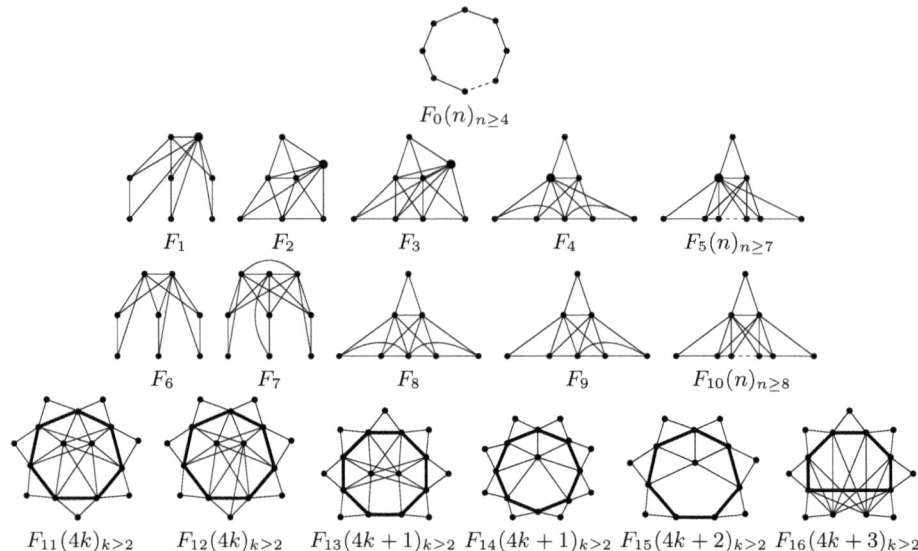

Fig. 2. Minimal forbidden induced subgraphs for path graphs (the vertices in the cycle marked by bold edges form a clique)

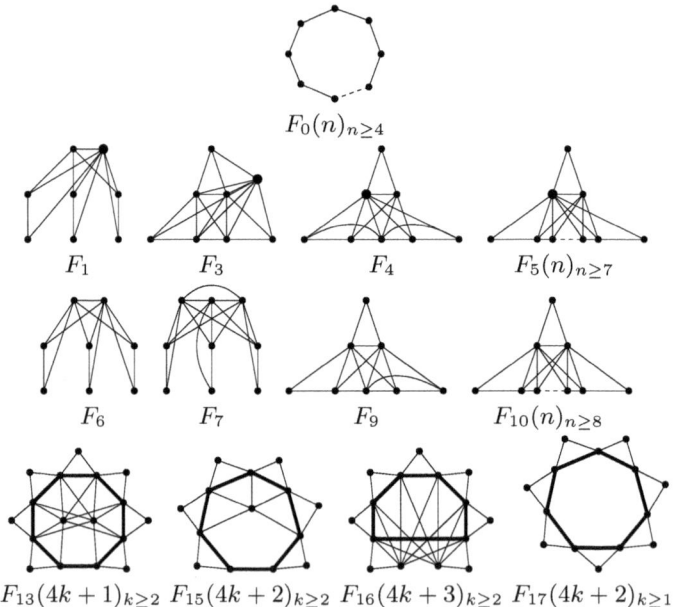

Fig. 3. Minimal forbidden induced subgraphs for directed path graphs (the vertices in the cycle marked by bold edges form a clique)

The algorithmic consequence of Theorem 2 and Lemma 1 is the following:

Theorem 7. *If there exists a polynomial algorithm that tests if a graph G is a path graph and returns a clique path tree of G when the answer is "yes." Then, there exists an algorithm with the same complexity to test if a graph is a directed path graph.*

Some efficient recognition algorithms for path graphs were given by Gavril [4], Schäffer [11] and Chaplick [1], whose complexity is respectively $O(n^4)$, $O(nm)$ and $O(nm)$ for graphs with n vertices and m edges. Another algorithm was proposed in [2] and claimed to run in $O(n + m)$ time, but it has only appeared as an extended abstract and is not considered to be complete or correct (see comments in [1, Section 2.1.4]). All of these algorithms can be extended to recognize directed path graphs with the use of Algorithm 2 without increasing the time complexity. Thus, the fastest recognition algorithm of directed path graphs with this method has complexity $O(nm)$. This is the fastest known algorithm to recognize directed path graphs. Previously, the fastest known algorithm to recognize directed path graphs was from Monma and Wei [9] and has complexity $O(n^2m)$.

References

1. Chaplick, S.: PQR-trees and undirected path graphs, M.Sc. Thesis, Dept. of Computer Science, University of Toronto, Canada (2008)
2. Dahlhaus, E., Bailey, G.: Recognition of path graphs in linear time. In: 5th Italian Conference on Theoretical Computer Science (Revello, 1995), pp. 201–210. World Sci., Publishing, River Edge (1996)
3. Gavril, F.: The intersection graphs of subtrees in trees are exactly the chordal graphs. J. Combin. Theory B 16, 47–56 (1974)
4. Gavril, F.: A recognition algorithm for the intersection graphs of paths in trees. Discrete Math. 23, 211–227 (1978)
5. Golumbic, M.C.: Algorithmic graph theory and perfect graphs. Annals Disc. Math. 57 (2004)
6. Fulkerson, D.R., Gross, O.A.: Incidence matrices and interval graphs. Pacific J. Math. 15, 835–855 (1965)
7. Lévêque, B., Maffray, F., Preissmann, M.: Characterizing path graphs by forbidden induced subgraphs. Journal of Graph Theory 62, 369–384 (2009)
8. McKee, T.A., McMorris, F.R.: Topics in intersection graph theory. SIAM Monographs on Discrete Mathematics and Applications, Philadelphia (1999)
9. Monma, C.L., Wei, V.K.: Intersection graphs of paths in a tree. Journal of Combinatorial Theory B 41, 141–181 (1986)
10. Panda, B.S.: The forbidden subgraph characterization of directed vertex graphs. Discrete Mathematics 196, 239–256 (1999)
11. Schäffer, A.A.: A faster algorithm to recognize undirected path graphs. Discrete Appl. Math. 43, 261–295 (1993)
12. Tondato, S.B.: Grafos Cordales: Árboles clique y Representaciones canónicas, Doctoral Thesis, Universidad Nacional de La Plata, Argentina (2009) (in spanish)

Connections between Theta-Graphs, Delaunay Triangulations, and Orthogonal Surfaces*

Nicolas Bonichon, Cyril Gavoille**, Nicolas Hanusse, and David Ilcinkas

LaBRI, CNRS & Université de Bordeaux, France
{bonichon,gavoille,hanusse,ilcinkas}@labri.fr

Abstract. Θ_k-graphs are geometric graphs that appear in the context of graph navigation. The shortest-path metric of these graphs is known to approximate the Euclidean complete graph up to a factor depending on the cone number k and the dimension of the space.

TD-Delaunay graphs, a.k.a. triangular-distance Delaunay triangulations, introduced by Chew, have been shown to be plane 2-spanners of the 2D Euclidean complete graph, i.e., the distance in the TD-Delaunay graph between any two points is no more than twice the distance in the plane.

Orthogonal surfaces are geometric objects defined from independent sets of points of the Euclidean space. Orthogonal surfaces are well studied in combinatorics (orders, integer programming) and in algebra. From orthogonal surfaces, geometric graphs, called geodesic embeddings can be built.

In this paper, we introduce a specific subgraph of the Θ_6-graph defined in the 2D Euclidean space, namely the half-Θ_6-graph, composed of the even-cone edges of the Θ_6-graph. Our main contribution is to show that these graphs are exactly the TD-Delaunay graphs, and are strongly connected to the geodesic embeddings of orthogonal surfaces of coplanar points in the 3D Euclidean space.

Using these new bridges between these three fields, we establish:

- Every Θ_6-graph is the union of two spanning TD-Delaunay graphs. In particular, Θ_6-graphs are 2-spanners of the Euclidean graph, and the bound of 2 on the stretch factor is the best possible. It was not known that Θ_6-graphs are t-spanners for some constant t, and Θ_7-graphs were only known to be t-spanners for $t \approx 7.562$.
- Every plane triangulation is TD-Delaunay realizable, i.e., every combinatorial plane graph for which all its interior faces are triangles is the TD-Delaunay graph of some point set in the plane. Such realizability property does not hold for classical Delaunay triangulations.

Keywords: Delaunay triangulation, theta-graph, orthogonal surface, spanner, realizability.

---

* All authors are partially supported by the ANR project "ALADDIN" and the INRIA project "CÉPAGE".
** Member of the "Institut Universitaire de France". Supported by the French-Israeli "Multi-Computing" project.

1   Introduction

A *geometric graph* is a weighted graph whose vertex set is a set of points of $\mathbb{R}^d$, and whose edge set consists of line segments joining two vertices. The weight of any edge is the Euclidean distance (L_2-norm) between its endpoints. The *Euclidean complete graph* is the complete geometric graph, in which all pairs of distinct vertices are connected by an edge.

Although geometric graphs are in theory specific weighted graphs, they naturally model many practical problems and in various fields of Computer Science, from Networking to Computational Geometry. Delaunay triangulations, Yao graphs, theta-graphs, β-skeleton graphs, Nearest-Neighborhood graphs, Gabriel graphs are just some of them [17]. A companion concept of geometric graphs is the *graph spanner*. A *t-spanner* of a graph G is a spanning subgraph H such that for each pair u, v of vertices the distance in H between u and v is at most t times the distance in G between u and v. The value t is called the *stretch factor* of the spanner.

Spanners have been independently introduced in Computational Geometry by Chew [7] for the complete Euclidean graph, and in the fields of Networking and Distributed Computing by Peleg and Ulman [31] for arbitrary graphs. Literature in connection with spanners is vast and applications are numerous. We refer to Peleg's book [30], and Narasimhan and Smid's book [28] for a comprehensive introduction to the topic.

1.1   Orthogonal Surfaces

With a point set M of $\mathbb{R}^d$ it is possible to associate other geometric objects. Assuming that M consists only of pairwise incomparable[1] points, the *orthogonal surface* of M is the geometric boundary of the set of points of $\mathbb{R}^d$ greater to at least one point of M (see Fig. 2 for $d = 3$).

Orthogonal surfaces are rich mathematical objects with connections to various fields, including order dimension, integer programming, and monomial ideals. Schnyder woods and orthogonal surfaces of coplanar points of $\mathbb{R}^3$ have been established by Miller [27], and Felsner and Zickfeld [15]. As a side effect, they gave an intuitive proof of a restricted version of the Brightwell-Trotter Theorem, which is an extension to multigraphs of Schnyder's characterization of planar graphs in terms of dimension of their incidence order [33].

The *geodesic embedding* of a point set $S \subset \mathbb{R}^2$ is a geometric graph with vertex set S. To define its edges, one considers a specific embedding $\phi : S \to \mathbb{R}^3$ such that the points of $\phi(S)$ are coplanar (see Section 2). There is an edge between the points $p, q \in S$ if the *join* point $\phi(p) \vee \phi(q)$ belongs to the orthogonal surface of $\phi(S)$, the join point being the point with maximum coordinates between $\phi(p)$ and $\phi(q)$ in each dimension.

---

[1] A point v is *greater* than u if, for each dimension i, v's ith coordinate is greater than u's ith coordinate.

1.2   Delaunay-Graphs

In his seminal paper [7], Chew has constructed plane spanners of the 2D Euclidean graph, namely planar subgraphs whose stretch factor is at most $\sqrt{10} \approx$ 3.162. His construction is based on the L_1-Delaunay graph, i.e., the dual of the Voronoi diagram for the Manhattan distance (L_1-norm). He conjectured that L_2-Delaunay graphs, i.e., classical Delaunay triangulations, are t-spanners for some constant t. This conjecture has been proved in [12], and the current best bounds on the stretch factor t of L_2-Delaunay graphs are $1.584 < t < 2.419$, proved respectively in [5] and [23]. Determining the exact stretch factor of this important class of geometric graphs is a challenging and open question. We refer to the recent survey [6].

More generally, for any given convex set Γ in the plane[2], one can define the Γ-Delaunay graphs as the dual of the Voronoi diagram of a set of points with respect to the convex distance function defined by Γ. Bose et al. [2] have shown that Γ-Delaunay graphs are plane t-spanners for some stretch factor t depending on the shape of Γ.

A natural question, widely open, is to determine whether L_2-Delaunay graphs are the "best" plane spanners in terms of stretch factor. It is known [13] that there are point sets for which no plane t-spanner can exist if $t < (1+10^{-11})\pi/2 \approx$ 1.570. On the upper bound side, Chew introduced in [7] the triangular distance-Delaunay graphs, *TD-Delaunay graphs* for short, whose convex distance function is defined from an equilateral triangle. He proved that TD-Delaunay graphs are plane 2-spanners. The stretch 2 is optimal with respect to the triangular distance because of some 3-gons.

1.3   Delaunay Realizability

Searching for the "best" plane spanner should be done, a priori, in the set of all planar graphs. Indeed, there is no advantage to limit the search to any restricted subclass, except maybe to plane triangulations. By *plane triangulation*, we mean a combinatorial plane graph in which all interior faces are triangles[3]. However, there are notorious plane triangulations that cannot be obtained from any L_2-Delaunay graphs, e.g., a K_4 for which a degree-3 vertex is added in each of its three interior faces [11,25]. This leads to the question of *realizability* of plane triangulations by L_2-Delaunay graphs, and more generally by Γ-Delaunay graphs for a convex distance function Γ. More formally, a plane triangulation G is Γ-Delaunay *realizable* if there exists a point set S such that the Γ-Delaunay graph of S is isomorphic to G.

Every triangulation of any polygon, i.e., every maximal outerplane graph, can be realized by a L_2-Delaunay graph [11]. Based on 3D hyperbolic geometry, Hodgson et al. [19] gave a combinatorial characterization of the graphs that

---

[2] To be more precise, Γ must be a compact and convex set with non-empty interior that contains its origin.

[3] Note that the outer face may not be necessarily a triangle nor the convex hull of the point set, see Figure 1(b).

are L_2-Delaunay realizable, leading to a polynomial-time recognition algorithm by the use of integer programming. The algorithm has been simplified later in [18,26]. Other connections between toughness, polyhedra of inscribable type, and L_2-Delaunay graphs have been developed in [10]. For an arbitrary convex distance function Γ, the Γ-Delaunay realizability has not yet been studied.

1.4   Theta-Graphs

Theta-graphs [9,22] and Yao graphs [34] are very popular geometric graphs that appear in the context of navigating graphs. Adjacency is defined as follows: the space around each point p is decomposed into $k \geqslant 2$ regular cones, each with apex p, and a point $q \neq p$ of a given cone C is linked to p if, from p, q is the "nearest" point in C. When the points are in general positions, the out-degree is at most k, and the points form a non-plane graph in general whenever $k > 6$.

Theta-graphs and Yao graphs differ in the way the nearest neighbor is defined. We focus on the 2D Euclidean space, so that each cone forms an angle of $\Theta_k = 2\pi/k$. For Yao graphs (Y_k-graphs for short), the nearest neighbor of p in the cone C is simply a point $q \neq p$ minimizing the L_2-distance between p and q. Whereas for theta-graphs (Θ_k-graphs for short), the nearest neighbor of p is the point whose orthogonal projection onto the bisector of C minimizes the L_2-distance.

Both graphs are known to be efficient spanners. The stretch factor of Θ_k-graphs and Y_k-graphs, proved respectively in [32] and in [34], is at most $1/(1 - 2\sin(\pi/k))$ for every $k > 6$. Very little is known for $k \leqslant 6$. For instance, it was known that Y_4-graphs are connected [16] and recently it has been shown that they are $8(29 + 23\sqrt{2})$-spanners[4] [4]. A very recent result [29] states that Y_6-graphs are 20.4-spanners. For $k = 7$, we observe that the current upper bound on the stretch of these graphs is larger than 7.562, and the upper bound drops under 2 only from $k \geqslant 13$.

Our main result relies on a specific subgraph of the Θ_k-graph, namely the *half-Θ_k-graph*, taking only half the edges, those belonging to non-consecutive cones in the counter-clockwise order (see Section 2 for a more formal definition). For even k, every Θ_k-graph is the union of two spanning half-Θ_k-graphs.

1.5   Our Results

Our main contribution is an unexpected connection between theta-graphs, TD-Delaunay graphs, and orthogonal surfaces. We stress that these objects come from rather different domains and can lead to new results. We show that (see Section 3 for a more precise statement):

> *For every point set $S \subset \mathbb{R}^2$ in general position, the half-Θ_6-graph of S,*
> *the TD-Delaunay graph of S, and the geodesic embedding of S are equal.*

Half-Θ_6-graph turns out to be a key ingredient of our result. This unification result implies that each of these objects can directly inherit of all the known properties from the others. In particular, we exhibit two important corollaries:

---

[4] This number is greater than 492.

1. *Every Θ_6-graph is the union of two spanning TD-Delaunay graphs.*

In particular, Θ_6-graphs are 2-spanners of the 2D Euclidean graph, because they contain a TD-Delaunay graph as spanning subgraph, which is a 2-spanner from [7]. Since the bound of 2 is optimal (by considering the apices of a quasi-equilateral triangle), we have therefore determined the stretch factor of Θ_6-graphs. Up to now, Θ_6-graphs were not known to be t-spanners for any constant t, and the best known bound on the stretch factor for Θ_7-graphs was larger than 7.562. Before this current paper, only Θ_k-graphs for $k \geqslant 13$ were known to be 2-spanners (see the previous best general upper bound [32]).

The other important consequence is:

2. *Every plane triangulation is TD-Delaunay realizable.*

We also show that the plane triangulation of K_4 is not L_1-Delaunay realizable, so that, to the best of our knowledge, the equilateral triangle is the only regular convex distance function Γ that is known to have the property that every plane triangulation is Γ-Delaunay realizable.

The paper is organized as follows. In Section 2 we precisely define all the objects we need, and in Section 3 we prove our main result. The corollaries are proved in Section 4. Due to space limitations, proofs are omitted.

2   Definitions

2.1   Half-Θ_6-Graph

A *cone* is the region in the plane between two rays that emanate from the same point, its apex. For each cone C, let ℓ_C be the bisector ray of C, and for each point p, let $C^p = \{x + p : x \in C\}$. Let us consider the rays obtained by a counter-clockwise rotation around the origin of the positive x-axis by angles of $2i\pi/k$ with integer i. Each pair of successive rays $2(i-1)\pi/k$ and $2i\pi/k$ defines a cone, denoted by A_i, whose apex is the origin. Let $\mathcal{A}_k = \{A_1, \ldots, A_k\}$.

The *directed Θ_k-graph* of a point set $S \subset \mathbb{R}^2$, denoted by $\overrightarrow{\Theta_k}(S)$, is defined as follows: (1) the vertex set of $\overrightarrow{\Theta_k}(S)$ is S; and (2) (p, r) is an arc of $\overrightarrow{\Theta_k}(S)$ if and only if there is a cone $A_i \in \mathcal{A}_k$ such that $r \in A_i^p \setminus \{p\}$ whose orthogonal projection onto ℓ_C^p is the closest to p.

This definition makes no assumptions on relative positions of points of S. In particular, it may happen that in $\overrightarrow{\Theta_k}(S)$ several arcs of a given cone have the same length (and the out-degree is larger than k), or some arc lie on the border of a cone. The notion of "general position" is discussed in Section 3. We now introduce a new graph, called *half-Θ_k-graph*, defined as follows:

Definition 1. *The directed half-Θ_k-graph of a point set $S \subset \mathbb{R}^2$, denoted by $\frac{1}{2}\overrightarrow{\Theta_k}(S)$, is the digraph induced by all the arcs (p, r) of $\overrightarrow{\Theta_k}(S)$ such that $r \in A_i^p$ for some even number i.*

Whenever $k \equiv 2 \pmod 4$, we denote by C_i the cone A_{2i}, and by $\overline{C}_i$ the opposite cone of C_i, i.e., $\overline{C}_i = A_{2i+k/2 \bmod 6}$ (observe that $2i + k/2$ is odd). An arc (p, r) such that $r \in C_i^p$ is said to be colored i.

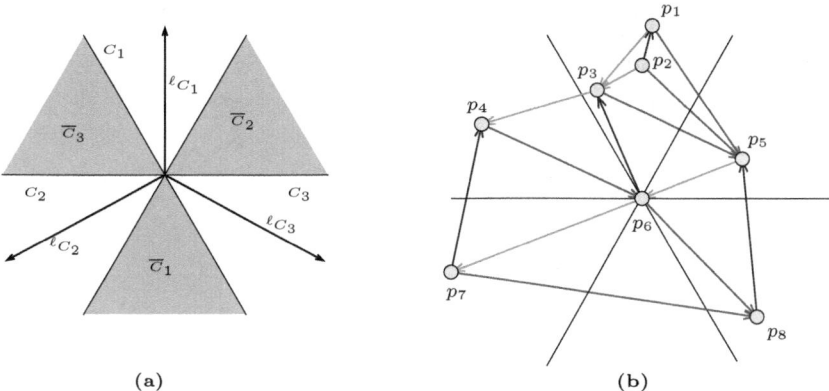

Fig. 1. (a) Illustration of notations for half-Θ_6-graphs. (b) An example of a directed half-Θ_6-graph.

In this paper, we focus on the half-Θ_6-graph. So, in counter-clockwise order starting from the positive x-axis, the six cones of $\mathcal{A}_6$ are encountered in the order $\overline{C}_2, C_1, \overline{C}_3, C_2, \overline{C}_1, C_3$ (see Fig. 1(a)). Fig. 1(b) shows an example of a directed half-Θ_6-graph on 8 points.

The set of points S is said to be *degenerate* if there exist two points p and q in S such that both (p,q) and (q,p) are arcs of $\frac{1}{2}\overrightarrow{\Theta_k}(S)$. The set S is said to be *non-degenerate* otherwise.

2.2   Geodesic Embeddings

Let $\mathcal{P}$ be a plane equipped with the standard basis $(\mathbf{e_x}, \mathbf{e_y})$, and let S be a finite set of points in the plane $\mathcal{P}$.

The following definitions are extracted from [27]. (Similar definitions can also be found in Felsner's book [14].) Let $(\mathbf{e_1}, \mathbf{e_2}, \mathbf{e_3})$ be the standard basis of $\mathbb{R}^3$. The plane $\mathcal{P}$ is now embedded in $\mathcal{P}' \subset \mathbb{R}^3$ where $\mathcal{P}'$ is the plane containing the origin of $\mathbb{R}^3$ with basis $(\mathbf{e_x'}, \mathbf{e_y'})$ where $\mathbf{e_x'} = (0, -1/\sqrt{2}, 1/\sqrt{2})$ and $\mathbf{e_y'} = (\sqrt{2/3}, -1/\sqrt{6}, -1/\sqrt{6})$. Observe that $\mathbf{e_1} + \mathbf{e_2} + \mathbf{e_3}$ is a normal vector[5] of $\mathcal{P}'$. Any point $p = (p_x, p_y) \in \mathbb{R}^2$ is mapped to $p' \in \mathcal{P}'$ with $p' = p_x \mathbf{e_x'} + p_y \mathbf{e_y'}$.

Consider the dominance order on $\mathbb{R}^3$: $p \succcurlyeq q$ if and only if $p_i \geqslant q_i$ for each $i \in \{1, 2, 3\}$. Note that any two different points of $\mathcal{P}'$ are incomparable. The *filter* generated by a set of points S of $\mathcal{P}$ is the set

$$\langle S \rangle = \left\{ \alpha \in \mathbb{R}^3 : \alpha \succcurlyeq v \text{ for some } v \in S \right\} .$$

The boundary $\mathfrak{S}_S$ of $\langle S \rangle$ is the *coplanar orthogonal surface* generated by S. Notice that in [27,15], the authors also consider *orthogonal surfaces*, a more general case where elements of S are pairwise incomparable but not necessarily

---

[5] I.e., $\forall p' = (p_1', p_2', p_3') \in \mathcal{P}'$, $p_1' \mathbf{e_1} + p_2' \mathbf{e_2} + p_3' \mathbf{e_3} = 0$.

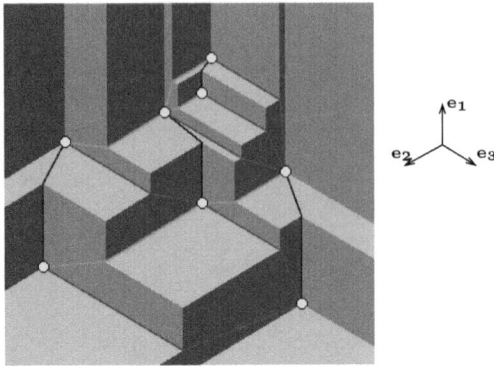

Fig. 2. A coplanar orthogonal surface with its geodesic embedding

in the same plane of normal vector $\mathbf{e_1} + \mathbf{e_2} + \mathbf{e_3}$. Fig. 2 shows an example of a coplanar orthogonal surface.

We denote by $p \vee q$ the point $(\max\{p_1, q_1\}, \max\{p_2, q_2\}, \max\{p_3, q_3\})$. If $p, q \in S$ and $p \vee q \in \mathfrak{S}_S$, then $\mathfrak{S}_S$ contains the union of the two line segments joining p and q to $p \vee q$. These lines are called *elbow geodesics* of $\mathfrak{S}_S$. The *orthogonal arc* of $p \in S$ in direction of the standard vector $\mathbf{e_i}$ is the piece of ray $p + \lambda \mathbf{e_i}, \lambda \geqslant 0$, which follows a crease of $\mathfrak{S}_S$. If $p \vee q$ is equal to $p + \lambda \mathbf{e_i}$, for some $\lambda \geqslant 0$, we say that it is an elbow of *type i*. The corresponding elbow geodesic is also said to be of type i. Observe that $p \vee q$ shares two coordinates (on the basis $(\mathbf{e_1}, \mathbf{e_2}, \mathbf{e_3})$) with at least one (and perhaps both) of p and q. We say that a geodesic elbow is *uni-directed* if its corresponding elbow $p \vee q$ shares two of its coordinates either with p or with q (but not with both).

An orthogonal surface $\mathfrak{S}_S$ is *uni-directed* if all the geodesic elbows are uni-directed.

Definition 2. *Let S be a set of points on $\mathcal{P}$ such that the orthogonal surface $\mathfrak{S}_S$ is uni-directed. The* geodesic embedding *of S is the directed graph $\overrightarrow{\mathbf{Geo}}(S)$ defined as follows:*

- *the vertices of $\overrightarrow{\mathbf{Geo}}(S)$ are the points of S.*
- *there is an arc from p to q colored i if and only if $p \vee q$ is an elbow of type i.*

2.3   TD-Delaunay Triangulation

We recall here the definition of TD-Delaunay graphs introduced in [7].

Let T (resp. $\tilde{T}$) be the equilateral triangle of side length 1 whose barycenter is the origin and one of its vertices is on the positive (resp. negative) y-axis . A *homothet* of T is obtained by scaling T with respect to the origin, followed by a translation: $p + \lambda T = \{p + \lambda z : z \in T\}$. The *triangular distance* between two points p and q is defined as follows:

$$d_T(p, q) = \min\{\lambda : \lambda \geqslant 0 \text{ and } q \in p + \lambda T\}$$

Notice that in general $d_T(p, q) \neq d_T(q, p)$.

Let S be a set of points in the plane $\mathcal{P}$. For each $p \in S$, we define the *TD-Voronoi cell of p* as:

$$V_T(p) = \{x \in \mathcal{P} : \text{ for all } q \in S, d_T(p, x) \leqslant d_T(q, x)\} \ .$$

Fig. 3(a) shows an example of a set of TD-Voronoi cells, also called TD-Voronoi diagram. Observe that the intersection of two TD-Voronoi cells may have a positive area. For instance, consider the following set $S = \{u = (-\sqrt{3}, 1), v = (\sqrt{3}, 1)\}$ (see Fig. 3(b)). The intersection $V_T(u) \cap V_T(v)$ contains the part of the plane below the two lines (o, u) and (o, v) where $o = (0, 0)$.

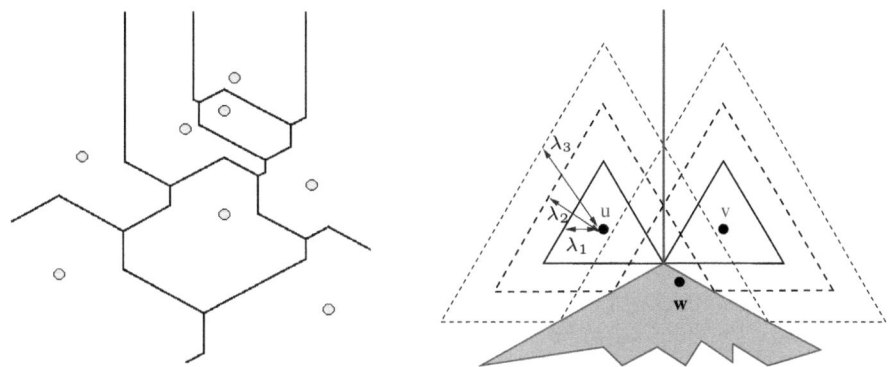

Fig. 3. (a) TD-Voronoi diagram. (b) $\lambda_1 < \lambda_2 < \lambda_3$ stand for three triangular distances. The set $\{u, v\}$ is an ambiguous point set, however $\{u, v, w\}$ is non-ambiguous.

We say that a set of points S is *non-ambiguous* if the intersection of any two TD-Voronoi cells of S is of null area[6].

Definition 3. *Let S be a non-ambiguous set of points of $\mathcal{P}$. The TD-Delaunay graph of S, denoted by* **TDDel**(S)*, is defined as follows:*

- *the vertex set of* **TDDel**(S) *is S; and*
- *(p, q) is an edge of* **TDDel**(S) *if and only if $V_T(p) \cap V_T(q) \neq \varnothing$.*

3   Unification of the Concepts

We will now prove that the three objects defined in Section 2 are essentially the same. Note that Fig. 1(b), 2, and 3(a) are based on the same set of points.

Given two points p and $q \in C_i^p$, we denote by $T_i(p, q)$ the set of points of $\mathcal{P}$ in $C_i^p \setminus \{p\}$ whose orthogonal projection onto $\ell_{C_i^p}$ is strictly closer to p than the

---

[6] For ambiguous set of points S, it is possible to have a partition of the plane by the interior of Voronoi cells plus the union of all boundaries, by ordering the elements of S to break ties. See for example [2].

orthogonal projection onto $\ell_{C_i^p}$ of q. Note that the boundary of $T_i(p,q)$ is an equilateral triangle. The interior of $T_i(p,q)$ is denoted by $T_i^\circ(p,q)$. Differently speaking, $T_i^\circ(p,q)$ is the set of points $T_i(p,q)$ deprived of the points lying on the axes of the cone C_i^p.

Lemma 1. *Let S be a set of points in the plane $\mathcal{P}$, and let p and q be two distinct points in this set. There is an arc (p,q) colored i in $\frac{1}{2}\overrightarrow{\Theta}_6(S)$ if and only if $q \in C_i^p$ and $T_i(p,q) \cap S = \varnothing$.*

Lemma 2. *Let S be a set of points in the plane $\mathcal{P}$, and let p and q be two distinct points in this set. $p \vee q$ is an elbow of type i if and only if $q \in C_i^p$ and $T_i^\circ(p,q) \cap S = \varnothing$.*

Lemma 3. *Let S be a set of points in the plane $\mathcal{P}$, and let p and q be two distinct points in this set. The Voronoi cells $V_T(p)$ and $V_T(q)$ share at least a point if and only if there exists $i \in \{1,2,3\}$ such that $q \in C_i^p$ and $T_i^\circ(p,q) \cap S = \varnothing$, or $p \in C_i^q$ and $T_i^\circ(q,p) \cap S = \varnothing$.*

Thanks to these technical lemmas, we first show the links existing between the different notions of "general position" corresponding to the three objects into consideration, and we then prove our main equivalence theorem.

Theorem 1. *Let S be a set of points in the plane $\mathcal{P}$.*
1. S is non-degenerate if and only if $\mathfrak{S}_S$ is uni-directed.
2. If S is non-degenerate, then S is non-ambiguous.

Theorem 2. *Let S be a non-degenerate point set in the plane $\mathcal{P}$. Let $\mathbf{Geo}(S)$, resp. $\frac{1}{2}\Theta_6(S)$, be the underlying undirected uncolored graph of $\overrightarrow{\mathbf{Geo}}(S)$, resp. $\frac{1}{2}\overrightarrow{\Theta}_6(S)$. We have*

$$\frac{1}{2}\Theta_6(S) = \mathbf{Geo}(S) = \mathbf{TDDel}(S) \ .$$

Moreover,

$$\overrightarrow{\mathbf{Geo}}(S) = \frac{1}{2}\overrightarrow{\Theta}_6(S) \ .$$

4   Applications

4.1   Spanner

In [7] it is shown that TD-Delaunay triangulations are plane 2-spanners. From Theorem 2, while observing that the Θ_6-graph is the union of two half-Θ_6-graphs (one using odd cones and the other even cones) we directly get the following corollary:

Corollary 1. *Every half-Θ_6-graph (and also Θ_6-graph) is a 2-spanner. Moreover the edges of the Θ_6-graph can be partitioned into two planar graphs.*

We observe that the bound of 2 is the best possible stretch for Θ_6-graphs and half-Θ_6-graphs as well. Indeed, the Θ_6-graph, and the half-Θ_6-graph as well, of some 3-gons (the apex of a quasi-equilateral triangle) has stretch arbitrarily close to 2.

4.2 Delaunay Realizability

Using the face counting algorithm introduced by Schnyder [33], Felsner and Zickfeld [15, Theorem 10] showed that for every plane triangulation G, a point set S such that $\mathbf{Geo}(S) = G$ can be computed in linear time[7]. Using the equivalence between geodesic embeddings and TD-Delaunay triangulations (Theorem 2) we directly get the following result (see Section 1.3 for the definition of realizability):

Corollary 2. *Every plane triangulation is TD-Delaunay realizable.*

This raises the following natural question: is there another distance function Γ such that every triangulation is Γ-Delaunay realizable? The first natural distance to be considered is the L_1-norm. The next theorem shows that not all triangulations are L_1-Delaunay realizable.

Theorem 3. *The plane triangulation of K_4 is not L_1-Delaunay realizable.*

To conclude on realizability, let us mention, that there are graphs that are realizable for a certain distance function and not for another and vice versa. For instance, Theorem 3 shows that K_4 is not L_1-Delaunay realizable but it is L_2-Delaunay realizable. On the other hand, there also exist graphs that are L_1-Delaunay realizable but not L_2-Delaunay realizable [11, Fig. 4].

5   Final Remarks

A Voronoi diagram is sometimes interpreted as a view from the top of a collection of cones whose apices lie on a plane and whose axes are oriented downward (see, e.g., [20]). Coplanar orthogonal surfaces are the exact generalisation of this idea for TD-Voronoi diagrams: the only difference lies on the shape of the base of the cones: circular (L_2-norm), square (L_1- and L_∞-norm) or triangular (triangular distance). Hence the TD-Voronoi cell of a point p of S is exactly the orthogonal projection on $\mathcal{P}$ of the points of $\mathfrak{S}_S$ dominated by p (see Fig. 2 and 3).

Among various generalizations of Voronoi diagrams, *Additively Weighted Voronoi Diagrams* have been widely studied (see, for example, [3,24,21]). In such a diagram, the point set is replaced by a set of weighted points $S = \{(p_1, w_1), \dots, (p_n, w_n)\}$. The distance between an element (p_i, w_i) of S and point x of the plane $\mathcal{P}$ is $d_{AW}((p_i, w_i), x) = d(p_i, x) - w_i$. The AW-Vonoroi cell of a weighted point $(p_i, w_i) \in S$ is naturally defined as follows:

$$V_{AW}(p_i, w_i) = \{x \in \mathcal{P} : \forall (p_j, q_j) \in S, i \neq j, d_{AW}((p_i, w_i), x) \leqslant d_{AW}((p_j, w_j), x)\}$$

An AW-Voronoi diagram can be interpreted as a view from the top of a collection of cones where the altitude of the apex of a cone is the weight of the corresponding element of S.

---

[7] Note that that this result holds also for 3-connected plane maps, but in this case the orthogonal surface is not uni-directed.

In our context we can define the *Additively Weighted Triangular Distance Voronoi Diagram* (or simply *AWTD-Voronoi diagram*) using the notion of distance: $d_{AWTD}((p_i, w_i), x) = d_T(p_i, x) - w_i$. From the previous remarks, one can see orthogonal surfaces (not necessarily coplanar) as AWTD-Voronoi diagrams.

The Yao graph [34] is very similar to the Θ-graph: in each cone of apex p, the selected neighbor of p is the nearest one in the cone instead of being the one with the nearest projection on ℓ_C. Half-Y_6-graphs can be defined as we did for half-Θ_6-graphs considering only 3 of the six cones. Unfortunately, half-Y_6-graphs do not have as nice structural properties. For instance, a half-Y_6-graph is not a plane graph in general.

Algorithms that compute Θ-graphs, geodesic embeddings and TD-Delaunay triangulations have been respectively proposed in [28,15,8]. It appears that the three proposed algorithms have the same time complexity of $O(n \log n)$ and are all essentially based on the "plane-sweep" algorithm. Hence, our connections between these objects do not give immediately a new insight from the algorithmic point of view.

Finally, the results in this paper have been recently used to construct a plane spanner of maximum degree 6 and stretch factor 6 of any Euclidean graph [1].

References

1. Bonichon, N., Gavoille, C., Hanusse, N., Perković, L.: Plane Spanners of Maximum Degree Six. In: Gavoille, C. (ed.) ICALP 2010. LNCS, vol. 6198, pp. 19–30. Springer, Heidelberg (2010)
2. Bose, P., Carmi, P., Collette, S., Smid, M.: On the Stretch Factor of Convex Delaunay Graphs. In: Hong, S.-H., Nagamochi, H., Fukunaga, T. (eds.) ISAAC 2008. LNCS, vol. 5369, pp. 656–667. Springer, Heidelberg (2008)
3. Bose, P., Carmi, P., Couture, M.: Spanners of Additively Weighted Point Sets. In: Gudmundsson, J. (ed.) SWAT 2008. LNCS, vol. 5124, pp. 367–377. Springer, Heidelberg (2008)
4. Bose, P., Damian, M., Douïeb, K., O'Rourke, J., Seamone, B., Smid, M.H.M., Wuhrer, S.: Pi/2-Angle Yao Graphs are Spanners. CoRR abs/1001.2913 (2010)
5. Bose, P., Devroye, L., Löffler, M., Snoeyink, J., Verma, V.: The spanning ratio of the Delaunay triangulation is greater than $\pi/2$. In: Proc. of 21st Canadian Conf. on Computational Geometry, CCCG 2009 (2009)
6. Bose, P., Smid, M.: On plane geometric spanners: a survey and open problems (2009) (submitted)
7. Chew, L.P.: There are planar graphs almost as good as the complete graph. Journal of Computer and System Sciences 39(2), 205–219 (1989)
8. Chew, P., Drysdale, R.L.: Voronoi diagrams based on convex distance functions. In: Proc. 1st Ann. Symp. on Computational Geometry, SCG 1985, pp. 235–244 (1985)
9. Clarkson, K.: Approximation algorithms for shortest path motion planning. In: Proc. 19th Ann. ACM Symp. on Theory of Computing, STOC 1987, pp. 56–65 (1987)
10. Dillencourt, M.B., Smith, W.D.: Graph-theoretical conditions for inscribability and Delaunay realizability. Discrete Mathematics 161(1-3), 63–77 (1996)

11. Dillencourt, M.B.: Toughness and Delaunay Triangulations. Discrete and Computational Geometry 5(1), 575–601 (1990)
12. Dobkin, D.P., Friedman, S.J., Supowit, K.J.: Delaunay graphs are almost as good as complete graphs. Discrete & Computational Geometry 5(4), 399–407 (1990)
13. Dumitrescu, A., Ebbers-Baumann, A., Grne, A., Klein, R., Rote, G.: On the geometric dilation of closed curves, graphs, and point sets. Computational Geometry: Theory and Applications 36(1), 16–38 (2007)
14. Felsner, S.: Geometric graphs and arrangements. Vieweg (2004)
15. Felsner, S., Zickfeld, F.: Schnyder Woods and Orthogonal Surfaces. Discrete and Computational Geometry 40(1), 103–126 (2008)
16. Fischer, M., Lukovszki, T., Ziegler, M.: Geometric searching in walkthrough animations with weak spanners in real time. In: Bilardi, G., Pietracaprina, A., Italiano, G.F., Pucci, G. (eds.) ESA 1998. LNCS, vol. 1461, pp. 163–174. Springer, Heidelberg (1998)
17. Goodman, J.E., O'Rourke, J. (eds.): Handbook of discrete and computational geometry. CRC Press, Inc., Boca Raton (1997)
18. Hiroshima, T., Miyamoto, Y., Sugihara, K.: Another Proof of Polynomial-Time Recognizability of Delaunay Graphs. IEICE Transactions on Fundamentals of Electronics, Communications and Computer Sciences E83-A(4), 627–638 (2000)
19. Hodgson, C.D., Rivin, I., Smith, W.D.: A Characterization of Convex Hyperbolic Polyhedra and of Convex Polyhedra Inscribed in the Sphere. Bulletin of the American Mathematical Society 27(3), 251–256 (1992)
20. Hoff III, K.E., Keyser, J., Lin, M., Manocha, D., Culver, T.: Fast computation of generalized Voronoi diagrams using graphics hardware. In: Proc. 26th Ann. Conf. on Comp. Graphics and Interactive Techniques, SIGGRAPH 1999, pp. 277–286 (1999)
21. Karavelas, M.I., Yvinec, M.: Dynamic Additively Weighted Voronoi Diagrams in 2D. In: Möhring, R.H., Raman, R. (eds.) ESA 2002. LNCS, vol. 2461, pp. 586–598. Springer, Heidelberg (2002)
22. Keil, J.M.: Approximating the complete Euclidean graph. In: Karlsson, R., Lingas, A. (eds.) SWAT 1988. LNCS, vol. 318, pp. 208–213. Springer, Heidelberg (1988)
23. Keil, J.M., Gutwin, C.A.: Classes of graphs which approximate the complete Euclidean graph. Discrete & Computational Geometry 7(1), 13–28 (1992)
24. Lee, D.T., Drysdale, R.L.: Generalization of Voronoi Diagrams in the Plane. SIAM Journal on Computing 10(1), 73–87 (1981)
25. Lenhart, W., Liotta, G.: Drawable and forbidden minimum weight triangulations. In: DiBattista, G. (ed.) GD 1997. LNCS, vol. 1353, pp. 1–12. Springer, Heidelberg (1997)
26. Lillis, K.M., Pemmaraju, S.V.: On the Efficiency of a Local Iterative Algorithm to Compute Delaunay Realizations. In: McGeoch, C.C. (ed.) WEA 2008. LNCS, vol. 5038, pp. 69–86. Springer, Heidelberg (2008)
27. Miller, E.: Planar Graphs as Minimal Resolutions of Trivariate Monomial Ideals. Documenta Mathematica 7, 43–90 (2002)
28. Narasimhan, G., Smid, M.: Geometric Spanner Networks. Cambridge University Press, Cambridge (2007)
29. O'Rourke, J.: The Yao Graph Y_6 is a Spanner. CoRR abs/1003.3713 (2010)
30. Peleg, D.: Distributed Computing: A Locality-Sensitive Approach. SIAM Monographs on Discrete Mathematics and Applications (2000)
31. Peleg, D., Ullman, J.D.: An optimal synchronizer for the hypercube. SIAM Journal on Computing 18(4), 740–747 (1989)

32. Ruppert, J., Seidel, R.: Approximating the d-dimensional complete Euclidean graph. In: 3rd Canadian Conference on Computational Geometry (CCCG), pp. 207–210 (1991)
33. Schnyder, W.: Planar Graphs and Poset Dimension. Order 5, 323–343 (1989)
34. Yao, A.C.C.: On constructing minimum spanning trees in k-dimensional spaces and related problems. SIAM Journal on Computing 11(4), 721–736 (1982)

Efficient Broadcasting in Random Power Law Networks*

Robert Elsässer and Adrian Ogierman

Institute for Computer Science
University of Paderborn
33102 Paderborn, Germany
{elsa,adriano}@upb.de

Abstract. We consider broadcasting in random power law graphs by using a simple modification of the so-called random phone call model introduced by Karp, Schindelhauer, Shenker, and Vöcking (FOCS 2000). In the phone call model, every time step each node calls on a randomly chosen neighbor, and establishes a communication channel to this node. The communication channels can then be used to transmit messages in both directions. We show that if we allow every node to choose ρ *neighbors* instead of one, where ρ is some constant, then the average number of message transmissions per node decreases exponentially in certain random power law graphs. Formally, we present an algorithm that completes broadcasting in time $\mathcal{O}(\log n)$ and uses $\mathcal{O}(n \log \log n)$ transmissions per message, with probability $1 - n^{-\Omega(1)}$, where n is the size of the underlying network.

Keywords: Broadcasting, Power Law Graphs.

1 Introduction

Information dissemination is one of the fundamental tasks in distributed computing. In this paper we consider the broadcasting problem, where a piece of information placed initially on a certain node in a network has to be disseminated to all nodes by using local communication only. Efficient broadcasting algorithms are very useful in various fields such as computer science, statistical physics, and sociology. As an example, consider the maintenance of replicated databases, in which every update on some data item has to be propagated to all nodes of the network deployed by the database. Another well-known application arises in the analysis of epidemic diseases. There, the question of interest is how fast the disease infects the whole population where contacts between individuals are modeled by the edges of the corresponding social network. However, in contrast to the broadcasting problem considered in this paper, spreaders are only active in a certain time window, and one is interested whether on networks modeling social contacts an epidemic outbreak occurs.

---

* Partly supported by the German Research Foundation under contract EL 399/2-1.

D.M. Thilikos (Ed.): WG 2010, LNCS 6410, pp. 279–291, 2010.
© Springer-Verlag Berlin Heidelberg 2010

There is a huge amount of work focusing on the (experimental and theoretical) analysis of the broadcasting problem in different communication models. In this paper we only consider randomized broadcasting on so called power law graphs. In e.g. [17] it has been observed that many real world networks such as the Internet, World-Wide-Web, as well as various social and biological networks have a so called power law degree distribution, i.e., the fraction of vertices with degree d is proportional to $d^{-\alpha}$, where α is a constant.

A huge amount of work concentrates on the design of random graph models which are well suited to describe the power law property observed in the networks mentioned above. Probably the most simple model is a generalization of the well-known Erdős-Rényi graphs. For a sequence $d = (d_1, \ldots, d_n)$ let $G(d)$ be the graph in which an edge is drawn between vertices i and j with probability $d_i d_j / \sum_{k=1}^{n} d_k$, independently. If d has a power law distribution, then the resulting graph is a power law graph, whp[1]. However, if the smallest degree is $o(\log n)$ (which is the case in most real world networks), then the graphs obtained by this model are not connected.

Another simple model generalizes the so called pairing or configuration model, which is often used to generate random regular graphs of small degree [17]. In this model, we start with an empty graph consisting of n nodes, where node i has d_i stubs. The graph is constructed in $\sum_{k=1}^{n} d_k / 2$ steps, and in each step we choose two unmatched stubs independently and uniformly at random. Then, these stubs are matched with each other, and an edge is created between the corresponding vertices.

In this paper we consider random graphs with power law degree distribution, in which the smallest degree is at least $\delta(\log \log n)^2$ with δ being a large constant. Using the techniques of this paper, our results can be generalized to the case when $d_{\min} \geq \delta \log \log n$. However, the proofs would require an elaborate case analysis and therefore we omit this case in the conference version of our paper.

We should mention that the real world networks mentioned above also have some other properties which are not fulfilled by random graphs. Nevertheless, overlays for peer-to-peer (P2P) networks are often constructed by using certain random graphs, e.g., Gnutella [1] or JXTA of Sun Microsystems [3], and it is worth noting that Gnutella obeys the power law degree distribution as pointed out in, e.g., [15]. Clearly, the broadcasting problem is one of the most fundamental questions in P2P systems.

In this paper, we are interested in the runtime and communication overhead produced by randomized broadcasting. One of the most simple and efficient communication models is the *random phone call model* introduced by Karp, Schindelhauer, Shenker, and Vöcking [16]. Given a graph, in each step every node opens a communication channel to a randomly chosen neighbor. The channels can then be used for bi-directional communication in this step, i.e, each node is able to send messages over every open channel incident to that node. At the end of each step, the open channels are closed. The phone call model is especially of interest in cases when plenty of different messages have to be spread in each

---

[1] With high probability or whp means with probability $1 - n^{-\Omega(1)}$.

communication step. Then, the nodes establish communication channels in each step anyway, and they decide which messages to transmit over the open channels, without knowing anything about the nodes at the other end of the channel. Thus, the cost of establishing communication channels amortizes over all transmissions, and we can consider the cost of each message separately (cf. [16]).

In the phone call model we distinguish between *push* and *pull* transmissions. If the message is forwarded over an open channel from the node which established the communication channel, then the transmission is called push. If the message is sent from the node to which the channel was opened, then we call it pull transmission. We consider randomized broadcasting protocols which can easily deal with certain kinds of edge/node failures in the network's topology [8]. We show that a simple modification of the random phone call model introduced in [9] leads to logarithmic runtime and (nearly) optimal communication overhead in the graphs described above.

1.1   Related Work

There is a huge amount of work considering epidemic type (broadcasting) algorithms on proper graph models for real world networks. Due to space constraints, we can only describe here the results which focus on the analytical study of push&pull algorithms.

Runtime. Most papers on randomized broadcasting analyze the runtime of algorithms where only push transmissions are allowed. For complete graphs of size n, Frieze and Grimmett [13] present an algorithm that broadcasts a message in time $\log_2(n) + \ln(n) + o(\log n)$ with a probability of $1 - o(1)$. Later, Pittel [18] shows that (with probability $1 - o(1)$) it is possible to broadcast a message in time $\log_2(n) + \ln(n) + f(n)$, where $f(n)$ can be any slow growing function. In [12], Feige et al. determine asymptotically optimal upper bounds for the runtime of the push algorithm on $G(n,p)$ graphs (i.e., traditional Erdös-Rényi random graphs [10,11]), bounded degree graphs, and Hypercubes. Boyd et al. consider the combined push&pull model in arbitrary graphs of size n, and show that the running time is asymptotically bounded by the mixing time of a corresponding Markov chain plus an $O(\log n)$ value [4]. Sauerwald shows that the same result also holds if only the push algorithm is considered [19].

Number of transmissions. Karp et al. [16] note that in complete graphs the pull approach is inferior to the push approach, until roughly $n/2$ nodes receive the message, and then the pull approach becomes superior. They present a push&pull algorithm, together with a termination mechanism, which reduces the number of total transmissions to $O(n \log \log n)$ (w.h.p.), and show that this result is asymptotically optimal. They also consider communication failures and analyse the performance of their method in cases where the connections are established using arbitrary probability distributions.

For sparser graphs it is not possible to get $O(n \log \log n)$ message transmissions together with a broadcast time of $O(\log n)$ in the standard phone call model.

In [7] Elsässer considers random $G(n, p)$ graphs, and shows a lower bound of $\Omega(n \log n / \log(pn))$ message transmissions for broadcast algorithms with a run-time of $O(\log n)$. On the positive side, for $p > \log^2 n/n$ he develops an algorithm that broadcasts in time $O(\log n)$ using $O(n \cdot (\log \log n + \log n / \log(pn)))$ transmissions, w.h.p. In [6] he generalizes these results to random power law graphs in which the smallest degree is bounded by $\Omega(\log^3 n)$.

In [9] Elsässer and Sauerwald consider a simple modification of the standard phone call model. In this model every node is allowed to open a channel to *four different* randomly chosen neighbors in every time step. For $G(n, p)$ graphs with $p > \log^2 n/n$, they show that this modification results in a reduction of the number of message transmissions down to $O(n \log \log n)$. In [2] Berenbrink et al. show similar results for random d-regular graphs with $d = O(\log n)$.

1.2   Our Results

In this paper we consider a simple modification of the random phone call model on random power law graphs. Our graphs are chosen uniformly at random from the space of all (simple) graphs with power law degree distribution, where the smallest degree is $t \geq \delta(\log \log n)^2$ with δ being a constant. Formally, if t denotes the smallest degree, then the number of nodes with degree $d > t$ is proportional to $(d - t)^{-\alpha}$, where $\alpha > 3$. Additionally, we assume that the largest degree in the network is $n^{1/\alpha} + t$. That is, every node which is assigned some degree larger than $n^{1/\alpha} + t$ (according to the distribution above) will have degree $n^{1/\alpha} + t$.

In each step, every node is allowed to call on ρ different neighbors chosen uniformly at random, and to establish communication channels to these nodes. The channels can be used for bi-directional communication, i.e., each node is allowed to send messages over the channels incident to it. At the end of each step, the channels are closed. We show that in this communication model there is an algorithm which completes broadcasting in time $O(\log n)$ by using $O(n \log \log n)$ transmissions of the message, whp. That is, by allowing each node to call on a constant number of different neighbors instead of one, the average number of message transmissions decreases exponentially compared to the random phone call model (cf. [6]).

To obtain the result described above, we adapt the techniques of [9] and [2] to our graphs. However, in the case of regular graphs, as considered in the papers above, one could use the fact that every node has the same behavior w.r.t. the broadcasting process in the underlying graph. In the case of power law graphs, the nodes have different degrees, and we cannot rely on similarities between nodes as in the regular case. Therefore, the main difficulty in our proofs is to cope with the different degrees during the execution of the broadcasting process.

2   Model and Annotation

In this section we describe our model, give an overview about the resulting network structure and introduce our broadcasting algorithm. The network contains

n nodes and every node $v \in V$ chooses his degree, defined as $deg(v)$, independently. The probability for a node degree $d > t$ is proportional to $\frac{1}{[d-t]^\alpha}$, with $\alpha > 3$, $t \geq \delta(\log \log n)^2$, and δ a large constant. In our proofs we assume that $t = \log^{o(1)} n$, however, our results also hold in the more general case when $t = n^{o(1)}$. If a node is assigned some node degree $d > n^{1/\alpha} + t$ according to the distribution above, then the degree of this node is set to $n^{1/\alpha} + t$. Every node has an estimation of n which is accurate to within a constant factor and all nodes have access to a global clock and work synchronously.

Before we state the results of this section, we need a few definitions. Let the set (or group) G_j contain all nodes with degree $2^j t$ to $2^{j+1}t - 1$, where $j \geq 0$. We assume for simplicity that $n^{1/\alpha} + t$ can be written as $2^k t$ for some k, and the last group only contains the nodes of degree $n^{1/\alpha} + t$. Therefore, in total we obtain less than $\log n$ different groups. The highest degree in a set G_j is defined by $deg(G_j)$. As we will see in the following subsections, it holds that the higher the node degree (of the nodes contained in a specific group), the lower the expected group size. We distinguish between the set of already informed nodes $I(i)$ before a specific round (or step) i, and its counterpart, the set of uninformed nodes $H(i)$. Furthermore, $I^+(i)$ is the set of nodes, which become informed in step $i - 1$ for the first time. $I_{G_k}(i)$ is the set of informed nodes of G_k before step i and $H_{G_k}(i)$ is the corresponding set of uninformed nodes. The set of uninformed nodes with at least l neighbors in $H(i)$ is given by $H_l(i) := \{v \in H(i) \mid |\Gamma_H(i)| \geq l\}$ where $\Gamma_H(v) := \{u \in H(i) \mid (v, u) \in E\}$.

In the following, we do not distinguish between a set and its size in our notations. That is, we always write S instead of $|S|$ in our analysis.

2.1 The Network Structure

Now we are going to provide some information about the resulting network structure. Due to space limitations some proofs are omitted.

Lemma 1. *The size of group $G_j, j \geq 1$, is $\mathcal{O}\left(\frac{n}{(2^{j+1}t)^{\alpha-1}} + \log n\right)$, whp.*

Next, we obtain an estimation on the number of nodes with degree $\mathcal{O}(t)$.

Lemma 2. *The number of nodes with degree smaller than $2t$ is $n(1 - o(1))$, whp.*

This lemma implies that the vast majority of all nodes are contained in the first group and have the minimum node degree up to a constant factor. We now state a lemma, which gives us an estimation on the average node degree of all nodes with degree higher then $2t$.

Lemma 3. *Let $deg(G_k)$ be the maximum node degree within G_k. Then, the average node degree among the nodes with degree $\geq 2t$ is bounded by t^2, whp.*

At this point we have an estimation of the resulting network structure. However, in our analysis we will consider the so called pairing (or configuration) model to construct a random graph with a certain degree distribution [17]. Thus, our results depend basically on random pairing decisions according to that model, which may lead to a multigraph. Our goal is to provide an efficient broadcasting algorithm for realistic networks, which are usually simple. Therefore, we will now state the probability for which a graph generated by the pairing model remains simple. Although the probability to achieve a simple graph may be small, it will be large enough to conclude that the results of this paper hold with high probability.

Theorem 1. *Let G be a random graph constructed according to the pairing model, and let the node degree follow a power law distribution with exponent $\alpha > 3$. Furthermore, let n denote the size of the graph and let t be the smallest degree. Then, G is a simple graph with probability $\frac{1}{e^{\mathcal{O}(t^2)}}$.*

Proof. Durrett [5, Theorem 3.1.2] showed that the number of self-loops and multiple edges are asymptotically independent Poisson random variables, which implies the statement of the theorem. □

2.2   The Broadcasting Model

The communication model used in this paper is based on the random phone call model introduced by Karp et al. [16]. In the phone call model, each node calls in every step on a neighbor chosen uniformly at random, and opens a communication channel to this node. Then, every channel can be used for bidirectional communication in this step. In our model, in each step every node calls on ρ different neighbors, and establishes communication channels to these nodes. If a channel is established between a pair of nodes, both of them are allowed to send messages over the channel. Then, the nodes have to decide which of the established channels they will use, and which messages they will send over the channel. We assume that the size of the messages exchange between a pair of nodes is not limited. However, if a specific message is really transmitted over an open channel, only then it will be counted in the communication overhead produced by that message. That is, establishing a communication channel is not taken into consideration when the total number of message transmissions are considered (cf. Introduction). The algorithm presented in this paper is *distributed*, i.e., the nodes use only local knowledge to make the decisions whether to send a message over an open channel or not. This local information can be, e.g., the age and number of broadcast messages, the time they arrived, or their own identifier. The algorithm we use is a simple modification of so called address-oblivious algorithms. An algorithm is called address-oblivious if decisions do not depend on the ID's of the nodes to which they were connected via an open channel in some previous step. That is, nodes are not allowed to remember with which nodes they communicated in the steps before (see [16]). In the following, we define some procedures which are frequently used by the nodes of the graph.

open choose ρ different neighbors, uniformly at random, and establish communication channels to them. These channels are called *outgoing* in the following. The procedure also establishes communication channels with all nodes which call on this node. These channels are called *incoming*.

push(M) send message M over all outgoing channels.
pull(M) send message M over all incoming channels.
receive receive and store all messages coming over open channels (if any).
close close all channels opened in the current round.

In each step i, any node $v \in V$ executes the procedure given in Algorithm 1 below. The algorithm will be run for every message. The nodes decide if a message has to be transmitted via push or pull, depending on the time the message has been generated. When several messages are to be considered, the node combines all messages which should be transmitted via push (pull) to a single message and forwards it over all open outgoing (incoming) channels. In the following we state the algorithm w.l.o.g. for one fixed message M and we assume that the message is created in time step 0. Hence, the age of the message is the same as the current time step i.

Algorithm 1 (i : step)

1	open
2	if $i \leq \lceil \beta \log n \rceil$ then {Phase 1}
3	if the message is created or received for the first time in the previous step then push(M) over all (ρ) outgoing channels
4	if $\lceil \beta \log n \rceil + 1 \leq i \leq \lceil \beta(\log n + \log\log n) \rceil$ then {Phase 2}
5	if the node is informed (has M) then push(M) over all (ρ) outgoing channels
6	if $\lceil \beta(\log n + \log\log n) \rceil \leq i \leq \lceil \beta \log n + 2\beta \log\log n \rceil$ then {Phase 3}
7	if the node is informed then pull(M) over all incoming channels
8	receive
9	close

In Algorithm 1, β is a large constant. The algorithm has 3 different phases, and each of these phases consists of several steps. In the first phase every informed node transmits the message exactly ρ times in a dedicated step. In the second phase the informed nodes perform push transmissions in all steps of this phase. In the third phase the informed nodes perform pull transmissions, i.e., send the message over all incoming channels in every step of this phase.

3   Analysis of the Algorithm

In this section, we analyze the behavior of Algorithm 1. For our analysis, we start with n nodes, and assign each node a certain number of stubs according to the power law distribution (as described in the introduction). At the beginning, the stubs are not paired with each other. In the first step (of the broadcasing algorithm), we consider ρ different stubs of the node on which the message is generated, and pair these stubs with unpaired stubs from the graph, according to the pairing model. To obtain a proper pairing, we divide the step into ρ substeps. In substep $j \leq \rho$, we choose (uniformly at random) a stub from the set of stubs of the first informed node, except the stubs chosen in substeps $1, \ldots, j-1$, and pair it with some unpaired stub from the graph. In some step i of the broadcasting algorithm, let $\tilde{V}(i) = \{1, \ldots, k\}$ be the set of nodes which are required to choose ρ different neighbors, and open communication channels to these nodes. The step is subdivided into $\rho \cdot k$ substeps. In substep j node $\lfloor (j-1)/\rho \rfloor + 1$ chooses a stub, uniformly at random, from the set of its stubs not considered so far in this step. If this stub is already paired, then it executes the procedure described in Algorithm 1 over the corresponding edge. Otherwise, it pairs this stub with an unpaired stub from the graph, and then it executes the procedure described in Algorithm 1. The process described above implies that in every step each node may generate up to ρ new edges in the network.

The analysis will start with one node generating the message. In *Phase 1* the message is spread via push transmissions, and we obtain a wide fundament for the remaining process.

Lemma 4. *(Phase 1.) Let G be a random network with n nodes and let the node degree follow a power law distribution with exponent $\alpha > 3$, where the smallest degree $t \geq \delta(\log \log n)^2$. Then, Algorithm 1 informs at least $\frac{n}{2}$ nodes in $\mathcal{O}(\log n)$ rounds, whp., by using $\mathcal{O}(n \log \log n)$ message transmissions.*

Due to space limitations and the similarity of this proof to the ones presented in [2], the proof is omitted. Unfortunately the same proof techniques as in Lemma 4 will not work to obtain $H(i) \leq \frac{n}{\log^{\mathcal{O}(1)} n}$. The next lemma points out that at least a constant fraction of the uninformed nodes will have at least a constant fraction of their connections in the informed group. Additionally we modify the spreading process. Obviously, a node with a high degree has a higher probability to get informed by push transmissions than a node with a low degree. This implies that the nodes in G_k will become faster informed with rising k. If now for a group G_k it holds that $|H_{G_k}(i)| < \frac{H(i)}{n} \cdot G_k$ in a step i, then we choose some nodes from this group and let them be uninformed, so that afterwards we obtain $|H_{G_k}(i)| = \frac{H(i)}{n} \cdot G_k$.

Lemma 5. *(Phase 2.) Let G be a random graph with n nodes and let the node degree follow the power law distribution with exponent $\alpha > 3$. Furthermore, let the smallest degree t be at least $\delta(\log \log n)^2$. Then, within additional $\mathcal{O}(\log \log n)$ rounds there are at most $\frac{n}{\log^{9+\Omega(1)} n}$ nodes uninformed, whp. Moreover, the number of message transmissions produced by the algorithm in these steps is $\mathcal{O}(n \log \log n)$.*

The proof is omitted due to space limitations. Until this point we only used push transmissions to spread the message. This implies that according to the process described at the beginning of this section, the stubs of the uninformed nodes are not paired yet. The next lemma considers the structure of the uninformed nodes at step s, which denotes the first time step right after Phase 2. This structural information will be used in Theorem 2.

Lemma 6. *Let G be a random network with n nodes and let the node degree follow the power law distribution with exponent $\alpha > 3$, where $t \geq \delta(\log \log n)^2$. Furthermore, let the number of uninformed nodes before step s be $\frac{n}{\log^c n}$, where $c = 9 + \Theta(1)$. Then, it holds that $\Omega\left(\frac{H(s)t}{\log^c n}\right) = H_1(s) = \mathcal{O}\left(\frac{H(s)}{\log^{q-1} n}\right)$ and $H_\rho(s+1) = \mathcal{O}\left(H(s)\left(\frac{1}{\log^{c-2} n}\right)^{\rho/2}\right)$, whp., where $2q < c$.*

Proof. Now we make an additional grouping of the nodes to achieve the desired result. There will be three groups. The first group is formed by nodes of degree between t and $\log^q n$, where $2q < c$ is a constant to be specified later. The second group contains the nodes of degree at least $\log^q n$ up to $\log^{c+1} n$. If v is such a node, then $\frac{\log n}{deg(v)} > \frac{H(s)}{n}$. The last group is the set of nodes having degree at least $\log^{c+1} n$. If v is such a node, then $\frac{\log n}{deg(v)} < \frac{H(s)}{n}$. The groups are called $H^{(1)}(s)$, $H^{(2)}(s)$, and $H^{(3)}(s)$, respectively. This grouping leads to three cases.

In the first group we know that the nodes are incident to less than $\delta \log n$ edges with both ends in $H(s)$, with probability at least $1 - n^{-2}$. These edges are called inner edges in the following. Thus, with high probability it holds that

$$H_1(s) \leq H(s)\left[1 - \left(1 - \frac{H(s)\delta \log n}{nt(1 - o(1))}\right)^{\log^q n}\right] + H^{(2)}(s) + H^{(3)}(s)$$

$$\leq H(s)\left[1 - e^{1/\left(\log^{c-1-q} n \, t\right)}\right] + H^{(2)}(s) + H^{(3)}(s)$$

$$\leq \frac{H(s)}{\log^{c-1-q} n} + H^{(2)}(s) + H^{(3)}(s).$$

In order to have an upper bound on $H_1(s)$, we need an estimate on $H^{(2)}(s) + H^{(3)}(s)$. Since the number of uninformed nodes in some group G_j is proportional to $H(s)/n$, we obtain that $H^{(2)}(s) + H^{(3)}(s) = O\left(\frac{H(s)}{(\log^q n)^{\alpha-1}}\right)$. Thus, for a properly chosen q with $2q < c$ and $q(\alpha - 1) > c$, we obtain the desired upper bound on $H_1(s)$. On the other hand, the lower bound is given by

$$H_1(s) \geq H(s)\left[1 - \left(1 - \frac{H(s)t}{nt}\right)^t\right] \geq H(s)\left[1 - e^{t/\log^c n}\right] \geq \frac{H(s)t}{\log^c n}.$$

For the second part we consider the same grouping. We know that in $H^{(3)}(s)$, each node is incident to at most $H_1(s) \log n/nt \leq H(s)/n$ inner edges. In $H^{(2)}(s)$, the nodes are incident to at most $\log n$ inner edges. In $H^{(1)}(s)$, a node is incident to ρ inner edges with probability at most $\binom{\log^q n}{\rho}\left(\frac{H_1(s)\log n}{nt(1-o(1))}\right)^\rho$. Thus, the expected number of nodes in $H_\rho(s+1)$ is bounded by

$$H^{(3)}(s)\left(\frac{H(s)\log n}{n}\right)^\rho + H^{(2)}(s) \cdot \log^{-(q-1)\rho} n + H^{(1)}(s)\binom{\log^q n}{\rho}\left(\frac{H_1(s)\log n}{nt(1-o(1))}\right)^\rho.$$

Applying standard Martingale techniques, we obtain the second statement. □

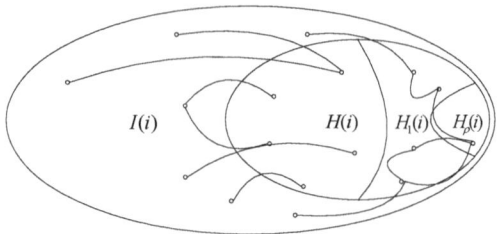

Fig. 1. Layout of the structure of the uninformed nodes at step i

Now we know how the structure of the uninformed nodes will be in step $s + 1$ and we are ready to analyze the remaining part.

Theorem 2. *Let G be a random network with n nodes and let the node degree follow the power law distribution with exponent $\alpha > 3$. Furthermore, let the smallest degree t be at least $\delta(\log \log n)^2$, where δ is a (large) constant. Then, Algorithm 1 informs all nodes of G within $\mathcal{O}(\log n)$ steps by using $\mathcal{O}(n \log \log n)$ message transmissions, whp.*

Proof. By Lemma 4, Algorithm 1 informs at least $\frac{n}{2}$ different nodes in $\mathcal{O}(\log n)$ rounds by using $\mathcal{O}(n \log \log n)$ message transmissions. After a few additional rounds and message transmissions we obtain $H(s) \leq \frac{n}{\log^{9 + \Omega(1)} n}$ (Lemma 5). The current situation is visualized in Figure 1. Here we have to keep in mind that the vast majority of nodes in $H(s)$ only has connections to nodes in $I(s)$. Thus, there is only a very small number of nodes in $H(s)$, which have at least one inner connection to some node of $H(s)$. Now, most these nodes will also become informed, since a node must have at least ρ stubs paired with nodes in $H(s)$ to have a chance to remain uninformed. With this in mind we continue our analysis. We will show that $\frac{H(s+i+2)}{H(s+i+1)} = o\left(\frac{H(s+i+1)^2}{H(s+i)^2}\right)$ for $i \geq 0$ by induction, unless $H(i + 1) \leq \log^q n$ for a (large) constant q.

For the further analysis we use a specific grouping. For each i, the nodes are divided into two sets $H'(s + i) = \left\{ v \in H(s + i) \mid \frac{1}{deg(v)} < \frac{H(s+i)}{n} \right\}$ and $H''(s + i) = \left\{ v \in H(s + i) \mid \frac{1}{deg(v)} \geq \frac{H(s+i)}{n} \right\}$. Then, the following claim holds.

Claim. Each node of $H''(s + i)$ is incident to at most $O(\log n)$ inner edges, whp., and each node of $H'(s+i)$ has at most a fraction of $O\left(\frac{H(s+i) \log n}{n}\right)$ of its neighbors in $H(s + i)$, whp.

This claim follows from a simple application of standard Chernoff bounds. We omit the proof due to space limitations. Then, for $i = 0$, we have

$$H(s + 2) \leq \mathcal{O}\left(H'(s + 1)\left(\frac{H(s + 1) \log n}{n}\right)^\rho\right) + H''(s + 1)\left(\frac{H(s + 1) \log n}{H_1(s)}\right)^\rho$$

which equals $o\left(\left(\frac{H(s+1)}{H(s)}\right)^2\right)$, for ρ large enough. Now we estimate the expected number of nodes in $H_1(s+i)$. Let c be defined as in Lemma 6. Then,

$$E[H_1(s+i)] \geq H''(s+i)\frac{H(s+i)}{\mathcal{O}(\log n) \cdot H_1(s+i-1)}$$
$$\geq \Omega\left(H(s+i)\frac{H(s+i)\log^{c-1} n}{H(s+i-1)}\right).$$

Now we derive a similar upper bound on $E[H_1(s+i)]$ and obtain that $E[H_1(s+i)]$ equals $\mathcal{O}\left(H''(s+i)\frac{H(s+i)\log^2 n}{H_1(s+i-1)}\right)+H'(s+i)$. However, since any node $v \in H'(s+i)$ has degree at least $n/H(s+i)$, we obtain

$$E[H_1(s+i)] = O\left(H(s+i)\frac{H(s+i)\log^2 n}{H_1(s+i-1)}\right) = \mathcal{O}\left(H(s+i)\frac{H_\rho(s+i-1)\log^2 n}{H_1(s+i-1)}\right).$$

By using Martingale techniques, we obtain that $H_1(s+i) = E[H_1(s+i)](1\pm o(1))$, whp. Now, $H_\rho(s+i+1)$ denotes here the set of nodes in $H''(s+i+1)$, which are incident to at least ρ inner edges. Then,

$$E[H_\rho(s+i+1)] \leq H(s+i+1)\left(\frac{H(s+i+1)\mathcal{O}(\log n)}{H_1(s+i)}\right)^\rho$$
$$\overset{(IH)}{\leq} H(s+i+1)\left(\frac{H(s+i)}{H(s+i-1)}\frac{\mathcal{O}(1)}{\log^{c-2} n\ (1-o(1))}\right)^\rho$$
$$\leq H(s+i+1)o\left(\frac{H(s+i+1)^2}{H(s+i)^2}\right),$$

whenever $\rho \geq 4$. As before, we may conclude that $H_\rho(s+i+1)$ equals $E[H_\rho(s+i+1)](1\pm o(1))$ with high probability, and the invariant follows.

Now we show that as soon as $|H(j)| \leq \log^q n$ for some j, then all nodes become informed within $\mathcal{O}(\log\log n)$ additional rounds, with high probability. For this, we are going to prove that a constant fraction of the uninformed nodes has less than ρ inner neighbors with a very high probability.

We assume that each node in $H(j)$ has degree $n^{1/\alpha}$. We consider a random subset S of size $|H(j)|$. A node has at least ρ neighbors in S with probability

$$p = \binom{n^{1/\alpha}}{\rho}\left(\frac{\log^q n\ n^{1/\alpha}}{\Theta(nt)}\right)^\rho \leq \mathcal{O}\left(\left(\frac{\log^q n}{n^{1-2/\alpha}t}\right)^\rho\right) \qquad \underbrace{\qquad}_{\text{with }\rho\text{ large enough}} \leq \qquad \frac{1}{n^{100}}.$$

Now, let A be defined as the event that *more then $\frac{7}{8}S$ nodes in S have more then ρ inner connections*. Then it follows that

$$Pr(A) = \sum_{k=\frac{7}{8}u}^u \binom{u}{k}p^k(1-p)^{u-k} \leq \left(\frac{p}{7/8}\right)^{\frac{7}{8}u}\left(\frac{1-p}{1/8}\right)^{\frac{1}{8}}u < \frac{1}{n^{50u}}.$$

The first inequality follows by [14]. Since this probability only holds for a random set S, we need to adapt this to our situation. We know that the set S has to be chosen very clumsy in order to let event A happen. However, since there are $\binom{n}{S}$ different sets of size S and $\binom{n}{S}\frac{1}{n^{50S}} \leq \frac{1}{n^{25S}}$, there is no set S in which event A occurs with probability $1 - n^{-25}$.   □

References

1. The Gnutella protocol specification v.0.4 (March 2001), http://www9.limewire.com/developer/gnutella_protocol_0.4.pdf
2. Berenbrink, P., Elsässer, R., Friedetzky, T.: Efficient randomised broadcasting in random regular networks with applications in peer-to-peer systems. In: Proceedings of PODC 2008, pp. 155–164 (2008)
3. Botros, S.M., Waterhouse, S.R.: Search in jxta and other distributed networks. In: Graham, R.L., Shahmehri, N. (eds.) Peer-to-Peer Computing, pp. 30–35. IEEE Computer Society Press, Los Alamitos (2001)
4. Boyd, S.P., Ghosh, A., Prabhakar, B., Shah, D.: Randomized gossip algorithms. Proceedings of IEEE 52(6), 2508–2530 (2006)
5. Durrett, R.: Random Graph Dynamics. Cambridge Series in Statistical and Probabilistic Mathematics. Cambridge University Press, New York (2006)
6. Elsässer, R.: On randomized broadcasting in power law networks. In: Dolev, S. (ed.) DISC 2006. LNCS, vol. 4167, pp. 370–384. Springer, Heidelberg (2006)
7. Elsässer, R.: On the communication complexity of randomized broadcasting in random-like graphs. In: Proceedings of SPAA 2006, pp. 148–157 (2006)
8. Elsässer, R., Sauerwald, T.: On the runtime and robustness of randomized broadcasting. In: Asano, T. (ed.) ISAAC 2006. LNCS, vol. 4288, pp. 349–358. Springer, Heidelberg (2006)
9. Elsässer, R., Sauerwald, T.: The power of memory in randomized broadcasting. In: Proceedings of SODA 2008, pp. 218–227 (2008)
10. Erdős, P., Rényi, A.: On random graphs I. Publicationes Mathematicae, 290–297 (1959)
11. Erdős, P., Rényi, A.: On the evolution of random graphs. Publication of the Mathematical Institute of the Hungarian Academy of Science, 17–61 (1960)
12. Feige, U., Peleg, D., Raghavan, P., Upfal, E.: Randomized broadcast in networks. In: Asano, T., Imai, H., Ibaraki, T., Nishizeki, T. (eds.) SIGAL 1990. LNCS, vol. 450, pp. 128–137. Springer, Heidelberg (1990)
13. Frieze, A., Grimmett, G.: The shortest-path problem for graphs with random-arc-lengths. Discrete Applied Mathematics 10, 57–77 (1985)
14. Hagerup, T., Rüb, C.: A guided tour of Chernoff bounds. Inf. Process. Lett. 33(6), 305–308 (1990)
15. Jovanovic, M., Annexstein, F., Berman, K.: Scalability issues in large p2p networks - a case study of Gnutella. Technical Report, University of Cincinatti (2003)
16. Karp, R., Schindelhauer, C., Shenker, S., Vocking, B.: Randomized rumor spreading. In: Proceedings of FOCS 2000, p. 565 (2000)
17. Newman, M.E.J.: The structure and function of complex networks. SIAM Review 45, 167–256 (2003)

18. Pittel, B.: On spreading a rumor. SIAM J. Appl. Math. 47(1), 213–223 (1987)
19. Sauerwald, T.: On mixing and edge expansion properties in randomized broadcasting. In: Tokuyama, T. (ed.) ISAAC 2007. LNCS, vol. 4835, pp. 196–207. Springer, Heidelberg (2007)

Graphs with Large Obstacle Numbers[*]

Padmini Mukkamala[1], János Pach[2], and Deniz Sarıöz[3]

[1] Rutgers, The State University of New Jersey
Piscataway, NJ 08904
padmini.mvs@gmail.com
[2] Ecole Polytechnique Fédérale de Lausanne
Station 8, CH-1015 Lausanne
pach@cims.nyu.edu
[3] The Graduate Center of the City University of New York
New York, NY 10016
sarioz@acm.org

Abstract. Motivated by questions in computer vision and sensor networks, Alpert et al. [3] introduced the following definitions. Given a graph G, an *obstacle representation* of G is a set of points in the plane representing the vertices of G, together with a set of connected obstacles such that two vertices of G are joined by an edge if an only if the corresponding points can be connected by a segment which avoids all obstacles. The *obstacle number* of G is the minimum number of obstacles in an obstacle representation of G. It was shown in [3] that there exist graphs of n vertices with obstacle number at least $\Omega(\sqrt{\log n})$. We use extremal graph theoretic tools to show that (1) there exist graphs of n vertices with obstacle number at least $\Omega(n/\log^2 n)$, and (2) the total number of graphs on n vertices with bounded obstacle number is at most $2^{o(n^2)}$. Better results are proved if we are allowed to use only *convex* obstacles or polygonal obstacles with a small number of sides.

1 Introduction

Consider a set P of points in the plane and a set of closed polygonal obstacles whose vertices together with the points in P are in *general position*, that is, no *three* of them are on a line. The corresponding *visibility graph* has P as its vertex set, two points $p, q \in P$ being connected by an edge if and only if the segment pq does not meet any of the obstacles. Visibility graphs are extensively studied and used in computational geometry, robot motion planning, computer vision, sensor networks, etc.; see [5], [15], [20], [21], [31].

Recently, Alpert, Koch, and Laison [3] introduced an interesting new parameter of graphs, closely related to visibility graphs. Given a graph G, we say that a set of points and a set of polygonal obstacles as above constitute an *obstacle representation* of G, if the corresponding visibility graph is isomorphic to G. A representation with h obstacles is also called an h-obstacle representation. The smallest number of obstacles in an obstacle representation of G is called the *obstacle number* of G and is denoted by $\mathrm{obs}(G)$. If

---

[*] Research supported by NSA grant 47149-00 01, NSF grant CCF-08-30272, and grants from BSF, OTKA, SNF.

we are allowed to use only *convex* obstacles, then the corresponding parameter $\mathrm{obs}_c(G)$ is called the *convex obstacle number* of G. Of course, we have $\mathrm{obs}(G) \leq \mathrm{obs}_c(G)$ for every G, but the two parameters can be very far apart.

A special instance of the obstacle problem has received a lot of attention, due to its connection to the Szemerédi-Trotter theorem on incidences between points and lines [28], [27], and other classical problems in incidence geometry [23]. It is an exciting open problem to decide whether the obstacle number of $\overline{K}_n$, the empty graph on n vertices, is $O(n)$ if the obstacles must be *points*. The best known upper bound is $n2^{O(\sqrt{\log n})}$; see Pach [22], Dumitrescu et al. [7], Matoušek [18], and Aloupis et al. [2].

Alpert et al. [3] constructed a bipartite graph G_1 and a split graph (a graph consisting of a clique and an independent set with possible edges between them) G_2 with obstacle number at least *two*. In Section 4, we complement their examples with a third one:

Theorem 1. *There is a graph G_3 that consists of two cliques with edges between them and satisfies* $\mathrm{obs}(G_3) \geq 2$.

Consequently, no graph of obstacle number *one* has an induced subgraph isomorphic to G_1, G_2, or G_3. The choice of these forbidden graphs may appear somewhat capricious at first glance. In Section 2, we will see that this set of graphs allows us to utilize some extremal graph theoretic tools developed by Erdős, Kleitman, Rothschild, Frankl, Rödl, Prömel, Steger, Bollobás, Thomason and others. They yield that the number of graphs with n vertices and bounded obstacle number is very small, compared to the total number of labeled graphs, which is $2^{\binom{n}{2}}$. More precisely, we obtain

Corollary 1. *For any fixed positive integer h, the number of graphs on n (labeled) vertices with obstacle number at most h is at most* $2^{o(n^2)}$.

Alpert et al. [3] raised the question whether there exist *bipartite* graphs with arbitrarily large obstacle number? Since the number of bipartite graphs with n labeled vertices is $\Omega(2^{n^2/4})$, it follows directly from Corollary 1 that the answer is yes.

Corollary 2. *For any fixed positive integer h, there exist bipartite graphs with obstacle number at least h.*

For every sufficiently large n, Alpert et al. constructed a graph with n vertices with obstacle number at least $\Omega(\sqrt{\log n})$. We also show in Section 2 how Theorem 1, combined with a result by Erdős and Hajnal [8], implies the existence of graphs with much larger obstacle numbers.

Corollary 3. *For every $\varepsilon > 0$, there exists an integer $n_0 = n_0(\varepsilon)$ such that for all $n \geq n_0$, there are graphs G on n vertices such that their obstacle numbers satisfy*

$$\mathrm{obs}(G) \geq \Omega\left(n^{1-\varepsilon}\right).$$

It turns out that for the proof of Corollary 3, in the place of Theorem 1 we can use the much simpler fact that there are graphs with obstacle number greater than *one*. In Section 3, we improve on the last two corollaries, using some estimates on the number of different *order types* of n points in the Euclidean plane, discovered by Goodman and Pollack [16], [17] (see also Alon [1]). We establish the following results.

Theorem 2. *For any fixed positive integer h, the number of graphs on n (labeled) vertices with obstacle number at most h is at most*

$$2^{O(hn\log^2 n)}.$$

Theorem 3. *For every n, there exist graphs G on n vertices with obstacle numbers*

$$\text{obs}(G) \geq \Omega\left(n/\log^2 n\right).$$

Note that Theorem 3 directly follows from Theorem 2. Indeed, since the total number of (labeled) graphs with n vertices is $2^{\Omega(n^2)}$, as long as $2^{O(hn\log^2 n)}$ is smaller than this quantity, there is a graph with obstacle number larger than h.

We prove a slightly better bound for convex obstacle numbers.

Theorem 4. *For every n, there exist graphs G on n vertices with convex obstacle numbers*

$$\text{obs}_c(G) \geq \Omega\left(n/\log n\right).$$

If we only allow segment obstacles, we get an even better bound. Following Alpert et al., we define the *segment obstacle number* $\text{obs}_s(G)$ of a graph G as the minimal number of obstacles in an obstacle representation of G, in which each obstacle is a straight-line segment.

Theorem 5. *For every n, there exist graphs G on n vertices with segment obstacle numbers*

$$\text{obs}_s(G) \geq \Omega\left(n^2/\log n\right).$$

In Section 4, we prove Theorem 1.

In the last section, we make some concluding remarks. In particular, we answer a question of Alpert et al. [3] by showing that for every positive integer h, there exists a graph with obstacle number *precisely* h. We also discuss possible extensions of the above notions to higher dimensions.

Given any placement (embedding) of the vertices of G in general position in the plane, a *drawing* of G consists of the image of the embedding and the set of *open segments* connecting all pairs of points that correspond to the edges of G. If there is no danger of confusion, we make no notational difference between the vertices of G and the corresponding points, and between the pairs uv and the corresponding open segments. The complement of the set of all points that correspond to a vertex or belong to at least one edge of G falls into connected components. These components are called the *faces* of the drawing. Notice that if G has an obstacle representation with a particular placement of its vertex set, then

(1) each obstacle must lie entirely in one face of the drawing, and
(2) each non-edge of G must be blocked by at least one of the obstacles.

2   Hereditary Properties, Universal Graphs, and Applications

The aim of this section is to review some results in extremal graph theory and then to apply them to establish Corollaries 1 and 3.

In 1985, Erdős, Kleitman, and Rothschild [11] proved that, as n tends to infinity, the number of all K_ℓ-free graphs on n vertices is asymptotically equal to the number of $(\ell - 1)$-partite graphs with n vertices with as equal vertex classes as possible. This result was soon generalized to graphs that do not contain some fixed (not necessarily induced) subgraph H [10].

Analogous questions based on the *induced* subgraph relation were investigated in [24], [26], and [25]. If a graph G does not contain an induced subgraph isomorphic to a fixed graph H, then the same is true for every induced subgraph of G. Therefore, this property is called *hereditary*. In order to formulate an Erdős-Kleitman-Rothschild type theorem valid for any hereditary graph property, we need some definitions and notations.

In notation, we do not distinguish between a graph property $\mathscr{P}$ and the set of all graphs that satisfy this property. In the same spirit, the set of all graphs on n labeled vertices, which satisfy property $\mathscr{P}$, is denoted by $\mathscr{P}^n$.

A graph is (r,s)-colorable if its vertex set can be partitioned into r blocks out of which s are cliques and every remaining block is an independent set. Let $\mathscr{C}(r,s)$ denote the set of all (r,s)-colorable graphs. A graph property which holds for all graphs is called *trivial*. Given any nontrivial hereditary graph property $\mathscr{P}$, define its *coloring number* as

$$r(\mathscr{P}) = \max\left\{r \mid \exists s : \mathscr{C}(r,s) \subseteq \mathscr{P}\right\}.$$

Since $r(\mathscr{P})$ is bounded from above by the number of vertices of any graph that does not satisfy $\mathscr{P}$, the parameter $r(\mathscr{P})$ exists and it is at least 1.

Theorem 6 (Bollobás, Thomason [6]). *For any nontrivial hereditary graph property* $\mathscr{P}$, *we have*

$$|\mathscr{P}^n| = 2^{\left(1 - \frac{1}{r(\mathscr{P})} + o(1)\right)\binom{n}{2}}.$$

Notice that if for some value r there is no s such that $\mathscr{C}(r,s) \subseteq \mathscr{P}$, then for every $r' > r$ there is no s for which $\mathscr{C}(r',s) \subseteq \mathscr{P}$. If there are $(2,0)$-colorable, $(2,1)$-colorable, and $(2,2)$-colorable graphs *none* of which is in $\mathscr{P}$, then by the preceding observations, $r(\mathscr{P}) = 1$. In that case, by Theorem 6, we have $|\mathscr{P}^n| = 2^{o(n^2)}$.

The familiar term for a $(2,0)$-colorable graph is bipartite. A $(2,1)$-colorable graph consists of a clique and an independent set, possibly with edges running between them; such a graph is often called a *split graph* [13], [30]. A $(2,2)$-colorable graph consists of two cliques, possibly with edges running between them—its complement is bipartite.

Proof of Corollary 1. We apply Theorem 6 to the hereditary property that a graph admits a 1-obstacle representation. The graphs G_1, G_2, and G_3 defined in the Introduction are $(2,0)$-, $(2,1)$-, and $(2,2)$-colorable. Thus, in view of the fact that, according to Alpert et al. and Theorem 1, none of them admits a 1-obstacle representation, we can conclude that the number of all graphs on n (labeled) vertices with obstacle number at most 1 is $2^{o(n^2)}$. In other words, Corollary 1 holds for $h = 1$.

For every fixed $h > 1$, consider a graph G on the vertex set $[n]$, which permits an h-obstacle representation on an n-element point set P in general position, with obstacles $O_1, \ldots, O_h$. Obviously, $E(G)$, the edge set of G, can be obtained as $\cap_{i=1}^{h} E(G_i)$, for suitable graphs G_i with obstacle number 1. Indeed, we can choose G_i to be the visibility graph of P in the presence of a single obstacle O_i ($i = 1, \ldots, h$). Therefore, the total number of labeled graphs on $[n]$ with obstacle number h can be bounded from above by the h-th power of the number of graphs with obstacle number 1. This completes the proof of Corollary 1.   □

Let G be a graph on n vertices and let k be a positive integer. We say that G is k-*universal* if it contains every graph on k vertices as an induced subgraph. Let $\hom(G)$ denote the maximum of the size of the largest independent set of vertices and the size of the largest complete subgraph in G. According to the quantitative form of Ramsey's theorem, due to Erdős and Szekeres [12], $\hom(G)$ is at least roughly $\frac{1}{2} \log n$. (In the sequel, all logarithms are taken modulo 2.)

In order to prove Corollary 3, we need the following result, which shows that if G avoids at least one induced subgraph with k vertices, for some $k \ll \log n$, then the Erdős-Szekeres bound on $\hom(G)$ can be substantially improved.

Theorem 7 (Erdős, Hajnal [8]). *For any fixed positive integer t, there is an $n_0 = n_0(t)$ with the following property. Given any graph G on $n > n_0$ vertices and any integer $k < 2^{c\sqrt{\log n}/t}$, either G is t-universal or we have $\hom(G) \geq k$. (Here $c > 0$ is a suitable constant.)*

Proof of Corollary 3. For the sake of clarity of the presentation, we systematically omit all floor and ceiling functions wherever they are not essential. Let H be a graph of t vertices that does not admit a 1-obstacle representation. Fix any $0 < \varepsilon < 1$, and choose an integer $N \geq n_0$, that satisfies the inequality

$$2^{c\sqrt{\varepsilon \log N}/t} > 2 \log N, \tag{1}$$

where c, n_0 are constants that appear in the previous theorem.

For any $n \geq N$, we set $m = n^{1-\varepsilon}$. According to a theorem of Erdős [9], there exists a graph G with n vertices such that

$$\hom(G) < 2 \log n < 2^{c\sqrt{\log(n/m)}/t}.$$

Consider an obstacle representation of G with the smallest number h of obstacles. Suppose without loss of generality that in our coordinate system all points of G have different x-coordinates. By vertical lines, partition the plane into m strips, each containing n/m points. Let G_i denote the subgraph of G induced by the vertices lying in the i-th strip ($1 \leq i \leq m$).

Obviously, we have

$$\hom(G_i) \leq \hom(G) < 2^{c\sqrt{\log(n/m)}/t},$$

for every i. Hence, applying Theorem 7 to each G_i separately, we conclude that each must be t-universal. In particular, each G_i contains an induced subgraph isomorphic to H. That is, we have $\mathrm{obs}(G_i) > 1$ for every i, which means that each G_i requires at least *two* obstacles.

As was explained at the end of the Introduction, each obstacle must be contained in an interior or in the exterior face of the graph. Therefore, in an h-obstacle representation of G, each G_i must have at least one internal face that contains an obstacle, and there must be at least one additional obstacle (which may possibly contained in the interior face of every G_i). At any rate, we have $h > m = n^{1-\varepsilon}$, as required. □

3   Encoding Graphs of Low Obstacle Number

The aim of this section is to prove Theorems 2–5. The idea is to find a short encoding of the obstacle representations of graphs, and to use this to give an upper bound on the number of graphs with low obstacle number.

We need to review some simple facts from combinatorial geometry. Two sets of points, P_1 and P_2, in general position in the plane are said to have the same *order type* if there is a one to one correspondence between them with the property that the orientation of any triple in P_1 is the same as the orientation of the corresponding triple in P_2. Counting the number of different order types is a classical task, see e.g.

Theorem 8 (Goodman, Pollack [16]). *The number of different order types of n points in general position in the plane is* $2^{O(n \log n)}$.

Observe that the same upper bound holds for the number of different order types of n labeled points, because the number of different permutations of n points is $n! = 2^{O(n \log n)}$.

In a graph drawing, the *complexity* of a face is the number of line segment sides bordering it. The following result was proved by Arkin, Halperin, Kedem, Mitchell, and Naor (see Matoušek, Valtr [19] for its sharpness).

Theorem 9 (Arkin et al. [4]). *The complexity of a single face in a drawing of a graph with n vertices is at most $O(n \log n)$.*

Note that this bound does not depend of the number of edges of the graph.

Proof of Theorem 2. For any graph G with n vertices that admits an h-obstacle representation, fix such a representation. Consider the visibility graph G of the vertices in this representation. As explained at the end of the Introduction, every obstacle belongs to a single face in this drawing. In view of Theorem 9, the complexity of every face is $O(n \log n)$. Replacing each obstacle by a slightly shrunken copy of the face containing it, we can achieve that every obstacle *is* a polygonal region with $O(n \log n)$ sides.

Let S be the point sequence starting with the vertices of G, followed by the vertices of every obstacle in cyclic order, one entire obstacle after another. Let I be the set of the starting positions of the h obstacles in S. G is completely determined by the (labeled) order type of S, together with I. To see this, first observe that I tells us which pairs in S are pairs graph vertices and which correspond to a side of some polygon. Now, notice that a given segment uv among graph vertices is blocked if and only if it meets some side ab of some polygon, for which a necessary and sufficient condition is that the ordered triples uav, avb, vbu, and bua have the same orientation.

If the length of S is N, then the number of possibilities for I is at most $\binom{N}{h} \le N^h$. Since $N \le n + c_1 h n \log n$ for some absolute constant $c_1 > 0$, according to Theorem 8 and our comment that follows it, the number of graphs with obstacle number at most h is at most

$$N^h \cdot 2^{O(N \log N)} = 2^{O(N \log N)} < 2^{chn \log^2 n},$$

for a suitable constant $c > 0$. This is a generous upper bound due to overcounting, and also because most pairs (S, I) do not encode obstacle representations.   □

If the average number of sides an obstacle can have is small, then we obtain

Theorem 10. *The number of graphs admitting an obstacle representation with at most h obstacles, having a total of at most hs sides, is at most*

$$2^{O(n \log n + hs \log(hs))}.$$

In particular, for segment obstacles ($s = 2$), Theorem 10 immediately implies Theorem 5. Indeed, as long as the bound in Theorem 10 is smaller than $2^{\binom{n}{2}}$, the total number of graphs on n labeled vertices, we can argue that there is a graph with segment obstacle number larger than h.

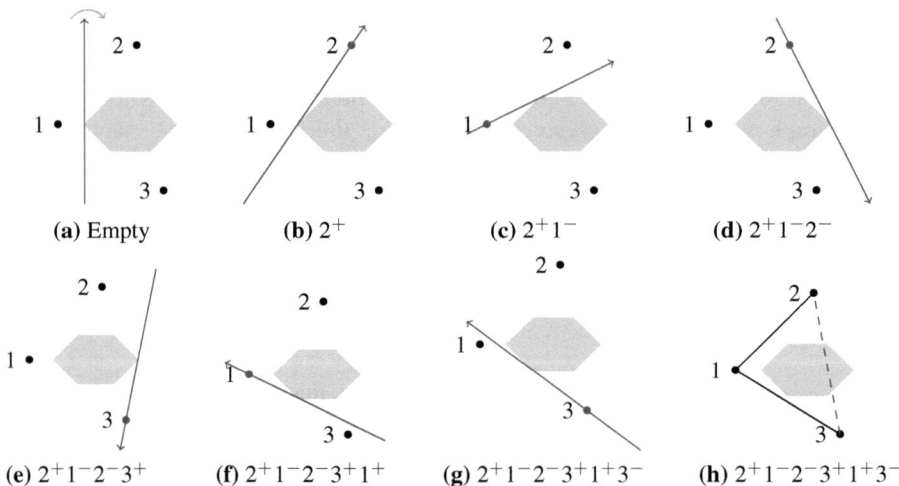

Fig. 1. Parts (a) to (g) show the construction of the sequence and (h) shows the visibilities. The arrow on the tangent line indicates the direction from the point of tangency in which we assign $+$ as a label to the vertex. The additional arrow in (a) indicates that the tangent line is rotated clockwise around the obstacle.

Proof of Theorem 4. As before, it is enough to bound the number of graphs that admit an obstacle representation with at most h convex obstacles. Let us fix such a graph G, together with a representation. Let V be the set of points representing the vertices, and let $O_1, \ldots, O_h$ be the convex obstacles. For any obstacle O_i, rotate an oriented tangent line ℓ along its boundary in the clockwise direction. We can assume without loss of generality that ℓ never passes through two points of V. Let us record the sequence of points met by ℓ. If $v \in V$ is met at the right side of ℓ, we add the symbol v^+ to the sequence, otherwise we add v^- (Fig. 1). When ℓ returns to its initial position, we stop. The resulting sequence consists of $2n$ characters. From this sequence, it is easy to reconstruct which pairs of vertices are visible in the presence of the single obstacle O_i. Observe that O_i blocks uv if and only if the subsequence induced on u and v has no consecutive pair with the same superscript. Hence, knowing these sequences for every obstacle O_i, completely determines the visibility graph G. The number of distinct sequences assigned to a single obstacle is at most $(2n)!$, so that the number of graphs with convex obstacle number at most h cannot exceed $((2n)!)^h/h! < (2n)^{2hn}$. As long as this number is smaller than $2^{\binom{n}{2}}$, there is a graph with convex obstacle number larger than h. $\qquad\square$

4   Proof of Theorem 1

Let the graph G_3 consist of a clique of blue vertices $B = \{b_i \mid i \in [4]\}$, a clique of red vertices $R = \{r_A \mid A \subseteq [4]\}$, and additional edges between every b_i and every r_A with $i \in A$. We say that a polygon is *solid* if all its edges are edges in G_3. For three distinct points p, q, and r, we denote by $\angle pqr$ the union of the rays $\overrightarrow{qp}$ and $\overrightarrow{qr}$. For a point set P, we denote by $\mathrm{conv}(P)$ the convex hull of P (the smallest convex set containing P).

Assume for contradiction that we are given a 1-obstacle representation of G_3. For a red vertex r_A, if there are points p and q such that $\angle pr_Aq$ strictly separates $\{b_i \mid i \in A\}$ from the remaining blue vertices, we say that r_A is *innocent*. If some red vertex r_A is not innocent, two obstacles will be required due to $\{r_A\} \cup B$, a contradiction.

Case 1: B is not in convex position. Without loss of generality, b_4 is inside triangle $\triangle b_1b_2b_3$.

Subcase 1a: The obstacle is in $\mathrm{conv}(B)$. Without loss of generality, the obstacle is inside $\triangle b_1b_4b_3$. Then $r_{\{1,4\}}$ is inside $\triangle b_1b_4b_3$, for the obstacle to block $b_2r_{\{1,4\}}$ and $b_3r_{\{1,4\}}$. Similarly, $r_{\{3,4\}}$ is inside $\triangle b_1b_4b_3$. For $r_{\{1,4\}}$ and $r_{\{3,4\}}$ to be innocent, the line through b_2 and b_4 separates $b_1r_{\{3,4\}}$ from $b_3r_{\{1,4\}}$. Without loss of generality, $r_{\{1,4\}}$ is inside $\triangle b_4r_{\{3,4\}}b_3$. Since $b_1r_{\{3,4\}}$ and $b_3r_{\{1,4\}}$ are separated by the solid $\triangle b_4r_{\{3,4\}}b_3$, two obstacles are needed, a contradiction.

Subcase 1b: The obstacle is outside of $\mathrm{conv}(B)$. Hence $r_{\{1,2,3\}}$ is outside of $\mathrm{conv}(B)$, and without loss of generality, in $\mathrm{conv}(\angle b_1b_4b_3)$. Therefore, the obstacle is inside the convex quadrilateral $Q = b_1b_4b_3r_{\{1,2,3\}}$. For b_1r_4 and b_3r_4 to be blocked, r_4 is inside Q. Then $\angle b_4r_4r_{\{1,2,3\}}$ separates $\mathrm{conv}(Q)$ into two regions with solid boundaries that respectively contain b_1r_4 and b_3r_4. Therefore, two obstacles are needed, a contradiction.

Case 2: B is in convex position. Without loss of generality, the bounding polygon of B is $b_1b_2b_3b_4$. In order for $r_{\{1,3\}}$ and $r_{\{2,4\}}$ to be innocent,

- $r_{\{1,3\}}$ and $r_{\{2,4\}}$ are outside of $\mathrm{conv}(B)$;
- for $r_{\{1,3\}}$: either $b_1, b_3 \in \mathrm{conv}(\angle b_2r_{\{1,3\}}b_4)$ or $b_2, b_4 \in \mathrm{conv}(\angle b_1r_{\{1,3\}}b_3)$; and
- for $r_{\{2,4\}}$: either $b_1, b_3 \in \mathrm{conv}(\angle b_2r_{\{2,4\}}b_4)$ or $b_2, b_4 \in \mathrm{conv}(\angle b_1r_{\{2,4\}}b_3)$.

Subcase 2a: $b_1, b_3 \in \mathrm{conv}(\angle b_2r_{\{1,3\}}b_4)$ and $b_2, b_4 \in \mathrm{conv}(\angle b_1r_{\{2,4\}}b_3)$. Without loss of generality, the quadrilateral $b_4b_1b_2r_{\{1,3\}}$ is convex and has b_3 inside, and without loss of generality, the quadrilateral $b_3b_4b_1r_{\{2,4\}}$ is convex and has b_2 inside. Hence,

$b_2 b_3 r_{\{1,3\}} r_{\{2,4\}}$ is a solid convex quadrilateral with $b_1 r_{\{2,4\}}$ outside and $b_3 r_{\{2,4\}}$ inside. Therefore, two obstacles are required, a contradiction.

Subcase 2b: $b_2, b_4 \in \text{conv}(\angle b_1 r_{\{1,3\}} b_3)$ or $b_1, b_3 \in \text{conv}(\angle b_2 r_{\{2,4\}} b_4)$. Due to symmetry, we proceed assuming the former. Without loss of generality, $Q = b_3 b_4 b_1 r_{\{1,3\}}$ is a convex quadrilateral. The obstacle is inside Q due to $r_{\{1,3\}} b_4$. In order for $b_1 r_{\{2,4\}}$ and $b_3 r_{\{2,4\}}$ to be blocked, $r_{\{2,4\}}$ is inside Q. Hence, $\angle r_{\{1,3\}} r_{\{2,4\}} b_4$ partitions $\text{conv}(Q)$ into two regions with solid boundaries that respectively contain $b_1 r_{\{2,4\}}$ and $r_{\{2,4\}} b_3$. Therefore, two obstacles are required, a contradiction.

This completes the proof of Theorem 1. $\qquad\square$

5   Concluding Remarks

A. First we answer a question from [3].

Proposition 1. *For every h, there exists a graph with obstacle number exactly h.*

Proof. Pick a graph G with obstacle number $h' > h$. (The existence of such a graph follows, e.g., from Corollary 1.) Let n denote the number of vertices of G. Consider a complete graph K_n on $V(G)$. Its obstacle number is *zero*, and G can be obtained from K_n by successively deleting edges. Observe that as we delete an edge from a graph G', its obstacle number cannot increase by more than *one*. This follows from the fact that by blocking the deleted edge with an additional small obstacle that does not intersect any other edge of G', we obtain a valid obstacle representation of the new graph. (Of course, the obstacle number of a graph can also *decrease* by the removal of an edge.) Since at the beginning of the process, K_n has obstacle number *zero*, at the end G has obstacle number $h' > h$, and whenever it increases, the increase is *one*, we can conclude that at some stage we obtain a graph with obstacle number precisely h. $\qquad\square$

The same argument applies to the convex obstacle number, to the segment obstacle number, and many similar parameters.

B. Let H be a fixed graph. According to a classical conjecture of Erdős and Hajnal [8], any graph with n vertices that does not have an induced subgraph isomorphic to H contains an independent set or a complete subgraph of size at least $n^{\varepsilon(H)}$, for some positive constant $\varepsilon(H)$. It follows that for any hereditary graph property there exists a constant $\varepsilon > 0$ such that every graph G on n vertices with this property satisfies $\text{hom}(G) \geq n^\varepsilon$.

Here we show that the last statement holds for the property that the graph has bounded obstacle number.

Proposition 2. *For any fixed integer $h > 0$, every graph on n vertices with $\text{obs}_c(G) \leq h$ satisfies $\text{hom}(G) \geq \frac{1}{2} n^{\frac{1}{h+1}}$.*

Proof. We proceed by induction on h. For $h = 1$, Alpert et al. [3] showed that all graphs with convex obstacle number *one* are so-called "circular interval graphs" (intersection graphs of a collection of arcs along the circle). It is known that all such graphs G whose maximum complete subgraph is of size x has an independent set of size at least $\frac{n}{2x}$; see [29]. Setting $x = \sqrt{n/2}$, it follows that $\text{hom}(G) \geq \frac{1}{2}\sqrt{n}$.

Let $h > 1$, and assume that the statement has already been verified for all graphs with convex obstacle number smaller than h. Let G be a graph that requires h convex obstacles, and consider one of its representations. Then we have $G = \cap_i G_i$, where G_i denotes the visibility graph of the same set of points after the removal of all but the i-th obstacle.

If the size of the largest independent set in G_1 is at least $\frac{1}{2}n^{\frac{1}{h+1}}$, then the statement holds, because this set is also an independent set in G. If this is not the case, then, by the above property of circular arc graphs, G must have a complete subgraph K of size at least $n^{\frac{h}{h+1}}$. Consider now the subgraph of $\cap_{i=2}^{h} G_i$ induced by the vertices of K. This graph requires only $h - 1$ obstacles. Thus, we can apply the induction hypothesis to obtain that it has a complete subgraph or an independent set of size at least $\frac{1}{2}(n^{\frac{h}{h+1}})^{\frac{1}{h}} = \frac{1}{2}n^{\frac{1}{h+1}}$.   □

It is easy to see that every graph G on n vertices with convex obstacle number at most h has the following stronger property, which implies that they satisfy the Erdős-Hajnal conjecture: There exists a constant $\varepsilon = \varepsilon(h)$ such that G contains a complete subgraph of size at least εn or two sets of size at least εn such that no edges between them belongs to G (cf. [14]).

C. Finally, we comment on higher dimensional representations. In *three* dimensions, every graph can be represented with one obstacle that is a polygonal chain.

Proposition 3. *In* three *dimensions, every* planar *graph can be represented with one convex obstacle.*

Proof. Given a planar graph G, triangulate a planar embedding of it to obtain the graph T. Now take a convex polyhedron C (no four vertices coplanar) with graph T. Let O be the convex hull of the set of midpoints of all pairs in $V(C)$ that do not correspond to edges in G. Clearly, $V(C)$ together with O (which can be perturbed to attain general position) constitute a 1-convex obstacle representation of G in *three* dimensions.   □

Proposition 4. *In dimensions* $d = 4$ *and higher, every graph can be represented with one convex obstacle.*

Proof. Let G be a graph with n vertices. Consider the moment curve

$$\{(t, t^2, t^3, t^4) : t \in \mathbb{R}\}.$$

Pick n points $v_i = (t_i, t_i^2, t_i^3, t_i^4)$ on this curve, $i = 1, \ldots, n$. The convex hull of these points is a *cyclic polytope* P_n. The vertex set of P_n is $\{v_1, \ldots, v_n\}$, and any segment connecting a pair of vertices of P_n is an edge of P_n (lying on its boundary). Denote the midpoint of the edge $v_i v_j$ by v_{ij}, and let O be the convex hull of the set of all midpoint v_{ij}, for which v_i and v_j are not connected by an edge in G. Obviously, the points v_i and the obstacle O (or its small perturbation, if we wish to attain general position) show that G admits a representation with a single convex obstacle.   □

References

1. Alon, N.: The number of polytopes, configurations and real matroids. Mathematika 33(1), 62–71 (1986)
2. Aloupis, G., Ballinger, B., Collette, S., Langerman, S., Por, A., Wood, D.R.: Blocking coloured point sets. In: 26th European Workshop on Computational Geometry (EuroCG 2010), Dortmund, Germany (March 2010), arXiv:1002.0190v1 [math.CO]
3. Alpert, H., Koch, C., Laison, J.: Obstacle numbers of graphs. Discrete and Computational Geometry, 27 (December 2009),
   http://www.springerlink.com/content/45038g67t22463g5 (viewed on 12/26/09)
4. Arkin, E.M., Halperin, D., Kedem, K., Mitchell, J.S.B., Naor, N.: Arrangements of segments that share endpoints: single face results. Discrete Comput. Geom. 13(3-4), 257–270 (1995)
5. de Berg, M., van Kreveld, M., Overmars, M., Schwarzkopf, O.: Computational Geometry. Algorithms and Applications, 2nd edn. Springer, Berlin (2000)
6. Bollobás, B., Thomason, A.: Hereditary and monotone properties of graphs. In: Graham, R.L., Nešetřil, J. (eds.) The Mathematics of Paul Erdős. Algorithms and Combinatorics 14, vol. 2, pp. 70–78. Springer, Berlin (1997)
7. Dumitrescu, A., Pach, J., Tóth, G.: A note on blocking visibility between points. Geombinatorics 19(1), 67–73 (2009)
8. Erdős, P., Hajnal, A.: Ramsey-type theorems. Discrete Appl. Math. 25(1-2), 37–52 (1989)
9. Erdős, P.: Some remarks on the theory of graphs. Bull. Amer. Math. Soc. 53, 292–294 (1947)
10. Erdős, P., Frankl, P., Rödl, V.: The asymptotic number of graphs not containing a fixed subgraph and a problem for hypergraphs having no exponent. Graph and Combinatorics 2, 113–121 (1986)
11. Erdős, P., Kleitman, D.J., Rothschild, B.L.: Asymptotic enumeration of K_n-free graphs. In: Colloq. Int. Teorie Comb., Roma, Tomo II, pp. 19–27 (1976)
12. Erdős, P., Szekeres, G.: A combinatorial problem in geometry. Compositio. Math. 2, 463–470 (1935)
13. Foldes, S., Hammer, P.L.: Split graphs having Dilworth number 2. Canadian Journal of Mathematics - Journal Canadien de Mathematiques 29(3), 666–672 (1977)
14. Fox, J., Pach, J.: Erdős–Hajnal-type results on intersection patterns of geometric objects. In: Horizons of Combinatorics, Bolyai Soc. Math. Stud., vol. 17, pp. 79–103. Springer, Berlin (2008)
15. Ghosh, S.K.: Visibility algorithms in the plane. Cambridge University Press, Cambridge (2007)
16. Goodman, J.E., Pollack, R.: Upper bounds for configurations and polytopes in $\mathbb{R}^d$. Discrete Comput. Geom. 1(3), 219–227 (1986)
17. Goodman, J.E., Pollack, R.: Allowable sequences and order types in discrete and computational geometry. In: New Trends in Discrete and Computational Geometry, Algorithms Combin., vol. 10, pp. 103–134. Springer, Berlin (1993)
18. Matoušek, J.: Blocking visibility for points in general position. Discrete & Computational Geometry 42(2), 219–223 (2009)
19. Matoušek, J., Valtr, P.: The complexity of lower envelope of segments with h endpoints. Intuitive Geometry, Bolyai Society of Math. Studies 6, 407–411 (1997)
20. O'Rourke, J.: Visibility. In: Handbook of Discrete and Computational Geometry. CRC Press Ser. Discrete Math. Appl, pp. 467–479. CRC, Boca Raton (1997)
21. O'Rourke, J.: Open problems in the combinatorics of visibility and illumination. In: Advances in Discrete and Computational Geometry, South Hadley, MA. Contemp. Math., vol. 223, pp. 237–243. Amer. Math. Soc., Providence (1999)
22. Pach, J.: Midpoints of segments induced by a point set. Geombinatorics 13(2), 98–105 (2003)

23. Pach, J., Agarwal, P.K.: Combinatorial geometry. Wiley-Interscience Series in Discrete Mathematics and Optimization. John Wiley & Sons Inc., New York (1995)
24. Prömel, H.J., Steger, A.: Excluding induced subgraphs: Quadrilaterals. Random Structures and Algorithms 2(1), 55–71 (1991)
25. Prömel, H.J., Steger, A.: Excluding induced subgraphs III: A general asymptotic. Random Structures and Algorithms 3(1), 19–31 (1992)
26. Prömel, H.J., Steger, A.: Excluding induced subgraphs II: extremal graphs. Discrete Applied Mathematics 44, 283–294 (1993)
27. Szemerédi, E., Trotter Jr., W.T.: A combinatorial distinction between the Euclidean and projective planes. European J. Combin. 4(4), 385–394 (1983)
28. Szemerédi, E., Trotter Jr., W.T.: Extremal problems in discrete geometry. Combinatorica 3(3-4), 381–392 (1983)
29. Tucker, A.: Coloring a family of circular arcs. SIAM Journal on Applied Mathematics 29(3), 493–502 (1975), http://www.jstor.org/stable/2100446
30. Tyškevič, R.I., Černjak, A.A.: Canonical decomposition of a graph determined by the degrees of its vertices. Vestsī Akad. Navuk BSSR Ser. Fīz.-Mat. Navuk 5(5), 14–26, 138 (1979) (in Russian)
31. Urrutia, J.: Art gallery and illumination problems. In: Handbook of Computational Geometry, pp. 973–1027. North-Holland, Amsterdam (2000)

The Complexity of Vertex Coloring Problems in Uniform Hypergraphs with High Degree

Edyta Szymańska*

Faculty of Mathematics and Computer Science,
Adam Mickiewicz University,
Poznań, Poland
edka@amu.edu.pl

Abstract. In this note we consider the problem of deciding whether a given r-uniform hypergraph H with minimum vertex degree at least $c\binom{|V(H)|-1}{r-1}$, has a vertex 2-coloring and a strong vertex k-coloring. Motivated by an old result of Edwards for graphs, we summarize what can be deduced from his method about the complexity of these problems for hypergraphs. We obtain the first optimal dichotomy results for 2-colorings of 3- and 4-uniform hypergraphs according to the value of c. In addition, we determine the computational complexity of strong k-colorings of 3-uniform hypergraphs for some c, leaving a gap which vanishes as $k \to \infty$.

1 Introduction

A *hypergraph* $H = (V, E)$ is a finite set of vertices V together with a family E of distinct, nonempty subsets of vertices called edges. In this paper we consider *r-uniform hypergraphs (r-graphs)* in which, for a fixed $r \geq 2$, each edge is of size r.

For an r-graph, the graph minimum degree $\delta(G)$ can be replaced by the *minimum (l-wise) degree* denoted by $\delta_l(H)$, for $1 \leq l \leq r - 1$, which is the largest integer d such that every l-element set of vertices of H is contained in at least d edges of H. One natural case is the *minimum vertex degree*, $\delta_1(H)$.

A k-coloring of a hypergraph $H = (V, E)$ is a function assigning colors from $\{1, 2, \ldots, k\}$ to vertices of H in such a way that no edge is monochromatic. The minimum number k such that H admits a k-coloring is called *the chromatic number* of H, denoted by $\chi(H)$.

We are studying the following decision problem.

Definition 1. *For fixed integers r, $1 \leq l \leq r - 1$, and k, and a real number $0 \leq c \leq 1$, define the problem $\Pi^{r,l}(k, c)$ as follows:*

Input: *r-uniform hypergraph $H = (V, E)$ with $|V(H)| = n$ and $\delta_l(H) \geq c\binom{n-l}{r-l}$*
Output: *Is H k-colorable $(\chi(H) \leq k)$?*

---

* Research supported by grant N206 017 32/2452.

D.M. Thilikos (Ed.): WG 2010, LNCS 6410, pp. 304–314, 2010.
© Springer-Verlag Berlin Heidelberg 2010

In particular, if we disregard the minimum degree condition, setting $c = 0$, we get $\Pi^{2,1}(2,0)$, the classical problem which is asking whether a given graph admits a 2-coloring, or, in other words, is bipartite. The problem becomes hard when more colors are allowed and it was shown by Lovász [10] that for every $k \geq 3$ $\Pi^{2,1}(k,0)$ is NP-complete.

In [6], Edwards considered the problem $\Pi^{2,1}(k,c)$ and found a deterministic polynomial time algorithm for the problem when $c > \frac{k-3}{k-2}$. His result is best possible in the sense that for $k \geq 3$ and $0 \leq c \leq \frac{k-3}{k-2}$ he also showed that $\Pi^{2,1}(k,c)$ is NP-complete.

We are extending this result to hypergraphs and determine the threshold value of the constant c for which 3- and 4-graphs are 2-colorable(bipartite). The hypergraph which is bipartite is also said to have *Property B*. The following theorems are the main results of this paper.

Theorem 2. $\Pi^{3,1}(2,c)$ is $\begin{cases} NP\text{-}complete \text{ for } c < \frac{1}{2}, \\ in \ P \text{ for } c > \frac{1}{2}. \end{cases}$

Theorem 3. $\Pi^{4,1}(2,c)$ is $\begin{cases} NP\text{-}complete \text{ for } c < \frac{3}{4}, \\ in \ P \text{ for } c > \frac{3}{4}. \end{cases}$

In the case of hypergraphs yet another version of coloring is being considered. A given k-coloring of H is called *strong* if for every edge of H, each color appears at most once in it. Analogously, the minimum number k such that H admits a strong k-coloring is called *the strong chromatic number* of H, denoted by $\chi_s(H)$. Notice that $\chi(H) \leq \chi_s(H)$. The related decision problem is formulated below.

Definition 4. *For fixed integers r, $1 \leq l \leq r-1$, and k, and a real number $0 \leq c \leq 1$, define the problem $\Pi_s^{r,l}(k,c)$ as follows:*
Input: *r-uniform hypergraph $H = (V,E)$ with $|V(H)| = n$ and $\delta_l(H) \geq c\binom{n-l}{r-l}$*
Output: *Is H k-strong colorable ($\chi_s(H) \leq k$)?*

This problem reduces to a graph coloring problem in the following sense.

Remark 1. **Strong coloring of an r-graph can be viewed as a vertex coloring of the clique graph $G_r(H)$ of the r-graph H, defined on the same set of vertices, with edge set $E(G_r(H)) = \{\{u,v\} : u,v \in e \text{ for some } e \in E(H)\}$. In this way, $\chi_s(H) = \chi(G_r(H))$, the ordinary chromatic number of the clique graph.**

Using the above relation we apply Edwards' graph result to the problem $\Pi_s^{3,1}(k,c)$ for $k \geq 3$ and complement it with the proof of NP-completeness for $k \geq 5$.

Theorem 5. *For $k \geq 5$ we have $\Pi_s^{3,1}(k,c)$ is* $\begin{cases} NP\text{-}complete \text{ for } c \leq \frac{(k-3)(k-4)}{(k-2)^2}, \\ in \ P \text{ for } c > \left(\frac{k-3}{k-2}\right)^2 \quad [6]. \end{cases}$

Moreover, we present an independent proof for $k = 4$ colors (Fact 15), where $\Pi_s^{3,1}(4,c)$ turns out to be polynomial for every $c > 0$.

Remark 2. For $k \geq 3$, $r \geq 3$ and $l \geq 2$ the problem $\Pi_s^{r,l}(k,c)$ is trivial for any $c > 0$. This is because in such a case the condition $\delta_l(H) \geq c\binom{n-l}{r-l}$ implies that every set of l vertices belongs to a hyperedge and all l vertices must have different colors, which imposes $\chi_s(H) = n$.

In what follows we will mainly consider the case of $l = 1$, and use a shorthand notation $\Pi^r(k,c) := \Pi^{r,1}(k,c)$ and $\Pi_s^r(k,c) := \Pi_s^{r,1}(k,c)$.

In the next section we give an overview of previous results. Section 3 contains the general framework of all proofs presented in this paper. The complexity results about the Property B in 3- and 4-graphs are presented in Sect. 4. Section 5 concerns the strong coloring problem of 3-graphs. The paper is concluded with some open problems.

2   Related Works

In the last ten years a series of papers appeared (see, e.g. [12], [13],[14]) where the structural properties of dense hypergraphs, satisfying the so called Dirac-condition were studied. They triggered further investigation of the computational aspects of such problems as matching, Hamilton cycle and packing (see e.g. [15], [8], [9]), which turned out to be polynomial under the restricted minimum degree condition.

The complexity of the hypergraph 2-coloring problem in dense hypergraphs was first addressed by Chen and Frieze in [5]. They showed, using the idea of Edwards, that every bipartite 3-uniform hypergraph with $\delta_2(H) > \alpha n$ can be 2-colored in $n^{O(1/\alpha)}$ time using a randomized algorithm. Their result, as stated there, relies on the assumption, that the input hypergraph is, indeed, 2-colorable. We eliminate this assumption by taking into account $\delta_1(H)$, i.e. the degree of a single vertex instead of a pair.

An expected polynomial time algorithm for coloring 2-colorable random 3-graphs was recently given in [11]. Other results concerning the inapproximability of 2-coloring bipartite hypergraphs were given by Guruswami, Dinur and others.

Strong colorings of general hypegraphs were studied by Agnarsson and Hall-dorssón in [1]. Their main motivation was to unify various coloring problems as strong colorings of appropriate hypergraphs and their results contain approxima-tion offline and online algorithms for strong colorings of arbitrary hypergraphs in terms of the size of a largest hyperedge and the number of hyperedges.

3   Preliminaries

3.1   General Framework

For all results presented in Sect. 4 we first show the NP-completeness of a prob-lem and this is followed by a polynomial time algorithm for the problem above the threshold. We will extensively use the notion of a link of a hypergraph. Sup-pose H is an r-graph and $v \in V(H)$. The *link (neighbourhood)* graph of v is an

$(r-1)$-graph $G(v) = \{\{x_1, \ldots, x_{r-1}\} : \{x_1, \ldots, x_{r-1}, v\} \in H\}$. If $r = 3$, then the link graph of every vertex in $V(H)$ is a graph.

All coloring algorithms given in this note share the following general framework.

1. Choose a suitable $(r-1)$-graph F and find in H a $O(\log n)$-size core K, i.e. a union of copies of F such that every vertex of H contains at least one copy of F in its link.
2. Check if $H[V(K)]$ is k-colorable.
3. For every admissible k-coloring of the core, check if it can be extended to $V(H) - V(K)$ (using properties of F and possibly 2-SAT).

The reduction to 2-SAT is possible if a coloring of the core leaves at most two colors available to every vertex outside it and if the constraints for hyperedges outside the core can be expressed by a 2-SAT formula.

The construction of the core relies on the following property of a bipartite graph.

Lemma 1. *For every bipartite graph $B = (X \cup Y, F)$ such that $|X| = n$ and $d_B(v) \geq \beta |Y|$ for all $v \in X$ there exists a set of vertices $D \subseteq Y$ of size $O(\log n)$ which dominates every vertex in X. Moreover, D can be constructed in $O(\log n)$ steps, each of them being polynomial in n.*

Proof. We will construct D by sequentially adding to it vertices of large degree. Since the degree of every vertex v in X is at least $\beta |Y|$, there exists a vertex $w \in Y$ such that $d_B(w) \geq \beta |X|$ and we add it to D. Observe that w covers at least $\beta |X|$ vertices of X, which we remove from X while w is removed from Y. In the remaining graph, every vertex in X will still have a degree of at least $\beta |Y|$ and the same argument can be repeated until all vertices in X are covered which happens after $O(\log n)$ steps. As a result, every vertex in X has a neighbor in D. $\qquad\square$

After appropriately defining the auxiliary bipartite graph, the set D will correspond to the core. In order to guarantee its existence the degree condition must be satisfied. This, in turn, will follow from the minimum degree bound of H and Turán numbers of certain subgraphs of link graphs. In the next subsection we will review some facts on Turán numbers for graphs and hypergraphs.

3.2   Turán Numbers for Graphs and Hypergraphs

Given an r-graph F, let $t_r(n, F)$, the Turán number of F, be the maximum number of edges in an n-vertex r-graph with no copy of F. The limit $d_r(F) = \lim_{n \to \infty} \frac{t_r(n,F)}{\binom{n}{r}}$ is referred to as the Turán density. In the case in which F is a complete graph on q vertices, $F = K_q$, we will denote the Turán density by $d_2(q)$. It is known that $d_2(3) = \frac{1}{2}$ and, in general, for $q \geq 3$ we have

$$d_2(q) = \frac{q-2}{q-1}. \tag{1}$$

We will recall a useful fact from [2, p.307], known as a supersaturation property, saying that graphs which exceed Turán density for a given subgraph contain already many copies of the subgraph.

Fact 6. *For all $q \geq 3$, and $\alpha > 1$, there exists $\gamma = \gamma(\alpha, q)$ such that a graph of order n with at least $\alpha d_2(q)\binom{n}{2}$ edges contains at least γn^q cliques of size q.*

Not much is known about Turán numbers for hypergraphs. One special case which we will use here is a 3-graph called Fano plane. The Fano plane F is the projective plane over the field with 2 elements. It has 7 vertices, which can be identified with the binary non-zero vectors of length 3. It has 7 edges, corresponding to the lines of the plane. A triple xyz is an edge if $x + y = z$. The Fano plane requires 3 colors for a proper coloring and therefore no 2-colorable hypergraph can contain it as a subgraph.

By the result of De Caen and Füredi [4] it is known that the Turán density of the Fano plane is $3/4$. On the other hand, from the supersaturation result by Erdős and Simonovits [3] we know that the next fact, similar to Fact 6 for graphs, is true.

Fact 7. *Let H be a 3-graph on $n \geq n_0$ vertices and F be the Fano plane. For every $\epsilon > 0$ there exists $\xi > 0$ such that if H contains at least $(\frac{3}{4} + \epsilon)\binom{n}{3}$ edges then H contains at least ξn^7 copies of F.*

4   Property B

Here we prove Theorems 2 and 3. The following two facts imply Theorem 2.

Fact 8. *For every $c < \frac{1}{2}$ the problem $\Pi^3(2, c)$ is NP-complete.*

Proof. To see the NP-completeness, consider the following reduction from $\Pi^3(2, 0)$. Let $H = (V, E)$ be a 3-graph on $|V| = n$ vertices. We fix $\epsilon > 0$ and construct a 3-graph $H' = (V', E')$ in such a way that $V' = V \cup V_1 \cup V_2$, where $|V_i| = N > \frac{n}{\epsilon}$ for $i = 1, 2$ and $E' = E \cup E_1 \cup E_2$, where

$$E_1 = \{xyz : x \in V, y \in V_1, z \in V_2\}$$

and

$$E_2 = \{xyz : x \in V_1, y \in V_2, z \in V_2 \text{ or } x \in V_1, y \in V_1, z \in V_2\}.$$

Notice that $|V'| = 2N + n$ and

$$\delta_1(H') = \min\left\{N^2 + \delta_1(H),\ nN + N(N - 1) + \binom{N}{2}\right\} \geq N^2.$$

Also, for every $\epsilon > 0$, the following inequality is true

$$N^2 > (\tfrac{1}{2} - \epsilon)\binom{2N + n}{2},$$

whenever $N > \frac{n}{\epsilon}$. Now, if $\chi(H) \leq 2$, i.e. H is 2-colorable, say in *red* and *blue*, then H' can be 2-colored by assigning blue to all vertices in V_1 and red to all vertices in V_2. On the other hand, if H' admits a 2-coloring, then, trivially, H has a 2-coloring.   □

Fact 9. *For every $c > \frac{1}{2}$ the problem $\Pi^3(2, c)$ is in P.*

Proof. We will describe the polynomial time algorithm for the problem in the case in which $c > \frac{1}{2}$. If $\delta_1(H) \geq (\frac{1}{2} + \epsilon)\binom{n-1}{2}$, then for every $v \in V$, its link graph $G(v)$ contains at least $(\frac{1}{2} + \epsilon)\binom{n}{2}$ edges. Since the Turán density for triangles $d_2(3) = \frac{1}{2}$, by Fact 6, $G(v)$ has enough edges to guarantee γn^3 triangles for some $\gamma > 0$. Thus, using Lemma 1 we can find a set K of $O(\log n)$ triangles constituting a core-like structure in H. To see this, consider an auxiliary bipartite graph $B = (X \cup Y, F)$ with $X = V(H)$ and Y corresponding to all triangles in the complete graph on n vertices. There is an edge $e = vT$ in F for $v \in X$ and a triangle T in Y whenever T is a subgraph of $G(v)$. By Fact 6 and Eq. (1), every vertex has many triangles in its link and therefore its degree satisfies the assertion of Lemma 1. Let $K := \bigcup_{T \in D} T$ be the set of triangles in the dominating set D.

Consequently, after constructing the core K, every vertex in $V(H) - V(K)$ has at least one triangle of K in its link. Next we sequentially consider all possible assignments of two colors to the vertices of K and for each of them check if it is a 2-coloring of the subhypergraph $H[V(K)]$. If none of them gives a bipartition of $H[V(K)]$ then it is not 2-colorable and so is H. Otherwise, we verify if a proper 2-coloring of $H[V(K)]$ can be extended to the vertices outside K. We can do this verification efficiently in this case because for every 2-coloring of the vertices of K there is one monochromatic edge in every $K_3 \subseteq K$. This leaves at most one color available for every vertex $v \in V(H) - V(K)$. If, after considering all triangles of the core, no color is available to a vertex or a hyperedge outside $H[V(K)]$ is monochromatic then we turn to the next 2-coloring of $H[V(K)]$. Otherwise, we return YES.   □

Facts 10 and 11 below yield Theorem 3.

Fact 10. *For every $c < \frac{3}{4}$, the problem $\Pi^4(2, c)$ is NP-complete.*

Proof. Fix $\epsilon > 0$ and for a 4-graph $H = (V, E)$ with $|V| = n$ construct a 4-graph $H' = (V', E')$ such that $V' = V \cup V_1 \cup V_2$, $|V_i| = N > \frac{n}{\epsilon}$ for $i = 1, 2$ and $E' = E \cup E_1 \cup E_2 \cup E_{12}$, where

$$E_1 = \{xyzt : x, y \in V, z \in V_1, t \in V_2\}$$

and

$$E_2 = \{xyzt : x \in V, y, z \in V_1, t \in V_2 \text{ or } x \in V, y \in V_1, z, t \in V_2\},$$

whereas E_{12} contains all quadruples intersecting only both V_1 and V_2. Observe that

$$\delta_1(H') =$$
$$\min\left\{2\binom{N}{2}N + (n-1)N^2 + \delta_1(H),\right.$$
$$\left.\binom{n}{2}N + (N-1)Nn + n\binom{N}{2} + (N-1)\binom{N}{2} + \binom{N}{3} + \binom{N-1}{2}N\right\}$$
$$\geq N^3 + o(N^3)$$

and for every $\epsilon > 0$,

$$N^3 > (\tfrac{3}{4} - \epsilon)\binom{2N+n}{3}$$

whenever $N > \frac{n}{\epsilon}$.

Thus, we have

$$\delta_1(H') > (\tfrac{3}{4} - \epsilon)\binom{|V'|}{3}.$$

If H is 2-colorable then, again, coloring V_1 red and V_2 blue results in a proper 2-coloring of H'. Conversely, if H is not 2-colorable, then there is no way to color H' with two colors. □

Fact 11. *For every $c > \frac{3}{4}$, the problem $\Pi^4(2,c)$ is in P.*

Proof. The main idea is similar to that used in the proof of Fact 9, but this time the argument relies on the Turán density of the Fano plane (see Sect. 3.2).

First we will construct a core in H using an auxiliary bipartite graph $B = (X \cup Y, F)$. Again $X = V(H)$, but Y will correspond this time to all Fano planes in the complete 3-graph on the vertices from H.

To construct a core using Lemma 1 we need to make a few observations first. By the degree assumption $\delta_1(H) \geq (\tfrac{3}{4} + \epsilon)\binom{n-1}{3}$, for every $v \in V$, its link graph $G(v)$ is a 3-graph such that $|E(G(v))| > (\tfrac{3}{4} + \epsilon)\binom{n-1}{3}$. By Fact 7, for every $v \in V(H)$ we are guaranteed at least ξn^7 copies of a Fano plane in its link graph $G(v)$ for some $\xi > 0$. Applying Lemma 1 to the graph B with $\beta = \xi$ we are able to find a $O(\log n)$-size subset D of Y which dominates every vertex in $V(H)$. This further implies that for every vertex $v \in V(H)$ there is at least one Fano plane in its link belonging to the core.

We will now proceed by sequentially considering 2-colorings of the subhypergraph induced by the vertices of the core $K := \bigcup_{F \in D} F$. If no proper 2-coloring can be found then H is not 2-colorable. Otherwise, if a proper 2-coloring of $H[V(K)]$ is encountered we will check if it can be extended to the vertices outside $V(K)$. Notice that for every 2-coloring of a Fano plane there is at least one monochromatic edge. In consequence, by our construction, every vertex in $V(H) - V(K)$ will have at most one color available. This implies that it can be verified in polynomial time if every proper coloring of the core with two colors can be extended to the remaining vertices and the fact follows. □

Remark 3. Since for all $c > 0$, $\delta_l(H) \geq c\binom{n-l}{r-l}$ implies $\delta_{l-1}(H) \geq c\binom{n-l+1}{r-l+1}$, the results in Facts 9 and 11 carry over to $l \geq 2$. It is not clear what kind of an $(r-1)$-graph would need to replace the Fano plane to get a similar result for $r \geq 5$.

5   Strong Coloring

In this section we prove Thereom 5. As mentioned in the Introduction (see Remark 1), a strong coloring of an r-graph can be seen as a proper coloring of the clique graph $G_r(H)$ obtained from $H = (V, E)$ by replacing each hyperedge with an r-clique. There is a relation stated below between the minimum vertex degree in a 3-graph H and the minimum vertex degree in its clique graph $G_3(H)$.

Fact 12. *For every $c > 0$ and a 3-graph $H = (V, E)$, if $\delta_1(H) \geq c\binom{n-1}{2}$ then $\delta(G_3(H)) \geq \sqrt{c}(n - 1)$.*

Proof. Let v be a vertex in $V(H)$ of the minimum degree $\delta_1(H)$ and let i be the number of isolated vertices in its link graph $G(v)$. Then $d_{G_3(H)}(v) = n-1-i := x$ and the smallest value of x satisfying $c\binom{n-1}{2} \leq \binom{x}{2}$ gives us a bound on the minimum degree in $G_3(H)$. Since $\frac{x-1}{n-2} \leq \frac{x}{n-1}$, we have $x^2 \geq c(n-1)^2$ and hence $\delta(G_3(H)) \geq \sqrt{c}(n-1)$.   □

In addition to Fact 12, we will need the following result of Edwards for graphs.

Theorem 13 (Th. 2.5,[6]). *For $k \geq 3$ we have*
$$\Pi^2(k, c) \text{ is } \begin{cases} \textit{NP-complete for } c \leq \frac{k-3}{k-2}, \\ \textit{in P for } c > \frac{k-3}{k-2}. \end{cases}$$

The statement below is directly implied by Theorem 13 and Fact 12.

Corollary 14. *For $k \geq 3$ and $c > \left(\frac{k-3}{k-2}\right)^2$ the problem $\Pi_s^3(k, c)$ is in P.*

Notice that if $k = 3$, then $\Pi_s^3(3, c)$ is in P for all $c > 0$. In the case of $k = 4$, Corollary 14 guarantees a polynomial algorithm for $\Pi_s^3(4, c)$ when $c > \frac{1}{4}$. We improve this and get the same result for all values of $c > 0$.

Fact 15. *For every $c > 0$, the problem $\Pi_s^3(4, c)$ is in P.*

Proof. We will follow the idea of the proof by Edwards [6] again. This time, in addition, we will reduce the problem $\Pi_s^3(4, c)$ for $c > 0$ to 2-SAT in polynomial time. Given a fixed strong 4-coloring f of the vertices of K we construct a 2-SAT formula $\phi(f)$ as follows. For every vertex $v \in V(H) - V(K)$ let $S(v)$ be the set of admissible colors for v. Observe that as a consequence of Lemma 1 and $\delta_1(H) \geq c\binom{n-1}{2}$, there is at least one edge in the link of every vertex of $V(H)$ and thus two colors are already used, so $|S(v)| \leq 2$. Let $V(H) - V(K) = \{v_1, \ldots, v_m\}$ and assign a variable x_{ij} to represent the fact that "vertex v_i has color j" for every $1 \leq i \leq m$, $1 \leq j \leq 4$. Now, $\phi(f)$ will contain the following variables and clauses:

- $\{x_{ij} | j \in S(v_i)\}$ for $1 \leq i \leq m$
- $\overline{x}_{ij} \vee \overline{x}_{ih}$ for $1 \leq i \leq m, 1 \leq j < h \leq 4$
- $\overline{x}_{ij} \vee \overline{x}_{hj}, \overline{x}_{ij} \vee \overline{x}_{lj}, \overline{x}_{hj} \vee \overline{x}_{lj}$ for $\{v_i, v_h, v_l\} \in H$ and $1 \leq j \leq 4$.

The pseudo-code of the algorithm is below.

ALGORITHM STRONGCOLOR

In: $k = 4$, a 3-graph H with $\delta_1(H) \geq c\binom{n-1}{2}$, where $0 < c \leq 1$.
Out: YES, if H has a strong $4-$coloring ; NO, otherwise

---

1. Build an auxiliary bipartite graph $B = (X \cup Y, F)$, where $X = V(H)$ and Y corresponds to all edges $T = K_2$ of a complete graph on n vertices. For every $v \in X$ and edge $T \in Y$ we add an edge $vT \in F$ if $T \subseteq G(v)$. (Note that for every $v \in X$, $d_B(v) \geq c|Y|$.)
2. Using Lemma 1 find a subset $D \subseteq Y$ which dominates every vertex in X. Let $K = \bigcup_{T \in D} T$.
3. For every function $f : V(K) \to \{1, 2, 3, 4\}$ do
   (a) If f is a strong 4-coloring of $H[V(K)]$ then check if f can be extended on $V(H) - V(K)$, i.e. construct an instance $\phi(f)$ of 2-SAT on the set of vertices $V(H) - V(K)$.
   (b) If $\phi(f)$ is satisfiable then return YES (and present the final strong 4-coloring).
4. If no strong 4-coloring found, return NO.

Observe that Steps 2 and 3 can be performed in polynomial time. Also, it is well known that 2-SAT is in P (see [7]).   □

It is worth noticing here that the above result does not necessarily imply that the problem of coloring clique graphs of hypergraphs with 4 colors is always polynomial.

On the hardness side, we have another, more general result.

Fact 16. *For $k \geq 5$ and $c \leq \frac{(k-3)(k-4)}{(k-2)^2}$, the problem $\Pi_s^3(k,c)$ is NP-complete.*

Proof. It is clear that $\Pi_s^3(k,c)$ is in NP. Now we will show that it is NP-complete for $c = \frac{(k-3)(k-4)}{(k-2)^2}$. The proof will use a reduction from 3-colorability, it is from $\Pi_s^3(3,0)$, which is NP-complete.

For this, given a 3-graph $H = (V, E)$, where $|V(H)| = n$ we construct a 3-graph $H' = (V', E')$ in the following way. For V' put $V_0 = V$ and add disjoint sets $V_1, \ldots, V_{k-3}$ of new vertices such that $|V_i| = n$ for $i = 1, \ldots, k-3$. Include E into E' and add an edge for every triple of vertices $\{x, y, z\}$, such that $x \in V_i$, $y \in V_j$, $z \in V_k$, where i, j, k are different indices from $\{0, 1, \ldots, k-3\}$. Observe that $N = |V'| = (k-2)n$ and the minimum degree

$$\delta_1(H') = \frac{(k-3)(k-4)n^2}{2} \geq \frac{(k-3)(k-4)}{(k-2)^2}\binom{N}{2}.$$

If H is 3-colorable in the strong sense then, since $V_1, \ldots, V_{k-3}$ are strongly independent sets in H', we can color each of them with a different color and hence H' admits a strong k-coloring this way. On the other hand, if H' can be colored using k colors, in a strong way, then the colors used for $V_1, \ldots, V_{k-3}$ must be all distinct. Therefore, only at most the same 3 colors are available for all vertices in V_0 and thus H is strongly 3-colorable.   □

Fact 16 and Corollary 14 together yield Theorem 5.

Concluding Remarks and Open Problems

Recently the author extended the results presented in this paper to $\Pi^{r,2}(2, c)$ and described polynomial time algorithms in the cases $r = 4, 5$ for $c < \frac{1}{2}$ and $c < \frac{3}{4}$, respectively. The computational complexity of the general problem $\Pi^{r,l}(k, c)$ with $1 \leq l \leq r - 1$ and $r > 5$ remains open. Moreover, it would be interesting to see what is the complexity of the strong coloring problem $\Pi_s^3(k, c)$ in the remaining interval for $c \in \left[\frac{(k-3)(k-4)}{(k-2)^2}, \left(\frac{k-3}{k-2} \right)^2 \right].$

Acknowledgements

I thank Marek Karpiński for bringing to my attention the result of Edwards.

References

1. Agnarsson, G., Halldorssón, M.: Strong colorings of hypergraphs. In: Persiano, G., Solis-Oba, R. (eds.) WAOA 2004. LNCS, vol. 3351, pp. 253–266. Springer, Heidelberg (2005)
2. Bollobás, B.: Extremal Graph Theory. Academic Press, London (1978)
3. Erdős, P., Simonovits, M.: Supersaturated graphs and hypergraphs. Combinatorica 3(2), 181–192 (1983)
4. De Caen, D., Füredi, Z.: The maximum size of 3-uniform hypergraphs not containing a Fano plane. J. Combinatorial Theory B 78, 274–276 (2000)
5. Chen, H., Frieze, A.: Coloring Bipartite Hypergraphs. In: Cunningham, W.H., Queyranne, M., McCormick, S.T. (eds.) IPCO 1996. LNCS, vol. 1084, pp. 345–358. Springer, Heidelberg (1996)
6. Edwards, K.: The complexity of colouring problems on dense graphs. Theoretical Computer Science 43, 337–343 (1986)
7. Even, S., Itai, A., Shamir, A.: On the Complexity of Timetable and Multicommodity Flow Problems. SIAM J. Comput. 5(4), 691–703 (1976)
8. Karpiński, M., Ruciński, A., Szymańska, E.: The Complexity of Perfect Matching Problems on Dense Hypergraphs. In: Dong, Y., Du, D.-Z., Ibarra, O. (eds.) ISAAC 2009. LNCS, vol. 5878, pp. 626–636. Springer, Heidelberg (2009)
9. Karpiński, M., Ruciński, A., Szymańska, E.: Computational Complexity of the Hamiltonian Cycle Problem in Dense Hypergraphs. In: López-Ortiz, A. (ed.) LATIN 2010. LNCS, vol. 6034, pp. 663–674. Springer, Heidelberg (2010)

10. Lovász, L.: Coverings and coloring of hypergraphs. In: Proceedings of the Fourth Southeastern Conference on Combinatorics, Graph Theory, and Computing, (Florida Atlantic Univ., Boca Raton, Fla., 1973), p. 312 (1973)
11. Person, Y., Schacht, M.: An expected polynomial time algorithm for coloring 2-colorable 3-graphs. In: Proceedings of EuroComb 2009. Electronic Notes In Discrete Mathematics, vol. 34, pp. 465–469 (2009)
12. Pikhurko, O.: Perfect matchings and K_4^3-tilings in hypergraphs of large codegree. Graphs Combin. 24(4), 391–404 (2008)
13. Rödl, V., Ruciński, A., Szemerédi, E.: Diracs theorem for 3-uniform hypergraphs. Combin. Probab. Comput. 15(1-2), 229–251 (2006)
14. Rödl, V., Ruciński, A., Szemerédi, E.: Perfect matchings in large uniform hypergraphs with large minimum collective degree. J. Combin. Theory, Ser. A 116, 613–636 (2009)
15. Szymańska, E.: The Complexity of Almost Perfect Matchings in Uniform Hypergraphs with High Codegree. In: Fiala, J., Kratochvíl, J., Miller, M. (eds.) IWOCA 2009. LNCS, vol. 5874, pp. 438–449. Springer, Heidelberg (2009)

The Number of Bits Needed to Represent a Unit Disk Graph

Colin McDiarmid[1] and Tobias Müller[2,*]

[1] University of Oxford
cmcd@stats.ox.ac.uk
[2] Centrum Wiskunde & Informatica
tobias@cwi.nl

Abstract. We prove that for sufficiently large n, there exist unit disk graphs on n vertices such that for every representation with disks in the plane at least $c^{\sqrt{n}}$ bits are needed to write down the coordinates of the centers of the disks, for some $c > 1$. We also show that d^n bits always suffice, for some $d > 1$.

1 Introduction and Statement of Results

A *unit disk graph* is the intersection graph of equal sized disks in the plane. That is, we can represent the vertices by disks $D_1, \ldots, D_n \subseteq \mathbb{R}^2$ of equal radius in such a way that $ij \in E$ if and only if $D_i \cap D_j \neq \emptyset$. Equivalently, we can represent G by a sequence of points $\mathcal{V} = (z_1, \ldots, z_n)$ in the plane such that $ij \in E(G)$ if and only if $\|z_i - z_j\| \leq 1$. We say that such a $\mathcal{V}$ *realizes* G.

Over the past 20 years or so, unit disk graphs have been the subject of a sustained research effort by many different authors. Partly because of their relevance for practical applications one of the main foci is the design of (efficient) algorithms for them.

One can of course store the unit disk graph G in a computer as an adjacency matrix or a list of edges, but for many purposes (algorithms) it is useful to actually store a representation as points in the plane. In this article we will study the number of bits that are needed to store such a representation. There are of course infinitely many realizations, but we will focus on a realization whose coordinates have the smallest possible bit size. Here we shall use the convention that a rational number is stored as a pair of integers (the denominator and numerator) that are relatively prime and those integers are stored in the binary number format (see for instance [7]).

We will denote the bit size of a rational number $q \in \mathbb{Q}$ by $\text{size}(q)$. The bit size of a point $z \in \mathbb{Q}^2$ will be the sum of the bit sizes of its coordinates $\text{size}(z) := \text{size}(z_x) + \text{size}(z_y)$, and the bit size of a realization $\mathcal{V} = (z_1, \ldots, z_n)$ of a unit disk graph G will be $\text{size}(\mathcal{V}) := \sum_{i=1}^{n} \text{size}(z_i)$. We are thus interested in the following quantity for G a unit disk graph:

* Research partially supported by a VENI grant from Netherlands Organisation for Scientific Research (NWO).

D.M. Thilikos (Ed.): WG 2010, LNCS 6410, pp. 315–323, 2010.

$$\text{size}(G) := \min_{\mathcal{V} \text{ realizes } G} \text{size}(\mathcal{V}).$$

Our main result in this paper is the following:

Theorem 1. *There exists a $\gamma > 1$ such that for each n, there exists a unit disk graph on n vertices with $\text{size}(G) > \gamma^{\sqrt{n}}$.*

Theorem 1 answers a question of Spinrad [8]. This question was also studied by Van Leeuwen and Van Leeuwen [6], who dubbed it the *Polynomial Representation Hypothesis* (PRH) for unit disk graphs. The PRH for unit disk graphs states that a unit disk graph can always be realized by points whose bit sizes are bounded by some polynomial in the number of vertices n. Theorem 1 above thus shows that this hypothesis is false. It is known that unit disk graph recognition is NP-hard [1], but membership in NP is still an open problem. Had the PRH been true, then this would have proved membership in NP, but as it is this remains an open problem.

Theorem 1 could be seen as bad news for those wishing to design algorithms for unit disk graphs. On the slightly positive side we offer the following upper bound:

Theorem 2. *There exists a constant γ such that for each n, each unit disk graph G on n vertices has $\text{size}(G) \leq \gamma^n$.*

Our results also hold for disk graphs (intersection graphs of disks not all of the same radius), but the proofs are more involved. We therefore postpone these proofs to the journal version of this paper.

2   Proofs

A *line arrangement* is a family $\mathcal{L} = (\ell_1, \ldots, \ell_m)$ of lines in the plane. A line arrangement is *simple* if every two lines intersect (there are no parallel lines), and there is no point on three lines.

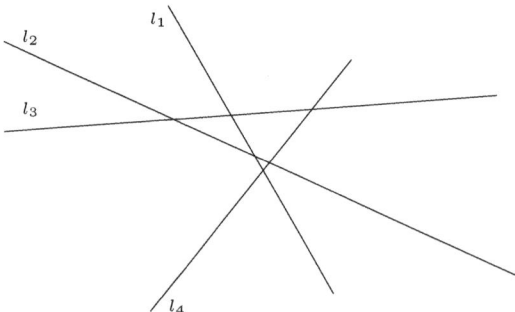

Fig. 1. A simple line arrangement

A *segment* is the portion of a line between two intersections with other lines. The *size* of a line arrangement $\mathcal{L}$ is the quotient of the length of a longest segment and a shortest segment.

The *combinatorial description* $\mathcal{D}$ of a line arrangement $\mathcal{L}$ is obtained as follows. We assume without loss of generality that no line other than l_0 is vertical. We add an auxiliary vertical line ℓ_0 to the left of all intersection points of the lines and we store the order in which each line $\ell_0, \ldots, \ell_k$ intersects the others from left to right (top to bottom for ℓ_0). For instance, the combinatorial description of the line arrangement in figure 1 is given by the sequences $(1, 2, 3, 4), (0, 3, 2, 4), (0, 3, 1, 4), (0, 2, 1, 4), (0, 1, 2, 3)$. If the line arrangement $\mathcal{L}$ has combinatorial description $\mathcal{D}$ then we say that $\mathcal{L}$ *realizes* $\mathcal{D}$.

We shall make use of the following impressive result of Kratochvil and Matousek [5] and independently Goodman, Pollack and Sturmfels [2].

Theorem 3 ([5],[2]). *For every k, there exists a combinatorial description of a simple line arrangement on $O(k)$ lines such that every realization of it has size at least 2^{2^k}.*

Here it should be mentioned that Kratochvil and Matousek's proof of Theorem 3 can only be found in the technical report version [4]. The following proposition allows us to encode a combinatorial description of a simple line arrangement into a unit disk graph.

Lemma 1. *Let $\mathcal{D}$ be a combinatorial description of a simple line arrangement on k lines. There exists a unit disk graph G on $O(k^2)$ vertices such that for every realization $\mathcal{V} = (z_0, \ldots, z_m)$ of it, up to isometry, the line arrangement $\mathcal{L} = \{\ell_1, \ldots, \ell_k\}$ where*

$$\ell_i := \{z : \|z - z_{2i}\| = \|z - z_{2i+1}\|\}, \quad i = 1, \ldots, k,$$

is a realization of $\mathcal{D}$. Moreover, all the segments defined by $\mathcal{L}$ will have length at most one.

Proof. Let $\mathcal{L}$ be a realization of $\mathcal{D}$, and let ℓ_0 be a vertical line to the left of all intersection points. We will call a connected component of $\mathbb{R}^2 \setminus (\bigcup_{i=0}^{k} \ell_i)$ a *cell*. Let c denote the number of cells and put $m = 2k + 1 + c$. It is easily seen that $c = 1 + 1 + 2 + 3 + \cdots + k = 1 + \binom{k+1}{2}$. So in particular $m = O(k^2)$. Let us arbitrarily label the cells as $C_1, \ldots, C_c$. For $i = 1, \ldots, c$, we place a point p_{2k+1+i} in the interior of C_i (we shall define points $p_0, \ldots, p_{2k+1}$ shortly).

Since none of the p_js that have been defined until now lie on the line l_i, for any sufficiently large radius R, we can place disks $D_i^0(R)$ and $D_i^1(R)$ of radius R on either side of ℓ_i such that all the p_js are contained in one of $D_i^0(R)$ and $D_i^1(R)$ (see figure 2). We now choose R big enough so that $D_i^0(R), D_i^1(R)$s can be constructed with this property for all $i = 0, \ldots, k$ and moreover we make sure that R is bigger than the distance between any of the p_js that have been defined until now. We let p_{2i} be the center of $D_i^0(R)$ and p_{2i+1} the center of $D_i^1(R)$ for $i = 0, \ldots, k$. To finish our construction, we set $z_i := \frac{1}{R} p_i$ for $i = 0, \ldots, m$

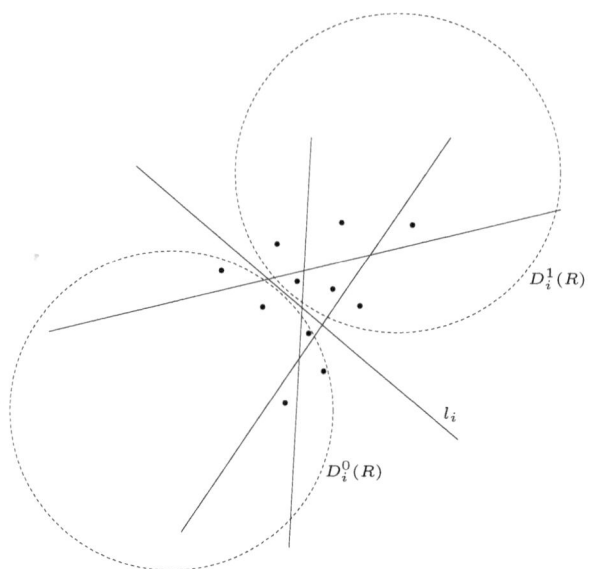

Fig. 2. The construction of $D_i^1(R), D_i^2(R)$

and $\mathcal{V} := (z_0, \ldots, z_m)$, and we let G be the corresponding unit disk graph on vertex set $\{0, \ldots, m\}$. To avoid confusion, let us stress that we use the second definition of a unit disk graph given in the introduction, i.e. $ij \in E(G)$ if and only if $\|z_i - z_j\| \leq 1$. This corresponds to the intersection graph of disks of radius $\frac{1}{2}$ centered on the z_is (or disks of radius $\frac{R}{2}$ centered on the p_is). Let us observe that the set of vertices $C = \{2k+2, \ldots, m\}$ forms a clique in G, and the neighbourhoods $N(2i), N(2i+1)$ partition C into two non-empty parts for all $i = 0, \ldots, k$.

We claim that G is as required. To this end, let $\mathcal{V}' := (z_0', \ldots, z_m')$ be an arbitrary realization of G. For $i = 0, \ldots, k$ we set

$$\ell_i' := \{z : \|z - z_{2i}'\| = \|z - z_{2i+1}'\|\}.$$

One of p_0, p_1 was to the left of ℓ_0 in our original construction, without loss of generality assume it was p_0. By applying a suitable isometry if needed, we can assume that ℓ_0' is vertical, and z_0' lies to the left of ℓ_0' (and z_1' to its right).

Now consider an arbitrary line ℓ_i for some $i \in \{0, \ldots, k\}$. In the original arrangement $\mathcal{L}$, the line ℓ_i intersects the other lines in some order $(i_1, i_2, \ldots, i_k)$ from left to right (top to bottom if $i = 0$ – in the next few paragraphs "left" should be replaced by "top" and "right" by "bottom" in case $i = 0$). We wish to show that ℓ_i' intersects the other ℓ_j's in the same order.

Let us relabel the points in the cells that are neighbouring ℓ_i as $t_0, \ldots, t_k$ and $b_0, \ldots, b_k$ where the t_js lie in the cells above ℓ_i and the b_j lie in the cells below ℓ_i, and t_0, b_0 lie in the leftmost cells, t_1, b_1 in the second leftmost cells and so on

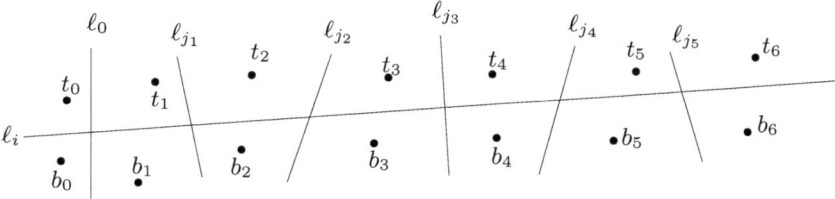

Fig. 3. The intersections of ℓ_i with the other lines

(see figure 3). For $r = 0, \ldots, k$, let t'_r denote the point of $\mathcal{V}'$ corresponding to t_r and let b'_r denote the point corresponding to b_r (by "corresponding to" we mean that both represent the same vertex of G).

We say that a line ℓ *separates* two sets $A, B \subseteq \mathbb{R}^2$ if A and B lie on different sides of ℓ. For $j = 1, \ldots, k$ let us set $A_j := \{t_r, b_r : r \leq j\}$, $B_j := \{t_r, b_r : r > j\}$ and $A'_j := \{t'_r, b'_r : r \leq j\}$, $B'_j := \{t'_r, b'_r : r > j\}$. By our choice of $p_0, \ldots, p_m$ above, all points in A_j have distance $< R$ from p_{2i_j} and distance $> R$ from p_{2i_j+1}, and these inequalities are reversed for the points in B_j (swapping the labels of p_{2i_j}, p_{2i_j+1} if necessary). By construction of G, we must then also have that all the points in A'_j have distance ≤ 1 to z'_{2i_j} and distance > 1 to z'_{2i_j+1} and these inequalities are reversed for the points in B'_j. Thus, $\ell'_{i_j} := \{z : \|z - z'_{2i_j}\| = \|z - z'_{2i_j+1}\|\}$ separates A'_j from B'_j for all $j = 1, \ldots, k$. For $r = 0, \ldots, k$, let u'_r be the intersection point of the segment $[t'_r, b'_r]$ with ℓ'_i and let v'_r be the intersection point of ℓ'_{j_r} with ℓ'_i (see figure 4). We must have that on ℓ'_i the points $\{u'_0, \ldots, u'_j\}$

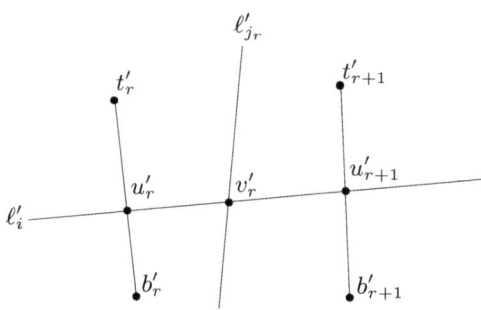

Fig. 4. The definition of u'_r and v'_r

and $\{u'_{j+1}, \ldots, u'_k\}$ lie on different sides of v'_j (here we use that the line segment $[t'_r, b'_r]$ stays on the same side of ℓ'_{i_j} that t'_r, b'_r are on). From this last observation it now follows that the order of the u'_rs and v'_rs on ℓ'_i from left to right is either $(u'_0, v'_1, u'_1, v'_2, \ldots, v'_k, u'_k)$ or the reverse order $(u'_k, v'_k, u'_{k-1}, v_{k-1}, \ldots, v'_1, u'_0)$.

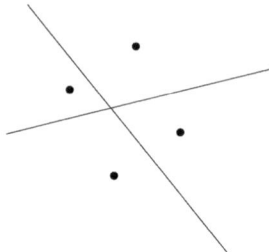

Fig. 5. An intersection point lies in the convex hull of the cell-points

Suppose that $i > 0$. In this case we must have that $i_1 = 0$, because the line ℓ_0 lies to the left of all intersection points between other lines in the original line arrangement $\mathcal{L}$. Note that t'_0, b'_0 and hence also u'_0 lie to the left of l'_0, since t'_0, b'_0 have distance ≤ 1 to z'_0 and distance > 1 to z'_1 and we have made sure that z'_0 lies to the left of l'_0 and z'_1 to the right. From this it follows that u'_0 is the leftmost point among the u'_js, which in turn shows that ℓ_i intersects the other lines in the desired order.

Suppose that $i = 0$. We have seen that ℓ'_0 intersects the other lines from top to bottom either in the correct order, or in the reverse of the correct order. In this last case we can reflect our point set $(z'_0, \ldots, z'_m)$ through the x-axis (notice this does not change the left-to-right orders on other lines) to fix it.

This proves that indeed, after applying a suitable isometry, $\mathcal{L}'$ has combinatorial description $\mathcal{D}$.

It remains to check that all the segments of the line arrangement $\mathcal{L}'$ have length at most 1. To this end, let p' be the intersection point of two lines ℓ'_i and ℓ'_j. Each of the four connected components of $\mathbb{R}^2 \setminus (\ell'_i \cup \ell'_j)$ must contain at least one element of $C' := \{z'_{2k+2}, \ldots, z'_m\}$. This is because the points of C' inside each of these regions corresponds with one of the non-empty sets of vertices $C \cap N(2i) \cap N(2j), C \cap N(2i+1) \cap N(2j), C \cap N(2i) \cap N(2j+1), C \cap N(2i+1) \cap N(2j+1)$ in G, where $C = \{2k+2, \ldots, m\}$ and $N(j)$ denotes the neighbourhood of j in G as before. It follows that p' lies in the convex hull $\text{conv}(C')$ of C' (see figure 5). Finally notice that, as C is a clique in G, the distance between any two points of C' is at most 1. Thus, if p', q' are two intersection points of the lines, then

$$\|p' - q'\| \leq \text{diam}(\text{conv}(C')) = \text{diam}(C') \leq 1.$$

This concludes the proof of Lemma 1   □

The last ingredient for the proof of Theorem 1 is the following elementary observation. For completeness we provide a proof in the appendix.

Lemma 2. *Let $a, b \in \mathbb{Q}$ be two rational numbers with bit sizes* $\text{size}(a), \text{size}(b) \leq B$. *Then* $\text{size}(a + b), \text{size}(a - b), \text{size}(ab), \text{size}(a/b) \leq 4B$.   □

Proof of Theorem 1. For an arbitrary $k \in \mathbb{N}$, let $\mathcal{D}$ be the combinatorial description from theorem 3, and let G be the unit disk graph that Lemma 1 constructs from it. It suffices to show that any realization of G has bit size at least $\Omega(2^k)$.

To this end, let $\mathcal{V} = (z_0, \ldots, z_m)$ be an arbitrary realization of G, and let B be such that each coordinate of each z_i is stored using at most B bits. Note that we can also write l_i as

$$l_i = \{z : (z_{2i+1} - z_{2i})^T z = (z_{2i+1} - z_{2i})^T (z_{2i+1} + z_{2i})/2\}.$$

This shows that the intersection point between two lines is the solution to a 2×2 linear system $Az = b$ where each entry of A and b can be obtained by applying a bounded number of arithmetic operations (i.e. addition, substraction, multiplication, division) to the coordinates of a bounded number of z_js. By Lemma 2, the entries of A and b thus all have bit size $O(B)$. From the familiar formula

$$\begin{pmatrix} a_{11} & a_{12} \\ a_{21} & a_{22} \end{pmatrix}^{-1} = \begin{pmatrix} \frac{a_{22}}{a_{11}a_{22}-a_{12}a_{21}} & \frac{-a_{12}}{a_{11}a_{22}-a_{12}a_{21}} \\ \frac{-a_{21}}{a_{11}a_{22}-a_{12}a_{21}} & \frac{a_{11}}{a_{11}a_{22}-a_{12}a_{21}} \end{pmatrix},$$

we see that the intersection point $z = A^{-1}b$ can be obtained by a bounded number of arithmetic operations from the entries of A and b. By another application of Lemma 2 we thus have that all intersection points have bit size $O(B)$.

By construction of G, there are two intersection points v, w with $0 < \|v-w\| \le 1/2^{2^k}$ (the longest segment has length at most 1 and the ratio of the longest to the smallest segment is at least 2^{2^k}). Let $s_x := |v_x - w_x|$, $s_y := |v_y - w_y|$. At least one of these numbers must be positive, assume without loss of generality it is s_x. Recall that rational numbers are stored as a pair (denominator, numerator) of integers. We see from $0 < s_x \le 1/2^{2^k}$ that its numerator must be at least 2^{2^k}, which gives size$(s_x) \ge 2^k$. On the other hand, we have already seen that size$(s_x) = O(B)$. Hence $B = \Omega(2^k)$, which concludes the proof. $\qquad\square$

Theorem 2 is a straightforward consequence of a result of Grigor'ev and Vorobjov. The following is a reformulation of Lemma 10 in [3]:

Lemma 3 ([3]). *For each $d \in \mathbb{N}$ there exists a constant $C = C(d)$ such that the following hold. Suppose that $h_1, \ldots, h_k$ are polynomials in n variables with integer coefficients, and degrees $\deg(h_i) < d$. Suppose further that the bit sizes of the all coefficients are less than B. If there exists a solution $(x_1, \ldots, x_n) \in \mathbb{R}^n$ of the system $\{h_1 \ge 0, \ldots, h_k \ge 0\}$, then there also exists one with $|x_1|, \ldots, |x_n| \le \exp[(B + \ln k)C^n]$.*

Proof of Theorem 2. Let G be a unit disk graph on n vertices. Consider the set of all $(x_1, y_1, \ldots, x_n, y_n, R) \in \mathbb{R}^{2n+1}$ that satisfy:

$$(x_i - x_j)^2 + (y_i - y_j)^2 \leq (R - 10)^2, \text{ for all } ij \in E(G),$$
$$(x_i - x_j)^2 + (y_i - y_j)^2 \geq (R + 10)^2, \text{ for all } ij \notin E(G),$$
$$R \geq 100.$$

This is a system of $1 + \binom{n}{2}$ polynomial inequalities of degree less than 3 in $2n + 1$ variables, with all coefficients small integers. It follows from the fact that any unit disk graph has a realization with all distances different from 1 (see for instance Proposition 1 of [6]) that the system has a solution, by "inflating" such a realization. Hence, by lemma 3 there exists a solution to this system with all numbers less than $\exp[\gamma^n]$ in absolute value for some γ (we absorb the factor $\ln(1 + \binom{n}{2}) + O(1)$ by taking $\gamma > C$). Let us now round down all numbers to the next integer. It is easily checked that we get a $(x_1', y_1', \ldots, x_n', y_n', R') \in \mathbb{Z}^{2n+1}$ that satisfies

$$(x_i' - x_j')^2 + (y_i' - y_j')^2 \leq (R')^2, \text{ for } ij \in E(G),$$
$$(x_i' - x_j')^2 + (y_i' - y_j')^2 > (R')^2, \text{ for } ij \notin E(G).$$

Hence, if we divide the x_i', y_i' by R' we get a realization of G that uses $O(\gamma^n)$ bits per coordinate.   □

Acknowledgements

We thank Erik Jan van Leeuwen for helpful discussions. We also thank Jan Kratochvíl and Jiří Matoušek for finding, scanning and sending us their 1988 preprint. We also thank the anonymous referees for helpful comments that have improved the presentation.

References

1. Breu, H., Kirkpatrick, D.G.: Unit disk graph recognition is NP-hard. Comput. Geom. 9(1-2), 3–24 (1998)
2. Goodman, J.E., Pollack, R., Sturmfels, B.: The intrinsic spread of a configuration in $\mathbf{R}^d$. J. Amer. Math. Soc. 3(3), 639–651 (1990)
3. Grigor'ev, D.Y., Vorobjov, N.N.: Solving systems of polynomial inequalities in subexponential time. J. Symbolic Comput. 5(1-2), 37–64 (1988)
4. Kratochvíl, J., Matoušek, J.: Intersection graphs of segments. KAM preprint series. Charles University, Prague (1988)
5. Kratochvíl, J., Matoušek, J.: Intersection graphs of segments. J. Combin. Theory Ser. B 62(2), 289–315 (1994)
6. van Leeuwen, E.J., van Leeuwen, J.: On the representation of disk graphs. Utrecht University Technical report number UU-CS-2006-037 (July 2006)
7. Schrijver, A.: Theory of linear and integer programming. Wiley-Interscience Series in Discrete Mathematics. John Wiley & Sons Ltd., Chichester (1986); A Wiley-Interscience Publication
8. Spinrad, J.P.: Efficient graph representations. Fields Institute Monographs, vol. 19. American Mathematical Society, Providence (2003)

A   The Proof of Lemma 2

Proof of Lemma 2. Let $a, b \in \mathbb{Q}$ be arbitrary with $\mathrm{size}(a), \mathrm{size}(b) \leq B$. Let us write $a = p_1/q_1, b = p_2/q_2$ with p_i, q_i relatively prime integers for $i = 1, 2$. Note that an integer $n \in \mathbb{Z}$ has bit size

$$\mathrm{size}(n) = 1 + \lceil \log_2(|n| + 1) \rceil, \tag{1}$$

(the extra one is for the sign). From (1) it follows that for two integers $n, m \in \mathbb{Z}$:

$$\begin{aligned} \mathrm{size}(nm) &\leq \mathrm{size}(n) + \mathrm{size}(m), \\ \mathrm{size}(n + m) &\leq 1 + \max(\mathrm{size}(n), \mathrm{size}(m)). \end{aligned} \tag{2}$$

From $ab = p_1 p_2 / q_1 q_2$ and (2), we see that

$$\begin{aligned} \mathrm{size}(ab) &\leq \mathrm{size}(p_1 p_2) + \mathrm{size}(q_1 q_2) \\ &\leq \mathrm{size}(p_1) + \mathrm{size}(p_2) + \mathrm{size}(q_1) + \mathrm{size}(q_2) \\ &= \mathrm{size}(a) + \mathrm{size}(b) \\ &\leq 2B. \end{aligned}$$

Completely analogously, $\mathrm{size}(a/b) \leq 2B$.

From $a + b = (p_1 q_2 + p_2 q_1)/q_1 q_2$ and (2) we see that

$$\begin{aligned} \mathrm{size}(a + b) &\leq \mathrm{size}(p_1 q_2 + p_2 q_1) + \mathrm{size}(q_1 q_2) \\ &\leq 1 + \max(\mathrm{size}(p_1) + \mathrm{size}(q_2), \mathrm{size}(p_2) + \mathrm{size}(q_1)) \\ &\quad + \mathrm{size}(q_1) + \mathrm{size}(q_2) \\ &\leq 1 + 3B \\ &\leq 4B. \end{aligned}$$

Completely analogously, $\mathrm{size}(a - b) \leq 4B$. $\qquad\square$

Lattices and Maximum Flow Algorithms in Planar Graphs

Jannik Matuschke* and Britta Peis

Technische Universität Berlin, Institut für Mathematik,
Straße des 17. Juni 136, 10623 Berlin, Germany
{matuschke,peis}@math.tu-berlin.de

Abstract. We show that the left/right relation on the set of s-t-paths of a plane graph [9] induces a so-called "submodular" lattice. If the embedding of the graph is s-t-planar, this lattice is even consecutive. This implies that Ford and Fulkerson's uppermost path algorithm for maximum flow in such graphs [4] is indeed a special case of a two-phase greedy algorithm on lattice polyhedra [2]. We also show that the properties submodularity and consecutivity cannot be achieved simultaneously by any partial order on the paths if the graph is planar but not s-t-planar, thus providing a characterization of this class of graphs.

1 Introduction and Preliminaries

The special case of flows in planar graphs has always played a significant role in network flow theory. The predecessor of Ford and Fulkerson's well-known path augmenting algorithm – and actually the first combinatorial flow algorithm at all – was a special version for s-t-planar networks, i.e., those networks where s and t can be embedded adjacent to the infinite face [4]. The basic idea of this *uppermost path algorithm* is to iteratively augment flow along the "uppermost" non-saturated s-t-path in the planar embedding of the network. In 2006, Borradaile and Klein [1] established an intuitive generalization of this algorithm to arbitrary planar graphs, which relies on a partial order on the set of s-t-paths in the graph, called the *left/right relation*.

Another area of combinatorial optimization, which has so far been unrelated to planar flow computations, is the optimization on lattice structures. In 1978, Hoffman and Schwartz introduced the notion of *lattice polyhedra* [6], a generalization of Edmond's polymatroids based on lattices, and proved total dual integrality of the corresponding inequality systems if certain additional properties hold. Later, several variants of two-phase greedy algorithms were developed, e.g., by Kornblum [10], Frank [5], and Faigle and Peis [2], to solve quite general classes of linear programs on these polyhedra efficiently.

---

* The author was supported by Berlin Mathematical School.

D.M. Thilikos (Ed.): WG 2010, LNCS 6410, pp. 324–335, 2010.

Our results. In this paper, we connect these two fields of research by showing that the left/right relation induces a lattice on the set of simple s-t-paths in a planar graph. If the network is s-t-planar, this lattice fulfills the two main properties required in Hoffman and Schwartz' framework, called *submodularity* and *consecutivity*. Our result implies that the uppermost path algorithm of Ford and Fulkerson is a special case of the two-phase greedy algorithm on lattice polyhedra, which, even more, can solve a variant of the flow problem with supermodular and monotone weights on the paths. However, the case of general planar graphs, i.e., not necessarily s-t-planar graphs turns out to be much more involved. In fact, we will characterize s-t-planar graphs as the only class of planar graphs that can be equipped with a lattice on the set of paths that is consecutive and submodular at the same time.

Outline. In the remainder of this section we will define lattice polyhedra (Subsection 1.1) and introduce the basic notions of graph structures (Subsection 1.2) and the left/right relation (Subsection 1.3) we need to present our results. We then will discuss the left/right relation in s-t-planar graphs and provide an intuitive characterization for the relation in this class of graphs, which leads to the insight that the relation induces a submodular and consecutive lattice on such graphs (Section 2). In Section 3, we discuss the case of general planar graphs and outline a considerably more involved proof to show that the left/right relation induces a submodular lattice in the general case. Finally, in Section 4 we show that consecutivity and submodularity cannot be achieved at the same time by any partial order in the non-s-t-planar case.

All proofs omitted in this paper due to space constraints can be found [12].

1.1   Lattice Polyhedra

Our interest in lattices is motivated by a two-phase greedy algorithm that can solve a primal/dual pair of very general linear programming problems on so-called lattice polyhedra, which have first been introduced by Hoffman and Schwartz [6]. More precisely, we are given a finite set E, a set system $\mathcal{L} \subseteq 2^E$, and two vectors $c \in \mathbb{R}^E$, $r \in \mathbb{R}^{\mathcal{L}}$, and consider the covering problem

$$(C) \quad \min \left\{ \sum_{e \in E} c(e)x(e) \ : \ x \in \mathbb{R}_+^E, \ \sum_{e \in S} x(e) \geq r(S) \ \forall S \in \mathcal{L} \right\}$$

and its dual, the packing problem

$$(P) \quad \max \left\{ \sum_{S \in \mathcal{L}} r(S)y(S) \ : \ y \in \mathbb{R}_+^{\mathcal{L}}, \ \sum_{S \in \mathcal{L}: e \in S} y(S) \leq c(e) \ \forall e \in E \right\}.$$

Observe that the packing problem (P) corresponds to an ordinary max flow problem if $\mathcal{L}$ is the set of s-t-paths of a given network and $r \equiv 1$. Before we can state Hoffman and Schwartz' main result, we need to introduce some definitions.

Definition 1 (Lattices, submodularity, consecutivity). *Let E be a finite set, $\mathcal{L} \subseteq 2^E$ and $\preceq$ be a partial order on $\mathcal{L}$. The pair $(\mathcal{L}, \preceq)$ is a* lattice *if for all $S, T \in \mathcal{L}$ the following two conditions are fulfilled.*

- *$\{L \in \mathcal{L} : L \preceq S, L \preceq T\}$ has a unique maximum element $S \wedge T$, called* meet.
- *$\{U \in \mathcal{L} : U \succeq S, U \succeq T\}$ has a unique minimum element $S \vee T$, called* join.

A function $r : \mathcal{L} \to \mathbb{R}$ is submodular *if $r(S \wedge T) + r(S \vee T) \leq r(S) + r(T)$ for all $S, T \in \mathcal{L}$. It is* supermodular, *if $r(S \wedge T) + r(S \vee T) \geq r(S) + r(T)$ for all $S, T \in \mathcal{L}$. A lattice $\mathcal{L}$ is* submodular, *if $(S \wedge T) \cap (S \vee T) \subseteq S \cap T$ and $(S \wedge T) \cup (S \vee T) \subseteq S \cup T$ for all $S, T \in \mathcal{L}$.[1] It is* consecutive *if $S \cap U \subseteq T$ for all $S, T, U \in \mathcal{L}$ with $S \preceq T \preceq U$.*

Theorem 1 (Hoffman and Schwartz [6]). *If $\mathcal{L}$ is a submodular and consecutive lattice and r is supermodular, then the inequality system defining (C) is totally dual integral. In this case, the corresponding polyhedron is called* lattice polyhedron.

Note that even if all requirements of Theorem 1 are fulfilled, no general algorithm is known that solves problems (C) and (P) in time polynomial in the cardinality of the ground set E. However, if we additionally require r to be monotone increasing w.r.t. $\preceq$ and assume that both r and the lattice are polynomially computable in the sense that the maximum element of any restricted sublattice of $\mathcal{L}$ can be found, an optimal and – in case of integral input – integral solution can be computed efficiently. The corresponding *two-phase greedy algorithm* goes back to Frank [5]. The following paragraph gives a brief sketch based on a refined presentation by Faigle and Peis [2].

The two-phase greedy algorithm. The first phase constructs a feasible solution to the dual problem (P). In each iteration, the maximum element M of the current lattice is obtained by an oracle call. From this set, a bottleneck element e with minumum residual capacity is chosen. Now the dual variable corresponding to M is increased as much as possible until the capacity of e is completely used up. The bottleneck element is removed from the ground set, and the lattice is restricted to the sublattice of sets not containing the removed elements. This procedure is re-iterated until the the remaining restricted lattice is empty and the first phase ends. In the second phase, the primal variables corresponding to the bottleneck elements are iteratively set in such a way that complementary slackness is fulfilled.

In a later section of this paper, we will show that ordering the set of *s-t*-paths in an *s-t*-plane graph "from top to bottom" yields a lattice that fulfills all requirements stated above. In this case, the first phase of the two-phase greedy algorithm corresponds to iteratively saturating the uppermost path of the network and erasing a corresponding bottleneck edge. This exactly coincides

---

[1] Submodularity of lattices is connected to submodularity of functions in the following way: A lattice is submodular if and only if all functions of the type $f(S) := \sum_{e \in S} x(e)$ for some vector $x \in \mathbb{R}^E_+$ are submodular.

with Ford and Fulkerson's uppermost path algorithm, even solving the problem for supermodular and monotone increasing weights on the paths.

1.2   Graphs

For the representation of graphs we will use a similar notation as used in [1]. Our results in Sections 2 and 3 will be valid for both directed and undirected graphs. We will assume that we are given a directed graph $G = (V, E)$ (if the graph is undirected, we can direct it arbitrarily), but we will allow paths to use all edges in arbitrary direction.[2] For this purpose equip every edge $e \in E$ with two antiparallel *darts*, a forward dart $\overrightarrow{e} := (e, 1)$ pointing in the same direction as the edge and a backward dart $\overleftarrow{e} := (e, -1)$ pointing in the opposite direction.

Definition 2. *We define* $\overleftrightarrow{E} := E \times \{1, -1\}$ *to be the set of all darts. For a dart* (e, i) *we use* $\mathrm{rev}((e, i)) := (e, -i)$ *to refer to its reverse. For* $e = (v, w)$*, we let* $\mathrm{tail}(\overrightarrow{e}) = v = \mathrm{head}(\overleftarrow{e})$ *and* $\mathrm{head}(\overrightarrow{e}) = w = \mathrm{tail}(\overleftarrow{e})$*. For* $D \subseteq \overleftrightarrow{E}$*, we define* $E(D) := \{e \in E : \overrightarrow{e} \in D \text{ or } \overleftarrow{e} \in D\}$*. We use* $G[\tilde{E}]$ *to refer to the subgraph that only contains the edges* $\tilde{E} \subseteq E$*.*

The basic notions of paths, cycles, and cuts are defined in the natural way except that all of these objects consist of darts rather than edges.

Definition 3 (Walk, path, cycle). *A* simple *x-y-path is a non-empty sequence of darts* $d_1, \ldots, d_k$ *such that* $\mathrm{head}(d_i) = \mathrm{tail}(d_{i+1})$ *for* $i \in \{1, \ldots, k-1\}$ *and* $x = \mathrm{tail}(d_1)$ *and* $y = \mathrm{head}(d_k)$*. If for all darts of an x-y-walk the underlying edges are pairwise distinct, then the walk is called x-y-path for $x \neq y$ or cycle if $x = y$. A path or cycle is called* simple *if the heads of all its darts are pairwise distinct.*

A *planar graph* is a graph that can be drawn on (or embedded in) the plane without any two edges intersecting. A graph together with such an embedding is called *plane graph*. The embedding partitions the plane into regions that are bordered by the edges. These regions are called *faces* and can be used to define the dual graph G^* as follows. The vertex set V^* of G^* is the set of all faces. For every edge in G, we introduce a corresponding edge in G^* that connects the faces that are separated by this primal edge, going from right to left. We refer to the faces left and right of a dart $d \in \overleftrightarrow{E}$ by $\mathrm{left}(d)$ and $\mathrm{right}(d)$, respectively. The face surrounding the drawing is called the *infinite face* f_∞.

1.3   The Left/Right Relation

Assumption. For the rest of this paper, let $G = (V, E)$ be a connected, planar graph, $s, t \in V$, with an embedding such that t is adjacent to f_∞. Furthermore, let $\mathcal{P}$ be the set of all simple s-t-paths in G.

---

[2] This helps to streamline the proofs, the resulting lattice can be restricted to directed paths later on by removing all paths that use backward darts. Note that removing elements from the ground set preserves submodular lattice structures.

In order to define a partial order on $\mathcal{P}$, we consider the vector space that is spanned by the edges of the graph and the subspace spanned by all cycles. It is well-known that the (clockwise) boundaries of the non-infinite faces comprise a basis of this cycle space.

Definition 4. *For a path or cycle $P \subset \overleftrightarrow{E}$ we define the vector $\delta_P \in \mathbb{R}^E$ by $\delta_P(e) := 1$ if $\overrightarrow{e} \in P$, $\delta_P(e) := -1$ if $\overleftarrow{e} \in P$ and $\delta_P(e) := 0$ otherwise. For a face $f \in V^*$ we define δ_f to be the vector corresponding to the set of darts in the clockwise boundary of f (with antiparallel darts canceling out). For a vector $\delta \in \mathbb{R}^E$, we let $\delta(\overrightarrow{e}) := \delta(e)$, $\delta(\overleftarrow{e}) := -\delta(e)$.*

Definition 5. *The subspace $\mathcal{S}_{cycle}(G) := \operatorname{span}\{\delta_C : C$ is a cycle in $G\}$ is called cycle space. Its elements are called circulations. The set $\{\delta_f : f \in V^* \backslash \{f_\infty\}\}$ is a basis of $\mathcal{S}_{cycle}(G)$. In particular, there is a unique linear mapping $\Phi : \mathcal{S}_{cycle}(G) \to \mathbb{R}^{V^*}$, such that $\Phi(\delta)(f_\infty) = 0$ and $\delta = \sum_{f \in V^*} \Phi(\delta)(f)\delta_f$ for all $\delta \in \mathcal{S}_{cycle}(G)$. The vector $\Phi(\delta)$ is called the face potential of δ.*

The left/right relation goes back to an order on circulations in planar graphs by Khuller et al. [8] that was generalized to flows by Weihe [13] and later specified for s-t-paths by Klein [9]. It yields useful applications for shortest path and maximum flow computations in planar graphs (cf. [9] and [1], respectively) and is based on the face potentials introduced above. Intuitively, the definition states that $P \preceq Q$ if and only if the circulation consisting of P and the reverse of Q is clockwise (as positive face potentials correspond to clockwise circulations).

Definition 6 (Left/right relation). *Let $P, Q \in \mathcal{P}$. If $\Phi(\delta_P - \delta_Q) \geq 0$, we say that P is left of Q and write $P \succeq Q$. If $\Phi(\delta_P - \delta_Q) \leq 0$, we say that P is right of Q and write $P \preceq Q$.*

Khuller et al. [8] showed that the left/right relation induces a lattice on the set of circulations in a planar graph corresponding to a min/max lattice on the space of face potentials. We will show that the relation also induces a lattice on the set of simple s-t-paths.

2   Uppermost Paths and the Path Lattice of an s-t-Plane Graph

Intuitively speaking, the uppermost path of an s-t-plane graph, is the s-t-path forming its "upper" boundary in a drawing where s is on the very left and t is on the very right of the drawing. The idea goes back to Ford and Fulkerson, who used it to introduce the uppermost path algorithm for the maximum flow problem in s-t-planar graphs [4], which iteratively saturates the uppermost residual path. We will give a definition of the uppermost path in combinatorial terms and use it to characterize the left/right relation in s-t-plane graphs. This yields all the desired lattice properties of the partial order and thus shows that the uppermost path algorithm corresponds to the two-phase greedy algorithm, which also saturates the maximum (w.r.t. $\preceq$) "residual" element of the lattice in each iteration [2].

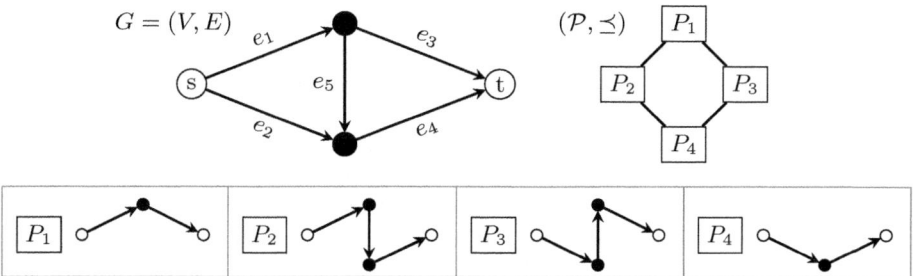

Fig. 1. An s-t-plane graph and the set of its simple s-t-paths ordered w.r.t. the left/right relation

Assumption. Throughout Section 2, we assume that the embedding of G is s-t-planar.

Theorem 2 (Uppermost and lowermost path). *There is a unique path $U \in \mathcal{P}$ such that* left$(d) = f_\infty$ *for all $d \in U$. It is called* uppermost path *of G. There also is a unique path $L \in \mathcal{P}$ such that* right$(d) = f_\infty$ *for all $d \in L$. This path is called* lowermost path *of G.*

Clearly, if U is the uppermost path of G, then U is also the uppermost path of any subgraph of G containing all edges of U. We will the following two helpful observations that are direct implications of the duality of cycles and cuts in planar graphs [14].

Lemma 1 (Orientation lemma). *Let U be the uppermost path of G and P any path of G. If $d \in U$ then* rev$(d) \notin P$.

Lemma 2 (Bridge lemma). *Let $d \in \overleftrightarrow{E}$. If* left$(d) =$ right(d), *then d is either contained in all simple s-t-paths or in none.*

This suffices to characterize the left/right relation in an s-t-plane graph in terms of the uppermost path property.

Theorem 3. *Let $P, Q \in \mathcal{P}$. Then the following statements are equivalent.*

1. *P is the uppermost path in $G[E(P \cup Q)]$.*
2. *Q is the lowermost path in $G[E(P \cup Q)]$.*
3. *P is left of Q.*

Proof. Without loss of generality, we can identify $\phi := \Phi(\delta_P - \delta_Q)$ with a potential function in $G[E(P \cup Q)]$, as edges not in P or Q do not affect the potentials.

$(1) \Leftrightarrow (2)$: Suppose P is the uppermost path of $G[E(P \cup Q)]$. Let L be the lowermost path of $G[E(P \cup Q)]$. Let $d \in L$. If d belongs to an edge of $E(P)$, the orientation lemma ensures that $d \in P$. Consequently, $d \in P \cap L$, and thus left$_{G[E(P \cup Q)]}(d) = f_\infty =$ right$_{G[E(P \cup Q)]}(d)$, implying $d \in Q$ by the bridge lemma. Hence $E(L) \subseteq E(Q)$, and as the two paths are simple, they are equal. The converse follows by symmetry.

(2) $\Rightarrow$ (3) : Suppose Q is the lowermost path of $G[E(P \cup Q)]$ and thus P is its uppermost path. Let f be a face of $G[E(P \cup Q)]$ and let d be a dart of that graph with right$(d) = f$. If rev$(d) \in P$ or $d \in Q$, then $f = $ right$(d) = f_\infty$, implying $\phi(f) = 0$. Otherwise, $d \in P$ or rev$(d) \in Q$, implying left$(d) = f_\infty$ and $\phi(f) = \phi(f_\infty) + \delta_P(d) - \delta_Q(d) \geq 0$. Thus, $\phi \geq 0$.

(3) $\Rightarrow$ (1) : Suppose $\phi \geq 0$. Let U be the uppermost path of $G[E(P \cup Q)]$ and $d \in U$. By the orientation lemma, $d \in P$ or $d \in Q$. But $d \in Q \setminus P$ is not possible, as $\delta_P(d) - \delta_Q(d) = \phi(\mathrm{right}(d)) - \phi(f_\infty) \geq 0$. So $d \in P$ for all $d \in U$, i.e., $U = P$.   □

Thus, the left/right order in s-t-plane graphs is in fact an uppermost/lowermost path order. Before we can show that this indeed yields a lattice, we need a final auxiliary result, which states that we can add a path to a subgraph without changing its uppermost path, as long as there already is a path above the path we add. This follows by simple elementary arguments using the orientation lemma and the bridge lemma.

Lemma 3. *Let $\bar{E} \subseteq E$ be an edge set such that $G[\bar{E}]$ is connected and let $Q \in \mathcal{P}$. If there is an s-t-path P in $G[\bar{E}]$ with $P \succeq Q$, then the uppermost path of $G[\bar{E} \cup E(Q)]$ is equal to the uppermost path of $G[\bar{E}]$.*

Finally, we can show the existence of a consecutive and submodular path lattice in an s-t-plane graph. We even get a nice characterization of meet and join of this lattice as the lowermost and uppermost path of $G[E(P \cup Q)]$.

Theorem 4. *$(\mathcal{P}, \preceq)$ is a consecutive and submodular lattice with $P \wedge Q$ being the lowermost path in $G[E(P \cup Q)]$ and $P \vee Q$ being the uppermost path in $G[E(P \cup Q)]$.*

Proof. Let $P, Q, R \in \mathcal{P}$.

Meet and join: Let U be the uppermost path in $G[E(P \cup Q)]$. Then $P, Q \preceq U$. Let $U' \in \mathcal{P}$ with $P, Q \preceq U'$. Then U' is the uppermost path of $G[E(P \cup U') \cup E(Q)]$ by Lemma 3. As U is contained in this graph, $U \preceq U'$. Thus, U is the least upper bound on P and Q with respect to $\preceq$. The characterization of the meet follows by symmetry.

Consecutivity: Suppose $P \preceq Q \preceq R$. Then P is the lowermost path and R is the uppermost path of $G' := G[E(P \cup R) \cup E(Q)]$ by Lemma 3. Thus, right$_{G'}(d) = f_\infty = $ left$_{G'}(d)$ for all $d \in P \cap R$. By the bridge lemma, this implies $d \in Q$.

Submodularity: As we have proven consecutivity, it suffices to show $P \wedge Q, P \vee Q \subseteq P \cup Q$. This immediately follows from the definition of $P \wedge Q$ and $P \vee Q$ as lowermost and uppermost path of $G[E(P \cup Q)]$ and the orientation lemma.   □

An example of a graph and its path lattice is depicted in Figure 1. Our result implies total dual integrality of the maximum flow problem in s-t-planar graphs, even when we introduce supermodular weights on the paths. Applying the two-phase greedy algorithm on the lattice yields an implementation of the uppermost

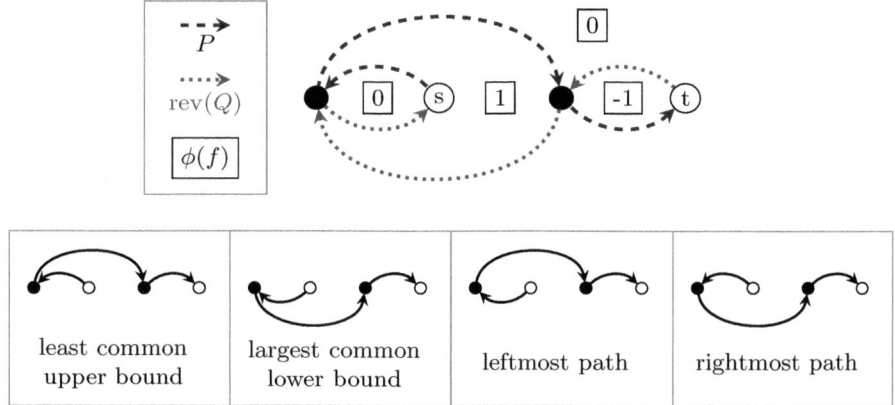

Fig. 2. This example shows that the rightmost path of $G[E(P \cup Q)]$ is not necessarily the largest common lower bound of the two paths in non-s-t-planar embeddings. However, subtracting the "positive" part of the circulation from P yields $P \wedge Q$.

path algorithm by Ford and Fulkerson that solves the maximum flow problem on s-t-planar graphs, also for the case of supermodular and monotone increasing path weights (cf. [12] for more details).

Running time. It can be shown that the general two-phase greedy algorithm can be implemented to run in time $\mathcal{O}(|E| \log(|E|) + T_{\mathcal{L}} + T_r)$ where $T_{\mathcal{L}}$ and T_r is the total time for computing the sequence of consecutive maximum elements of the lattice restrictions and the corresponding values of r, respectively. In the special case of the path lattice in s-t-planar graphs, successive uppermost paths occurring throughout the course of the algorithm can be computed in amortized constant time, yielding an overall running time of $\mathcal{O}(|V| \log(|V|))$, the same time that was shown by Itai and Shiloach [7] for the uppermost path algorithm. Details on this implementation of the two-phase greedy algorithm in general and the uppermost path algoritm in particular can be found in [12].

3   The Path Lattice of a General Plane Graph

We will now show that the left/right relation also defines a lattice in general plane graphs. However, the proof will require significantly more effort this time. In contrast to the s-t-planar case, meet and join of two paths $P, Q \in \mathcal{P}$ are not always the minimum (rightmost) and maximum (leftmost) path in $G[E(P \cup Q)]$ in the general case (cf. Figure 2 for an example). The intuitive idea for constructing the meet is the following: We subtract the "positive part" of the circulation $\delta_P - \delta_Q$ (i.e., those faces that prevent P from being right of Q) from the path P. In this way we obtain a set of darts $D^{P \wedge Q} \subseteq P \cup Q$ that contains the meet $P \wedge Q$, as we shall see later. We formalize this idea in the following lemma.

Lemma 4. *Let $P, Q \in \mathcal{P}$ and $\phi := \Phi(\delta_P - \delta_Q)$.*

- *Let $S^+ := \{f \in V^* : \phi(f) > 0\}$ and $\delta^{P \wedge Q} := \delta_P - \sum_{f \in S^+} \phi(f)\delta_f$. Then $\delta^{P \wedge Q} \in \{-1, 0, 1\}^E$ and $D^{P \wedge Q} := \{d \in \overleftrightarrow{E} : \delta^{P \wedge Q}(d) = 1\} \subseteq P \cup Q$.*
- *Let $S^- := \{f \in V^* : \phi(f) < 0\}$ and $\delta^{P \vee Q} := \delta_P - \sum_{f \in S^-} \phi(f)\delta_f$. Then $\delta^{P \vee Q} \in \{-1, 0, 1\}^E$ and $D^{P \vee Q} := \{d \in \overleftrightarrow{E} : \delta^{P \vee Q}(d) = 1\} \subseteq P \cup Q$.*

It is straightforward to check that if $P \preceq Q$, then $P = D^{P \wedge Q}$ and $Q = D^{P \vee Q}$. Unfortunately, $D^{P \wedge Q}$ and $D^{P \vee Q}$ are not s-t-paths in general. However, it can be shown that each of these sets consists of a unique simple s-t-path and some cycles and that these paths are meet and join of P and Q, respectively. The proof of Lemma 4 can be obtained by a simple case distinction, and, as a by-product, leads to the following additional result.

Lemma 5. *Let R be a simple path and C be a simple cycle in $D^{P \wedge Q}$ or $D^{P \vee Q}$, respectively. If $R \cap C = \emptyset$, then R does not cross C, i.e., all darts of R are either in the interior of C or none of them is.*

The following lemma is the key insight on our way to proving the desired result. Its proof, however, is rather lengthy and involves many minor details. The key idea is that such a cycle must consist of edges of P and Q and that these paths can only enter or leave the cycle from or to the left, i.e., from or towards its interior, thus being "trapped" inside the cycle.

Lemma 6. *There are no counterclockwise simple cycles in $D^{P \wedge Q}$. There are no clockwise cycles in $D^{P \vee Q}$.*

Theorem 5. *$(\mathcal{P}, \preceq)$ is a submodular lattice with $P \wedge Q$ being the unique simple s-t-path contained in $D^{P \wedge Q}$ and $P \vee Q$ being the unique simple s-t-path contained in $D^{P \vee Q}$.*

Proof. As $\delta^{P \wedge Q} \in \{-1, 0, 1\}^E$ is the sum of δ_P and some circulations, $\delta^{P \wedge Q}$ is a unit s-t-flow of value 1. Thus, there is a flow decomposition $\delta_R + \sum_{i=1}^k \delta_{C_i} = \delta^{P \wedge Q}$ for a simple s-t-path $R \subseteq D^{P \wedge Q}$ and some – by Lemma 6 clockwise – simple cycles $C_1, \ldots, C_k \subseteq D^{P \wedge Q}$ with $E(R), E(C_1), \ldots, E(C_k)$ pairwise disjoint.

By linearity of Φ, we deduce that $\Phi(\delta_P - \delta_R) = \Phi(\delta_P - \delta^{P \wedge Q}) + \sum_{i=1}^k \Phi(\delta_{C_i})$ and $\Phi(\delta_P - \delta_R) = \Phi(\delta_Q - \delta^{P \wedge Q}) + \sum_{i=1}^k \Phi(\delta_{C_i})$. Since $\Phi(\delta_P - \delta^{P \wedge Q}) \geq 0$ and $\Phi(\delta_Q - \delta^{P \wedge Q}) \geq 0$ by construction of $\delta^{P \wedge Q}$ and $\Phi(\delta_{C_i}) \geq 0$ for the clockwise cycles C_i, this implies $R \preceq P$ and $R \preceq Q$.

Now let $S \in \mathcal{P}$ be a path with $S \preceq P$ and $S \preceq Q$. We show $S \preceq R$. First, we consider only faces incident to R. So let $\bar{f} \in \{\text{left}(d) : d \in R\} \cup \{\text{right}(d) : d \in R\}$. By Lemma 5, R cannot cross any of the cycles C_i, and, as t is on the exterior of any such cycle, R cannot have any darts in the interior of a cycle. This implies that $\bar{f}$ is not in the interior of any of the cycles C_i and $\Phi(\delta_{C_i})(\bar{f}) = 0$. Thus, $\Phi(\delta_R - \delta_S)(\bar{f})$ is equal to $\Phi(\delta_P - \delta_S)(\bar{f})$ if $\bar{f} \notin S^+$ and equal to $\Phi(\delta_Q - \delta_S)(\bar{f})$ if $\bar{f} \in S^+$. In both cases $\Phi(\delta_R - \delta_S)(\bar{f})$ is non-negative.

Now let $\hat{f} \in V^*$ be any face. As S does not contain a cycle in the primal graph, it does not contain a cut in the dual graph. Thus, there is a path in the dual that leads from $\hat{f}$ to some face $\bar{f}$ incident to R and does not intersect S or R, i.e., the potential does not change along the path. Thus $\Phi(\delta_R - \delta_S)(\hat{f}) = \Phi(\delta_R - \delta_S)(\bar{f}) \geq 0$. Consequently, $S \preceq R$.

We thus have shown that R is the meet of P and Q (R is unique by anti-symmetry). Likewise, we can show that $D^{P \vee Q}$ contains a unique s-t-path that is the least common upper bound of P and Q. Thus, $(\mathcal{P}, \preceq)$ is a lattice with meet and join as described above. Submodularity follows from the same case distinction that can be used to prove Lemma 4.   □

Complete versions of the proofs in this section can be found in [12].

4   A Characterization of s-t-Planar Graphs

It is easy to observe that the path lattice induced by the left/right relation in general planar (but non-s-t-planar) graphs is not necessarily consecutive (cf. the paths P_1, P_2, P_4 in Figure 3). Of course one might ask whether this property can be achieved by a different partial order on the paths. Indeed, one can show that no partial order in any planar but not s-t-planar graph can induce a lattice that is submodular and consecutive at the same time.

The central idea to proving this negative result is to show it for two graphs that comprise the s-t-planar equivalent to the famous Kuratowski graphs $K_{3,3}$ and K_5 [11]. So let $K_{3,3}^{s-t}$ and K_5^{s-t} be the graphs that arise from the respective Kuratowski graphs by deleting the edge connecting s and t ($K_{3,3}^{s-t}$ is depicted in Figure 3).

Lemma 7. *Let* $P, Q \in \mathcal{P}$ *such that the subgraph* $G[E(P \cup Q)]$ *contains only the two paths* P *and* Q. *If* $\preceq$ *is a partial order that induces a submodular lattice, then* $P \preceq Q$ *or* $P \succeq Q$.

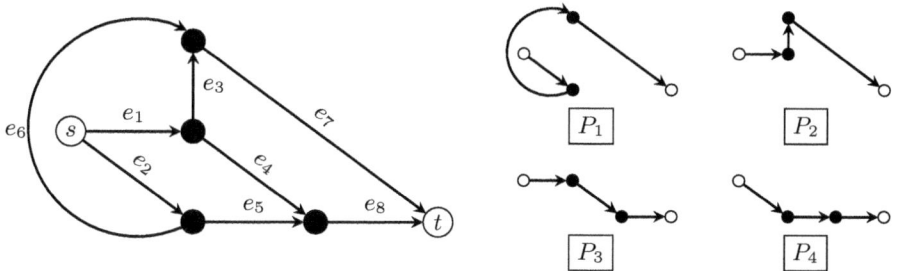

Fig. 3. The graph $K_{3,3}^{s-t}$ and four of its s-t-paths, which suffice for showing that there is no partial order that induces a submodular and consecutive lattice on the set of paths

Proof. By submodularity, $P \vee Q \subseteq P \cup Q$ and thus $P \vee Q$ is a path in $G[E(P \cup Q)]$. This implies $P \vee Q = P$ or $P \vee Q = Q$. $\square$

Lemma 8. *The set of s-t-paths in $K_{3,3}^{s-t}$ or K_5^{s-t} (or a subdivision of these graphs) cannot be equipped with a partial order $\preceq$, such that $(\mathcal{P}, \preceq)$ is a consecutive and submodular lattice.*

Proof. Assume by contradiction $\preceq$ is defined such that $(\mathcal{P}, \preceq)$ is a consecutive and submodular lattice with meet $\wedge$ and join $\vee$ on the set of s-t-paths in $K_{3,3}^{s-t}$. Consider the four paths P_1, P_2, P_3, P_4 depicted in Figure 3. It can easily be checked that P_i and P_j are the only two s-t-paths in $G[E(P_i \cup P_j)]$ for $i \neq j$. Thus, by Lemma 7, the paths form a chain w.r.t. $\preceq$. Since P_1 and P_2 are the only two of the paths sharing $\vec{e_7}$ as common dart, and P_2 and P_3 are the only two of the paths sharing $\vec{e_1}$, and P_3 and P_4 are the only two of the paths sharing $\vec{e_8}$, consecutivity demands that either $P_1 \succ P_2 \succ P_3 \succ P_4$ or $P_1 \prec P_2 \prec P_3 \prec P_4$. In both cases $\vec{e_2} \in P_1 \cap P_4 \setminus P_2$ yields a contradiction to consecutivity. Note that all arguments used in this proof are invariant under the operation of subdividing edges. The result for K_5^{s-t} can be derived by a similar line of argumentation. $\square$

Theorem 6. *A graph is s-t-planar if and only if it is planar and there is a partial order on the set of its s-t-paths that induces a consecutive and submodular lattice.*

Proof. Necessity follows from Theorem 5. For sufficiency, let G be a graph that is planar but not s-t-planar. Let G' be the graph that is obtained from G by adding an edge e from s to t. As G is not s-t-planar, G' is not planar and thus, by Kuratowski's theorem, it contains a subdivision K' of $K_{3,3}$ or K_5. As G is planar, one of the subdivided edges in K' must contain e. Let s' and t' be the endpoints that are connected by this subdivided edge. Clearly, G contains a subdivision K of $K_{3,3}^{s'-t'}$ or $K_5^{s'-t'}$, respectively. By Lemma 8, there is a set of s'-t'-paths in K (and thus in G) that cannot be equipped with any partial order that induces a consecutive and submodular lattice. These paths can all be extended to s-t-paths in G by using the s-s'-path and the t'-t-path contained in the subdivided edge connecting s' and t' in K. Thus G contains a set of s-t-paths that cannot be equipped with a partial order of a consecutive and submodular lattice. $\square$

5   Conclusion and Outlook

We have established a connection between optimization on lattice polyhedra and planar network flow theory by showing that the structure exploited by two important flow algorithms corresponds to a submodular lattice. For the s-t-planar case, this implies that the uppermost path algorithm of Ford and Fulkerson is a special case of a more general two-phase greedy algorithm, which allows for certain types of weights on the lattice sets. Thus, a closer study of the weighted maximum flow problem is of obvious interest. First results in this direction can be found in [12]. Future research could also deal with the question in how far the structural result presented here lead to new insights for existing or new planar graph algorithms. Finally, it might be interesting to investigate whether the path lattice is distributive and our results extend a line of distributive lattices in planar graph structures connected to the left/right relation [3].

References

1. Borradaile, G., Klein, P.: An $O(n \ log \ n)$ algorithm for maximum st-flow in a directed planar graph. In: Proceedings of the Seventeenth Annual ACM-SIAM Symposium on Discrete Algorithms, pp. 524–533 (2006)
2. Faigle, U., Peis, B.: Two-phase greedy algorithms for some classes of combinatorial linear programs. In: Proceedings of the Nineteenth Annual ACM-SIAM Symposium on Discrete Algorithms, pp. 161–166 (2008)
3. Felsner, S., Knauer, K.: ULD-lattices and Δ-bonds. Combinatorics, Probability and Computing 18(5), 707–724 (2009)
4. Ford Jr., L.R., Fulkerson, D.R.: Maximal flow through a network. Canadian Journal of Mathematics. Journal Canadien de Mathématiques 8, 399–404 (1956)
5. Frank, A.: Increasing the rooted-connectivity of a digraph by one. Mathematical Programming 84(3, Ser. B), 565–576 (1999)
6. Hoffman, A.J., Schwartz, D.E.: On lattice polyhedra. In: Hajnal, A., Sós, V.T. (eds.) Proceedings of the Fifth Hungarian Colloquium on Combinatorics. Colloquia mathematica Societatis János Bolyai, vol. I, vol. 18, pp. 593–598 (1978)
7. Itai, A., Shiloach, Y.: Maximum flow in planar networks. SIAM Journal on Computing 8, 135 (1979)
8. Khuller, S., Naor, J., Klein, P.: The lattice structure of flow in planar graphs. SIAM Journal on Discrete Mathematics 6(3), 477–490 (1993)
9. Klein, P.N.: Multiple-source shortest paths in planar graphs. In: Proceedings of the Sixteenth Annual ACM-SIAM Symposium on Discrete Algorithms, pp. 146–155 (2005)
10. Kornblum, D.F.: Greedy algorithms for some optimization problems on a lattice polyhedron. Ph.D. thesis, Graduate Center of the City University of New York (1978)
11. Kuratowski, K.: Sur le probleme des courbes gauches en topologie. Fundamenta Mathematicae 15(271-283), 79 (1930)
12. Matuschke, J.: Lattices and maximum flow algorithms in planar graphs. Diploma thesis, TU Berlin (2009)
13. Weihe, K.: Maximum (s,t)-flows in planar networks in $O(|V| \log |V|)$ time. Journal of Computer and System Sciences 55(3), 454–475 (1997)
14. Whitney, H.: Non-separable and planar graphs. Transactions of the American Mathematical Society 34(2), 339–362 (1932)

Author Index

GPSR Compliance

The European Union's (EU) General Product Safety Regulation (GPSR) is a set of rules that requires consumer products to be safe and our obligations to ensure this.

If you have any concerns about our products, you can contact us on ProductSafety@springernature.com

In case Publisher is established outside the EU, the EU authorized representative is:

Springer Nature Customer Service Center GmbH
Europaplatz 3
69115 Heidelberg, Germany

Batch number: 09474011

Printed by Printforce, the Netherlands